Mycotechnology

Present Status and Future Prospects

Mycotechnology

Present Status and Future Prospects

Editor
MAHENDRA RAI
Professor and Head
Department of Biotechnology
S.G.B. Amravati University
Maharashtra, India

I.K. International Publishing House Pvt. Ltd.
NEW DELHI • BANGALORE

Published by

I.K. International Publishing House Pvt. Ltd.
S-25, Green Park Extension, Uphaar Cinema Market
New Delhi – 110 016 (India)
E-mail : info@ikinternational.com

ISBN 81-89866-08-7

10 9 8 7 6 5 4 3 2

Published by Krishan Makhijani for I.K. International Publishing House Pvt. Ltd., S-25, Green Park Extension, Uphaar Cinema Market, New Delhi – 110 016 and Printed by Rekha Printers Pvt. Ltd., Okhla Industrial Area, Phase II, New Delhi – 110 020.

Foreword

Fungi are extremely diverse life forms that share an unusual, osmotrophic nutritional strategy. The essential nature of fungi is to secrete enzymes, organic acids and other metabolites across their cell walls and into the surrounding environment thereby degrading polymeric molecules that are then reabsorbed as "pre-digested" foodstuffs. Because fungi feed on organic matter made by other organisms, they are heterotrophs. However, unlike animals, they literally live in their food, transforming their surrounding habitat in a singular and profound way.

In nature, fungi decompose organic materials of all kinds. They are particularly efficient plant pathogens, causing diseases on a wide range of wild and cultivated species. They are dominant components of soil microbial populations where they are agents of biodegradation and serve as major players in the great recycling processes of nature. In the human habitat they cause economic damage by destroying wood, paper, fabrics and other polymeric materials. Although most biologists are aware of their environmental importance, the position of fungi in modern biotechnological processes is often understimated. This edited volume on *Mycotechnology: Present Status and Future Prospects* helps rectify the under appreciation of mycology in the human economy and highlights the way in which secreted fungal metabolites can be exploited in commerce.

What is biotechnology? What is mycotechnology? There are far more published definitions for the first term than for the second. For example, the United Nations Convention on Biological Diversity uses an extremely broad definition: "Biotechnology means any technological application that uses biological systems, living organisms, or derivatives thereof, to make or modify products or processes for specific use" (Article 2, Use of Terms). Another widely used sense of the term is: "The use of microorganisms or biological substances to perform specific commercial processes, often using the techniques of genetic engineering." Biotechnology also has been defined aphoristically as "biology making money." By extrapolation, mycotechnology is the use of fungi to or fungal components to perform specific industrial processes, or, more succintly, as "mycology making money." As in other fields, contemporary innovations have not emerged *de novo*. Rather, these developments are firmly grounded in the history of industrial mycology. Yeasts and molds have been used since antiquity in food (e.g., bread, cheese), beverage (beer, wine, sake) and condiment (soy sauce, kimchee) fermentations. Fungal systems were also pivotal in basic science. Modern biochemistry could be said to begin with the publication of Buchner's famous 1897 paper on "Alcoholic fermentation without yeasts." The "one gene-one enzyme" theory came out of work on the bread mold *Neurospora crassa*. In the early 20th century, a number of fungal products such as diastase (amylase) and citric acid became important commercial products. Perhaps the most celebrated example of the potential of fungi to make money is the largescale production of penicillin and other β-lactam antibiotics. Fungi have also been used to produce many other medicinally active metabolites of which the ergot alkaloids, produced by species of *Claviceps*, have the longest history. More recently, bioactive sesquiterpenes, polyketides and nonribosomal peptides also have been exploited. Further, the nutritional flexibility of fungi has been mobilized for use in the bioconversion of toxic xenobiotics of which the bioremediation of pentachlorophenol is an example.

Like all fields of biology, mycology has been transformed by the dual revolution in molecular biology and computer technology. Databases are growing at logarithmic rates. In addition to making the scientific literature almost universally available, the new databases associated with genomic and proteomic research aid experimentalists in myriad ways. SPPADBASE, for example, is the first on line searchable database of PCR primers for phytopathogenic fungi.

Dr. Rai has put together a stellar group of authors to elucidate several important topics in fungal biotechnology. These readings will serve as a useful starting point for both teaching and research. Many challenges remain in mycotechnology. Unexploited fungal products and processes offer promises to pharmaceutical and agrochemical companies. Mycotechnology has a bright future in the 21st century.

Joan Bennett, Ph.D.
Professor of Plant Biology and Pathology
Cook College of Rutgers University
&
Associate Vice President
Rutgers University
59 Dudley Road
New Brunswick, NJ 08901, USA
promycogirl@yahoo.com

Preface

Fungi are eukaryotes and are now included in a separate kingdom. They play a key role in shaping human life. The present book *Mycotechnology: Present Status and Future Prospects* is an effort to cover the application of fungi in the present biotechnology era. The life-saving drug Penicillin, and alcohol are, after all, the important gifts of Penicillium and the yeasts fermentation. The yeast is the model eukaryotic organism, and the first eukaryote to have its genome completely sequenced.

Fungi produce a wide variety of pharmaceuticals, like Penicillin, Cyclosporin, Griseofulvin, and Mevinolin, etc. In addition to this, they also form compounds like the ergot alkaloids, steroid derivatives and antitumor agents. Ergot toxins in low concentration are valuable drugs that decrease blood pressure. The biotransformations by fungi is immensely useful, and most of the steroids in chemical use today are transformed or modified by fungi during their manufacture. Studies have proved that many fungal products are known to inhibit cancer. In Japan and China enormous work has been done on the anticancerous properties of mushrooms. The fungi in fermentation cultures are also use to produce many commercial enzymes, most of which are for use in processing foods. Citric acid production by fungi is well known and is a major commercial chemical used by the people.

In the field of sustainable agriculture also, fungi play crucial role as biocontrol agents of weeds and other pests like nematodes, insects, and mites. They are growth promoters like mycorrhizal fungi, which form symbioses in 90 per cent plants on the earth, are the center of attraction for the researchers. The fermentation of industrial waste products, e.g., molasses, straw, and wood pulps can be useful for the production of alcohol as an alternative source of biofuel, which is the need of the hour.

The present book includes current and rare topics on Mycotechnology, such as, molecular techniques (for analyses of soil fungi, diagnosis of ochratoxin-A producing fungi, for identification of ectomycorrhizal fungi), SPPADBASE, bioactive sesquiterpenes, mycological applications of Raman spectroscopy, etc.

I am thankful to all the contributors for submission of valuable manuscripts, especially Aniket Gade for his help in editing. I wish to thank my students – Alka Karwa, Ravindra Ade, Prajakta Deshmukh and Ajit Chande for their contribution in checking manuscripts and help in type-setting.

Finally, and most importantly, I would like to thank Shivangi, Shivani and Aditya for their unconditional love, support, and patience during the editing work.

Editor

Contents

Contributors

Adriana P.D. Silveira
Department of Microbiology
Institute of Agronomy (IAC)
Campinas, Av.
Barao de Itapura 1.480, C.P. 28-13001-970
State of São Paulo, Brazil

Alexander Urban
Department for Systematic and Evolutionary Botany,
Faculty Center Botany, University of Vienna,
Rennweg 14, A-1030 Vienna
E-mail: alexander.urban@univie.ac.at

Annemieke Verbeken
Department of Biology
Ghent University, K.L. Ledeganckstraat 35
B-9000 Gent, Belgium

Archna Mehta
Microbiology Laboratory, Department of Botany,
Dr. H. S. Gour University,
Sagar-470 003, Madhya Pradesh, India
E-mail: archuunisagar@yahoo.com

Carlos García-Estrada
Departamento de Ecología
Genética y Microbiología, Área de Microbiología
Facultad de Ciencias Biológicas y Ambientales
Universidad de León
Campus de Vegazana s/n, 24071, León, Spain

Cesar Arriagada Escamilla
Estación Experimental del Zaidín
Departamento de Microbiología del Suelo
Profesor Albareda 1. CP-18008. Granada-España
E-mail: cesar.arriagada@eez.csic.es

Chris I. Ford
Lincoln University
Centre for Soil and Environmental Quality
PO Box 84, Lincoln, New Zealand, Australia

Donát Magyar
Plant Protection Institute of the
Hungarian Academy of Sciences
Department of Plant Pathology
H-1022 Budapest, Herman Ottó u. 15. Hungary
E-mail: magyard@freemail.hu

Emily J. Parker
Massey University
Institute of Fundamental Sciences, Private Bag 11 222
Palmerston North, New Zealand, Australia

Guangi Wang
Hawaii Natural Energy Institute
Department of Oceanography
School of Ocean and Earth Science and Technology
University of Hawaii at Manoa, Honolulu, HI 96822
E-mail: guangyi@hawaii.edu

Hong J. Di
Lincoln University
Centre for Soil and Environmental Quality
PO Box 84, Lincoln, New Zealand, Australia

Jenness M. Guthrie
Massey University,
Institute of Fundamental Sciences, Private Bag 11 222,
Palmerston North, New Zealand, Australia

Jiří Kučera
Czech Republic Food Research Institute Prague
(Výzkumný ústav potravinářský Praha)
Radiová 7, CZ 10231 Praha 10, Czech Republic
Tel: +420 296 792 364 Fax: +420 272 701 983
E-mail: j.kucera@vupp.cz

Juan Francisco Martín
Instituto de Biotecnología de León (INBIOTEC)
Parque Científico de la Granja,
Av. Real, nº1, 24006, León, Spain

Keith C. Cameron
Lincoln University
Centre for Soil and Environmental Quality
PO Box 84, Lincoln, New Zealand, Australia

Kirsty S. H. Boyd-Wilson
Hort Research, Environment
and Risk Management Group
Gerald St, PO Box 51
Lincoln, New Zealand, Australia

Kris De Gussem
Department of Analytical Chemistry
Ghent University, Proeftuinstraat 86
B-9000 Gent, Belgium

L. Niessen
Lehrstuhl für Technische Mikrobiologie,
Technische Universität München, Weihenstephaner Steig 16, D-85350 Freising, Germany
Phone +49 8161 715496, fax +49 8161 713327
E-mail: niessen@wzw.tum.de

López, Clara E.,
CEREMIC (Mycologal Work Reference Center)
Suipacha 531- (2000) Rosario, Argentina
Tel. and Fax: +54-341-4307624
E-mail: clopez@fbioyf.unr.edu.ar

Luc Moens
Department of Analytical Chemistry
Ghent University, Proeftuinstraat 86
B-9000 Gent, Belgium

M. Carol Carter
Biology Department, Portland State University,
P.O. Box 751, Portland, Oregon 97207
E-mail: b7mc@pdx.edu

Mahendra Rai
Department of Biotechnology
SGB Amravati University, Amravati-444602
Maharashtra, India
E-mail: mkrai123@rediffmail.com or
pmkrai@hotmail.com

Matthew Gordon
Biology Department, Portland State University
P.O. Box 751, Portland, Oregon 97207
E-mail: b7mc@pdxi.edu

Monika Walter
Hort Research, Environment and
Risk Management Group
Gerald St, PO Box 51, Lincoln, New Zealand, Australia
E-mail: mwalter@hortresearch.co.nz

Neetu Dahiya
Department of Biotechnology, Panjab University
Chandigarh-160014, India
E-mail: ineetudahiya@yahoo.com

Peter Vandenabeele
Department of Biology, Ghent University
K.L. Ledeganckstraat 35, B-9000 Gent, Belgium

Pradeep Mehta
Microbiology Laboratory, Department of Botany
Dr. H. S. Gour University
Sagar-470 003, Madhya Pradesh, India
E-mail: pradeepunisagar@hotmail.com

Quanzi Li
Hawaii Natural Energy Institute
Department of Oceanography
School of Ocean and Earth Science and Technology
University of Hawaii at Manoa, Honolulu, HI 96822
E-mail: guangyi@hawaii.edu

Quirico Migheli
Department of Plant Protection
Center for Biotechnology Development and
Biodiversity Research, University of Sassari,
Via E. De Nicola 9, I-07100 Sassari, Italy
Tel: +39 079 229295
E-mail: qmigheli@uniss.it

Sanghwal Yoon
Hawaii Natural Energy Institute
Department of Oceanography
School of Ocean and Earth Science and Technology
University of Hawaii at Manoa, Honolulu, HI 96822
E-mail: guangyi@hawaii.edu

Soraya C. Franca
Department of Microbiology
Institute of Agronomy (IAC)
Campinas, Av., Barao de Itapura 1.480
C.P. 28-13001-970, State of São Paulo, Brazil

Stefano Ghignone
Plant Protection Institute
Sect. Turin – CNR, V.le P.A. Mattioli
25, I-10125 Turin, Italy
Tel: +39 011 6705982 Fax: +39 011 6705962
E-mail: stefano.ghignone@unito.it

Wolf-Rainer Abraham
GBF-National Research Centre for Biotechnology
Chemical Microbiology, Mascheroder
Weg 1, D-38124 Braunschweig, Germany
Tel: +49-531-6181419 Fax: +49-531-6181411
E-mail: wab@gbf.de

Zacchino, Susana A
Pharmacognosy
School of Biochemical and Pharmaceutical Sciences
Suipacha 531- (2000)
Rosario, Argentina

1 Molecular Methods for Analysis of Soil Fungi

M. Carol Carter and Matthew Gordon

Biology Department, Portland State University, P.O. Box 751, Portland, Oregon 97207.

INTRODUCTION

Soil is perhaps our most important natural resource. Our food, shelter, and water depend on soil quality and quantity. Evaluating soil ecosystems and understanding how the system components interact are critical tasks that must be done to effectively manage and preserve the soil resource. To understand soil, it is important to consider the both nonliving, physical aspect of soil and the living, biotic component.

The physical aspects of soil are diverse. Moisture, minerals and organic components vary in three dimensions. Additionally, difference in the texture, pH, and climatic exposure gives particular character to the soil. Yet despite these many variables, detailed characterizations of soils based on physical features are routinely conducted and extensive databases describing soil features over vast geographic areas of the world are available. http://sis.agr.gc.ca/cansis/nsdb/intro.html; http://www.ftw.nrcs.usda.gov/ssur_data.html).

In comparison to the body of knowledge that has been generated on the physical attributes of soils, our information on the biotic component of soils is less. A shovel full of soil is likely to contain thousands of different organisms integrated with each other and with the organic and mineral components of the soil. Yet, in only a few cases substantiated estimations of biotic diversity of particular soils have been made, let alone specific taxonomic determinations (Söderstrom, 1975; Wicklow and Whittingham, 1978). Requisite to understanding the biota of soil is the ability to determine the identity, abundance, and distribution of organisms in the soil. However, it is a problem to separate the organisms from each other and from their physical environment for qualitative and quantitative assessment. In the past, surveys of soil organisms had been based on dilution, culture, and isolation. But the isolation of many organisms was not successfully achieved by routine culture methods and their identity was not recognized in such studies (Borneman and Hartin, 2000; Heredia *et al.*, 2001). Because of the limitations of isolation through cultures (Dix and Webster, 1995), progress in characterizing the biotic component of soil has been slow. Molecular approaches offer new strategies for studying soil organisms, but methods that work well on organisms above ground, may not be suitable for those below.

This chapter reviews some approaches to surveying soil microorganisms, particularly fungi. Our goal is not to present an exhaustive survey of techniques that could have been used to deal with the particular challenges of defining the biotic contents of the lithosphere ecosystem. Rather, we review selected molecular approaches for identifying or characterizing fungi. Some emphases have been made on techniques that allow characterization of fungi without initial culture, either from sporocarps or directly from soil. In addition, we present a new strategy for developing models for estimating the distribution of specific taxa.

IDENTIFYING SOIL FUNGI

Morphological Methods for Identification of Fungi

Fungi in soil can produce a variety of structures, the morphology of which has been used for classification and identification. Four types of structures that may be associated with soil are mycelia, mycorrhizae, sporocarps, and spores.

Mycelia are formed of filamentous hyphae that extend through the soil, secrete enzymes, and absorb nutrients and water. Resources are transported through the mycelium, often over great distances, in response to internal source-sink gradients or environmental stimuli (Rayner, 1991). Although mycelium can be the largest part of fungus in terms of biomass, it is typically difficult to observe since the invisibly thin hyphae may be distributed throughout a comparatively large volume of soil. In some soil fungi, a portion of the mycelium complex forms a fungal mat. The mat is a complex of aggregated cord-like strands of hyphae (rhizomorphs), soil particles, and fungal exudates that heavily colonizes a limited area of soil. Fungal mats are distributed in patches at various depths within the soil and are visible to the unaided eye upon removal of upper soil layers. Most of the fungal mats are similar in morphology (Ogawa, 1985) and it is not possible to determine fungal species from mat characteristics. The true extent and distribution of a mat-forming fungus cannot be determined from the size of the mat since fine hyphae extend out from the mat region.

Mycorrhizae are the extensions of mycelia that form on and extend into the fine root tips of plants. Actually, the term mycorrhizae is applied to the assemblage of both plant and fungal cells. The mycorrhizae is the region of exchange between the plant and the fungus, where the fungus takes carbon compounds from the plant and the plant takes nutrients and water from the fungus. In the past, mycorrhizal morphological features were thought to be useful for visual taxonomic identification of the fungus partner and useful for determining distribution patterns of the fungus within the soil (Agerer, 1988). However, mycorrhizal morphology has been found to be variable within species depending on host, environmental conditions, age, and season (Egger, 1995; Lefevre and Muller, 1998; Giomaro *et al.,* 2000), and different species can have the same morphology (Kraigher *et al.,* 1995; Bertini *et al.,* 1998). This lack of consistency between taxonomy and morphology makes identification of mycorrhizal fungi, by their morphology, difficult and ambiguous.

The sporocarp (or fruiting body) is the part of the fungus where genetic recombination and meiosis takes place and spores are formed. Traditionally, the fruiting body and associated spores are used for taxonomic identification of ecosystem fungi because these reproductive bodies usually have morphologically stable features. While the presence of a particular species of fungus in an area can be determined by the presence of the sporocarps of that species, the absence of sporocarps of a species is not evidence of the absence of a species from the soil. Fruiting bodies are ephemeral and the time of fruiting of most fungi is not well defined. A researcher studying fungal ecology, using sporocarps as indicators of the presence of species, must make many visits to the site of study over the course of several fruiting seasons (O'Dell *et al.,* 1996). Sporocarps may be eaten by animals or harvested by people between checks, leading to false negative results. A mature fungal individual which extensively colonizes a region of soil may not produce sporocarps every year and the lack of sightings lead to false negatives in short term studies. Lastly, in the "imperfect fungi", the sporocarp is unknown.

Fungal spores can be isolated from soil relatively easily with density gradient and differential centrifugation (http://www.inra.fr/Internet/Centres/Dijon/bbceipm/ Mychintec/ Protocole Workshop_Procedures.html). There are general guidelines for classifying certain types of spores based on morphology (Castellano *et al.,* 1989), but it is impossible to identify most of the fungal species from their spores alone.

Because of the difficulty of observing fungi *in situ*, isolating them from other soil components, and identifying them by visual means, our knowledge of the ecology of soil fungi is severely limited. Methods of identification that are less tedious and less ambiguous than those based on the morphology and culture characteristics will allow researchers to make more rapid progress in understanding the ecology of these organisms.

Molecular Methods for Identification of Fungi

Molecular biological techniques, particularly polymerase chain reaction (PCR) technology (Mullis and Faloona, 1987), have revolutionized biology. The main advantage of the PCR is to offer a way to generate relatively large amounts of a desired stretch of DNA from very small amounts of starting DNA. Once a sufficient quantity of the target DNA is produced, other procedures such as DNA sequencing or fingerprinting can be performed for establishing identity.

Amplification in PCR is accomplished from replication of a targeted portion of single-stranded DNA by a thermostable polymerase. Replication proceeds from sites where complimentary oligonucleotides (primers) have annealed. After replication, the DNA is heat-denatured to single strands, cooled, re-annealed with the primers and replicated again. Through repeated cycles of replication and denaturation, exponential production of DNA may occur. The discretely sized DNA products produced through PCR amplification are called amplicons. Amplicon size, approximated by comparison to molecular weight standards by gel electrophoresis, is often used as evidence that the desired target has been amplified, especially if the original primers were derived from sequence information and the exact number of nucleotide basepairs (bp) flanked by the chosen primer sequences is known. The optimal length for the DNA stretches targeted by primers is usually in the range of 200 to 2000 bp—longer or shorter regions may require special protocols. (For the detailed description of the process see Campbell and Reece (2002), or other recent college-level cell biology text.)

Soon after the PCR technique was described, biologists began using sequence information generated from pre-PCR molecular biology studies to design PCR primers for molecular taxonomic studies. Some primers are designed from conserved regions of genomes and are valuable for characterizations at high taxonomic levels. These *characterizing* primers are sometimes called universal primers. Other *specific* primers, selected from less conserved regions of genomes, are useful to study lower-level taxa such as within genera or species. A third type of primer, called a *random primer* or arbitrary primer (Welch and McClelland, 1990; Williams *et al.,* 1993) targets unknown regions of a genome and produces several amplicon size classes per PCR reaction.

Analysis of amplicons produced by characterizing primers require additional tests to make identifications. Tests, such as restriction fragment length polymorphism (RFLP) analysis, denaturing gel electrophoresis, or DNA sequencing provide additional information for characterizations, and may or may not establish identity. Although use of characterizing primers requires more steps beyond the PCR, over time the process may be efficacious since many of the procedures, from primer selection to post-PCR analyses, may be based on previously published work. Also, often a group of related organisms may be studied from one set of characterizing primers or nested primer sets. The universal rDNA primers are probably the most widely used characterizing primers for the study of both prokaryotic and eukaryotic organisms. Applications of rDNA primers to the study of fungi are discussed later.

Specific primers are usually desired for targeting DNA of a single taxon, usually at the species level. Under appropriately stringent amplification conditions, it is assumed that no other organism's DNA will produce amplicons with the primers, or, if non-target DNA is amplified, it will be readily distinguishable from the target DNA (Differences such as significant amplicon length differences or obvious quantitative differences in PCR product are the examples of non-specificities that may be tolerated).

Specific primers require detailed sequence knowledge of a portion of the DNA of a species. For studies on a species without previous relevant sequence information, the process of producing a set of specific primers can be arduous. Initially, an appropriate DNA target must be selected from genomic DNA, isolated, and purified in sufficient quantity. This process may involve PCR with characterizing primers from a sporocarp or pure culture to gain sufficient DNA for sequencing. Alternatively, random primers (discussed below) may be used to find an appropriate portion of DNA. The isolated and purified DNA is sequenced and appropriate primer regions are selected with the aid of primer design computer software. Freeware primer design programs are available at the National Center for Bio-Information (NCBI) website (http://www.ncbi.nlm.nih.gov/) as well as other internet sources. Primers that flank a region of 400-700 basepairs are practical since the size range is small enough to amplify well, yet the fragment is big enough to further characterize by RFLP, if desired. The primer pairs, in their opposite strand orientation, should be unique in reference to other reported DNA sequences. This information is obtained by database searches utilizing the search capabilities provided by the NCBI database and using the basic local alignment search tool (Altschul *et al*., 1990; http://www.ncbi.nlm.nih.gov/BLAST) within the NCBI site.

Even if the BLAST search reveals that sequences identical or nearly identical to those of the proposed primers occur in the DNA of other organisms, the primer pair may still be suitable for a specific study. For example, if the matching sites in the non-target organism are thousands of basepairs apart, one can assume that amplification under ordinary PCR conditions would not occur. Also, if matches are found in an organism that would not normally be associated with the target organism the primer pair may still be used (e.g., if the desired target is a soil fungus and the non-target DNA is associated with a marine mammal, the primer sequences will still be acceptable). One or more primer pairs are selected with respect to the above criteria, the oligonucleotides are synthesized and testing of the primers can begin.

The testing involves performing PCR with the primers on the target organism to see if the expected fragment is produced. If amplicons other than the expected target are produced, this may indicate that the primers are associated with the regions of DNA that include multiple genes with sequence variability from insertions, deletions, or introns. Sometimes weak bands are produced in addition to the expected amplicon product. Weak bands may result from imperfect primer matches to non-target sites. These bands can sometimes be eliminated by increasing the stringency of the PCR reactions, such as by increasing the annealing temperature during the reaction. To confirm that the expected target has indeed, been amplified, the PCR product is subjected to sequencing, either directly from the initial PCR reaction or from an agarose plug from the gel band of interest. If the sequencing result produces readouts with high noise and multiple peaks per position, it may be that the primer pair allowed amplification at multiple sites in the genome. Although poor sequence output may be caused by technical problems, lack of sequence confirmation indicates the primer pair may not be suitable for use without further troubleshooting. The goal of producing a good specific primer set is to allow the PCR to produce an amplicon product, which on electrophoresis gives a gel image with a single or predictable number of strong bands, with few if any minor bands. If this goal is reached with the known target species, related species are tested and if no amplicon bands are produced, the study involving taxa identification or detection can proceed.

Sometimes *nested* primer sets are used in order to identify and characterize an organism. Nested priming requires two or more consecutive PCR amplifications. In nested PCR, the first and second amplifications are often conducted under different conditions, such as different annealing temperatures and cycle numbers. Additionally, the primer pairs may be selected for their differential specificity to conserved or variable regions of DNA, for example, the first primer pair may be characterizing primers that select for phyla, while the second, specific primer pair selects for species. Nested PCR may be useful in mixed genome samples because it may increase sensitivity and specificity of the PCR. Nested PCR strategies are limited by length of the targeted DNA, since the second PCR priming must fit within the region defined by the first set of primers.

A random primer is a single oligonucleotide, often about 10 nucleotides long that primes both sides of a length of DNA (Williams *et al.,* 1993). Random primers are initially selected by criteria of purine/pyrimidine balance and nucleotide number. Because random primers anneal to multiple sites in a DNA sample, multiple amplicon kinds are usually produced. When separated by electrophoresis these amplicons produce a series of bands that collectively form a fingerprint of the organism. Random primers are useful in studies of sub-specific taxonomic relationships such as strains.

Molecular probes based on ribosomal DNA sequences

Ribosomal DNA (rDNA) is ideal in many ways for developing both specific and characterizing primers; and robust PCR reactions are typical. In fungi there are four types of ribosomal genes which produce 25-28S, 18S, 5.8S, and 5S subunits of eukaryotic ribosomes [the size of the large subunit varies among eukaryotic taxa (Gutell and Fox, 1988)]. In fungi, 18S, 5.8S, and 25-28S ribosomal genes are organized in a series separated by spacer regions (Fig. 1.1). The 5S genes may be associated with other rDNA but are frequently dispersed (Ganley and Scott, 2002). Usually there are hundreds of copies of the ribosomal gene series per cell organized in tandem (Long and

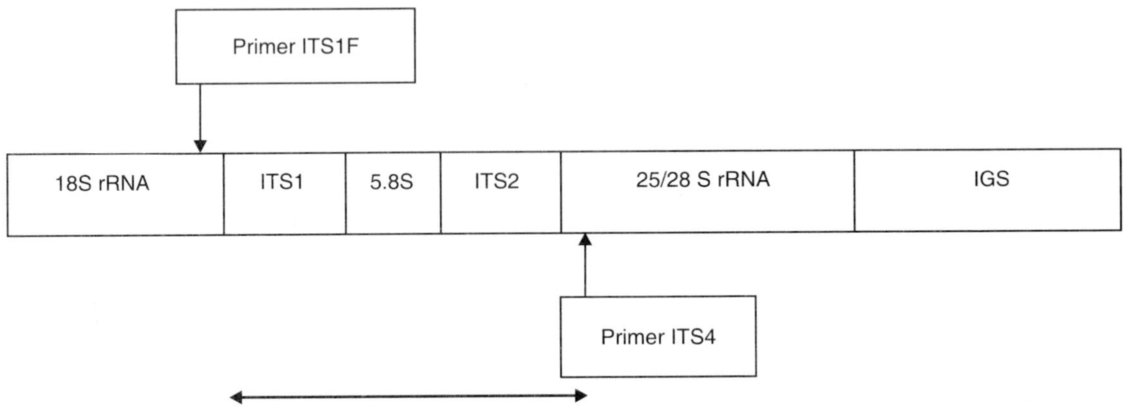

Fig. 1.1. A diagram of the ribosomal RNA gene cluster showing the positions of primers ITS1F and ITS4 that delineate the ITS amplicon (double-headed arrow). The amplicon contains the internal transcribed spacers ITS1 and 2 and the 5.8S ribosomal gene (Segment lengths are not to scale).

Dawid, 1980). Despite the large number of copies, there is thought to be relatively little sequence variation among the gene copies. There seems to be some mechanism that allows evolutionary change to occur in concert within rDNA (Elder and Turner, 1995; Ganley and Scott, 2002). Because there are hundreds of rDNA genes per cell, and copies of each type of gene all have virtually the same sequence, there are a large number of primable sites at the beginning of exponential amplification. PCR reactions set up with rDNA primers generally work well and produce generous amounts of amplicon product. Even if the quality of the DNA sample is compromised, rDNA primers may still produce adequate amplicon amounts for characterization, since some primable targets may persist.

The RNA transcripts from rDNA are, after some post-transcriptional modification, used directly in the formation of ribosomes, that is, no translation occurs. Thus, while the genetic code allows codon redundancy for amino acids, no such redundancy occurs among rDNA. Portions of the DNA sequences of ribosomal genes are among the most evolutionarily conserved of all genes (Hillis and Dixon, 1991) at the nucleotide level. Conservation may extend to the Kingdom level in some regions. Other regions of the ribosomal genes or their spacer regions are moderately conserved; while still others may be quite variable. The multiple degrees of sequence conservation associated with rDNA have been exploited for determining evolutionary relationships at several taxonomic levels (Hillis and Dixon, 1991; Soltis *et al.,* 1999; Campbell and Reece, 2002). Sequence variation is also found in the spacer regions between genes. The intergenic spacer (IGS) region (Fig. 1.1) is useful in classifications among the prokaryotes, but the comparable region in eukaryotes is rather large and often so hypervariable that it is not as favored for PCR studies (but see Jackson *et al.,* 1999). On the other hand, ITS regions in eukaryotes are generally moderately variable, and in the fungi the variation seems just about right for studying genera and species by PCR (White *et al.,* 1990; Jackson *et al.,* 1999).

White *et al.* (1990) described PCR primers designed from conserved regions of the 25-28S and 18S genes that flank ITS regions and were therefore useful in the amplification of ITS regions of ribosomal RNA (rRNA) gene in fungi and other organisms. Later, Gardes and Bruns (1993) described primers flanking the ITS region that are specific to fungi and fungi in the phyla Basidiomycota and

Ascomycota. These apparently kingdom- and phylum-specific primers allow the amplification of parts of the fungal rRNA gene in mixed-genomic samples such as mycorrhizae of soil. Since the publication of these important studies, many additional potential primers have been tested and their sequences are available for use by others when references to the original investigators are maintained (e.g., the Rytas Vilgalys's web site, http://www.biology.duke.edu/fungi/mycolab/).

Nested priming within ribosomal genes can be used to generate species-specific probes for fungi. The technique starts with the amplification of a region of rDNA from several strains or sporocarps from different populations of the species of interest (SOI), using a universal or characterizing primer pair suitable for fungi (as described above). The amplicons produced are then sequenced and compared to sequences in the rDNA database (*e.g.*, Michigan State University Ribosomal Database Project II, http://rdp.cme.msu.edu/html/index.html; The European Ribosomal DNA Database at the University of Gent, Belgium, http://rdp.cme.msu.edu/html/index.html) and to sequences available through NCBI or other general sequence repositories (*e.g.*, the European Bioinformatic Institute, http://www.ebi.ac.uk/embl/; the DNA Data Bank of Japan, http://www.ddbj.nig.ac.jp). When the sequences of the SOI are aligned and compared to other sequences in the database, there may be areas that are identical within all the samples sequenced of the SOI, but are not conserved in other species. An example of a potentially useful primer is one which is composed of 18-23 nucleotides with a sequence that is at least 70% unique to the SOI alone and with the majority of the unique sequence positions on the 3' end of the primer (a few mismatches within the 5' end of the primer do not seriously affect the polymerase's ability to initiate second strand synthesis). In addition, a primer pair must flank a length of DNA of a convenient length for PCR amplification, preferably greater than 300 bps, so additional tests such as RFLP analysis on the amplicon can be performed easily. Primer pairs should be well matched in terms of length and GC content so that annealing temperature will affect each primer. Similarly, synthesized primers should be tested against DNA extracted from other fungi including closely related species to determine their specificity. If the primers do not cause the amplification in a PCR even with related species, the primers will act as a specific probe for the target species when used in an appropriate PCR (*e.g.*, Kikuchi *et al.*, 2000, Millner *et al.*, 2001). If the primers produce amplicons in related species, they may still be valuable tools for characterizing a genus or related organisms (*e.g.*, Simon *et al.*, 1992).

Molecular Probes based on RAPDs

A technique of randomly amplified polymorphic DNA (RAPD) was developed by Williams *et al.* (1990) and has been successfully used as a DNA fingerprinting technique. Arbitrarily-primed (AP) PCR (Welsh and McClelland, 1990) is a similar technique developed about the same time as the RAPD technique. The techniques have very high resolution of genomic variability especially when multiple fingerprints are generated by a series of primers. The RAPD or AP-PCR methods can identify individuals or a vegetatively reproducing individual's genets. In order to identify the genets, however, the starting DNA sample must not be multigenomic, so the fungus must be cultured from samples of soil taken throughout an area or obtained from sporocarps or other structures that can be isolated. RAPD is the technique by which the extents of giant subterranean *Armillaria bulbos* individuals were documented from fruiting bodies and rhizomorphs (Smith *et al.*, 1992).

The RAPD technique can be used as a starting point to generate highly specific primers for identification at the species level. With highly specific primers, detection may be possible directly from a soil sample—without having to obtain sporocarps or pure cultures. A possible advantage of using specific primers derived from RAPD amplicons, rather than from rDNA, is that cross reactivity to species other than the SOI, may be less.

To develop specific primers from RAPD amplicons, a selection process must be applied to eliminate amplicons that are individually variable and to identify amplicons present in all individuals of the species or strain of interest, but not present in related species. With these tasks in mind, the following procedure can be used to produce a specific molecular probes from RAPD amplicons:

1. DNA is isolated from each of several sporocarps or pure cultures of the SOI collected from multiple populations, so that species-specific or strain-specific DNA markers can be screened.
2. RAPD is conducted on the samples using a variety of random primers. Variation of banding patterns in the RAPD products can result from experimental inconsistencies or sample DNA quality. These sources of non-genomic variation have to be minimized when conducting surveys for species specific probes. In our experience with hundreds of samples, "homemade" RAPD reaction mixtures are not very reliable for the screening of possible species specific markers. Investigators should invest in using proprietary RAPD reaction mixtures such as "Ready to Go" RAPD beads (Amersham Biosciences). With proprietary mixtures, all that is added is water, sample DNA and primer, minimizing the impact of experimental variation that often plagues RAPD studies (*e.g.*, Penner *et al.,* 1993). However, even with the RAPD beads, DNA quality and quantity must be carefully controlled.
3. The RAPD PCR products of the samples are electrophoresed and the banding patterns are compared. From a survey of several random primers, a small set of amplicons for developing taxon-specific primers are selected based on commonality to all populations of interest, amplicon size, and band intensity. A selected amplicon is isolated from gels by taking a tiny core of gel at the band position of interest. The core can be used directly in the PCR reaction mixture and amplicon is used as template DNA for reamplification with the random primer. Electrophoresis of an aliquot of the reamplification product should show a single band of the expected size.
4. Isolated RAPD amplicons must be sequenced. However, they cannot be sequenced in the easy way by initiating sequencing reactions with their own primers because their flanking ends are identical. Forward and reverse sequencing products would be produced simultaneously giving scrambled sequence information. Therefore, in most situations, RAPD fragments must be cloned prior to sequencing (Comes *et al.,* 1997).
5. Sequences from cloned fragments are subjected to database sequences and primer pairs are designed appropriately, as described above, and synthesized.
6. Primer pairs are tested for specificity against isolated DNA of the SOI and for cross-reactivity to other non target species. Bertini *et al.,* (1998) used primers derived from RAPDs to create specific primers for the ectomycorrhizal fungus *Tuber sp.* for life history studies. Weber *et al.,* (2002) successfully used modified RAPD primers to detect a specific strain of *Laccaria bicolor* in the field. In the present chapter, we report the use of specific primers for *Tricholoma magnivelare* for the detection and distribution of its mycelium in forest soil.

Other Methods

Sometimes fungal identity or taxa comparisons can be assessed without resorting to sequence analysis or development of highly specific primers. Taxon characterization by restriction fragment length polymorphisms (RFLP) analysis of universal PCR products is often very useful for screening a number of unknown but possibly related individuals. DNA from individual sporocarps, pure cultures, or mycorrhizae are used as PCR templates. The PCR is set up with universal primers such as those described by Gardes and Bruns (1993) for targets in the ITS regions in the rRNA gene series of fungi or other more specific primers such as those that have been found to target many AV mycorrhizae (Kjøller and Rosendahl, 2001). The amplicons obtained from individual PCRs are first categorized for length variations. Then, amplicons are subjected to digestion with a series of DNA restriction enzymes. Usually four base cutters are used to increase the likelihood that the relatively short sequence length of amplicons will contain a restriction site. The exception to this is EcoRI, a six base cutter. The 5.8S gene (Fig. 1.1) contains a conserved EcoRI site in most fungi surveyed (White *et al.*, 1990; Jobes and Thien, 1997). The EcoRI restriction will usually divide amplicons targeted by primers flanking the 5.8S gene into two pieces, the sizes of which can be compared among samples. Furthermore, the EcoRI digest can serve as a positive control for the digestibility of the amplicon samples.

In general, if there are insertions or deletions, there will be size differences (polymorphisms) in the digestion products from different restriction enzymes. Lack of restriction in a sample (when other samples cut or when the EcoRI digest works) indicates a sequence variation precisely at the restriction site. Restriction fragment size classes from unknown fungi can be compared to databases of fragment size classes from identified organisms (Cole *et al.*, 2003, and see the University of Minnesota Herbarium database at http://fungi.umn.edu/odbinfo.html). However, it is usually better to compile one's own RFLP database and to use other databases as supplementary information. This is because sizing of fragments from gel positions by comparison to a molecular weight marker is only an approximation. Different techniques of estimating, and different gel conditions give different results.

Henrion *et al.* (1994) used RFLP analysis of the ITS and inter-genic spacer (IGS) regions of the rRNA genes to distinguish DNA from 7 different *Tuber* species. Erland *et al.* (1994) used RFLP analysis of the ITS regions of the rRNA gene to identify the ectomycorrhizal fungus *Tylospora fibrillosa* in mycorrhizae from Sitka spruce plantations. Kraigher *et al.* (1995) used RFLP analysis of the ITS region to distinguish the morphologically similar mycorrhizae of two species of the ectomycorrhizal genus *Lactarius*. Kernaghan (2001) used RFLP analysis of rRNA amplified from DNA obtained from mycorrhizae from the alpine and subalpine zones in an area of the Canadian Rockies to identify the fungal component of 6 of 18 mycorrhizal types collected.

DIRECT ANALYSIS OF FUNGAL DNA FROM SOIL

Working with soil directly for studying fungi rather than working with individual sporocarps or soil fungal cultures presents special problems. DNA extracted directly from soil will be multi-genomic and investigators will need to be resourceful to target just the fungal component in the soil. Two approaches are discussed below. The first approach describes methods for characterizing the fungal community as a whole. The second approach describes methods for characterizing a specific fungus directly from soil, without subculture.

Community Fingerprinting

Community fingerprinting is a form of DNA fingerprinting used for comparative studies of populations of organisms that occupy certain ecological niches. The organisms targeted are expected to be related only at high taxonomic levels, such as phyla or kingdom. For community fingerprinting of soil without culturing, ribosomal gene complexes, with their combinations of conserved and variable regions, are targeted for PCR amplification from a single, multiple-genome sample (Borneman and Hartin, 2000; Peters *et al.*, 2000). The acronym RISA (ribosomal intergenic spacer analysis) was used to describe the first rDNA community fingerprinting studies for bacterial (Ranjard *et al.*, 2000) and fungal communities (Ranjard *et al.*, 2001). Bacterial RISA uses primers that flank the spacer region between the 16S and 23S subunits of the *rrn* operon (which is comparable to the non-transcribed intergenic spacer or IGS of eukaryotic organisms). For analysis of fungal communities by RISA, primers in the 18S and 25-28S ribosomal genes that flank the ITS1 and the ITS2 are more appropriate (Gardes and Bruns, 1993), since IGS regions tend to be long and hypervariable in eukaryotes, complicating routine PCR. Ranjard *et al.*, (2001) used a primer pair designed from *Penicillium* ITS flanking sequences to profile fungal communities. While these primers were designed from *Penicillium* sequences, and are perfectly homologous to many fungi, our review of NCBI database matches indicates that the primer pairs are not fungal specific and are in fact homologous to sequences reported for other groups of organisms, such as nematodes, that could be present in soil. For amplification of fungal sequences, the primer ITS1F (fungal specific) coupled with the "universal" primer ITS4 can be used (Gardes and Bruns, 1993). However, even this combination of primers could also amplify DNA of non-fungal organisms. For example, *Chlorella*, a green alga species that can be found in soil, has a nearly perfect match to primer ITS1F in its 18S DNA (Yeh *et al.*, 2002).

With RISA, multiple amplicon size classes are expected, representing the size variation of the spacer regions among taxa. A high resolution gel acrylamide is used to resolve the bands. Denaturing gradient gel electrophoresis (DGGE) or single strand conformational polymorphism (SSCP, Hongyo *et al.,* 1993) techniques are methods of choice of many investigators to resolve sequence differences in amplicons generated from mixed genomic samples (*e.g.*, Kjöller and Rosendahl, 2000; Koschinsky *et al.,* 2000; Ranjard, 2001). Gel band images are scanned and converted to electrophoregrams to facilitate sample comparisons. The profiles obtained by the electrophoregrams have peaks representing various fragment lengths (in basepairs) and peak height measured from band intensity represents relative abundance. Ranjard *et al.,* (2001) used principal component analysis and found that the profiles presented a fairly consistent picture of populations at sites, where the most variability was observed among plots (as expected) than among replicate DNA extractions and PCR runs of the individual samples. Ranjard's RISA technique, whereby amplicons were labeled by fluorescence-tagged primers and analysis was by automated sequencing gel technology, produced 82 to 115 peaks (bands) per profile of his "fungal" samples, ranging in size from 300 to 1100 bp. Using similar methods profiles of bacteria communities tended to produce even more peaks than were produced in "fungal" profiles. Positive controls to determine the number of peaks generated from known mixed genome samples (for example PCR amplicons from measured mixtures of pure cultures) amplified by ribosomal primers and subjected to DGGE have not been reported, as far as we know, but such controls would be helpful to show that how the number and sizes of peaks correlates with the number and proportion of taxa DNA in the starting multigenomic sample. Conformational variations of DNA strands, inconsistencies of the gel lattice, polymerase fidelity factors, concatenate primer-dimer formations, chimerical strands from sheared and hybridized fragments, or variation among the rDNA copies within an organism (Kjøller and Rosendahl, 2000; Kim and Breuil 2001) may contribute to the generation of multiple electrophoregram peaks. Therefore, an assumption that multiple peaks represent multiple species in the community seems unsupportable. Furthermore, because of the reaction kinetics of the PCR, the starting proportions of target templates are skewed by the final cycle because, in general, smaller and more abundant starting templates (flanked by the primers) are preferentially amplified over less abundant, larger templates. Therefore, it should be expected that some DNA size classes in the starting multigenomic DNA sample will be over represented in at the end of the cycling regime. It is possible that PCR conditions could be adjusted to compensate somewhat for amplification bias.

Even though the RISA technique may not really be indicating actual numbers and proportions of species within a soil sample, investigators have found that it is a useful technique for describing differences between sites or documenting changes in community structure over time if the same extraction and processing procedures are strictly followed (Kozdroj and van Elsas, 2000).

A variation on the RISA technique termed T-RFLP (Liu *et al.,* 1997; Dickie *et al.,* 2002) may be an improvement on the community RISA fingerprinting method since it may minimize the "noise" and over-estimation of taxa that is produced by DGGE. This genetic profiling method begins with amplification from universal ribosome primers (as in RISA) from a mixed genome DNA extracted directly from soil. The universal primers are 5-prime-tagged with different fluorescent moieties, so amplicons end up with strand-specific tags. After amplification the PCR product is subjected to restriction digest, the reaction mixture is resolved on an acrylamide sequencing gel, and the restriction products of each strand are detected by fluorescence spectrophotometry. Investigators say that by using this technique with fungal specific universal primers and a series of restriction enzymes, some fungi can be identified to genus or species by matching up restriction patterns with a database of known restriction patterns (Dickie *et al.*, 2002). It seems essential that pattern analysis software be used for the matching process and that the "known" fungal RFLP database is produced in the same laboratory and preferably by the same technician to avoid size discrepancies from experimental variables and technician technique. The more fungal species in the soil sample the harder it will be to sort out the patterns. Also, as in RISA, the proportions of low abundance species will be even lower after PCR and may be undetectable by T-RFLP. However, with both RISA and T-RFLP a general concept of character and complexity of a fungal community of fungi can be obtained, and with T-RFLP, perhaps a more accurate qualitative and quantitative comparison among samples can be made.

IDENTIFYING AND MODELING THE SUBTERRANEAN DISTRIBUTION OF A SINGLE FUNGAL SPECIES

Many uncharacterized organisms are found in soil and an extraction of total DNA from soil will produce a complex, multigenomic sample. Only very specific molecular probes could be expected to allow the precise identification of a SOI from such a complex mixture. The following gives an overview of how specific primers were developed and successfully used to identify and model the underground distribution of *Tricholoma magnivelare* (Gordon, 2003). *T. magnivelare* is an ectomycorrhizal fungus that produces sporocarps valued for their culinary use and harvested commercially.

1. Sample collection: Soil samples were taken beneath the duff layer throughout a study site that included two small fruiting patches of *T. magnivelare.* Sporocarp tissue was also collected. The locations of collections were determined with a global positioning systems (GPS) unit.

2. Extraction: A standard DNA extraction protocol can be used with grinding or freezing to extract DNA from sporocarps (Gardes and Bruns, 1993). However, DNA extraction from soil presents significant problems especially for soils rich in humic substances (Yeates *et al.,* 1998; Ogram, 2000; Burgmann, 2001). In our experience, even kits that may be designed specifically for soil extraction may not give satisfactory results for certain soils. In Gordon's (2003) study, an extraction buffer containing bovine serum albumin, polyvinylpyrolidine, and sodium dodecylsulfate was most successful. It is important to develop an assay to test against false positives related to PCR inhibition from co-extracted contaminants. Spiking aliquots of study site soil with known amounts of target DNA or finely ground sporocarp tissue is a way to test for PCR inhibition.

3. Primer design: Gordon (2003) designed primers from a cloned and sequenced RAPD amplicon common to all sporocarps tested from several widely spaced Oregon populations of *T. magnivelare* (Carter, unpublished study). A BLAST search and extensive testing with related species indicated that the primers were suitably specific.

4. PCR conditions: Like the extraction procedure, PCR reaction mixes must be optimized. Non-specificity can be improved using hot-start or antibody-mediated PCR techniques and high (65°C) annealing temperature.

5. Detection: After PCR, gel electrophoresis, and staining with ethidium bromide, the presence of the *T. magnivelare* DNA amplicon was detected and quantified by a fluorescence imaging system, adjusted by an internal DNA mass standard included in the gel. The intent of this procedure was to acquire relative fluorescence value for the amplicon band of interest that was independent of the characteristics of each individual gel so that bands from different gels could be compared in a meaningful way.

6. Assay: Once standard conditions were established from optimization of extraction, PCR, and detection conditions, it was possible to develop an assay for estimating the concentration of *T. magnivelare* in individual soil samples. A standard curve of *T. magnivelare* concentration was developed by "dilutions" of known amounts of finely-ground sporocarp tissue with soil. The diluent soil was from a nearby site that had no detectable *T. magnivelare.* The enriched soil samples, ranging from 2000 ppm to 10 ppm, were subjected to DNA extraction, PCR, and detection according to optimized protocols. The relative fluorescence values of the enrichment series was graphed against concentration. A linear regression of all points, excluding the 2000 ppm point which showed saturation of the fluorescence detection, showed a high correlation ($r = 0.92$) between the known concentration of *T. magnivelare* tissue in the soil and the fluorescence of the specific band after PCR of the DNA extracted from the soil. The range of values was grouped into four categories (none, low, medium, and high). These categories were used to assign ranked concentration values to fluorescence results obtained from actual soil samples at the study site.

7. Distribution modeling: In Gordon's study (2003) the continuous distribution of *T. magnivelare* in the soil was inferred from the amplicon band fluorescence value (none to high) corresponding to the soil sample position. The data points for amplicon concentration value along with the corresponding GPS coordinates was analyzed with the "Interpolate Grid" function in the Geographic Information Systems Spatial Analyst Program (ArcView GIS 3.2, 1992-1999, Environmental Systems Research Institute). The analytical procedure divides the study area into cells and, for each cell, whether or not sampled, a value of intensity is calculated based on the nearest cells that have actual values, giving heavier weight to closer data points. A continuous plane representing a model of the distribution and concentration of *T. magnivelare* over a subset of the study site is shown in Fig. 1.2. In the model, the darker the shade, the higher is the predicted concentration of the fungus.

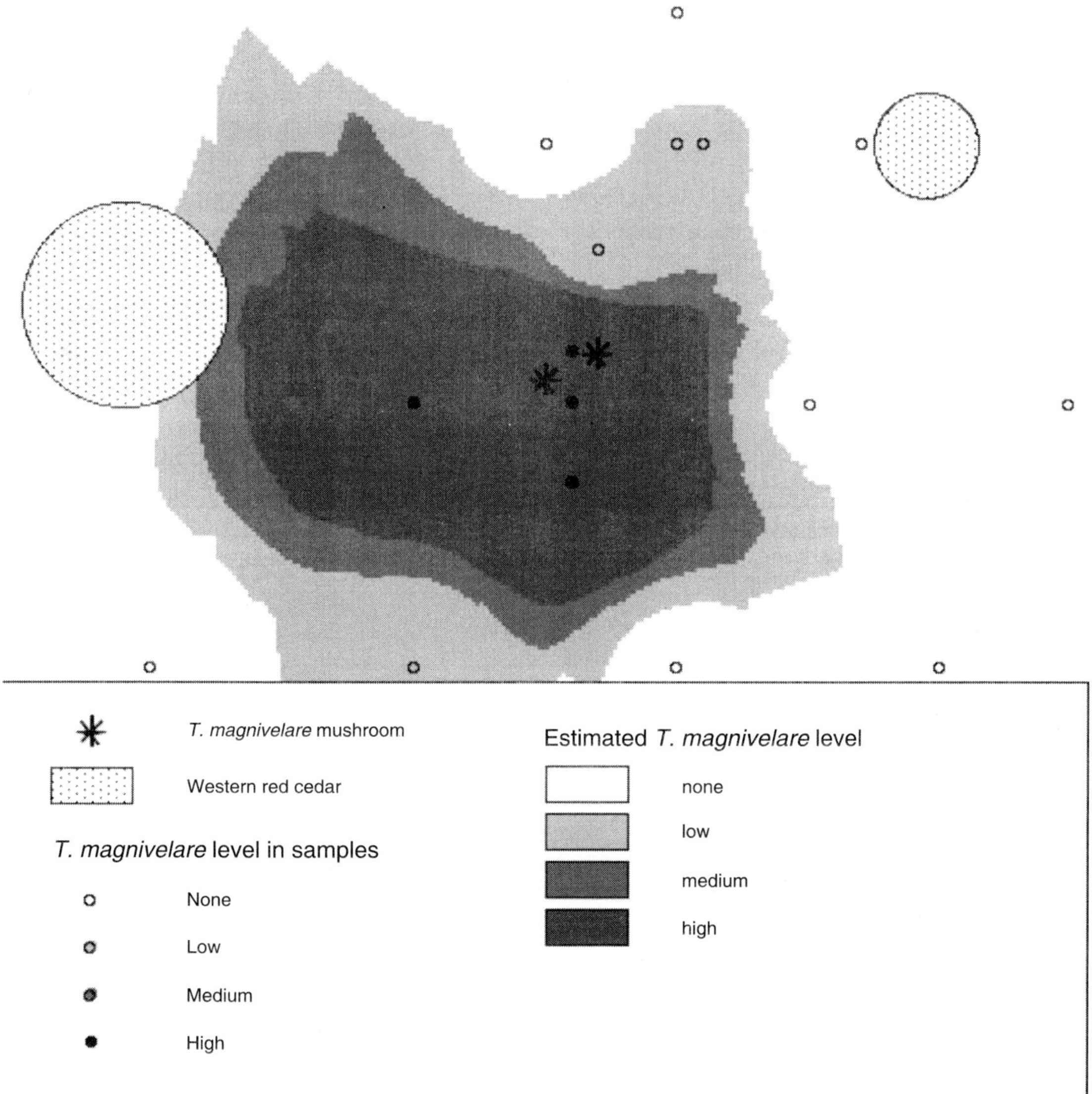

Fig. 1.2. Model of the distribution of *Tricholoma magnivelare* in soil at a forest site, based on an estimation of *T. magnivelare* DNA concentration in soil derived from a semiquantitative PCR assay.

FUTURE PERSPECTIVES AND CONCLUSIONS

Studies of soil organisms present significant challenges. Soil organisms are difficult or impossible to observe *in situ*, and, if one attempts to isolate them from the soil milieu, their character and the relationship to the whole may be altered. Molecular techniques can be used to study soil organisms, albeit with some obstacles, to yield significant information on the largely undescribed world of the soil biota.

Two types of molecular methods have been emphasized in this chapter—those that are based on molecular probes that elucidate community structure and those that are based on probes that allow detection, distribution, and temporal monitoring of a single organism.

To elucidate community structure, characterizing probes (*e.g.*, universal PCR primers) are used to generate complex community fingerprints. Individual species within the community are not usually identified, but community and temporal changes can be documented. Organisms that are present in relatively low concentration in soil samples will appear (in electrophoregrams) to be even lower after PCR with the characterizing primers used for community fingerprinting, since the higher concentration targets tend to be preferentially amplified in typical PCR reactions.

To study the biology of a single type of organism within soil, investigators can use specific molecular probes such as species-specific PCR primers. Species-specific PCR primers allow the presence, absence, and even the concentration of a species of interest (SOI) to be determined within a soil sample. It is likely that more accurate estimations of organism abundance in a mixed genome sample could be achieved by using real time PCR (RT-PCR) or quantitative PCR (QPCR) technology. RT-PCR and QPCR are relatively recent innovations that are mainly used to quantify gene expression levels but can be used to estimate concentrations of any DNA target in a sample at the start of a PCR. RT-PCR uses specific primers with fluorescent molecular probes and special PCR instrumentation that allow the monitoring of primer-targeted amplicon production, cycle by cycle. The advantage to real time monitoring rather than using end point determinations (such as described in the above *T. magnivelare* distribution study), to estimate starting concentrations of target DNA within samples, is that with RT PCR, calculations can be derived from the linear (exponential) phase of each PCR. RT-PCR has been used to monitor fungal populations in soil (Filion *et al.*, 2002; Lees *et al.*, 2002). The present drawback to RT-PCR technology is that the instrumentation is expensive and may be unavailable to many investigators.

Future studies may enlarge upon the ability of geoprocessing software such as that available through GIS to perform spatial analyses of the biotic component of soil. In addition to modeling the distribution of a single organism, multiple organisms could be modeled and displayed in site maps. Fungal community profiles may be subjected to cluster analysis and other statistical treatments (Dickie *et al.*, 2002) for determination of taxon distribution patterns significantly different from random. Emerging patterns in two or three dimensions can be displayed with geoprocessing software. Using the layering capabilities of the software, physical site characteristics can be integrated into models. Finally, distribution of an SOI can be considered in relationship to the fungal community as a whole, and other organisms such as trees and shrubs can be integrated into the analysis along with the physical features of sites.

REFERENCES

Agerer R. (1988). *Colour Atlas of Ectomycorrhizae.* Einhorn-Verlag Eduard Dietenberger, Munich.

Altschul S.F., Gish W., Miller W., Myers E.W. and Lipman D.J. (1990). Basic local alignment search tool. *Journal of Molecular Biology* 215: 403-410.

Bertini L., Agostini D., Potenza L., Rossi I., Zeppa S., Zambonelli A. and Stocchi V. (1998). Molecular markers for the identification of the ectomycorrhizal fungus *Tuber borchii. New Phytologist* 139: 565-570.

Borneman J. and Hartin R.J. (2000). PCR primers that amplify fungal rRNA genes from environmental samples. *Applied and Environmental Microbiology* 66 (10): 4356-4360.

Burgmann H., Pesaro M., Widmer F. and Zeyer J. (2001). A strategy for optimizing quality and quantity of DNA extracted from soil. *Journal of Microbiological Methods* 45: 7-20.

Campbell N.A. and Reece J.B. (2002). *Biology, Sixth Ed.*, Benjamin Cumming, San Francisco.

Castellano M.A., Trappe J.M., Maser Z. and Maser C. (1989). *Key to Spores of the Genera of Hypogeous Fungi of North Temperate Forests.* Mad River Press, Eureka, CA.

Cole J.R., Chai B., Marsh T.L., Farris R.J., Wang Q., Kulam S.A., Chandra S., McGarrell D.M., Schmidt T.M., Garrity G.M.and Tiedje J.M. (2003). The Ribosomal Database Project (RDP-II): previewing a new autoaligner that allows regular updates and the new prokaryotic taxonomy. *Nucleic Acids Research* (1): 442-3.

Comes A.M., Humbert J.F. and Laurent F. (1997). Rapid cloning of PCR-derived RAPD probes. *BioTechniques* 23: 210-212.

Dickie I.A., Bing X. and Koide R.T. (2002). Vertical niche differentiation of ectomycorrhizal hyphae in soil as shown by T-RFLP analysis. *New Phytologist* 156: 527-535.

Dix N.J. and Webster J. (1995). *Fungal Ecology.* Chapman & Hall, London.

Egger K.N. (1985). Molecular analysis of ectomycorrhizal fungal communities. *Canadian Journal of Botany* 73(suppl. 1): S1415-S1422.

Elder J.F. Jr. and Turner B.J. (1995). Concerted evolution of repetitive DNA sequences in eukaryotes. *The Quarterly Review of Biology* 70: 297-320.

Erland S., Henrion B., Martin F., Glover L.A. and Alexander I.J. (1994). Identification of the ectomycorrhizal basidiomycete *Tylospora fibrillosa* Donk by RFLP analysis of the PCR-amplified ITS and IGS regions of ribosomal DNA. *New Phytologist* 126: 525-532.

Filion M., St.-Arnaud M., Hamel C. and Jabaji-Hare S.H. (2002). Use of quantitative real-time PCR to monitor fungal populations in soil. *Phytopathology* 92(6 suppl.): S143.

Ganley A.R.D. and Scott B. (2002). Concerted evolution in the ribosomal RNA genes of an *Epichloe* endophyte hybrid: comparison between tandemly arranged rDNA and dispersed 5S rrn genes. *Fungal Genetics and Biology* 35: 39-51.

Gardes M. and Bruns T.D. (1993). ITS primers with enhanced specificity for basidiomycetes—application to the identification of mycorrhizae and rusts. *Molecular Ecology* 2: 113-118.

Giomaro G, Zambonelli A., Sisti D., Cecchini M., Evangelista V. and Stocchi V. (2000). Anatomical and morphological characterization of mycorrhizas of five strains of *Tuber borchii* Vittad. *Mycorrhiza* 10: 107-114.

Gordon M. (2003). A species-specific molecular probe for *Tricholoma magnivelare*, the American matsutake. Masters Thesis, Portland State University, Portland, OR 97207.

Gutell R.R. and Fox G.E. (1988). A compilation of large subunit RNA sequences presented in a structural format. *Nucleic Acids Research* 16: r175-r269.

Henrion B., Chevalier G. and Martin F. (1994). Typing truffle species by PCR amplification of the ribosomal DNA spacers. *Mycological Research* 98(1): 37-43.

Heredia G., Reyes M., Arias R.M. and Bills G.F. (2001). *Talaromyces ocotl* sp. nov. and observations on *T. rontundus* from conifer forest soils of Veracruz State, Mexico. *Mycologia* 93(3): 528-540.

Hillis D.M. and Dixon M.T. (1991). Ribosomal DNA: molecular evolution and phylogenetic inference. *The Quarterly Review of Biology* 66(4): 411-453.

Hongyo T., Buzard G.S., Calvert R.J. and Weghorst C.M. (1993). 'Cold SSCP': a simple, rapid and non-radioactive method for optimized single-strand conformation polymorphism analyses. *Nucleic Acids Research* 21(16): 3637-3642.

Jackson C.J., Barton R.C. and Evans *E.G.*V. (1999). Species identification and strain differentiation of dermatophyte fungi by analysis of ribosomal-DNA intergenic spacer regions. *Journal of Clinical Microbiology* 37(4): 931-936.

Jobes D.V. and Thien L.B. (1997). A conserved motif in the 5.8S ribosomal RNA r(RNA) gene is a useful diagnostic marker for plant internal transcribed spacer (ITS) sequences. *Plant Molecular Biology Reporter* 15: 326-334.

Kernaghan G. (2001). Ectomycorrhizal fungi at tree line in the Canadian Rockies II. Identification of ectomycorrhizae by anatomy and PCR. *Mycorrhiza* 10: 217-229.

Kikuchi K., Matsushita N., Guerin-Laguette A., Ohta A. and Suzuki K. (2000). Detection of *Ticholoma matsutake* by specific ITS primers. *Mycological Research* 104(12): 1427-1430.

Kim S.H. and Breuil, C. (2001). Common nuclear ribosomal internal transcribed spacer sequences occur in the sibling species *Ophiostoma piceae* and *O. quercus. Mycological Research* 105(3): 331-337.

Kjöller R. and Rosendahl S. (2001). Molecular diversity of glomalean (arbuscular mycorrhizal) fungi determined as distinct *Glomus* specific DNA sequences from roots of field grown peas. *Mycological Research* 105(9): 1027-1032.

Koschinsky S., Peters S., Schwieger F. and Tebbe C.C. (2000). Applying molecular techniques to monitor microbial communities in composting processes. In*: Microbial Biosystems: New Frontiers* (Proceedings of the 8th International Symposium on Microbial Ecology, Bell C.R., Brylinsky M. and Johnson-Green P., Eds., Atlantic Canada Society for Microbial Ecology, Halifax, Canada, pp. 461-468.

Kozdroj J. and van Elsas J.D. (2000). Application of polymerase chain reaction-denaturing gradient gel electrophoresis for comparison of direct and indirect extraction methods of soil DNA used for microbial community fingerprinting, *Biology and Fertility of Soils* 31(5): 372-378.

Kraigher H., Agerer R. and Javornik B. (1995). Ectomycorrhizae of *Lactarius lignyotus* on Norway spruce, characterized by anatomical and molecular tools. *Mycorrhiza* 5: 175-180.

Lees A.K., Cullen D.W., Sullivan L. and Nicolson M.J. (2002). Development of conventional and quantitative real-time PCR assays for the detection and identification of *Rhizoctonia solani* AG-3 in potato and soil. *Plant Pathology* 51(3): 293-302.

Lefevre C. and Muller W. (1998). *Tricholoma magnivelare* (Peck) Redhead. In: *Mycologue Publications and Forest Resource Development Agreement.* Canadian Forest Service, Victoria, Canada, CDE18.1-CDE18.5.

Liu W.T., Marsh T.L., Cheng H. and Forney L.J. (1997). Characterization of microbial diversity by determinng terminal restriction fragment length polymorphiosms of genes encoding 16S rRNA. *Applied Environmental Microbiology* 63: 4516-4522.

Long E.O. and Dawid I.B. (1980). Repeat genes in eukaryotes. *Annual Review of Biochemistry* 49: 727-764.

Millner P.D., Mulbry W.W. and Reynolds S.L. (2001). Taxon-specific oligonucleotide primers for detection of *Glomus etunicatum. Mycorrhiza* 10: 259-265.

Mullis K.B. and Faloona F.A. (1987). Specific synthesis of DNA *in vitro* via a polymerase-catalysed chain reaction. *Methods in Enzymology* 155: 335-350.

O'Dell T.E., Smith J.E., Castellano M. and Luoma D.L. (1996). Diversity and conservation of forest fungi. In: Managing forest ecosystems to conserve fungus diversity and sustain wild mushroom harvests, Pilz D. and Molina R., Eds., USDA Forest Service Gen. Tech. Rep. PNW-GTR-371. Pac. Northwest Res. Stn., Portland, OR. Pp. 5-18.

Ogawa M. (1985). Ecological characters of ectomycorrhizal fungi and their mycorrhizae. *Japanese Agricultural Research Quarterly* 18(4): 305-314.

Ogram A. (2000). Soil molecular microbial ecology at age 20: methodological challenges for the future. *Soil Biology & Biochemistry* 32: 1499-1504.

Penner G.A., Bush A., Wise R., Kim W., Domier L., Kasha K., Laroche A., Scoles G., Molnar S.J. and Fedak G. (1993). Reproducibility of random amplified polymorphic DNA (RAPD) analysis among laboratories. *PCR Methods and Applications* 2: 341-345.

Peters S., Koschinsky S., Schwieger F. and Tebbe C.C. (2000). Succession of microbial communities during hot composting as detected by PCR-single-strand-conformation polymorphism-based genetic profiles of small-subunit rRNA genes. *Applied and Environmental Microbiology* 66(3): 930-936.

Ranjard L., Poly F., Combrisson J., Gourbiere F., Richaume A., Thioulouse J. and Nazaret S. (2000). Heterogeneous cell density and genetic structure of bacterial pools associated with various soil microenvironments as determined by enumeration and DNA fingerprinting approach (RISA). *Microbial Ecology* 39: 263-272.

Ranjard L., Poly F., Lata J.-C., Mougel C., Thioulouse J. and Nazaret S. (2001). Characterization of bacterial and fungal soil communities by automated ribosomal intergenic spacer analysis fingerprints: biological and methodological variability. *Applied and Environmental Microbiology* 67(10): 4479-4487.

Rayner A.D.M. (1991). The challenge of the individualistic mycelium. *Mycologia* 83(1): 48-71.

Simon L., Lalonde M. and Bruns T.D. (1992). Specific amplification of 18S fungal ribosomal genes from vesicular-arbuscular endomycorrhizal fungi colonizing roots. *Applied and Environmetal Microbiology* 58: 291-295.

Smith M.L., Bruhn J.N. and Anderson J.B. (1992). The fungus *Armillaria bulbosa* is among the largest and oldest of living organisms. *Nature* 356: 428-431.

Soderstrom B.E. (1975). Vertical distribution of microfungi in a spruce forest soil in the south of Sweden. *Transactions of the British Mycological Society* 65: 419-25.

Soltis P.S., Soltis D.E., Wolf P.G., Nickrent D.L., Chaw S.M. and Chapman R.L. (1999). The phylogeny of land plants inferred from 18S rDNA sequences: pushing the limits of rDNA signal?. *Molecular Biology and Evolution* 16: 1774-1784.

Weber J., Diez J., Selosse M.-A. Tagu D. and LeTacon F. (2002). SCAR markers to detect mycorrhizas of an American *Laccaria bicolor* strain inoculated in European Douglas-fir plantations. *Mycorrhiza* 12: 19-27.

Welsh J. and McClelland M. (1990). Fingerprinting genomes using PCR with arbitrary primers. *Nucleic Acids Research* 18: 7213-7218.

White T.J., Bruns T.D., Lee S. and Taylor J. (1990). Amplification and direct sequencing of fungal ribosomal RNA genes for phylogenetics. In: *PCR Protocols: A Guide to Methods and Applications*, Innis M.A., Gelfand D.H., Sninsky J.J. and White T.J., Eds., Academic Press, New York, 315-322.

Wicklow D.T. and Whittingham W.F. (1978). Comparison of soil microfungal populations in disturbed and undisturbed forests in Northern Wisconsin. *Canadian Journal of Botany* 56: 1702-1709.

Williams J.G.K., Hanafey, M.K., Rafalski J.A. and Tingey S.V. (1993). Genetic analysis using random amplified polymorphic markers. *Methods of Enzymology* 218: 704-740.

Williams J.G.K., Kubelik A.R., Livak K.J., Rafalski J.A. and Tingey S.V. (1990). DNA polymorphisms amplified by arbitrary primers are useful as genetic markers. *Nucleic Acids Research* 18: 6531-6535.

Yeates C., Gillings M.R., Davison A.D., Altavilla N. and Veal D.A. (1998). Methods for microbial DNA extraction from soil for PCR amplification. *Biological Procedures Online* 1(1): 40-47, (www.biologicalprocedures.com).

Yeh W.-J., Len S.-C and Su H.M. (2002). Accession AF479758, NCBI Database, http://www.ncbi.nlm.nih.gov/.

2 Bioremediation of Aged Pentachlorophenol Residues by *Trametes* Spp.

Monika Walter[1], Chris I. Ford[1,2], Jenness M. Guthrie[1,3], Kirsty S. H. Boyd-Wilson[1], Hong J. Di[2], Keith C. Cameron[2] and Emily J. Parker[3]

[1]HortResearch, Environment and Risk Management Group, Gerald St, Lincoln, New Zealand [2]Lincoln University, Centre for Soil and Environmental Quality, Lincoln, New Zealand, [3]Massey University, Institute of Fundamental Sciences, Private Bag 11 222, Palmerston North, New Zealand

INTRODUCTION

Bioremediation of xenobiotics *in vitro* or in spiked soil has been widely studied. Bioremediation of aged soil residues, however, is not well understood. Chlorophenols, such as pentachlorophenol (PCP), can undergo complex physico-chemical reactions in soil, which are temporally dependent, for example, on soil type, organic matter content and pH. This, in turn, affects the amount of potential PCP and congeners in solution, the risk of leaching and/or availability for microbial degradation processes. To understand bioremediation of aged PCP residues by white-rot fungi, knowledge of the PCP behaviour in soil and the induction of the biodegrading enzyme system is required as well as survival of the augmented organism in the contaminated substrate and maintenance of its remedial activity, are discussed.

Pentachlorophenol is the main pollutant considered with other relevant compounds such as polycyclic aromatic hydrocarbons (PAHs), as this class of compounds are often co-contaminants with PCP and lower substituted chlorophenols (Lamar *et al.*, 1993a). These are all regarded as priority pollutants by the United States Environmental Protection Agency (EPA) and are a threat to the environment and to human health (Czaplicka, 2004). They are persistent organic pollutants (POPs), which are largely insoluble, or sparingly soluble, in water at normal soil pH. Their chemical and biocidal nature also renders them resistant to microbial degradation at even low concentrations. For example, PCP acts as a biocide in soil/aqueous systems at concentrations as low as 4 mgL^{-1} (Davis *et al.*, 1994).

The aim of this chapter is to review and describe the fate of PCP in the soil environment, such as sorption and sequestering that directly influence its bioavailability for biodegradation. The enzyme systems involved in bioremediation of PCP by white-rot fungi, as well as the growth substrate requirements for successful white-rot fungal soil colonisation and induction of these enzymes have been described. We have also discussed the costs and obstacles faced in commercial bioremediation.

PCP USAGE IN NEW ZEALAND

Technical grade PCP (Table 2.1) was widely used as an antifungal treatment in the timber industry between 1950 and 1990. An estimated 5000 tonnes of PCP was used at timber treatment sites throughout New Zealand over a period of approximately 40 years (Finnbogasson, 1994). PCP was also used domestically for moss and algae control, for home treatment of timber by farmers, industrially to control slime in cooling waters in pulp and paper production, and in mushroom culture as a soil and timber steriliser (Stevenson, 1992). Although the use of PCP by the forestry industry voluntarily ceased in 1988, it was not deregistered and import-prohibited until December 1991. This widespread use of the chemical resulted in a variety of "waste streams" including contaminated soils from drippage and spills. As a result, there are about 800 contaminated timber treatment sites throughout New Zealand (Nadebaum *et al.*, 1992a; 1992b; Ellis *et al.*, 1993; Szabo, 1993).

Table 2.1. Generic technical grade pentachlorophenol formulation (Davis *et al.*, 1994)

Constituents	*Fraction (%; w/w)*
Pentachlorophenol	85-90
Tetrachlorophenol	4-8
Trichlorophenol	0.1
Chlorinated phenoxyphenols, dioxins and furans	2-6

Technical grade PCP is a toxic biocide and carcinogenic to laboratory animals (NTP, 1989). The New Zealand Ministry for the Environment has published soil acceptance criteria for acceptable PCP levels in soils (Table 2.2). These restrictions mean that amelioration and remediation of heavily contaminated sites with PCP residues is necessary before these sites can be used for other purposes.

Table 2.2. Soil acceptance criteria for PCP levels in New Zealand soils (Ellis *et al.*, 1997)

Future land use	*Soil PCP levels (mg kg^{-1}; dry weight basis)*
Industrial (unpaved)	570
Industrial (paved)	1000
Agricultural land	0.7
Residential land (rural)	1.4
Residential land (average)	7

While technical grade PCP has been the main fungicide used in timber treatment in New Zealand, a wide variety of chemicals was used alone and/or in combination with PCP (Nadebaum *et al.*, 1992a; 1992b). These include boron for timber framing for houses, plus copper, chrome and arsenic treatments for ground contact posts, decking and marine piles. In addition, creosote was used for railway sleepers and ground contact fence posts, and PCP as an antifungal treatment (aqueous) for freshly sawn timber for domestic use (0.5% w/v) and timber for export (0.75% w/v). The use of PCP in oil consisted of a ~ 4% w/v solution in fuel oil (Nadebaum *et al.*, 1992a) to treat fence posts, and poles of species such as Douglas fir that were untreatable by aqueous based solutions. The boron solutions, utilised for the boron diffusion process, also contained 0.2% NaPCP (Nadebaum *et al.*, 1992a). The main source of polycyclic aromatic hydrocarbon (PAH) contamination in New Zealand is from the use of creosote in timber treatment, as well as the sites of old coal gas plants. Globally, PAH contamination is generated by combustion of fossil fuels, coal gasification and asphalt production (Aamot *et al.*, 1996; Knox *et al.*, 1999).

FATE OF PCP IN THE ENVIRONMENT

The solubility of PCP at pH 10 is 478 000 mg L^{-1} compared with 15 mg L^{-1} at pH 5.0 (Christodoulatos and Mohiuddin, 1996). Therefore, PCP was applied to untreated timber in a buffered solution with sodium tetraborate at pH 10 (Nadebaum *et al.*, 1992a). The anionic form of PCP is more mobile in soils than the ionic forms. Movement of PCP in soil is mainly by advection with the draining water (Boyd *et al.*, 1989; Davis *et al.*, 1994; Mills *et al.*, 2006). Fugacity models for PCP movement off site have been reported (Pollard *et al.*, 1993; Chi and Huang, 2002a; 2002b). In New Zealand, e.g., the result of movement off site is that of dibenzo-*p*-dioxin (a technical grade PCP impurity). Octa and hepta substituted congeners have been found at levels in Lake Rotorua higher than reported in the lower reaches of the Rhine (Szabo, 1993).

The environmental fate of PCP has been described by Boyd *et al.*, (1989), and transformations of chlorophenols in the environment have recently been reviewed by Czaplicka (2004). Briefly, upon entering the environment, PCP may volatilise, photo-decompose, be taken up by plants, sorbed onto soil particles, oxidatively coupled onto soil humus and/or decomposed by microorganisms via methylation, dechlorination, hydroxylation and/or mineralization. Bioremediation of aged PCP residues in soil is largely influenced by its bioavailability. Therefore, the processes governing its bioavailability (sorption, sequestration and aging) are further described in detail. A lack of understanding, for example, of sorption processes has led to some failures of *in situ* remediation projects (Weber Jr *et al.*, 1999).

SORPTION

The term sorption is used to denote reactive exchange of a solute onto soil without reference to a specific mechanism. Hydrophobic organic chemicals mainly bind to the organic matter in the soil and also, to a lesser extent, the fine mineral fraction (clay and silt) (Chiou, 1989; Hasset and Banwart, 1989; Pignatello, 1989; Aislabie *et al.*, 1997). Two opposing hypotheses exist for the sorption of hydrophobic organic chemicals to soil. One view maintains that the process is by physical adsorption to the organic matter, in which physical binding of the solute to the organic matter occurs (Hasset and Banwart, 1989; Pignatello, 1989). The other view holds that the hydrophobic moieties exist in the soil organic phase, because they diffuse or partition into the solid organic matter, much as a hydrophobic compound partitions between aqueous solution and an organic solvent in which it is highly soluble (Chiou, 1989).

Partition is defined as "the distribution of a compound between different bulk phases by forces common to solution" (i.e., a solubilisation effect) (Smith *et al.*, 1988). This relationship is expressed by the organic carbon partitioning coefficient (K_{oc})[1]. The degree of partitioning can be evaluated experimentally from the slope of the adsorption isotherm and calculated from the organic fraction of the soil (f_{oc}) (Karickhoff, 1984; Chiou, 1989; Hasset and Banwart, 1989; Pignatello, 1989). A partition process gives a linear isotherm[2], whereas any deviation from linearity is usually an indication of a slower diffusion, or a kinetic adsorption mechanism, rather than simple partitioning (Pignatello, 1989; Weber Jr *et al.*, 1999; Divincenzo and Sparks, 2001).

SORPTION TO ORGANIC MATTER

The carbon content in soil organic matter has been reported generically as 56% (Pollard *et al.*, 1993). However, earlier studies (Lee *et al.*, 1990; Christodoulatos *et al.*, 1994; Christodoulatos and Mohiuddin, 1996) generally did not factor in the heterogeneous nature of soil organic matter when generating empirical models. The sorptive nature of heterogeneous soil organic matter is further enhanced by the presence of creosote and carrier oils as forms of NAPLs used for timber preservatives like PCP (Fall *et al.*, 2001). The term NAPL (non aqueous phase liquid) encompasses both DNAPL (D = dense) and LNAPL (L = Light). The effect of sorption of PCP in NAPLs was found to be similar to soil organic matter (Fall *et al.*, 2001). However, it was reported elsewhere (Pollard *et al.*, 1993) that oils present in soil, whether they are creosotes or the carrier oils used for delivery of the timber preservative, can significantly sequester hydrophobic organic compounds; their partition coefficients can be an order of magnitude greater than for soil organic matter.

The most commonly reported figure for the pKa of PCP is 4.75 (Shiu *et al.*, 1994; Christodoulatos and Mohiuddin, 1996; Fall *et al.*, 2001). Below this pH, PCP is predominantly in the neutral form and behaves as a neutral hydrophobic compound. Partitioning onto soil organic matter is the more dominant mechanism of sorption. The partition coefficients for the neutral form are far higher than for the more soluble anionic form (Shiu *et al.*, 1994; Christodoulatos and Mohiuddin, 1996; Wu *et al.*, 2000; Fall *et al.*, 2001).

A significant number of studies generated empirical mathematical models describing PCP sorption to soil organic matter (Davis *et al.*, 1994; Christodoulatos and Mohiuddin, 1996; Fall *et al.*, 2000; Fall *et al.*, 2001; Wu *et al.*, 2001). The K_{oc} values for a number of these have been presented for comparative purposes (Christodoulatos and Mohiuddin, 1996; Wu *et al.*, 2000). These K_{oc} values for PCP are sensitive to changes in pH in the range between the pKa values of 4.75 and 6.0. This range corresponds to the change from PCP being predominantly in the neutral phenol form, to being predominantly in the anionic form.

A major flaw with empirical mathematical models is the tendency to ignore sorption dynamics. They assume that the sorbate rapidly reaches equilibrium between the sorbent and aqueous phase (Pignatello, 1989). Sorption effects can produce apparent hysteresis curves. However, some useful results have been presented from the use of multiple linear regression (Fall *et al.*, 2001), in particular the isolation of the different effects that contribute to PCP sorption. Soil organic matter (f_{oc}) and pH, when included in the model together, described the majority of the differences observed for PCP

1 Normalised partition coefficent (K_{oc}, with respect to organic carbon), $K_{oc} = K_d/f_{oc}$, f_{oc} = organic carbon content.
2 Linear model, $q_e = K_dC_e$, q_e = amount of solute sorbed per unit weight of soil, K_d = partition coefficient of the solute and C_e = the solute equilibrium concentration in the liquid phase.

sorption between the eight soils tested, and pH and f_{oc} were not highly correlated with each other (Fall *et al.*, 2001). Only the presence of oil and surfactant (triton) had significant effects on PCP sorption. However, in a solubility study, ionic strength was found to be significant for the anionic form of PCP above pH 8 (Stapleton *et al.*, 1994).

SORPTION TO CLAYS AND MINERALS

Soils with organic matter contents of less than 0.1% exhibit a different sorption mechanism. The more polar trichlorophenol adsorbs more strongly to the mineral components than the less polar pentachlorophenol or tetrachlorophenol (Chiou, 1989; Hasset and Banwart, 1989; Zielke *et al.*, 1989; Davis *et al.*, 1994; Wu *et al.*, 2000). It has been suggested that this sorption mechanism is masked in the presence of soil organic matter. Another theory is that in water-saturated soils these adsorption sites are taken up by the strong dipole interaction between water and the mineral surfaces (Miller *et al.*, 1989).

Clays have negatively charged and hydrophilic surfaces that tend to bind cations in solution, and repel hydrophobic organic compounds (Boyd *et al.*, 1988a, 1988b; McClaren and Cameron, 1996). Non-clay minerals generally have a low hydrophobic organic compound sorptive capacity. The 1:1 clays such as kaolinite tend to have low cation exchange capacity (CEC) values and therefore the sorptive capacity for hydrophobic organic compounds is limited. The 2:1 smectite clays (swelling clays) expand in the presence of water because of the physical and chemical nature of the clay sheets (McClaren and Cameron, 1996). This also allows cations to penetrate the interior of the clay. The cations in the interior, and on the surface of the clay, can be replaced with charged organic cations such as alkyl ammonium compounds. The hydrophobic part of these organic cations can be superior sorbents when compared with sorption for hydrophobic compounds and PCP on soil organic matter (Brixie and Boyd, 1994). These compounds are termed organoclays, organosmectites, organobentonites, HDTMA-clay or hexadecyltrimethylammonium-exchanged clays (Jaynes and Boyd, 1991; Brixie and Boyd, 1994; Stapleton *et al.*, 1994; Soule and Burns, 2001).

Smectite clays such as montmorillonite can act as concentrating agents by sorbing organic molecules to their surfaces. This can result in the increased utilisation by microorganisms or the sequestering of these compounds, resulting in a reduced bioavailability of the organic compounds (Alexander, 1999). Chlorophenols adsorbed on to montmorillonite (Miller *et al.*, 1989) when the clay was treated with hexadecyltrimethylammonium (HDTMA). The chlorophenol homologous series was also tested for its affinity to the clay alkyl ammonium complex. The order was, phenol < 3-chlorophenol < 3,5-dichlorophenol < 3,4,5-trichlorophenol and 2,4,6-trichlorophenol (Zielke *et al.*, 1989). At times, sorption stimulates microbial activity and at other times sorption retards microbial breakdown (Alexander, 1999).

Studies have investigated PCP sorptive partitioning into the hydrophobic alkyl chains from HDTMA, di-octadecyl-dimethyl ammonium (DODMA), and several other ammonium ions, cationically bound to bentonite-producing organoclays (Boyd *et al.*, 1988a, 1988b; Miller *et al.*, 1989). The less hydrophobic, lower molecular ammonium ions sorbed considerably less PCP than the more hydrophobic quaternary alkyl ammonium compounds (HDTMA and DODMA). The resulting PCP sorption for HDTMA and DODMA organoclays was reported as a partition process and 10 times greater than the PCP sorption to soil organic matter (pH 5.5) (Boyd *et al.*, 1988a, 1988b; Miller *et al.*, 1989). In addition, further studies have reported up to 30 times greater PCP sorptive capacity of organoclays than with soil organic matter (Brixie and Boyd, 1994). This partition value for Dicocodimethyl-ammonium (DICOCO) organoclay was similar to the octanol-water partitioning coefficient for PCP at the given pH (Shiu *et al.*, 1994). Therefore, sorption of PCP closely approximated the hypothesis of partitioning into the more hydrophobic compounds of the series.

The increased sorption of anionic PCP at higher pH values cannot be explained by hydrophobic partitioning. Two mechanisms of anionic PCP sorption to organoclay complexes are presented in the literature (Li *et al.*, 2000): (1) The hydrophobic part of the anionic PCP is sorbed on to the hydrophobic tail of the QUAT (quaternary alkyl ammonium compounds). (2) Some QUAT moieties form a double layer by their hydrophobic tails partitioning onto the QUAT compounds sorbed on the clay surfaces. The anionic head group of PCP is then free to bind with cationic ammonium ion in the second layer. The limiting factor in both the mechanisms is the mutual repulsion of the anionic PCP head groups.

The use of organoclay complexes to reduce leachable PCP concentrations from contaminated soils has been studied in detail (Brixie and Boyd, 1994). It was discovered that a 20% w/w organoclay complex in soil, incorporating DICOCO as the QUAT compound, reduced the leachable PCP levels in the soil to below 0.2 mg L^{-1}. Lower amounts of this organoclay complex still had significant effects on the level of PCP in solution of all the soils tested. At a level of 3% w/w, 60-80% of PCP was removed from solution in the soil contamination range of 3000-5000 mg kg^{-1}.

Organoclay-modified soils contaminated with 2,4-D, phenanthrene and naphthalene (Nye *et al.*, 1994) can enhance biodegradation processes and have aided biodegradation *in vitro* of 2,4-dichlorophenol (Witthuhn *et al.*, 2005). This is achieved by reversibly sorbing the chlorophenol that would otherwise inhibit biodegradation. As concentrations in solution drop, the sorbed chlorophenol is desorbed back into solution to re-establish equilibrium. Therefore, potential exists to detoxify and enhance white-rot fungal biodegradation of more highly PCP contaminated soil than previously reported (Lamar and Dietrich, 1990a, 1990b; Leštan and Lamar, 1996; Walter *et al.*, 2005a, 2005b). Organoclays and their properties were first discovered in 1950 and are routinely utilised in industrial applications such as thickeners in paints, greases and oil drilling muds, and to gel various organic liquids (Murray, 1995). The sorption of POPs by organoclays can be regarded as nanotechnology (Yuan, 2004) because the distance between the swelling clay sheets is typically 2.0nm (McClaren and Cameron, 1996).

AGEING AND SEQUESTERING: HYSTERESIS

The organic compounds that sorb onto soil organic matter rarely desorb at the same rate and with identical partition coefficients. This difference is commonly called hysteresis (Pignatello, 1989). The reasons reported for hysteresis are:

1. Changes effected by biological agents: conversion into biomass; degradation into simpler compounds; polymerisation into more complex compounds and hydrogen and/or covalent bonding to the soil organic matter (Bollag *et al.*, 1992).

2. Abiotic mechanisms, such as competitive adsorption from natural adsorbents and a failure to establish equilibrium.

Irreversible PCP sorption, or extreme hysteresis, has been reported at low pH in podzol soils (Fall *et al.*, 2000). The explanation for this extreme hysteresis is that the partition coefficients for the neutral form of PCP, which is the dominant form at low pH, are orders of magnitude higher than for the anionic form (Divincenzo and Sparks, 1997). This phenomenon appears to contradict partition theory. However, one argument for this extreme case is that a potential energy barrier exists because of the resulting decrease of entropy that would be observed if a hydrophobic sorbate were desorbed back into solution (Pignatello, 1989). The added effect of condensed aged organic matter, as explained previously, could also be a major contributing factor to this extreme hysteresis. The neutral form of PCP has a higher affinity for this material than the anionic form. However, in other studies, the opposite has been recorded where there is little hysteresis at pH 4.0 (Divincenzo and Sparks, 2001), yet an increased level of hysteresis at pH 8.0. This sorption of the ionised form was adequately described by the Langmuir[3] isotherm (Barrow, 1979; Atkins and de Paula, 2002), which suggests a surface binding mechanism. Also, the K_d for the ionised species was significantly reduced at higher temperature (Divincenzo and Sparks, 2001).

SEQUESTERING

In addition to hysteresis, non-sorbed compounds can be sequestered by slowly diffusing into soil micropores and nanopores (Pignatello, 1989). The release of organic compounds sequestered in this way can be extremely slow and limit bioremediation rates. In some soils, the portion of the void volume that exists in nanopores (ϕ < 200 nm) can be as high as 30% and they have also been found to a depth of 1.0 m (Alexander, 1999). This slow diffusion process is described by the concept of tortuosity (McClaren and Cameron, 1996; Alexander, 1999), because transport pathways are not always straight and the molecule typically may undergo several sorption/desorption steps before it enters void spaces large enough to be available for biodegradation. Slow sorption is concentration dependent and considerably higher diffusion rates occur at elevated concentrations of PCP in the aqueous phase (Divincenzo and Sparks, 1997).

Some organic pollutants can be readily extracted from the soil and aquifer environment by conventional procedures. The evidence of reduced bioavailability of these compounds is a decreased rate of biodegradation. This can reach the stage where only vigorous extraction techniques can remove the compound in question. These compounds are termed aged or weathered residues. Such aged compounds are often referred to as being sequestered because the compound has become inaccessible, or unavailable for biodegradation. Some compounds reach the stage where they cannot be removed by even the most vigorous extraction techniques. Yet the compound remains in a form that is recognisable as the parent molecule or a metabolite derived from it. These compounds are generally known as bound residues. The mechanism of binding is not necessarily covalent bonding as complexing of the parent compound with humic materials in soils is also a common mechanism (Rüttimann-Johnson and Lamar, 1997; Alexander, 1999; Tuomela *et al.*, 1999). In general, the

3 Langmuir equation, $\theta = k_2(k_1+k_2P)$, θ = fraction of surface sorption sites occupied, k_1 = desorption rate, k_2 = adsorption rate, P = concentration.

processes of ageing and sequestering are poorly understood. The following mechanisms are presented most often for non-sorbed compounds.

Partitioning into a solid absorbent (Chiou, 1989; Hasset and Banwart, 1989), as mentioned previously, results in some compounds sorbing rapidly into the less condensed high O/C ratio humic material. However, the more condensed form of humic material tends to sequester compounds by a slow diffusion process (Weber Jr *et al.*, 1999). There is evidence that both mechanisms are responsible for the sequestering process, through solids from laboratory scale studies, on defined sorbents via diffusion into micro/nanopores and partitioning (Alexander, 1999).

Oxidative coupling is a process that determines the fate of chlorophenols in soils and is responsible for deviations from isotherm linearity. This process is analogous to the polymerisation of natural phenolic compounds into soil organic matter. The degree of chlorophenol coupling appears to be inversely proportional to the amount of chlorine substitution due to the electron withdrawing effects of chlorine on the benzene ring (Pignatello, 1989; Roy-Arcand and Archibald, 1991). The reaction can be catalysed by phenol oxidases (laccases) produced by bacteria, white-rot fungi and plants, or by various types of soil minerals. It can also be auto-oxidative. The end product of oxidative coupling is a covalently bound residue that is strongly immobilised and resistant to biological biodegradation. This has been suggested as a process for decontaminating polluted soils (Bollag, 1992). These polymerised compounds in solution are insoluble and precipitate out. Oxidative coupling is thought responsible for the dehalogenation of chloride in chlorophenols rather than the enzymatic processes promoted by microbes that are required to initiate the reaction (Dec and Bollag, 1994).

Bound residues are not considered pollutants in the USA and therefore removal is not regulated there. Therefore, it has been suggested that the formation of bound pollutant residues produced by bioremediation processes should be promoted (Bollag *et al.*, 1992; Alexander, 1999). However, compounds that have diffused into micropores and nanopores and are still solvent extractable by vigorous techniques, are considered to remain pollutants (Alexander, 1999). Therefore some researchers have suggested that this highly resistant fraction, which is not bioavailable, should not be subject to removal, as is presently advocated by regulatory authorities such as USEPA (White *et al.*, 1997, 1998, 1999a, 1999b; Robertson and Alexander, 1998; Tang *et al.*, 1998; Chung and Alexander, 1998; 2002; Liste and Alexander, 2002). The resistance of this fraction to biodegradation has been reported as the cause of the failure of some bioremediation projects to meet regulatory standards and the cause of incomplete bioremediation reported for many different compounds at a variety of contaminated sites (Alexander, 1999; White *et al.*, 1999a, 1999b). Techniques are being developed to extract the portion of a sequestered compound that is available for biodegradation. These methods include extraction with n-butanol and supercritical fluid extraction (Liste and Alexander, 2002).

Soil type has been reported to have a considerable effect on ageing and sequestration processes (Chung and Alexander, 1998, 2002). The effects found to be significant to sequestering of organic chemicals in soil (Boyle, 1998; Boyle *et al.*, 1998; White *et al.*, 1999a; Kottler *et al.*, 2001) are:

1. Soil organic matter content which increases sequestering
2. The drying and wetting cycles of the soil which decreases sequestering
3. The proportion of the soil void volume that exists as micro and nanopores which increases sequestering

WOOD DECAY BY WHITE-ROT FUNGI

White-rot fungi are very effective agents in nature for degrading the cell walls of woody tissue. The extra-cellular enzyme systems of white-rot fungi which are responsible for lignin degradation are oxidative, non-stereo selective, and non-specific. These characteristics are at least partly due to the free-radical mechanism of degradation used by white-rot fungi, which is considered ideal for the biodegradation of persistent organic pollutants (POPs) in the environment. Wood degrading fungi and their enzyme systems involved in the degradation processes are further described below as an introduction to discuss their place in bioremediation systems.

THE CELL WALL STRUCTURE OF WOODY TISSUES

The cell walls of woody tissues are made up of three main components: cellulose, hemi-cellulose and lignin. The proportions of each component vary from to species. Softwoods (gymnosperms, coniferous trees) generally have a higher proportion of lignin than hardwoods (angiosperms) (Betts *et al.*, 1991). Cellulose is the major component comprising 45-50% and 45-55% of woody tissues in softwoods and hardwoods, respectively. Hemicellulose makes up 24-40% and 25-35% of hardwoods and softwoods, respectively, and lignin comprises 18-25% of hardwoods and 25-35% of softwoods. The lignin polymer has an irregular structure containing a mixture of *R*- and *S*-enantiomeric forms, which makes it difficult to degrade (Betts *et al.*, 1991).

TAXONOMY OF WOOD DECAY FUNGI

The majority of fungi that are capable of degrading wood belong to the phylum Basidiomycota and are placed in the class Hymenomycetes. Fungi in this phylum produce club-shaped structures which produce spores known as basidia, hence the name Basidiomycota. Wood decay hymenomycetes are arbitrarily divided into two groups by the type of wood decay caused. One group of hymenomycetes decays the cellulose in the primary walls of woody tissues leaving behind the brown-coloured lignin. These fungi are known as brown-rot fungi. Other hymenomycetes degrade the lignin leaving behind the cellulose and hemicellulose. These fungi are known as white-rot fungi because of the bleached appearance of wood decayed by these fungi. Some white-rot fungi can also degrade the cellulose and hemicellulose in wood, although the lignin is generally degraded first.

The group of fungi known as white-rot fungi includes edible mushrooms such as *Lentinus edodes* (Shiitake), *Agaricus bisporous* (common button mushrooms) and *Pleurotus ostreatus* (oyster mushrooms). Other examples include bracket fungi such as *Trametes versicolor* (Turkey tail fungi) and *Pycnoporus coccineus*. White-rot fungi do not generally produce conidia. An exception to this is *Phanerochaete chrysosporium* which has one or more anamorphic states.

White-rot fungi produce non-selective, extra-cellular enzymes capable of complete mineralisation of lignin through degradation to carbon dioxide. In *P. chrysosporium,* ligninolytic enzymes are generally produced only during secondary metabolism under nutrient deficient conditions such as low carbon or nitrogen. More recently it has been shown that ligninolytic enzymes of *P. chrysosporium* can be induced in high nitrogen or carbon medium in the presence of lignocellulose compounds such as wheat straw (Kapich *et al.*, 2004). Other fungi, such as *T. versicolor*, produce ligninolytic enzymes under nutrient sufficient conditions. Lignin degradation of *Pinus radiata* tracheids by *T. versicolor* is shown in Fig. 2.1.

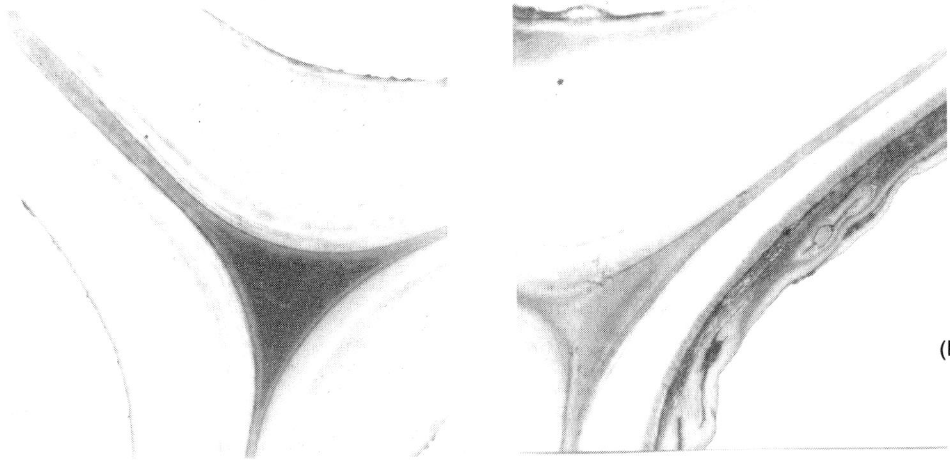

(A) (B)

Fig. 2.1. Intact *Pinus radiata* tracheids prior to (A) and after (B) inoculation with the New Zealand *Trametes versicolor* isolate HR131 (× 13,000). *P. radiata* sawdust is the basis of our *T. versicolor* growth substrates used for bioremediation of aged PCP residues (Walter *et al.* 2005a, 2005b, 2004). Photographs by Siva Sivakumaran, Hort Research.

LIGNINOLYTIC ENZYMES PRODUCED BY WHITE-ROT FUNGI

The ligninolytic enzymes produced by white-rot fungi have been categorised into two groups: peroxidases and laccases. The peroxidases are further subdivided based on their reactions. Lignin peroxidase (LiP) oxidises a variety of substrates that have high redox potentials. Manganese peroxidase (MnP) requires manganese to complete its catalytic cycle and its main function appears to be oxidation of chelated manganese (II) to manganese (III). More recently discovered peroxidases, known as versatile peroxidases (VPs), oxidise substrates of both LiP and MnP (de la Rubia *et al.*, 2002). Laccases are also known as phenol oxidases as this appears to be the primary reaction that they catalyse. White-rot fungi produce one or more of these types of enzyme, depending on the species and the environmental conditions.

For each type of enzyme there are many different isomeric forms (isozymes) produced. This is largely dependent upon the environmental conditions. Different fungi can produce a number of different isozymes of each type of enzyme. For example, six lignin peroxidase isozymes (H1, H2, H6, H7, H8 and H10) and four manganese dependent peroxidase isozymes (H3, H4, H5 and H9) have been isolated from the white-rot fungus *P. chrysosporium* (Farrell *et al.*, 1989). The assignment, H1-H10, is based on the order in which the isozymes elute from a chromatographic Mono Q column. The conditions to produce the different types of ligninolytic enzymes can vary from species to species. For example LiP is predominantly produced by *P. chrysosporium* under secondary metabolic conditions, such as low nitrogen or carbon. However, in *Coriolus versicolor* (*T. versicolor*) maximal LiP production was attained under carbon rich conditions (Collins and Dobson, 1996).

Manganese is important for manganese peroxidase production, but it can also affect laccase production. Laccase production by the white-rot fungi *T. versicolor* and *Pleurotus sajor-caju* has been shown to be induced by manganese at the level of gene transcription (Collins and Dobson, 1997). Manganese is also thought to influence the production of LiP indirectly. Mester *et al.*, (1995) studied the effect of manganese on veratryl alcohol production by *P. chrysosporium* and *Bjerkandera* sp. strain BOS55 under nitrogen-regulated and deregulated conditions, respectively. They found that the presence of manganese decreased the amount of veratryl alcohol produced by these fungi. Because LiP concentrations were increased with increasing levels of veratryl alcohol, they concluded that the level of manganese has an indirect effect on LiP concentration. Copper has been shown to induce laccase production by *P. sajor-caju*, *T. versicolor*, *Trametes trogii*, and *P. ostreatus* (Collins *et al.*, 1997).

Several aromatic compounds have been shown to enhance the production of laccase in white-rot fungi. This induction is thought to be part of the mechanism used by the fungi to reduce the potentially toxic effects of such compounds (Collins and Dobson, 1997). Common inducers include 2,5-xylidine, veratryl alcohol, ferulic acid and 1-hydroxybenzotriazole (Dodson *et al.*, 1986; Collins and Dobson, 1996, 1997). PCP was found to induce a laccase production (Sederati *et al.*, 2003) which was confirmed by our own work studying *Trametes* sp. isolate HR577 (Guthrie, 2006).

PEROXIDASES

There are several classes of peroxidase based on sequence homology, structural similarity and organism of origin. Well-known examples include horseradish peroxidase (a class III plant peroxidase) and yeast cytochrome *c* peroxidase (a class I peroxidase). LiP, MnP and VP are monomeric class II peroxidases – secretory fungal peroxidases (Larrondo *et al.*, 2005). Class II peroxidases have four conserved disulfide bridges (MnPs have an additional conserved disulfide bond near the C-terminal), two structural Ca^{2+} ions, a secretion sequence and are glycosylated. The disulfide bonds are involved in enzyme stability, and *O*-glycosylation protects the C-terminal of the peroxidases from attack by proteolytic enzymes (Reading and Aust, 2001). Glycosylation of these peroxidases is not believed to be necessary for enzyme activity as deglycosylated enzymes have slightly higher activity than glycosylated enzymes. The calcium ions are structural and maintain the integrity of the active site (Poulos *et al.*, 1993).

Peroxidases require hydrogen peroxide to complete their catalytic cycle (Gold *et al.*, 1984; Tien and Kirk, 1984; Higson, 1991). White-rot fungi generate hydrogen peroxide with extracellular enzymes such as glyoxyl oxidase and aryl alcohol oxidase and the intracellular enzyme pyranose oxidase (also known as glucose-2-oxidase) (Reading *et al.*, 2003). *P. chrysosporium* is the most extensively studied white-rot fungus. It produces LiP and MnP but not laccase. The only hydrogen peroxide generating enzyme that is secreted under experimental conditions for this fungus is glyoxyl oxidase (Cullen and Kersten, 1996). In contrast, *Bjerkandera* sp. strain BOS55 uses extracellular aryl alcohol oxidase to produced hydrogen peroxide (de Jong *et al.*, 1994). While pyranose oxidase is an intracellular enzyme, it is also believed to be involved in the lignin degradation process (Cullen and Kersten, 1996).

The typical peroxidase catalytic cycle has been described by Aust (1995). The resting enzyme reacts with hydrogen peroxide in a 2-electron reaction to give a stable radical cation Fe (IV) complex (compound I) that can carry out a 1-electron oxidisation to convert the substrate RH to produce radical R•. The final step in the cycle reduces the heme complex back to its resting state with the subsequent 1-electron oxidation of another molecule of substrate. In the presence of excess hydrogen peroxide compound II is converted into compound III, an inactive form of the enzyme.

The iron in the heme protoporphyrin IX prosthetic group is the active site for peroxidases. Reactions occur at an exposed heme-edge. Access to the heme channel in lignin is more restricted for manganese peroxidases than in other peroxidases. Despite this restriction these enzymes are involved in degradation of lignin, a large complex polymer. The crystal structures of manganese peroxidase and lignin peroxidase and several mutants have been determined and these, combined with other studies, reveal some clues as to how these enzymes can degrade lignin.

White-rot fungi produce a variety of extracellular products other than ligninolytic enzymes. Some of these are also implicated in lignin degradation. One compound synthesised by white-rot fungi is veratryl alcohol (VA, 3,4-dimethoxybenzyl alcohol). VA is a substrate for LiP and is believed to play a role in protecting some isozymes from conversion to compound III. It does not protect all isozymes from inactivation by H_2O_2 (Farrell *et al.*, 1989). But it has also been shown that tryptophan can stabilise LiP isozymes (Collins *et al.*, 1997). In the LiP crystal structure, a VA-binding site has been found (Choinowski *et al.*, 1999). Sequence alignment of known lignin peroxidases shows a conserved tryptophan residue (Trp 171). This residue is only found in LiPs and VPs. The tryptophan residue has been modified at the C_β atom by an autocatalytic post-translational modification. Timofeevski *et al.*, (2000) created a MnP mutant that replaced a serine with tryptophan at the position equivalent to Trp 171 in lignin peroxidases. This mutant mimicked *Pleurotus* VP and acted as a hybrid between MnP and LiP. Mutants of lignin peroxidase that lack the tryptophan 171 residue are incapable of oxidising VA (Doyle *et al.*, 1998; Blodig *et al.*, 2001). Some researchers have proposed that the VA radical cation (VA•), produced by LiP oxidation of VA, acts like a diffusible mediator compound to oxidise lignin and lignin-like compounds. However, $VA^{\bullet+}$ is not stable enough to act as a diffusible mediator. The current theory is that a LiP-$VA^{\bullet+}$ complex acts as a redox partner (Doyle *et al.*, 1998). The crystal structure of MnP shows a Mn^{2+} binding pocket where chelated Mn^{2+} is oxidised to Mn^{3+} (Martínez, 2002). The MnP compound II can only be returned to its resting state by oxidising complexed Mn^{2+} to Mn^{3+} (Sutherland and Aust, 1996).

Chelated Mn^{3+} is a powerful oxidant that can diffuse away from the enzyme where it can then act as a free radical mediator for polymer degradation. White-rot fungi produce a number of different organic acids that can act as chelators. Oxalate is the only one of these organic acids that is produced in sufficient quantities for this purpose (Gold *et al.*, 2000). Versatile peroxidases have only recently been described. They have a Mn^{2+} binding site and a C_α-modified tryptophan residue at the equivalent position to Trp 171 in LiP. They function as hybrids of MnP and LiP.

LACCASE

Laccases (phenol oxidase EC 1.10.3.2) are glycosylated phenol oxidases that exist as monomers, homodimers, or homotetramers (Solomon *et al.*, 1996). The molecular weight of laccases is generally around 50-80 kDa per monomeric unit. Laccase is a part of the group of enzymes known as multicopper oxidases. Examples of other multicopper oxidases include ascorbate oxidase, ceruloplasmin, cytochrome *c* oxidase, rustacyanin, azurin and Fet3. These enzymes catalyse the 4-electron reduction of oxygen to water with the concomitant 1-electron oxidation of substrate. Multicopper oxidases have a minimum of four copper atoms per monomeric unit. The copper atoms have been categorised into three groups, types 1-3, based on their spectroscopic and geometric properties. Type 1 (T1) copper atoms have a conserved cysteine ligand, two histidine ligands and either methionine (most multicopper oxidases) or phenylalanine or leucine (fungal laccases). The copper-sulphur (cysteine) bond has a charge transfer at about 600 nm giving the protein a blue appearance. The EPR spectrum for type-1 copper displays narrow parallel hyperfine splitting. The type 2 (T2) copper has a large parallel hyperfine splitting in the EPR spectrum, but lacks strong absorption bands in the UV-visible spectrum. Type 3 (T3) copper atoms form a binuclear couple. They are EPR silent but have an intense charge transfer band at 330 nm due to a bridging hydroxide. In laccases, the T2 and T3 copper atoms form a trinuclear copper cluster.

Laccase production is not restricted to white-rot fungi. It was originally discovered in 1883 in the Japanese lacquer tree *Rhus vernicifera* (Nitta *et al.*, 2002). Other fungi such as the ascomycete *Botrytis cinerea* (Yaropolov *et al.*, 1994) and bacteria such as *Azospirillum* sp. (Gianfreda *et al.*, 1999) also produce laccase. Laccase-like multicopper oxidases have also been found in some insects such as *Manduca sexta* and *Anopheles gambiae* (Dittmer *et al.*, 2004). Plant laccases are involved in reactions such as polymerisation of *p*-hydroxycinnamyl alcohols to form lignin. But fungal laccases from white-rot fungi are thought to be involved in lignin degradation.

Unlike plant laccases and other multi copper oxidases, fungal laccases do not have an axial-bonded methionine to the type 1 copper. Fungal laccases generally have higher T1 copper redox potentials than other multicopper oxidases including plant laccases. Amongst fungal laccases the T1 copper redox potential can vary widely between species. This has led to a further division of fungal laccases into so called "high" and "low" redox potential laccases. Before the laccase crystal structures were known, many theories were proposed to try to explain these differences. One reason for the difference in redox potential between fungal laccases and other multicopper oxidases is the ligation of the T1 copper centre. Leucine and phenylalanine (found only in fungal laccases) do not bind to the copper, and they increase the hydrophobicity surrounding the T1 copper. Because there is no bond to the axial ligand, the copper-S (cysteine) bond is much stronger. This accounts, in part, for an increase in redox potential of fungal laccases of about 100 mV compared with other multicopper oxidases (Palmer *et al.*, 1999; Xu *et al.*, 1999; Kanbi *et al.*, 2002). Phenylalanine and leucine appear to have the same effect on redox potential. Xu *et al.,* (1998) proposed that laccases with phenylalanine in the axial position would have a higher T1 copper redox potential than those with leucine as the axial ligand. Mutants of laccase from *Rhizoctonia solani* and *Myceliophthora thermophila* were made where the axial leucine was changed to phenylalanine. This mutation did not affect the redox potential of the T1 copper site. However, a mutant of *Myceliophthora thermophila* where the axial ligand was changed to a histidine resulted in a decrease in the T1 copper redox potential of 30 mV (Palmer *et al.*, 2003). In this mutant, the axial histidine had a strong bond to the T1 copper, while one of the other histidine residues present in the T1 site was no longer bound to the copper. This mutant had distorted geometry around the copper and the colour of the enzyme changed from blue to green. Since changing the axial ligand of the T1 copper cannot account for the differences in redox potentials, other amino acids near the T1 binding site were compared. Xu *et al.,* (1998) noted that in 'high' redox potential laccases there was a conserved Leu-Glu-Ala tripeptide motif, whereas 'low' redox potential laccases had a conserved Val-Ser-Gly motif at the equivalent position. This motif is close to the T1 copper. A LEA-VSG mutant of *Rhizoctonia solani* and a VSG-LEA mutant of *Myceliophthora thermophila* laccase were made to see if this would affect the

T1 copper redox potential. No effect was seen on the redox potential. But the mutations did affect the K_M, k_{cat} and fluoride inhibition of the mutants. It also resulted in a change in the pH optimum of the enzymes. The tripeptide motifs are now believed to be involved in substrate binding near the T1 copper site.

Although many laccase isozymes have been purified to apparent homogeneity by SDS-PAGE, the IEF gels of purified isozymes often show a series of bands with slightly different pIs (isoelectric points). This difference in bands is not believed to be due to different isozymes, but rather to differing degrees of glycosylation of the purified isozymes. Because of this variation in the glycosylation it has been difficult to obtain crystals suitable for X-ray diffraction. The first crystal structure of a laccase protein was published in 2001 (Ducros *et al.*, 2001). Laccase from *Coprinus cinereus* was deglycosylated prior to crystallisation to obtain suitable crystals for X-ray diffraction. Unfortunately, the deglycosylation procedure resulted in the loss of T2 copper thus giving the structure of an inactive protein. However, some useful information could still be obtained from the crystal structure. Although the T2 copper was absent, the trinuclear copper site appeared to be very similar to other multicopper oxidase trinuclear copper sites (Ducros *et al.*, 2001).

Solomon *et al.*, (1996) noted that while most multicopper oxidases are very stereospecific, fungal laccases are not. In addition, the K_M values for fungal laccases are generally in the range of 1-10 mM compared with other multicopper oxidases, where the K_M values are generally less than 1 mM. They suggested that this indicates the absence of a substrate-binding pocket in fungal laccases. Bertrand *et al.*, (2002) published the first crystal structure of an active *T. versicolor* laccase isozyme (isozyme IIIb) complexed with 2,5-xylidine to 2.4 Å. They observed that the xylidine was bound in a wide cavity and suggested that there was a ligand-induced fit of the T1 binding site to the substrate. They suggested that this allows the laccase to oxidise a variety of different compounds. The second crystal structure of a *T. versicolor* laccase isozyme (isozyme I) was published by Piontek *et al.*, (2002). This structure was 1.90 Å and did not have any substrate bound in the active site. *C. cinereus* laccase is classified as a 'low' redox potential enzyme while *T. versicolor* laccase is considered a 'high' redox potential enzyme. Piontek *et al.*, (2002) compared the crystal structure of both these laccases and suggested a reason for the high redox potential of the *T. versicolor* laccase compared to *C. cinereus*. They noticed that the T1 copper of the histidines had a longer bond distance in *T. versicolor* (0.17 Å). They looked for a cause and found a hydrogen bond between a glutamic acid (Glu-460) that is on the same helix as the histidine and a serine (Ser-113) that was highly conserved in high E° enzymes. This hydrogen bond pulls the helix away from the T1 site causing a lengthening of the Cu-N (His) bond resulting in an increase in redox potential. The equivalent position to the glutamic acid in *C. cinereus* is a methionine, which cannot form the hydrogen bond with serine.

The most studied white-rot fungus, *P. chrysosporium*, produces a multiplicity of LiP and MnP isozymes but does not produce laccase (Martinez *et al.*, 2004). Although multicopper oxidase-like sequences have been found following the publication of the *P. chrysosporium* genome. Because this white-rot fungus has been studied extensively, the cofactors involved in MnP and LiP degradation of lignin have been well characterised. Purified laccase can oxidise simple phenolic compounds in buffered aqueous solution without the addition of cofactors such as H_2O_2, VA, oxalate or manganese. The highest redox potential for a laccase is 800 mV (Li *et al.*, 1999) while for LiP the redox potential is around 1-1.1 V (Call and Mücke, 1997) and for the chelated Mn^{2+}/Mn^{3+} couple from MnP oxidation the redox potential is around 0.9-1.2 V (Call and Mücke, 1997). Due to its relatively low redox potential, compared with LiP and MnP, laccase on its own does not degrade non-phenolic lignin model compounds. There has been much debate over its role in lignin degradation. Evans (1985) used antibodies to study whether *T. versicolor* could degrade lignin in the absence of laccase. *In vitro* studies suggested that laccase could degrade lignin under certain conditions although this did not happen in stationary *in vivo* cultures. Morohoshi *et al.* (1987) studied lignin degradation using purified laccase isozymes from *T. versicolor*. They showed that laccase I was able to polymerise lignin, while laccase III depolymerised it. Since the substrate ranges of LiP and MnP can be increased by addition of mediator compounds, researchers have looked for compounds to mediate laccase reactions. Several synthetic compounds such as 2,2'-azino-bis(3-ethylbenzthiazoline-6-sulfonic acid) (ABTS) have been shown to act as mediators. ABTS is generally used as a substrate in laccase assays to determine the amount of laccase activity in a sample. Laccase oxidises ABTS to $ABTS^{\bullet +}$, an intensely blue-green coloured cation radical. Bourbonnais and Paice (1990) showed that oxidation of non-phenolic compounds by either laccase or the $ABTS^{+}$ radical alone does not occur. This suggests that both components are needed for the oxidation reaction. Collins and Dobson (1996) showed that oxidation of anthracene by laccase purified from *T. versicolor* was enhanced in the presence of ABTS. No increase in oxidation of anthracene was observed when the ultrafiltrate from the enzyme purification was used instead of a buffer. However, use of the crude enzyme concentrate caused significant oxidation of anthracene. Collins and Dobson (1996) suggested that the crude enzyme concentrate contained a compound that acted as a mediator in anthracene oxidation by the fungal laccase. They determined that the potential mediator was less than 10 kDa in size and heat stable. Johannes and Majcherczyk (2000) compared several natural mediators in the oxidation of polycyclic aromatic hydrocarbons. They found that the most effective natural mediators were 4-hydroxybenzoic acid (HBA) and 4-hydroxybenzyl alcohol.

If the white-rot fungus is grown on a solid substrate such as wheat straw (Leontievsky *et al.*, 1997), the colour of the laccases purified is described as 'yellow' or 'yellow-brown'. These laccases lack the characteristic 'blue' absorption at 600 nm. Comparison of N-terminal amino acid sequences showed high homology between blue and yellow laccase forms for laccases purified from *T. versicolor*, *Panus tigrinus*, *Phlebia radiata*, *Phlebia tremellosa* and *Agaricus bisporus* (Leontievsky *et al.*, 1997). They suggested that 'yellow' laccases produced by the white-rot fungus *P. tigrinus* are lignin-modified 'blue' laccases. These so-called 'yellow' laccases have been shown to degrade non-phenolic lignin compounds without the need for mediators.

TREATMENT TECHNOLOGIES FOR PCP AND CHLOROPHENOL REMOVAL FROM CONTAMINATED SOIL

There are hundreds of soil and solid waste remediation technologies reported for contaminated site clean up, utilising physical, chemical, thermal and bioremediation strategies or combinations thereof (USEPA, 2004). A recent review paper (Khan *et al.*, 2004) covered the general topic of site remediation technologies. A summary of remediation technologies and associated costs is presented in Table 2.3.

Table 2.3. Summary of remediation technologies and cost data

Process	References	Technology	$US/tonne[1]	
			Lower	Upper
Landfill	Aislabie *et al.*,1997; Semple *et al.*, 2001; Watanabe, 2001	Excavate and dump	79	230
Solidification and Stabilisation	Semple *et al.*, 2001	Cement, pozzolan or lime based	39	118
Vitrification	Semple *et al.*, 2001		79	828
Soil washing	Semple *et al.*, 2001		39	237
Solvent Extraction	Khodadoust *et al.*, 1999	Ethanol/water	160	313
Physiochemical Washing	Semple *et al.*, 2001		79	276
Vapour Extraction	Semple *et al.*, 2001; USEPA, 2001		12	414
Chemical treatment	Aislabie *et al.*, 1997; Semple *et al.*, 2001	Dehalogenation Chemical decomposition In-situ flushing Surface amendments	39	710
Thermal treatment	Semple *et al.*, 2001	Thermal desorption	39	355
Thermal treatment	Aislabie *et al.*, 1997; Semple *et al.*, 2001; USEPA, 2001	Thermal desorption	59	300
Bioremediation	Aislabie *et al.*, 1997; Semple *et al.*, 2001; USEPA, 2001	Land farming	16	493
Bioremediation	Semple *et al.*, 2001; USEPA, 2001	Bioventing	5	118
Bioremediation	Semple *et al.*, 2001; USEPA, 2001	Slurry phase	79	162
Bioremediation	Semple *et al.*, 2001	Bio pile	24	55
Bioremediation	Aislabie *et al.*, 1997	Engineered soil cell	138	246
Bioremediation	Alexander, 1999	Soil pile	97	108
Bioremediation	Semple *et al.*, 2001; Watanabe, 2001	In-situ	49	276
Bioremediation	Watanabe, 2001	Phytoremediation	10	34
Bioremediation	Glaser, 2003	White-rot Fungi	75	150
Bioremediation	Alexander, 1999	Composting	43	108

[1] The cost data presented have been adjusted from different unit systems to be presented as US$/tonne of material treated. All the cost data are post 1996 and were not adjusted for inflation. The conversion from different national currencies reflects the exchange rate given on the 6th March 2003.

Some remediation technologies may cause the formation of products that are more recalcitrant, toxic or carcinogenic than the parent compounds (McGrath and Singleton, 2000). For example, landfills, which have traditionally provided the cheapest method of disposal in New Zealand (Aislabie *et al.*, 1997), do not remediate and toxic wastes may leach into the surrounding soil and ground water. Moreover, some remediation treatments such as incineration may be so harsh that treated soils lose their normal capability to serve again as a natural resource (Cheng and Mulla, 1999). Bioremediation has been proposed as an alternative to physical and chemical remediation technologies because of high public acceptance, lower cost structures and increased understanding of the underlying science involved (Pointing, 2001; Watanabe, 2001).

BIOREMEDIATION BY WHITE-ROT FUNGI

Laboratory investigations (Bumpus and Aust, 1985; Bumpus *et al.*, 1985; Mileski *et al.*, 1988; Chaudry, 1994; Walter *et al.*, 2003, 2004, 2005a) demonstrated white-rot fungal bioremediation potential. This has led to bioaugmentation of contaminated sites *in situ* (Lamar and Dietrich, 1990a, 1990b; Lamar *et al.*, 1993a, 1993b; Holroyd and Caunt, 1995; Walter *et al.*, 2005b). New Zealand biosecurity law prevents the use of the widely studied fungi *P. chrysosporium* Burds. and *Phanerochaete sordida* (Karst.) Erikss. & Ryv. because they are not native to New Zealand. Therefore, New Zealand studies have been restricted to native white-rot fungi and their ability to degrade PCP first *in vitro* (Walter *et al.*, 2003, 2004) and then in aged residues (Walter *et al.*, 2004, 2005a). Selected isolates from the genus *Trametes* were used. An overview of the research is presented in Walter and Boyd-Wilson (2006).

For white-rot fungal remediation, an initial growth substrate or, specifically, a co-substrate is required for fungal biomass production before contact with PCP (Leštan and Lamar, 1996; Walter *et al.*, 2004). This requirement has led to the incorporation of mushroom growing techniques for large-scale production of fungi for delivery[4], inoculation or bioaugmentation into contaminated sites (Stamets and Chilton, 1983; Stamets, 2000). However, the resulting inocula production is not always of sufficient quality (Leštan and Lamar, 1996) to provide consistent results necessary for commercial bioremediation. This has resulted in the development of hydro-gels and pelletising for white-rot fungi encapsulation purposes (Leštan and Lamar, 1996, 1999; Leštan *et al.*, 1996, 1998). In contrast to other areas of white-rot fungal research, for example, genomics (Martinez *et al.*, 2004), biodegradation pathways (Reddy and Gold, 2000) and enzymology (Gianfreda *et al.*, 1999; Hofrichter, 2002), there was initially less focus on delivery system development for bioaugmentation into contaminated sites (Leštan and Lamar, 1996). More recently, research has begun to address this issue (Leštan and Lamar, 1998, 1999; Walter *et al.*, 2003, 2004, 2005a; Schmidt *et al.*, 2005).

The important factors when screening white-rot fungi for bioremediation are tolerance to xenobiotics, growth temperature and pH range, nutritional requirements and ability to express the appropriate extracellular enzyme systems. Also important is the ability to colonise a range of soil types to degrade the target xenobiotics (Lamar and Dietrich, 1990a; Lamar *et al.*, 1999; Walter *et al.*, 2003). There are major intra-species differences as well as inter-species differences between white-rot fungi (Lamar and Dietrich, 1990a). These manifest themselves when screening for the above traits. Generally, the fungi with the highest tolerance to PCP and the most powerful lignin degrading enzyme systems tend to be the most useful for PCP degradation. Most strains tend to have fairly broad tolerances to pH and temperature, even though their optima vary considerably (Lamar *et al.*, 1999; Walter *et al.*, 2003). PCP tolerance can be measured by spiking agar plates and measuring mycelial extension over time. The ability of the lignin degrading system can be assayed most conveniently by the use of colormetric dyes such as Poly R-478, which change colour as their chromophores are modified by lignolytic enzymes (Walter *et al.*, 2003) along with the capability of the fungus to degrade softwood lignin as mentioned previously. Exceptions include some *T. versicolor*

[4] This term does not refer to physical transportation, but rather a system of production which results in the inoculation of the particular organism into the contaminated site in a metabolic state to facilitate biodegradation. However, physical transportation may be part of this process.

strains which produce considerable amounts of laccase, and can also make very effective PCP degraders, even though they may not degrade softwood lignin (Lamar and Dietrich, 1990a). However, considering that a broader spectrum of xenobiotics have non-phenolic functional groups, the extent of the lignin degrading ability is paramount. The genus *Phanerochaete* also has the ability to produce undesirable intermediate compounds such as pentachloroanisole (Lamar and Dietrich, 1990a, 1990b; Lamar, 1992; Lamar *et al.*, 1993a, 1993b).

PCP DEGRADATION BY WHITE-ROT FUNGI IN LIQUID CULTURE

The pathways for biodegradation of PCP by white-rot fungi and bacteria have been studied extensively (McAllister *et al.*, 1996; Reddy and Gold, 2000). The most recent *in vitro* white-rot pathway reported for purified LiP and MnP from *P. chrysosporium* suggests (Reddy and Gold, 2000) the initial oxidative dechlorination of PCP by the *p* chlorine to tetrachlorobenzoquinone (TCBQ), which is in equilibrium with tetrachlorodihydroxybenzene (TCHB) in solution. This is followed by a series of reductive dechlorinations and hydroxylations to the point where the benzene ring is cleaved with the ultimate products being carbon dioxide and water. Pentachloroanisole, in trace amounts, is also produced.

The first step in PCP biodegradation by white-rot fungi showed that conditions of nitrogen limitation were required initially to degrade PCP. However, this appears to be true for only the first step in the process of PCP metabolism, since degradation of TCHB occurs more rapidly under nitrogen sufficient conditions (Reddy and Gold, 2000). The remaining steps in the pathway are carried out in both nitrogen sufficient and nitrogen limiting conditions. Under initial nitrogen sufficient conditions, LiP promoted the O-methylation of PCP to pentachloroanisole.

PCP appears to have been partially resistant to *T. versicolor* laccase degradation in some earlier studies (Roy-Arcand and Archibald, 1991), since high chlorine substitution of the benzene ring resulted in a decrease in biodegradability. However, in the presence of a mixture of lower chlorophenols, pentachlorophenol was rapidly degraded in the presence of laccase. Phenoxy radicals produced from other chlorophenols in the chlorophenol mixture were thought to be responsible for this. The mechanism of easily degradable laccase substrates catalysing the degradation of more recalcitrant compounds has been confirmed (Gianfreda *et al.*, 1999). Furthermore, other compounds have been observed to promote PCP degradation in the presence of laccase, such as 2,6-dimethoxyphenol (Gianfreda *et al.*, 1999). PCP was found to induce laccase or MnP activity (Sederati *et al.*, 2003).

The main role of laccase enzyme systems in nature appears to be the polymerisation of various monomers, dimers and oligomers which are the breakdown products of lignin depolymerisation (Yaropolov *et al.*, 1994) to humic matter. This has been observed *in vitro* where polymerised compounds have been detected in the presence of laccase and PCP (Ullah *et al.*, 2000). White-rot fungal laccase was found to polymerise PCP metabolites by covalently binding to soil humic matter (Rüttimann-Johnson and Lamar, 1997; Tuomela *et al.*, 1999). White-rot fungi (*P. ostreatus* and *T. versicolor*) have been reported to degrade chlorinated organic compounds with up to six substituted chlorine atoms, in particular, highly substituted PCBs and dieldrin (Zeddal *et al.*, 1993).

An alternative pathway for PCP degradation is the production of PCA, a general microbial detoxification response (Lamar and Dietrich, 1990a, 1990b; Hill *et al.*, 1995). This reaction appears to be more prevalent for the genus *Phanerochaete* than the genus *Trametes*

(Leštan and Lamar, 1996). The production of this compound is undesirable from a bioremediation perspective, because it is more hydrophobic and recalcitrant than PCP (Lamar and Dietrich, 1990a; Lamar *et al.*, 1990). However, this can result in an increased tendency to bioaccumulate (Lamar *et al.*, 1990; Shaw and Chadwick, 1999), mitigating the effect of its reduced toxicity (Renner *et al.*, 1986). Conversely, being less soluble it is less likely to contaminate ground water (Chung and Aust, 1995). The production of PCA from species other than in the genus *Phanerochaete* appears to be minimal (Rüttimann-Johnson and Lamar, 1997). The PCP methylation process is promoted by a transmembrane methyl transferase (Aust *et al.*, 2004), rather than an extracellular enzyme.

PCP DEGRADATION IN SOIL

The fate of PCP in soil is typically assessed with radiolabelled ^{14}C-PCP (Rüttimann-Johnson and Lamar, 1996, 1997; Tuomela *et al.*, 1999; Walter *et al.*, 2003). The amount of PCP mineralisation to carbon dioxide in soil (Lamar *et al.*, 1990) is generally lower than in liquid culture (Mileski *et al.*, 1988; Reddy and Gold, 2000). Mineralisaton values given for *P. chrysosporium* in liquid cultures varied between 26% and 42% (Mileski *et al.*, 1988). The comparative values given for PCP mineralisation in soil by *Phanerochaete* spp. were approximately 2% (Lamar *et al.*, 1990), with values of 30-40% reported for *T. versicolor* (Tuomela *et al.*, 1999; Walter *et al.*, 2004). The mineralisation of PCP by *T. versicolor, P. ostreatus, Irpex lacteus* (Fr.:Fr.) Fr. and *Bjerkandera adusta* (Wild.:Fr.) Karst ranged from 0.9-8.8% for a Marshan sandy loam soil (pH 6.4) (Rüttimann-Johnson and Lamar, 1997). The solvent extractability of the PCP metabolites is dependent on the pH, and the amount of organic matter present in the soil. At pH 4.8, PCP metabolites covalently bind to the soil humic materials irreversibly (Lamar *et al.*, 1990). An increased fraction of extractable PCP metabolites has been observed at pH 7.0.

The quinonoid intermediates (TCBQ) from PCP biodegradation appear to react preferentially typically by polymerisation with soils' humin, fulvic and humic acids (Lamar and Dietrich, 1990a, 1990b; Tuomela *et al.*, 1999; Morimoto *et al.*, 2000). This is by oxidative coupling (Bollag *et al.*, 1992) rather than the further mineralisation steps reported for liquid culture (Reddy and Gold, 2000). In addition, all three extracellular enzyme systems from white-rot fungi seemed to be able to promote this polymerisation reaction (Rüttimann-Johnson and Lamar, 1996; 1997). Moreover, it has been suggested that no toxic compounds are released from natural humus decomposition (Vertraete and Top, 1999). Therefore, once covalently bound to humic substances, the bioavailablity and toxicity of PCP and lower substituted chlorophenol metabolites is considerably reduced (Rüttimann-Johnson and Lamar, 1997). The natural turnover of humic materials is estimated to be 2.5×10^5 times slower than the easily decomposable organic matter of sugars and amino acids, and 1.0×10^5 slower than the slowly degraded organic matter such as hemicelluloses (Rüttimann-Johnson and Lamar, 1997). Any release of PCP metabolites by degradation from humic materials would also be over extended periods, with the final degradation product of humic materials being carbon dioxide. The turnover of organic matter in the soil is a complex process (McClaren and Cameron, 1996), with some classes being stable for several thousand years while others are degraded after one year. It has been reported, generally, that soil organic matter turnover has been estimated to range between 10 years for tropical forests, and 40 years for arable crop lands.

BIOAUGMENTATION

The majority of the white-rot fungal field-scale bioaugmentation research was first carried out between 1990-1994 by Richard Lamar and his associates working for the US Forest Service Laboratories in Madison, Wisconsin (Lamar and Dietrich, 1990a, 1990b; Davis *et al.*, 1993; Lamar *et al.*, 1993a, 1993b). Probably the largest field-scale PCP soil remediation, with 6000 m^3 of soil, was conducted with *P. chrysosporium* in Finland with average PCP levels of 203 mg kg^{-1} (Holroyd and Caunt, 1995). Since then, further research was carried out on the bioremediation potential of other white-rot fungi (McAllister *et al.*, 1996; Walter *et al.*, 2003). Research efforts increased on delivery systems for the white-rot inocula into the contaminated sites (Leštan and Lamar, 1996, 1999; Leštan *et al.*, 1996, 1998; Schmidt *et al.*, 2005). In spite of promising reports of laboratory-scale remediation, very few field studies have been published since then on PCP remediation (Walter *et al.*, 2005b). Commercial white-rot fungal field remediation has been limited.

SUCCESSES AND FAILURES OF BIOAUGMENTATION

The highest success rate for microbial inoculation processes has been achieved with aboveground bioreactors. These can be specially designed vessels for treatment of liquid waste streams, or solid phase slurries, with either a single species of bacterium or a mixed consortium. The cells are usually immobilised, encapsulated or incorporated as biofilms. These processes can be designed for specific compounds or waste streams. Successes have also been obtained with inoculations for solid phase treatments, *ex situ* or *in situ*, where designed aboveground solid phase reactors have been used (Lamar and Dietrich, 1990a; Aislabie *et al.*, 1997; Alexander, 1999).

The success record for *in situ* inoculation treatments in soils, aquifers and surface waters is inconsistent. Populations of microorganisms grown in favourable laboratory conditions are subject to a variety of abiotic and biotic stresses when inoculated *in situ*. These stresses have to be mediated for the enzyme systems responsible for biodegradation of a particular compound to be expressed. Complication factors include: more than one priority pollutant; competition; inability to move through the soil; temperature; the need of a carbon source or co-substrate; limited nutrients; pH; salinity and natural toxins.

MULTIPLE POLLUTANTS

The presence of multiple pollutants (Lamar *et al.*, 1993b; Alexander, 1999) can seriously inhibit the growth of the introduced microorganisms. This is more of a problem for substrate-specific detoxifying bacteria than for white-rot fungi. White-rot extracellular enzyme systems can degrade a wide range of xenobiotics. Metal ions as co-contaminants did not affect white-rot fungi (Glaser, 2003). Bioremediation enzymes expressed by white-rot fungi are generally not triggered by the pollutant itself, but occur as a function of enzymes expressed during secondary metabolism triggered by nutrient limitation, or as induction by a stress response (Gianfreda *et al.*, 1999; Pointing, 2001). White-rot fungi need a co-substrate and will rarely utilise the target pollutant as a carbon source. Creosote is one possible exception (Lamar *et al.*, 1993b). White-rot fungi will grow on a wide range of co-substrates yet not all will induce the required enzyme systems.

COMPETITION

Competition between white-rot fungi and indigenous soil microorganisms needs to be managed carefully (Stamets, 2000; Schmidt *et al.*, 2005), particularly where the pollutant or pollutant levels have little or no microbial toxicity. Boyle (1995) suppressed indigenous microorganisms when studying colonisation of soil by white-rot fungi to reduce "microbial contamination". In one of our studies, examining the effect of 11 different soil types (non-contaminated) on colonisation by three *T. versicolor* isolates, we needed to inhibit indigenous microorganisms when adding non-colonised growth substrate with the fungal inoculum. In contrast, when the fungal growth substrate was pre-colonised, even if only for 3 days, prior to non-contaminated soil inoculation with the *T. versicolor* isolates, "microbial contamination" did not occur. The white-rot fungi readily colonised the soil (Schmidt *et al.*, 2005). We tested the effect of nutrient soil amendments on *T. versicolor* colonisation and "microbial contamination". All nutrient amendments, including non-colonised growth substrate, resulted in proliferation of indigenous fungi (Table 2.4). However, "microbial contamination" was never noted in any of our studies when working with spiked or aged PCP residues (Walter *et al.*, 2004, 2005a). PCP is an effective biocide and can seriously inhibit microorganisms, including the augmented white-rot fungus (Davis *et al.*, 1994). Therefore, PCP tolerant white-rot strains have been identified in the laboratory (Walter *et al.*, 2003). A selected isolate was able to reduce aged PCP residue levels of up to 1500 mg kg^{-1} PCP (Walter *et al.*, 2004, 2005a, 2005b). Current formulation research and changes to the growth substrate allowed us to decontaminate aged PCP residues up to 6000 mg kg^{-1} (Ford, 2006).

Table 2.4. Total mycelial cover for three *T. versicolor* isolates and indigenous soil fungi after 21 days growth at 18°C on nutrient amended soil. Fungi were inoculated using a 11-day old sawdust-corn meal-starch (SCS, 25% v/v) growth substrate (Schmidt *et al.*, 2005).

Treatment (n = 9; 3 isolates × 3 replicates)	Amount of nutrient amendment per kg dry soil (non-contaminated Temuka deep clay)	Total mycelial cover (%)	Presence of other fungi
No nutrient amendment		7	-
Carbon	9 g glucose	10	+
Nitrogen	5 g ammonium tartrate	11	+
Phosphorus	1 g di-sodium hydrogen orthophosphate 2-hydrate	11	+
Calcium	6 g calcium chloride 2-hydrate	15	+
C + N[a]		19	+
P + Ca[a]		18	+
C + P + Ca[a]		19	+
N + P + Ca[a]		20	+
C + N + P + Ca[a]		24	+
SCS amendment	42.5 g (dry weight) non-colonised, sterilised SCS	28	+
LSD (10%)	Fisher's LSD test	8	

[a] Amounts of each nutrients are as specified for single amendments

SOIL COLONISATION

An inability to move through the soil is more of a problem for bacteria than white-rot fungi, because of hyphal extension in the latter. Hyphal penetration of soil particles is limited by the size of micropores. However, the diameter of the extracellular enzymes (5 nm) (Archibald *et al.*, 1997) is an advantage for white-rot fungi over bacteria that do not secrete extracellular enzymes themselves. Soil types also have a major influence on the ability of white-rot fungi to colonise soils as shown in Table 2.5 (Walter, 2004). In this study the only correlation that occurred for all three *T. versicolor* isolates was between mycelial cover and percentage base saturation. Percentage base saturation measures the proportion of the cation exchange capacity that is occupied by the cations Ca^{2+}, Mg^{2+}, K^{+} and Na^{+}. The Ca^{2+} cation occupies 80% of the total base saturation in neutral mineral soils. Therefore, correlations between Ca^{2+} and fungal growth, as well as between the pH and fungal growth, could be expected (Walter, 2004). However, this was not the case, and confirms the findings of Lamar *et al.* (1987) who found no relationship between the growth of *P. chyrsosporium* and the pH of 3 soils. The pH of soil tends to be more neutral than the optimum conditions for growth of white-rot fungi. This is not a serious limitation because white-rots can survive in soils (Fig. 2.2) and still degrade PCP and other xenobiotics over a wide range of pH values (pH 3.8-9.6) (Lamar *et al.*, 1993b).

The pH and the organic carbon content of the soil are the factors that determine PCP sorption (Christodoulatos and Mohiuddin, 1996; Fall *et al.*, 2001). How pH affects the upper level of PCP tolerances to white-rot fungi is not known. Some assumptions can be made from sorption studies, since it is known that the neutral and more toxic form of PCP at lower pH is less soluble. It binds to organic matter more strongly than the anionic form. However, as the pH increases, the less

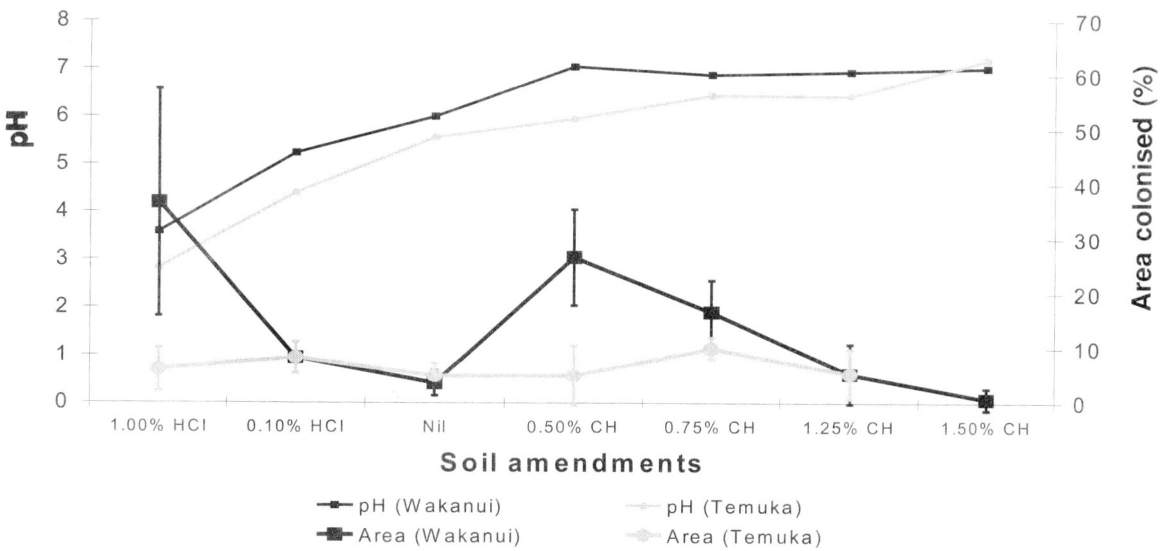

Fig. 2.2. Effect of soil pH on the growth (area colonised, %) of *T. versicolor* isolate HR131 in Wakanui silt loam (Wakanui) and Temuka deep clay loam (Temuka) after 7 days incubation at 30°C in the dark. Soil pH was adjusted over a 4-week period with hydrochloric acid (HCl) and calcium hydroxide (CH) according to Bolan *et al.* (1996). Soils were non-contaminated and non-sterile, without nutrient/growth substrate amendments. The soil microcosms (20 mL, 60% field capacity) were surface inoculated with a fungal mycelial plug (ϕ 10 mm).

Table 2.5. Mean mycelial cover for three *T. versicolor* isolates for 11 New Zealand soils (non-contaminated, sterile) amended with sterile growth substrate (SCS, 25% v/v) and surface inoculated with a fungal mycelial plug (f 8 mm) after 7 days incubation at 30°C in the dark.

Soils	New Zealand soil classification[a] (USDA Great Group)[b]	pH	BS %	OM %	CEC (cmol·kg^{-1})	Ca	Mg	K	Olsen P µg·ml^{-1}	S µg·g^{-1}	N µg·g^{-1}	Mycelial coverage %
Bealey silt loam	Allophanic brown (Dystrudept)	4.8	9	14.2	16.1	0.7	0.40	0.34	7	30	6	14.2
Cass soil – high country yellow brown earth	Acid brown (Dystrudept)	5.4	23	16.3	15.8	2.3	0.67	0.57	18	6	6	55.8
Kaipaki peaty loam	Mesic organic (Medihemist)	5.4	41	16.7	13.4	3.6	1.18	0.55	8	7	6	47.5
Hamilton clay loam	Orthic granular (Haplohumult)	6.0	63	12.2	14.0	3.7	2.80	2.03	6	26	71	54.2
Horotiu sandy loam	Orthic allophonic (Hapludand)	6.1	63	17.3	13.1	4.7	1.8	1.67	9	38	93	34.2
Pakaroa hill soil	Acidic orthic brown (Dystrudept)	6.1	64	13.9	14.2	5.1	2.76	1.05	10	38	33	75.8
Takahe silt loam	Fragic pallic (Haplustept)	5.5	55	8.0	11.6	3.1	2.42	0.46	5	5	6	69.2
Tamahere soil	Artifact fill anthropic (Haplanthrept)	5.4	42	9.9	18.0	1.7	0.61	0.87	10	8	9	59.2
Temuka deep clay loam	Typic gley (Humaquept)	5.7	82	11.8	16.5	9.6	2.83	0.82	35	30	6	92.5
Unnamed	Fluvial raw (Ustifluvent)	5.8	100	3.1	2.2	1.6	0.34	0.23	6	2	6	88.3
Wakanui silt loam	Mottled immature pallic (Aquic Haplustepts)	5.8	63	10.8	11.2	5.3	0.61	0.99	45	5	6	100.0
LSD (5%)	Fisher's LSD test											17.7

Soil property analyses include SCS amendment; BS = Base saturation; OM = Organic matter; CEC = Cation exchange capacity

[a] Hewitt (1993)

[b] Soil Survey Staff (1998)

toxic and more soluble anionic form of PCP predominates. The toxic effect of protonated PCP is mitigated, to some extent, by sorption to organic matter at a mildly acidic pH (pH 3.8-4.5). The extra solubility of the anionic form is balanced by its lower toxicity at more neutral pH. The use of organoclays may possibly aid in the immobilisation of PCP over a wider pH range (Wu *et al.*, 2000). This will mean white-rot fungi can tolerate higher levels of PCP in soils (Nye *et al.*, 1994; Witthuhn *et al.*, 2005).

TEMPERATURE

Temperature affects the growth rate of all microorganisms. Extreme temperatures can inhibit growth or even kill the fungus. Mushroom substrate beds grown outdoors are obviously subject to fluctuating temperatures on both a diurnal and seasonal basis (Stamets and Chilton, 1983). While lower temperatures may result in slow growth of the organism (Alexander, 1999), fungi studied in bioremediation were found to grow and biodegrade xenobiotics over a wide temperature range (Lamar and Dietrich, 1990a; Walter *et al.*, 2003, 2005b).

IMPORTANCE OF GROWTH SUBSTRATE

The techniques of mushroom cultivation requires much attention. Mushrooms are considerably more difficult to produce than ordinary agricultural or horticultural produce (Stamets and Chilton, 1983). Mushroom cultivation techniques were initially utilised for producing white-rot fungal inocula (Stamets and Chilton, 1983; Stamets, 2000) for bioaugmentation with some success in remediating PCP and creosote contaminated sites (Lamar and Dietrich, 1990a, 1990b; Lamar *et al.*, 1993a, 1993b). These techniques usually involved a lignocellulosic or supplementing a lignocellulosic substrate with grains, starch or cornmeal, for both white-rot fungi inoculum production and as substrate amendments to the contaminated soil being treated (Stamets and Chilton, 1983; Stamets, 2000). The lignocellulosic substrate was usually wood chips and sawdust from angiosperm species (hardwoods) such as willow (*Salix* spp.), poplar (*Populus* spp.) and oak (*Quercus* spp.). In addition to sawdust and grains, composted sawdust and bark were tested as white-rot fungal substrates, for *T. versicolor* and *P. chrysosporium.* These were found to be inferior to non-composted substrates for white-rot fungal growth (Walter *et al.*, 2004). Information has been reported on the nutritional composition of processed and unprocessed grains, that are typically used to supplement the chip/sawdust amendments (Stamets, 2000).

The economics of white-rot fungal bioaugmentation precludes the use of sophisticated control techniques for inoculum production (Leštan and Lamar, 1999). In some cases temperature control is absent, because the substrate beds for mushroom growth are often located outdoors (Stamets and Chilton, 1983). The manufacture of large substrate-inoculum volumes for bioaugmentation purposes necessitates the use of shallow substrate beds and levels of supplements that promote rapid mycelial growth. However, excess heat generation needs to be avoided. Excess nitrogen supplementation (Stamets and Chilton, 1983) can also cause the temperature of the substrate bed to rise to the point where thermophilic organisms become established, composting the substrate bed.

The mushroom cultivation techniques did not produce consistent inoculum quality. Contamination by non white-rot fungi was a constant problem (Stamets and Chilton, 1983; Leštan and Lamar,

1996). The generation of excess metabolic heat (flashing) during the transportation of the inocula for *in situ* remediation is another problem (Leštan and Lamar, 1996, 1999). Leštan and Lamar (1996) reported that delivery systems for inoculation of white-rot fungi into soil had received little attention. Accordingly, recent research has attempted to address these issues (Leštan *et al.*, 1998; Leštan and Lamar, 1999; Walter *et al.*, 2003, 2004, 2005a; Schmidt *et al.*, 2005). For example, one such important area covers measurement techniques for fungal inocula potency. This resulted in fluorescein diacetate (FDA) hydrolysing ability being utilised to test for "biological potential" (Leštan and Lamar, 1996; Walter *et al.*, 2004). High FDA activity has been reported as being positively correlated with ergosterol and fungal dry matter (Leštan and Lamar, 1996), laccase expression and fungal colonisation of soil (Schmidt *et al.*, 2005). Fungal enzyme activities and soil colonisation are influenced by substrate formulation, structure, and temperature, soil conditions and soil type (Leštan and Lamar, 1996, 1999; Leštan *et al.*, 1996, 1998; Schmidt *et al.*, 2005).

The influence of the various formulation factors contribute to the extent of proliferation of the fungus through soil (Leštan and Lamar, 1996; Schmidt *et al.*, 2005). For example, *T. versicolor* grown on *Pinus radiata* sawdust-grain substrate mixtures colonised uncontaminated soil better than a *Populus* sp. sawdust growth substrate with equivalent grain supplementation, as measured by biological potential. This was supported by laccase results and visual observation (Schmidt *et al.*, 2005).

FORMULATION OF GROWTH SUBSTRATES

Microorganisms in their physical environment can exist as discrete entities, such as free cells or mycelial fragments, or they can be attached or linked to each other, and be sorbed or encapsulated by inorganic and/or organic matter. Immobilisation techniques for environmental purposes have been reviewed (Cassidy *et al.*, 1996). White-rot fungal growth on lignocellulosic substrates can be considered an immobilisation technique, however, the nature of the immobilisation is uncontrolled. Therefore, inoculant consistency is important for a commercial bioaugmentation process (Trevors *et al.*, 1992; Leštan and Lamar, 1996; Leštan *et al.*, 1996).

The use of immobilisation techniques has been demonstrated to be superior to the use of free cells and mycelia for their long-term viability when introduced into both soil and aqueous environments for the purpose of degrading xenobiotics. This applies for not only environmental applications, but also in such areas as biological control in agriculture and horticulture, medicine, pharmaceutical production and forestry (Leštan and Lamar, 1996; Cassidy *et al.*, 1996; Jen *et al.*, 1996; Rudzinski *et al.*, 2002). For example, mycelial fragments of white-rots are more susceptible to growth inhibition by PCP in soil, and other xenobiotics, than when immobilised in pelleted and encapsulated formulations (Leštan and Lamar, 1996).

In general, there are six methods of immobilisation of microbial cells (Cassidy *et al.*, 1996). These are: flocculation, adsorption to surfaces, covalent bonding to a carrier, crosslinking of cells, encapsulation in a polymer gel, and entrapment in a matrix. Encapsulation in a polymer gel for bacteria is reported to be superior to other forms of immobilisation techniques (Cassidy *et al.*, 1996). In particular, hydro gels are used for this purpose. Because of their hydrophilic nature, they are benign for the growth and multiplication of bacteria and other microorganisms. The hydrogel structure allows small molecules such as ethanol and glucose to diffuse through the gel at levels up to 90% of their diffusion rate in water (Smidsrod and Skjak-Braek, 1990); their defined nature assists in optimising a microbial inoculant for a particular end use (Leštan and Lamar, 1996, 1999; Leštan *et al.*, 1996, 1998; Leštan and Lamar, 1999).

An immobilisation process for *T. versicolor* on nylon (Sederati *et al.*, 2003) allowed this white-rot fungus to tolerate step-wise additions of pentachlorophenol and 2,4-dichlorophenol totalling 3400 mg L^{-1} and 2000 mg L^{-1}, respectively, in 30% aqueous ethanol culture. The maximum PCP removal rate was 7.9 $mgL^{-1}h^{-1}$. The first formulation of a pelleted substrate for white-rot fungi, with or without alginate encapsulation, was based on mushroom growing techniques (Leštan and Lamar, 1996; Leštan *et al.*, 1996). For example, *T. versicolor* and *P. chrysosporium* pellet formulations have used a mixture of a hardwood sawdust (oak or aspen poplar) starch, cornmeal and lignosulfonate. These pellets, without alginate encapsulation, were demonstrated to be better for promoting white-rot fungal colonisation of soils over alginate beads without amendments (Leštan and Lamar, 1996). The white-rot is inoculated on to the surface of the pellet. There is no encapsulation involved. However, hyphae may penetrate into the centre of the pellet during growth. The pelletising process is well documented (Leštan and Lamar, 1996, 1999; Leštan *et al.*, 1996, 1998).

WHITE-ROT FUNGAL PELLETS

The ratio of sawdust to grain amendments is typically 75:25 on a dry weight basis. Higher sawdust to grain ratios (87.5:12.5) tend to lower biological potential, which results in a slower colonisation of the pellet or substrate, and increases the potential of contaminant proliferation (Stamets and Chilton, 1983; Leštan and Lamar, 1996; Walter *et al.*, 2004). Nitrogen levels in most woods are low (Stamets and Chilton, 1983). Therefore, colonisation by white-rot fungi is retarded. Carbon and nitrogen supplementations in the form of grains can aid rapid growth and expression of the appropriate extra-cellular enzymes (Leštan and Lamar, 1996). Unfortunately, supplementation can also aid microbial contaminant proliferation.

ENCAPSULATION OF WHITE-ROT FUNGI IN HYDROGELS

In the 1980s, technology began to be developed for encapsulation of biological control agents (Boyette *et al.*, 1991) for the use of mycorrhizae inoculation in forestry (Mauperin *et al.*, 1987). However, it was not until 1996 that encapsulation techniques were developed for white-rot fungal remediation of solid phase sites. Alginate, agar, gelatin, κ-carrageenan, chitosan, PVAL (poly vinyl alcohol) and agarose have been used as white-rot fungal encapsulation techniques that have produced viable white-rot fungal inocula (Leštan and Lamar, 1996, 1999; Leštan *et al.*, 1996, 1998; Loomis *et al.*, 1997; Leidig *et al.*, 1999; Arica *et al.*, 2001). Furthermore, hydrogel encapsulation of pellets has also been reported (Leštan and Lamar, 1996, 1999; Leštan *et al.*, 1998). This produces the advantage of isolating the water required for successful microbial growth on the surface of the pellet, while concurrently maintaining a lower moisture content in the pellet to maintain its mechanical strength. This is necessary for the pellet to maintain its structural integrity to survive the delivery process into solid phase environments.

Ca-alginate and other hydrogels have recently been studied more intensively as delivery systems of microorganisms for environmental purposes because of their defined nature. This produces greater inoculant consistency than previous mushroom growing techniques. In New Zealand, the cheapest hydrogels available are Na-alginate and κ-carrageenan (cheapest). Alginate is being used extensively in the food industry (Cassidy *et al.*, 1996). Their non-toxic nature and cost make them suitable for microbial growth and environmental bioaugmentation purposes. Agar and agarose are regarded as suitable encapsulation hydrogels for white-rot fungi. However, they are produced almost exclusively for scientific research requirements and their current cost structures eliminate their consideration for commercial purposes (Leštan *et al.*, 1998). In addition, the encapsulating material should also be biodegradable. This criterion excludes polymers such as PVAL, even though PVAL has been used to successfully encapsulate *T. versicolor* mycelial fragments and was a useful slow release formulation (Leidig *et al.*, 1999).

The use of coated pellets made from hydrogel of medium-viscosity alginate allows the moisture required for fungal growth to be concentrated on the surface of the pellet. The interior of the pellet can be formulated at lower moisture contents to increase the mechanical strength of the pellet to a level where it could survive the delivery into soil (Leštan and Lamar, 1996). In contrast, for pellets not coated with alginate to support white-rot fungal growth, the moisture content of the pellets had to be at a level that would not provide sufficient mechanical strength to allow the pellets to be delivered into contaminated soils in the field. Subsequently, Ca-alginate coated pellets were tested with five white-rot fungi, *P. chrysosporium, P. sordida, T. versicolor, I. lacteus* and *B. adusta* (Leštan and Lamar, 1996). The results demonstrated that this method of immobilising white-rot fungi was practical for delivery into contaminated sites. In addition, the lower moisture content of the interior of the pellets (35.5% for *T. versicolor*) provided the necessary mechanical stability, even when being introduced into sandy soils.

The influence of humidity on the formation of mycelial coats on alginate-coated pellets showed that the key to successful inoculum production is rapid growth (Leštan and Lamar, 1999). Generation of metabolic heat can cause the relative humidity in the immediate surroundings to fall. This results in the drying out of the hydrogel surface and the incomplete colonisation of the surface of the pellet by the mycelia. Pellets not fully colonised by white-rot fungal mycelia are not suitable for inoculation into contaminated soil (Leštan and Lamar, 1996). Competition from soil indigenous fungi on such pellets has been shown to limit the white-rot fungal proliferation in soils. The lower biological potential of such pellets also reduces white-rot fungal tolerance to PCP.

Pelletised white-rot fungal formulation was better than alginate for the proliferation of *P. chrysosporium* in soil (Leštan and Lamar, 1996). The reason given was the mass transfer limitations with the alginate pellet retarding growth. The most effective formulation reported was this same pellet formulation, but with alginate encapsulation (Leštan and Lamar, 1996). The PCP degradation observed was 86%, 82% and 90% for *I. lacteus, B. adusta* and *T. versicolor*, respectively. The inoculation of these pellets was into unamended soil that had been PCP spiked. The survival of white-rot fungi in soil was enhanced considerably with complete colonisation of the alginate coat on the pellet surface. Fungal contaminants, such *Trichoderma* and *Fusarium* spp., grew on any pellets that had incomplete surface colonisation. Leštan and Lamar (1996) reported on the effect of mixing different types of pellets with different soil on the weight and mechanical structure of the pellets. Their mechanical strength was substantially higher than what was required for successful inoculation into soil.

Amendments to alginate are frequently reported in the literature (Trevors *et al.*, 1992; Bennett *et al.*, 1996; Cassidy *et al.*, 1996; Leštan and Lamar, 1996; Loomis *et al.*, 1997; Childress *et al.*, 1998). These have taken the form of clay fillers for hydrogels, or supplements such as corn-cob grits or skim-milk powder. In all cases, addition of these two amendments increased the survival of encapsulated bacteria, or white-rot fungi in soil storage. These amendments may also provide a buffering effect (Cassidy *et al.*, 1996). The addition of clay to hydrogel increases the mechanical strength of the beads and acts as a bulking agent. Clay type can affect survival rate of the encapsulated organisms (Cassidy *et al.*, 1996).

CONCLUSIONS

White-rot fungi do not survive in soil unless it is amended with a carbon source. Nitrogen supplementation is also beneficial (Lamar and Dietrich, 1990a; Morgan *et al.*, 1993; Boyle, 1994; Walter *et al.*, 2004). Typical amendments that have been used in successful field-scale trials are usually hardwood chips or sawdust from species such as willow or poplar. These are usually added to soil without further amendments, since the fungal inoculum itself is usually nutrient fortified. There is only a single report in the literature where soil supplementation was not used in a successful bioaugmentation field trial. In this case, creosote was a co-contaminant with PCP (Lamar *et al.*, 1993b) and probably acted as a carbon source. Laboratory studies have also reported the growth of white-rot fungi on aged creosote contaminated soil (Eggen and Majcherczyk, 1998).

The success of bioremediation is dependent on various factors. Bioavailability of the pollutant and toxicity of the pollutant to the degrading organisms being paramount. Bioavailability of an organic compound is not only affected by water solubility, but also by aging. For example, spiked PCP levels were found to be less toxic and more readily degraded by white-rot fungi than an equivalent aged PCP soil residue (Walter *et al.*, 2005a). A recalcitrant compound has considerable time to interact with physical or chemical components of soils. Thus, its physical, chemical or biological behaviour becomes altered, although the original identity (molecular structure) of the compound is still intact and unchanged. This 'aging' also results in sequestering or complexing. Sequestered compounds are still solvent extractable in comparison with complexed substrates which are non-extractable. These are known as bound residues. Sequestration and complexing mechanisms depend on interactions with the soil organic matter and clay particles (Benoit *et al.*, 1999). The processes are reversible, albeit at a slower rate, particularly under increased microbial activity (Cort and Bielefeldt, 2000) and are also observed in bioremediation applications. For example, we have observed the extractability of aged PCP residues increased during the first few weeks of fungal treatment with *T. versicolor*, or compost treatments (Miller *et al.*, 2004). However, little is known about the microbial mechanisms involved in the desorption processes of the recalcitrant compound.

Microbial metabolism is largely derived from studies of water-soluble substrates. However, recalcitrant organic compounds are generally of low molecular weight and are water-insoluble. The mechanism of microbial mediated metabolism of these recalcitrant compounds, found in NAPLs, is quite different from that of water-soluble substrates because of their reduced bioavailability (Alexander, 1999). Biodegradation of recalcitrant compounds can occur intracellularly and/or extracellularly, depending on the organism, substrate and enzymes involved. Three mechanisms are known whereby microorganisms access recalcitrant compounds (Alexander, 1999):

1. Spontaneous partitioning of the recalcitrant compound from the NAPL to the water phase and subsequent utilisation of the low water-soluble fraction by the organism
2. Direct attachment of the cell to the NAPL fraction and compound diffusion into the cell
3. Pseudo-solubilisation by excretion of microbial surfactants.

In the case of PCP, which was often added in creosote, the creosote itself formed part of the NAPL fraction in the soil solution.

The cost of the white-rot fungal bioremediation is affected by the amount of fungal inoculum required, and to a lesser extent by the quantities of other wood-based amendments. The amount of fungal inoculum required is in turn determined by the its ability to colonise the soil, and the induction of detoxifying enzymes by the growth substrate. The "right growth substrate and soil amendment" is also needed. For example, in some soil types, such as the Bealey silt loam (Table 2.5), colonisation was still severely limited in sterile non-contaminated soils, even when 75% (v/v) growth substrate was added. In contrast, by further developing the "right growth substrate and soil amendment" we now can successfully bioremediate aged PCP residues at levels up to 6000 mg kg^{-1} with a *T. versicolor* isolate (Ford, 2006). This is a 3-fold increase of our earlier reported upper limit (Walter *et al.*, 2005a, 2005b).

Bioremediation feasibility and assessment studies on actual field polluted soils are essential. Remediation potential will vary between isolates, soil types, presence of co-contaminants, contamination levels and the composition and age of the growth substrate, and the soil amendments used. Bioremediation pathways will vary between isolates. For example, PCA was reported to be an intermediate metabolite of PCP remediation by *P. chrysosporium* (Lamar and Dietrich, 1990b; Chung and Aust, 1995). It was never considered an intermediate metabolite in bioremediation of PCP by *T. versicolor* (Tuomela *et al.*, 1999; Walter *et al.*, 2004, 2005a, 2005b). Equally there are no data available on the pathway of degradation of aged PCP residues. Our own research has reported the mineralization of PCP by *T. versicolor in vitro* and in soil spiking aged soil residues with ^{14}C-PCP (Walter *et al.*, 2004). More recent research indicates that maybe both mineralization and polymerisation may occur. Which pathway appears to be dependent on the enzyme systems triggered by the growth substrate, environmental conditions and/or pollutants present? Research on degradation pathways of aged PCP residues by white-rot fungi is urgently needed.

In spite of successful field remediation using white-rot fungi, extensive commercial uptake and widespread application have been limited. The reasons include economic drivers and socio-legal frameworks (Clothier *et al.*, 2004). While remediation treatments such "excavate and dump" or disposal into landfill are permissible and still more cost-efficient, there will be few financial incentives for companies or landowners to consider other treatment options for the clean-up of polluted sites. The ownership of the pollution-problem is also highly debated, particularly when companies acted within the legal guidelines and frameworks and/or government advice at the time when the pollution occurred. In the US there are over 300,000 contaminated sites, with an estimated annual clean-up bill of US$9 billion. Successful remediation technologies met with resistance (Gibb, 1999), as then regulators may force polluters to clean up their sites within five years (Clothier *et al.*, 2004). As Gibb (1999) noted that "...given the choice between spending $25 million on a risky but fast clean-up that will deplete a company's cash reserves and hurt its stock price, or paying lawyers $1 million a year to delay, any sensible CEO will choose the latter". However, as economy develops, societal values change, this eventually affects legislation within a jurisdiction. In New Zealand, the growth of environmentalism, or the "green" movement, has led to passing of laws to better protect the environment and a greater demand for remediation of contaminated sites (Clothier *et al.*, 2004).

REFERENCES

Aamot E., Steinnes E. and Schmid R. (1996). Polycyclic aromatic hydrocarbons in norwegian forest soil: Impact of long range atmospheric transport. *Environmental Pollution* 92 (3): 275-280.

Aislabie J., Smith D., Wickstrom M., Loyd-Jones G. and Huser B. (1997). Bioremediation: Treatment option for contaminated sites, Manaaki-Whenua Press, Lincoln, New Zealand. Pp 60.

Alexander M. (1999). Biodegradation and Bioremediation. Second Edition, Academic Press, London. Pp 453.

Archibald F.S., Bourbonnais R., Jurasek L., Paice M.G. and Reid I.D. (1997). Kraft pulp bleaching and delignification by Trametes versicolor. *Journal of Biotechnology* 53(2-3): 215-236.

Arica M.Y., Kacar Y. and Genc O. (2001). Entrapement of white-rot fungus *Trametes versicolor* in Ca-alginate beads: Preparation and biosorption kinetic analysis for cadmium removal from an aqueous solution. *Bioresource Technology* 90(2): 121-129.

Atkins P. and de Paula J. (2002). Processes at solid surfaces. In: Physical Chemistry, Oxford University Press, Oxford, 977-1013.

Aust S.D. (1995). Mechanisms of degradation by white-rot fungi. *Environmental Health Perspectives* 103(5): 59-61.

Aust S.D., Swaner P.R. and Stahl J.D. (2004). Detoxification and metabolism of chemicals by white-rot fungi. *Pesticide Decontamination and Detoxification* 863: 3-14.

Barrow G.M. (1979). Surfaces and heterogeneous catalysis. In: Physical Chemistry. Eds. Jakson D.C. and Wagely S., Kosaido Printing Co., Ltd., Tokyo, 731-748.

Bennett J.W., Turner A.J., Loomis A.K. and Connick W.J.J. (1996). Comparison of alginate and Pesta for formulation of *Phanerochaete chrysosporium. Biotechnology Techniques* 10(1): 7-12.

Benoit P., Barriuso E. and Soulas G. (1999). Organic compounds in the environment. *Journal of Environmental Quality* 28: 1127-1135.

Bertrand T., Jolivalt C., Briozzo P., Caminade E., Joly N., Madzak C. and Mougin C. (2002). Crystal structure of a four-copper laccase complexed with an arylamine: insights into substrate recognition and correlation with kinetics. *Biochemistry* 41(23): 7325-7333.

Betts W.B., Dart R.K., Ball A.S. and Pedlar S.L. (1991). Biosynthesis and structure of lignocellulose. In: Biodegradation: Natural and Synthetic Materials. Ed. Betts W.B., Springer-Verlag, London, 91-117.

Blodig W., Smith A.T., Doyle W.A. and Piontek K. (2001). Crystal structures of pristine and oxidatively processed lignin peroxidase expressed in *Escherichia coli* and of the W171F variant that eliminates the redox active tryptophan 171. Implications for the reaction mechanism. *Journal of Molecular Biology* 305(4): 851-861.

Bolan, N.S., Syers, J.K. and Tillman, R.W. (1996). Ionic strength effects on surface charge and adsorption of phosphate and sulphate by soils. *Journal of Soil Science* 37: 379-388.

Bollag J.-M. (1992). Decontaminating soil with enzymes. *Environmental Science and Technology* 26(10): 1876-81.

Bollag J.-M., Myers C.J. and Minard R.D. (1992). Biological and chemical interactions of pesticides with soil organic matter. *The Science of the Total Environment* 123/124: 205-217.

Bourbonnais R. and Paice M.G. (1990). Oxidation of non-phenolic substrates. An expanded role for laccase in lignin biodegradation. FEBS Letters 267(1): 99-102.

Boyd S.A., Mortland M.M. and Chiou C.T. (1988a). Sorption characteristics of organic compounds on hexadecyltrimethylammonium-smectite. *Soil Science Society of America Journal* 52: 652-657.

Boyd S.A., Mikesell M.D. and Lee J.-F. (1989). Chorophenols in soils. In: Reactions and Movements of Organic Chemicals in Soils. Ed. Kissel D.E., Soil Science Society of America and American Society of Agronomy, Madison, Wisconsin, 209-228.

Boyd S.A., Shaobai S., Lee J.-F. and Mortland M.M. (1988b). Pentachlorophenol sorption by organo-clays. *Clays and Clay Minerals* 36(2): 125-130.

Boyette C.D., Quimby P.C.J., Connick W.J.J., Daigle D.J. and Fulgham F.E. (1991). Progress in production, formulation and application of mycoherbicides. In: Microbial Control of Weeds. Ed. TeBeest D.O., Chapman and Hall, New York, 209-222.

Boyle D.C. (1994). Development of a practical method for Inducing white rot fungi to grow into and degrade organopollutants in soil. *Candian Journal of Microbiology* 41: 345-353.

Boyle D.C. (1998). Nutritional factors limiting the growth of *Lentinula edodes* and other white-rot fungi in wood. Soil Biology and Biochemistry 30(6): 817-823.

Boyle D.C., Wiesner C. and Richardson A. (1998). Factors affecting the degradation of polyaromatic hydrocarbons in soil by white-rot fungi. *Soil Biology and Biochemistry* 30(7): 873-882.

Brixie J.M. and Boyd S.A. (1994). Organic chemicals in the environment: Treatment of contaminated soils with organo clays to reduce leachable pentachlorophenol. *Journal of Environmental Quality* 23(6): 1283-1290.

Bumpus J.A. and Aust S.D. (1985). Biodegradation of environmental pollutants by the white rot fungus *Phanerochaete chrysosporium*: involvement of the lignin degrading system. *BioEssays* 6(4): 166-170.

Bumpus J.A., Tien M., Wright D. and Aust S.D. (1985). Oxidation of persistent environmental pollutants by a white rot fungus. *Science* 228: 1434-1436.

Call H.-P. and Mücke I. (1997). History, overview and applications of mediated lignolytic systems, especially laccase-mediator-systems (Lignozyme®-process). *Journal of Biotechnology* 53: 163-202.

Cassidy M.B., Lee H. and Trevors J.T. (1996). Environmental applications of immobilised microbial cells: A review. *Journal of Industrial Microbiology* 16: 70-101.

Chaudry G.R., ed. (1994). Biological degradation and bioremediation of toxic chemicals. Discorides Press: Portland Oregon, USA.

Cheng H.H. and Mulla D.J. (1999). The Soil Environment. In: Bioremediation of Contaminated Soils. Eds. Kral D.M. and Viney M.K., American Society of Agronomy, Crop Society of America and Soil Science Society of America, Madison,1-13.

Chi J. and Huang G.L. (2002a). A multimedia fugacity river model of Pentachlorophenol in South Drainage Canal, China. *Journal of Environmental Science and Health Part a-Toxic/Hazardous Substances & Environmental Engineering* 37(1): 113-125.

Chi J. and Huang G.L. (2002b). Modelling air-water exchange process of pentachlorophenol in the aquatic environment. *Water Quality Research Journal of Canada* 37(2): 445-458.

Childress A.M., Bennett J.W., Connick W.J.J. and Daigle D.J. (1998). Formulation of filamentous fungi for bioremediation. *Biotechnology Techniques* 12(3): 211-214.

Chiou C.T. (1989). Theortical considerations of the partition uptake of nonionic organic compounds by soil organic matter. In: Reactions and Movements of Organic Chemicals in Soils. Ed. Kissel D.E., Soil Science Society of America, Madison, Wisconsin, USA,1-30.

Choinowski T., Blodig W., Winterhalter K.H. and Piontek K. (1999). The crystal structure of lignin peroxidase at 1.70 A resolution reveals a hydroxy group on the Cβ of tryptophan 171: A novel radical site formed during the redox cycle. *Journal of Molecular Biology* 286(3): 809-827.

Christodoulatos C. and Mohiuddin M. (1996). Generalised models for prediction of pentachlorophenol adsorption by natural soils. *Water Environment Research* 68(3): 370-378.

Christodoulatos C., Korfiatis G.P., Talimcioglu N.M. and Mohammed M. (1994). Adsorption of pentachlorophenol by natural soils. *Journal of Environmental Science and Health* A 29(5): 883-898.

Chung M. and Alexander M. (1998). Differences in sequestration and bioavialability of organic compounds aged in dissimilar soils. *Environmental Science and Technology* 32(7): 855-860.

Chung M. and Alexander M. (2002). Effect of soil properties on bioavailability and extractability of phenanthrene and atrazine sequestered in soil. Chemosphere 48(1): 109-115.

Chung N. and Aust S.D. (1995). Degradation of pentachlorophenol in soil by *Phanerochaete chrysosporium. Journal of Hazardous Materials* 41: 177-183.

Clothier B.E., Green S.R., Vogeler I., Robinson B.H., Mills T.M.M., Duwig C., Roygard J.F.K., Walter M., van den Dijssel C.W. and Scotter D.R. (2004). Contaminants in the rootzone: transport, uptake and remediation, In: Waste Management. Eds. Juhasz A.L., Magesan G. and Naidu R. Science Publishers, Inc., Plymouth, UK, 3-34.

Collins P.J. and Dobson A.D.W. (1996). Oxidation of fluorene and phenanthrene by Mn(II) dependent peroxidase activity in whole cultures of *Trametes (Coriolus) versicolor. Biotechnology Letters* 18(7): 801-804.

Collins P.J. and Dobson A.D.W. (1997). Generation of species specific DNA probes for the lignin peroxidase genes of white rot fungi. In: Methods in Biotechnology volume 2: Bioremediation Protocols. Ed. Sheehan D., Humana Press Inc, Totowa, NJ, 177-185.

Collins P.J., Field J.A., Teunissen P. and Dobson A.D.W. (1997). Stabilisation of lignin peroxidases in white-rot fungi by tryptophan. *Applied and Environmental Microbiology* 63(7): 2543-2548.

Cort T. and Bielefeldt A. (2000). Organic compounds in the environment. *Journal of Environmental Engineering* 126: 635-643.

Cullen D. and Kersten P.J. (1996). Enzymology and molecular biology of lignin degradation. In: The Mycota. Eds. Esser K. and Lemke P.A., Springer-Verlug, Berlin Heidelberg, 295-312.

Czaplicka M. (2004). Sources and transformations of chlorophenols in the natural environment. *Science of The Total Environment* 322(1-3): 21-39.

Davis A., Campbell C., Colbert C., Ruby M.V., Bennet M. and Tobin S. (1994). Attenuation and biodegradation of chlorophenols in ground water in a former wood treating facility. *Ground Water* 32(2): 248-257.

Davis M.W., Glaser J.A., Evans J.W. and Lamar R.T. (1993). Field evaluation of the lignin-degrading fungus *Phanerochaete sordida* to treat creosote-contaminated soil. *Environmental Science and Technology* 27(12): 2572-2576.

de Jong E., Cazemier A.E., Field J.A. and de Bont J.A.M. (1994). Physiological role of chlorinated aryl alcohols biosynthesized de novo by the white rot fungus *Bjerkandera* sp. strain BOS55. *Applied and Environmental Microbiology* 60(1): 271-277.

de la Rubia T., Ruiz E., Pérez J. and Martínez J. (2002). Properties of a laccase produced by *Phanerochaete flavido-alba* induced by vanillin. *Archives of Microbiology* 179(1): 70-73.

Dec J. and Bollag J.-M. (1994). Dehalogenation of chlorinated phenols during oxidative coupling. *Environmental Science and Technology* 28(3): 484-490.

Dittmer N.T., Suderman R.J., Jiang H., Zhu Y-C.,Gorman M.J., Kramer K.J. and Kanost, M.R. (2004). Characterization of cDNAs encoding putative laccase-like multicopper oxidases and developmental expression in the tobacco hornworm, *Manduca sexta*, and the malaria mosquito. Anopheles gambiae. *Insect Biochemistry and Molecular Biology* 34(1): 29-41.

Divincenzo J.P. and Sparks D.L. (1997). Slow sorption kinetics of pentachlorophenol on soil: Concentration effects. *Environmental Science and Technology* 31(4): 977-983.

Divincenzo J.P. and Sparks D.L. (2001). Sorption of the neutral and charged forms of pentachlorophenol on soil: Evidence for different mechanisms. *Archives of Environmental Contamination and Toxicology* 40(4): 445-450.

Dodson A.P.J., Harvey P.J., Evans C.S. and Palmer J M. (1986). Properties of extracellular ligninase from *Trametes versicolor.* Proceedings of the Biotechnology in the Pulp and Paper Industry, Swedish Forest Products Research Laboratory and The Swedish Association of Pulp and Paper, Stockholm, Sweden, 185-186.

Doyle W.A., Blodig W., Veitch N.C., Piontek K. and Smith A.T. (1998). Two substrate interaction sites in lignin peroxidase revealed by site-directed mutagenesis. Biochemistry 37(43): 15097-15105.

Ducros V., Brzozowski A.M., Wilson K.S., Østergaard P., Schneider P., Svendson A. and Davies G.J. (2001). Structure of the laccase from *Coprinus cinereus* at 1.68 Å resolution: evidence for different "type 2 Cu-depleted" isoforms. *Acta Crystallographica* D 57(2): 333-336.

Eggen T. and Majcherczyk A. (1998). Removal of polycyclic aromatic hydrocarbons (PAH) in contaminated soil by white rot fungus. *Pleurotus ostreatus. International Biodeterioration & Biodegradation* 41(2): 111-117.

Ellis H., Shaw C., de Bazin M. and Huser B. (1993). Steering Committee on sites contaminated by timber treatment and antisapstain chemicals. Ministry for the Environment, Wellington, New Zealand, 1-10.

Ellis H., de Bazin M., Husser B., Shaw C. and Waters J. (1997). Health and Environmental Guidelines for Selected Timber Treatment Chemicals: Soil Acceptance Criteria, Ministry for the Environment, Wellington, New Zealand, 1-55.

Evans C.S. (1985). Laccase activity in lignin degradation by *Coriolus versicolor in vivo* and *in vitro* studies. *Microbiology Letters* 27: 339-343.

Fall C., Chavarie C. and Chaouki J. (2000). Desorptive behaviour of pentachlorophenol and phenanthrene in soil-water systems. *Water Environment Research* 72(2): 162-169.

Fall C., Chavarie C. and Chaouki J. (2001). Generalised model of pentachlorophenol distribution of amended soil-water systems. Water Environment Research 73(1): 110-117.

Farrell R.L., Murtagh K.E., Tien M., Mozuch M.D. and Kirk T.K. (1989). Physical and enzymatic properties of lignin peroxidase isoenzymes from *Phanerochaete chrysosporium. Enzyme and Microbial Technology* 11(6): 322-328.

Finnbogasson T.W. and St. Quintin O.N.C. (1994). Review and assessment of available pentachlorophenol (PCP) and dioxin treatment technologies. Report prepared for: New Zealand Ministry for the Environment and Timber Industry Environmental Council. Ministry for the Environment, Wellington, New Zealand. Pp. 123.

Ford C.I. (2006). White-rot fungi: formulation for inoculation and biodegradation of pentachlorophenol in contaminated sites, PhD Thesis. Lincoln University, Lincoln, New Zealand. Pp. 227.

Gianfreda L., Xu F. and Bollag J.-M. (1999). Laccases: a useful group of oxidoreductive enzymes. *Bioremediation Journal* 3(1): 1-25.

Gibb W.W. (1999). Not cleaning up. *Scientific American* 280(2): 22-23.

Glaser J.A. (2003). Treatment technology overview: Bioremediation *ex-situ* solid phase: Fungal based bioremediation, USEPA, online at http://www.epareachit.org/

Gold M.H., Youngs H.L. and Gelpke M.D.S. (2000). Manganese peroxidase. *Metal Ions in Biological Systems* 37: 559-586.

Gold M.H., Kuwahara M., Chiu A.A. and Glenn J.K. (1984). Purification and characterization of an extracellular H_2O_2 requiring diarylpropane oxygenase from the white rot basidiomycete. *Phanerochaete chrysosporium, Archives of Biochemistry and Biophysics* 234(2): 353-362.

Grey R., Hoffer C. and Schlosser D. (1998). Degradation of 2-chlorophenol and formation of 2-chlorophenol-1,4 benzoquinone by mycelia and cell-free Crude culture liquids of *Trametes versicolor* in relation to extra-cellular laccase activity. *Journal of Basic Microbiology* 38(5-6): 371-382.

Guthrie J.M. (2006). Degradation of PCP by laccases of the white-rot fungus *Trametes* sp. HR577, PhD Thesis. Massey University, Palmerston North, New Zealand. Pp. 103.

Hasset J.J. and Banwart W.L. (1989). The sorption of nonpolar organics by soils and sediments. In: Reactions and Movement of Organic Chemicals in Soils. Ed. Kissel D.E., Soil Society Science of America, American Society of Agronomy, Madison, Wisconsin, Madison, Wisconsin, 41-44.

Hewitt A.E. (1993). New Zealand soil classification. Landcare Research Science Series No. 1., Manaaki-Whenua Press, Lincoln, New Zealand. Pp. 131.

Higson F.K. (1991). Degradation of xenobiotics by white rot fungi. *Reviews of Environmental Contamination and Toxicology* 122: 111-152.

Hill J.L., Hocking A.D. and B W.F. (1995). The role of fungi in the production of chloroanisoles in general purpose freight containers. *Food Chemistry* 54: 161-166.

Hofrichter M. (2002). Review: lignin conversion by manganese peroxidase (MnP), Enzyme and Microbial Technology 30(4): 454-466.

Holroyd M.L. and Caunt P. (1995). Large-scale soil bioremediation using white-rot fungi. Third International *In situ* and On-site Bioreclamation Symposium, San Diego, California. 181-187.

Jaynes W.F. and Boyd S.A. (1991). Clay mineral type and organic compound sorption by hexadecyltrimethylammonium-exchanged clays. *Soil Science Society of America Journal* 55: 43-48.

Jen A.C., Conley Wake M. and Mikos A.G. (1996). Review: Hydro gels for cell immobilisation,.*Biotechnology and Bioengineering* 50: 357-364.

Johannes C. and Majcherczyk A. (2000). Laccase activity tests and laccase inhibitors. *Journal of Biotechnology* 78(2): 193-199.

Kanbi L.D., Antonyuk S., Hough M.A., Hall J.F., Dodd F.E. and Hasnain S.S. (2002). Crystal structures of the Met148Leu and Ser86Asp mutants of rusticyanin from *Thiobacillus ferrooxidans*: insights into the structural relationship with the cupredoxins and the multi copper proteins. *Journal of Molecular Biology* 320(2): 263-275.

Kapich A.N., Prior B.A., Botha A., Galkin S., Lundell T. and Hatakka A. (2004). Effect of lignocellulose-containing substrates on production of ligninolytic peroxidases in submerged cultures of *Phanerochaete chrysosporium* ME-446. *Enzyme and Microbial Technology* 34(2): 187-195.

Karickhoff S.W. (1984). Organic pollutant sorption in aquatic systems. *Journal of Hydraulic Engineering* 110(6): 707-735.

Khan F.I., Husain T. and Hejazi R. (2004). An overview and analysis of site remediation technologies. *Journal of Environmental Management* 71(2): 95-122.

Khodadoust A.P., Sorial G.A., Wilson G.J., Siudan M.T., Griffiths R.A. and Brenner R.C. (1999). Integrated system for remediation of contaminated soils. *Journal of Environmental Engineering* 125(11): 1033-1041.

Knox A.S., Gamerdinger A.P., Adriano D.C., Kolka R.K. and Kaplan D.I. (1999). Sources and practices contributing to soil contamination. In: Bioremediation of Contaminated Soils. Ed. Kral D.M., American Society of Agronomy, Crop Science Society of America and Soil Science Society of America, Madison, 53-87.

Kottler B.D., White J.C. and Kelsey J.W. (2001). Influence of soil moisture on the sequestration of organic compounds in soil. *Chemosphere* 42(8): 893-898.

Lamar R.T. (1992). The role of fungal lignin-degrading enzymes in xenobiotic degradation. *Current Opinion in Biotechnology* 3: 261-266.

Lamar R.T. and Dietrich D.M. (1990a). Use of lignin-degrading fungi in the disposal of pentachlorophenol treated wood. *Journal of Industrial Microbiology* 9: 181-191.

Lamar R.T. and Dietrich D.M. (1990b). *In situ* depletion of pentachlorophenol from contaminated soil by *Phanerochaete* spp. *Applied and Environmental Microbiology* 56(10): 3093-3100.

Lamar R.T., Glaser J.A. and Kent Kirk T. (1990). Fate of pentachlorophenol (PCP) in sterile soils inoculated with the white-rot basidomycete *Phanerochaete chrysosporium*: Mineralisation volatilisation and depletion of PCP. *Soil Biology and Biochemistry* 22(4): 433-440.

Lamar R.T., Evans J.W. and Glaser J.A. (1993a). Solid-phase treatment of a pentachlorophenol-contaminated soil using lignin-degrading fungi. *Environmental Science and Technology* 27(12): 2566-2571.

Lamar R.T., Davis M.W., Dietrich D.M. and Glaser J.A. (1993b). Treatment of a pentachloropenol and creosote contaminated soil using the lignin-degrading fungus *Phanerochaete sordida*: A field demonstration. *Soil Biology and Biochemistry* 26(12): 1603-1611.

Lamar R.T., Main L.M., Dietrich D.M. and Glaser J.A. (1999). Screening of fungi for soil remediation potential. In: Bioremediation of Contaminated Soils. Eds. Kral D.M. and Viney M.K., American Society of Agronomy, Crop Science Society of America, Soil Science Society of America, Madison, 437-456.

Larrondo L.F., González A., Perez-Acle T., Cullen D. and Vicuña R. (2005). The nop gene from *Phanerochaete chrysosporium* encodes a peroxidase with novel structural features. *Biophysical Chemistry* 116(2): 167-173.

Lee L.S., Rao P.S.C., Nkedi-Kizza P. and Delfino J.J. (1990). Influence of solvent and sorbent characteristics on distribution of pentachlorophenol in octanol-water and soil-water systems. *Environmental Science and Technology* 24(5): 654-661.

Leidig E., Prusse U., Vorlop K.-D. and Winter J. (1999). Biotransformation of Poly R-478 by continuous cultures of PVAL-encapsulated *Trametes versicolor* under non-sterile conditions. *Bioprocess Engineering* 21: 5-12.

Leontievsky A.A., Vares T., Lankinen P., Shergill J.K., Pozdnyakova N.N., Myasoedova N.M., Kalkkinen N., Golovleva L.A., Cammack R., Thurston C.F. and Hatakka A. (1997). Blue and yellow laccases of ligninolytic fungi *FEMS. Microbiology Letters* 156(1): 9-14.

Leštan D. and Lamar R.T. (1996). Development of fungal inocula for bioaugmentation of contaminated soils. *Applied and Environmental Microbiology* 62(6): 2045-2052.

Leštan D. and Lamar R.T. (1999). Influence of humidity on production of pelleted fungal inoculum. *World Journal of Microbiology and Biotechnology* 15(3): 349-357.

Leštan D., Leštan M. and Lamar R.T. (1998). Growth and viability of mycelial fragments of white-rot fungi on some hydrogels. *Journal of Industrial Microbiology & Biotechnology* 20(3-4): 244-250.

Leštan D., Leštan M., Chapelle J.A. and Lamar R.T. (1996). Biological potential of fungal inocula for bioaugmentation of contaminated soils. *Journal of Industrial Microbiology* 16: 286-294.

Li K., Xu F. and Eriksson K.-E. (1999). Comparison of fungal laccases and redox mediators in oxidation of a nonphenolic lignin model compound. *Applied and Environmental Microbiology* 65(6): 2654-2660.

Li Z., Burt D. and Bowman R.S. (2000). Sorption of ionizable organic solutes by surfactant-modified zeolite. *Environmental Science and Technology* 34(17): 3756-3760.

Liste H.-H. and Alexander M. (2002). Butanol extraction to predict bioavailability of PAHs in soil. *Chemosphere* 46(7): 1011-1017.

Loomis A.K., Childress A.M., Daigle D.J. and Bennett J.W. (1997). Alginate encapsulation of the white rot fungus *Phanerochate chrysosporium. Current Microbiology* 34: 127-130.

Martínez Á.T. (2002). Molecular biology and structure-function of lignin-degrading heme peroxidases. *Enzyme and Microbial Technology* 30(4): 425-444.

Martinez D., Larrondo L.F., Putnam N., Gelpke M.D.S., Huang K., Chapman J., Helfenbein K.G., Ramaiya P., Detter J.C., Larimer F., Coutinho P.M., Henrissat B., Berka R., Cullen D. and Rokhsar D. (2004). Genome sequence of the lignocellulose degrading fungus *Phanerochaete chrysosporium* strain RP78. *Nature Biotechnology* 22(6): 695-700.

Mauperin C., Mortier F., Garbaye J., Le Tacon F. and Garr G. (1987). Viability of an ectomycorrhizal inoculum produced in a liquid medium and entrapped in a calcium alginate gel. *Candian Journal of Botany* 65: 2326-2329.

McAllister K.A., Lee H. and Trevors J.T. (1996). Microbial degradation of pentachlorophenol. *Biodegradation* 7: 1-40.

McClaren R.G. and Cameron K. (1996). Soil science: sustainable production and environmental protection. 2 Edition. Oxford University Press: Auckland, New Zealand. Pp 277.

McGrath R. and Singleton I. (2000). Pentachlorophenol transformation in soil: A toxicological assessment. *Soil Biology and Biochemistry* 32(1): 1311-1314.

Mester T., de Jong E. and Field J.A. (1995). Manganese regulation of veratryl alcohol in white rot fungi and its indirect effect on lignin peroxidase. *Applied and Environmental Microbiology* 61(5): 1881-1887.

Mileski G.J., Bumpus J.A., Jurek M.A. and Aust S.D. (1988). Biodegradation of pentachlorophenol by the «white rot fungus *Phanerochaete chrysosporium. Applied and Environmental Microbiology* 54(12): 2885-2889.

Miller G.C., Herbert V.R. and Miller W.W. (1989). Effect of sunlight on organic contaminants at the atmosphere soil interface. In: Reactions and Movements of Organic Chemicals in Soils. Ed. Kissel D.E., Soil Science Society of America, American Society of Agronomy, Madison, Wisconsin, USA, 99-110.

Miller M.N., Stratton G.W. and Murray G. (2004). Effects of nutrient amendments and temperature on the biodegradation of pentachlorophenol contaminated soil. *Water, Air and Soil Pollution* 151: 87-101.

Mills T., Arnold B., Sivakumaran S., Northcott G., Vogeler I., Robinson B., Norling C. and Leonil D. (2006). Phytoremediation and long-term site management of soil contaminated with pentachlorophenol (PCP) and heavy metals. *Journal of Environmental Management* 79: 232-241.

Morgan P., Lee S.A., Lewis S.T., Sheppard A.N. and Watkinson R.J. (1993). Growth and biodegradation by white rot fungi inoculated into soil. *Soil Biology and Biochemistry* 25(2): 279-287.

Morimoto K., Tatsumi K. and Kuroda K.-I. (2000). Peroxidase catalysed co-polymerisation of pentachlorophenol and a potential humic precursor. *Soil Biology and Biochemistry* 32: 1071-1077.

Morohoshi N., Fujita K., Wariishi H. and Haraguchi T. (1987). Degradation of lignin by the extracellular enzymes of Coriolus versicolor V. Degradation of isolated lignin fractions by laccases I, II, and III. *Mokuzai Gakkaishi* 33(4): 310-315.

Murray H.H. (1995). Clays in industry and the environment. Clays controlling the environment: 10th International clays conference, Adelaide, Australia. CSIRO Publishing, Collingwood, Victoria, Australia, 49-55.

Nadebaum P., Gifford J. and Bingham A. (1992a). The report on the national task group investigating site contamination from the use of timber treatment chemicals. Ministry for the Environment and Ministry of Health, Wellington, New Zealand, 1-66.

Nadebaum P., Gifford J. and Bingham A. (1992b). New Zealand national task group on site contamination from the use of timber treatment chemicals: Pentachlorophenol risk assessment pilot study. Camp Scott Furphy, Melbourne, Australia, 2-9.

Nitta K., Kataoka K. and Sakurai T. (2002). Primary structure of a Japanese lacquer tree laccase as a prototype enzyme of multicopper oxidases. *Journal of Inorganic Biochemistry* 91(1): 125-131.

NTP (1989). Technical Report-349 (abstract): Toxicology and carcinogenesis of two technical grade pentachlorophenol mixtures (Cas No. 87-86-5) in B6C3F$_1$ mice (feed studies), National Institute of Environmental Health Services (NIEHS), online at http://ntp-server.niehs.nih.gov/index.cfm?objectid=070867E3-04A7-5F1D-4163D6EB3859804D

Nye J.V., Guerin W.F. and Boyd S.A. (1994). Heterotrophic Activity of Microorganisms in Soils Treated with Quaternary Ammonium-Compounds. *Environmental Science and Technology* 28(5): 944-951.

Palmer A.E., Randall D.W., Xu F. and Solomon E.I. (1999). Spectroscopic studies and electronic structure description of the high potential type 1 copper site in fungal laccase: insight into the effect of the axial ligand. Journal *of the American Chemical Society* 121(30): 7138-7149.

Palmer A.E., Szilagyi R.K., Cherry J.R., Jones A., Xu F. and Solomon E.I. (2003). Spectroscopic characterization of the Leu513His variant of fungal laccase: effect of increased axial ligand interaction on the geometric and electronic structure of the type 1 Cu site. *Inorganic Chemistry* 42(13): 4006-4017.

Pignatello J.J. (1989). Sorption dynamics of organic compounds in soils and sediments. In: Reactions and Movements of Organic Chemicals in Soils. Ed. Kissel D.E., Soil Science Society of America and American Society of Agronomy, Madison, Wisconsin, 45-80.

Piontek K., Anotorini M. and Choinowski T. (2002). Crystal structure of a laccase from the fungus *Trametes versicolor* at 1.90-Å resolution containing a full complement of coppers. *The Journal of Biological Chemistry* 277(No. 40): 37663-37669.

Pointing S.B. (2001). Feasiblity of bioremediation by white-rot fungi. *Applied Microbiology and Biotechnology* 57: 20-33.

Pollard S.J.T., Hoffman R.E. and Hrudery S.E. (1993). Screening of risk management options for abandoned wood-preserving sites in Alberta, Canada. *Canadian Journal of Civil Engineering* 20: 787-800.

Poulos T.L., Edwards S.L., Wariishi H. and Gold M.H. (1993). Crystallographic refinement of lignin peroxidase at 2 A. *Journal of Biological Chemistry* 268(6): 4429-4440.

Reading N.S. and Aust S.D. (2001). Role of disulfide bonds in the stability of recombinant manganese peroxidase. *Biochemistry* 40(27): 8161-8168.

Reading N.S., Welch K.D. and Aust S.D. (2003). Free radical reactions of wood-degrading fungi. *ACS Symposium Series 845* (Wood Deterioration and Preservation): 16-31.

Reddy G.V.B. and Gold M.H. (2000). Degradation of pentachlorophenol by *Phanerochaete chrysosporium*: intermediates and reactions involved. *Microbiology* 146(2): 405-413.

Renner G., Hopper C. and Gokel J.M. (1986). Acute toxicities of pentachlorophenol, pentachloroanisole, tetrachlorohydroquinone, tetrachlorocatechol, tetrachlororesorcinol, tetrachlorodimethoxybenzenes and tetrachlorobenzendiol diacetates administered to mice. *Toxicological and Environmental Chemistry* 11: 37-50.

Robertson B.K. and Alexander M. (1998). Sequestration of DDT and dieldrin in soil: Disappearance of acute toxicity but not the compounds. *Environmental Toxicology and Chemistry* 17(6): 1034-1038.

Roy-Arcand L. and Archibald F.S. (1991). Direct dechlorination of chlorophenolic compounds by laccases from *Trametes (Coriolus) versicolor. Enzyme and Microbial Technology* 13(3): 194-203.

Rudzinski W.E., Dave A.M., Vaishnav U.H., Kumbar S.G., Kulkarni A.R. and Aminabhavi T.M. (2002). Hydro gels as controlled release devices in agriculture. *Designed Monomers and Polymers* 5(1): 39-65.

Rüttimann-Johnson C. and Lamar R.T. (1996). Polymerization of pentachlorophenol and ferulic acid by fungal extracellular lignin-degrading enzymes. *Applied and Environmental Microbiology* 62(10): 3890-3893.

Rüttimann-Johnson C. and Lamar R.T. (1997). Binding of pentachlorophenol to humic substances in soil by the action of white rot fungi. *Soil Biology and Biochemistry* 29(7): 1143-1148.

Schmidt K.R., Chand S., Gostomski P.A., Boyd-Wilson K.S.H., Ford C. and Walter M. (2005). Fungal inoculum properties and its effect on growth and enzyme activity of *Trametes versicolor* in soil. *Biotechnology Progress* 21(2): 377-385.

Sederati M.R., Keshavarz T., Leontievsky A.A. and Evans C.S. (2003). Transformation of high concentrations of chlorophenols by the white-rot basidiomycete *Trametes versicolor* immobilized on nylon mesh. *Electronic Journal of Biotechnology* 6(2): 104-114.

Semple K.T., Reid B.J. and Fermor T.R. (2001). Impact of composting strategies on the treatment of soils contaminated with organic pollutants. *Environmental Pollution* 112(2): 269-283.

Shaw I.C. and Chadwick J. (1999). Principles of Environmental Toxiclogy. 1 Edition. Taylor & Francis: London. Pp 216.

Shiu W.-Y., Ma K.-C., Varhanickova D. and MacKay D. (1994). Chlorophenols and alkylphenols: A review and correlation of the environmentally relevant properties and fate in an evaluative environment. *Chemosphere* 29(6): 1155-1224.

Smidsrod O. and Skjak-Braek G. (1990). Alginate as immobilization matrix for cells, *Trends in Biotechnology* 8: 71-78.

Smith J.E., Witoski P.J. and Chiou C.T. (1988). Partition of nonionic organic compounds in aquatic systems. *Reviews of Environmental Contamination and Toxicology* 103: 127-151.

Soil Survey Staff. (2003). Keys to Soil Taxonomy. Ninth Edition. USDAC Natural Resources Conservation Service, Washington DC. Pp. 325.

Solomon E.I., Sundaram U.M. and Machonkin T.E. (1996). Multicopper Oxidases and Oxygenases. *Chemical Reviews* 96(7): 2563-2605.

Soule N.M. and Burns S.E. (2001). Effects of organic cation structure on behaviour of organobentonites. *Journal of Geotechnical and Geoenvironmental Engineering* 127(4): 363-370.

Stamets P. (2000). Growing gourmet and medicinal mushrooms. 2nd Edition. Ten Speed, Berkley, California, USA. Pp. 553.

Stamets P. and Chilton J.S. (1983). The mushroom cultivator. Agarikon Press, Olympia, Washington, USA. Pp. 408.

Stapleton M.G., Sparks D.L. and Dentel S.K. (1994). Sorption of pentachlorophenol to HDTMA-clay as a function of ionic strength and pH. *Environmental Science and Technology* 28: 2330-2335.

Stevenson P. (1992). The scandal of the PCP dumps. *Terra Nova* June:10-15.

Sutherland G.R.J. and Aust S.D. (1996). The Effects of Calcium on the Thermal Stability and Activity of Manganese Peroxidase. *Archives of Biochemistry and Biophysics* 332(1): 128-134.

Szabo M. (1993). New Zealand's poisoned paradise. *New Scientist* July: 29-33.

Tang J., Carroquino M.J., Robertson B.K. and Alexander M. (1998). Combined effect of sequestration and bioremediation of polycyclic aromatic hydrocarbons in soil. *Environmental Science and Technology* 32(22): 3586-3590.

Tien M. and Kirk T.K. (1984). Lignin-degrading enzyme from *Phanerochaete chrysosporium*: Purification, characterization, and catalytic properties of a unique H_2O_2-requiring oxygenase. *Proceedings of the National Academy of Sciences of the United States of America* 81(7): 2280-2284.

Timofeevski S.L., Nie G., Reading N.S. and Aust S.D. (2000). Substrate specificity of lignin peroxidase and a S168W variant of manganese peroxidase. Archives *of Biochemistry and Biophysics* 373(1): 147-153.

Trevors J.T., van Elsas J.D., Lee H. and van Overbeek L.S. (1992). Use of alginate and other carriers for encapsulation of microbial cells for use in soil. *Microbial Releases* 1: 61-69.

Tuomela M., Lyytikainen M., Oivanen P. and Hatakka A. (1999). Mineralisation and conversion of pentachlorophenol (PCP) in soil inoculated with the white-rot fungus *Trametes versicolor. Soil Biology and Biochemistry* 31(1): 65-74.

Ullah M.A., Bedford C.T. and Evans C.S. (2000). Reactions of pentachlorophenol with laccase from *Coriolus versicolor. Applied Microbiology and Biotechnology* 53(2): 230-234.

USEPA (2001). Remediation technology cost compendium, United States Environmental Protection Agency (USEPA): Office of Solid Waste and Emergency Response, Report EPA-542-R-01-099, 1-55. Available online at http://www.clu-in.org/

USEPA (2004). Treatment technologies for site clean up: Annual status report (eleventh edition), United States Environmental Protection Agency (USEPA): Office of Solid Waste and Emergency Response, Report EPA-542-R-09-009, 1-50. Available online at http://www.clu-in.org/asr/

Vertraete W. and Top E.M. (1999). Soil clean-up: lessons to remember. *International Biodeterioration & Biodegradation* 43(3): 147-153.

Walter M. (2004). Towards optimisation of white-rot fungi bioremediation, PhD Thesis, University of Canterbury. Pp 209.

Walter M. and Boyd-Wilson K.S.H. (2006). Development of a biotechnology tool using New Zealand white-rot fungi to degrade pentachlorophenol in soil – a summary. In: Recent Research Developments in Multidisciplinary Applied Microbiology. Understanding and Exploiting Microbes and Their Interactions. Biological, Physical, Chemical and Engineering Aspects, Wiley-VCH. In press.

Walter M., Boul L., Chong R. and Ford C. (2004). Growth substrate selection and biodegradation of PCP by New Zealand white-rot fungi. *Journal of Environmental Management* 71(4): 361-369.

Walter M., Boyd-Wilson K.S.H., McNaughton D. and Northcott G. (2005a). Laboratory trials on the bioremediation of aged pentachlorophenol residues. *International Biodeterioration & Biodegradation* 55(2): 121-130.

Walter M., Guthrie J.M., Sivakumaran S., Parker E., Slade A., McNaughton D. and Boyd-Wilson K. (2003). Screening of New Zealand native white-rot isolates for PCP degradation. *Bioremediation Journal* 7(2): 119-128.

Walter M., Boyd-Wilson K., Boul L., Ford C., McFadden D., Chong B. and Pinfold J. (2005b). Field-scale bioremediation of pentachlorophenol by Trametes versicolor. *International Biodeterioration & Biodegradation* 56(1): 51-57.

Watanabe M.E. (2001). Can bioremediation bounce back. Nature Biotechnology 19: 1111-1115.

Weber Jr W.J., Huang W. and LeBoeuf E.J. (1999). Geo-sorbent organic matter and its relationship to the binding and sequestration of organic contaminants. *Colloids and Surfaces A: Physiochemical and Engineering Aspects* 151(1-2): 167-179.

White J.C., Quinones-Rivera A. and Alexander M. (1998). Effect of wetting and drying on the bioavailability of organic compounds sequestered in soil. *Environmental Toxicology and Chemistry* 17(12): 2378-2382.

White J.C., Alexander M. and Pignatello J.J. (1999a). Enhancing the bioavailability of organic compounds sequestered in soil and aquifer solids. *Environmental Toxicology and Chemistry* 18(2): 182-187.

White J.C., Kelsey J.W., Hatzinger P.B. and Alexander M. (1997). Factors affecting sequestration and bioavailability of phenanthrene in soils. *Environmental Toxicology and Chemistry* 16(10): 2040-2045.

White J.C., Hunter M., Nam K., Pignatello J.J. and Alexander M. (1999b). Correlation between biological and physical availabilities of phenanthrene in soils and soil humin in ageing experiments. *Environmental Toxicology and Chemistry* 18(8): 1720-1727.

Witthuhn B., Klauth P., Klumpp E., Narres H.-D. and Martinius H. (2005). Sorption and biodegradation of 2,4-dichlorophenol in the presence of organoclays. *Applied Clay Science* 28(1-4): 55-66.

Wu C.T., Gumpertz M.L. and Boos D.D. (2001). Comparison of GEE, MINQUE, ML, and REML estimating equations for normally distributed data. *American Statistician* 55(2): 125-130.

Wu Q., Blume H.P., Rexilius L., Folcshow M. and Schleuss U. (2000). Sorption of atrazine, 2,4-D, nitrobenzene and pentachlorophenol by urban and industrial wastes. *European Journal of Soil Science* 51(2): 335-344.

Xu F., Palmer A.E., Yaver D.S., Berka R.M., Gambetta G.A., Brown S.H. and Solomon E.I. (1999). Targeted mutations in a *Trametes villosa* laccase. Axial perturbations of the T1 copper. *Journal of Biological Chemistry* 274(18): 12372-12375.

Xu F., Berka R.M., Wahleithner J.A., Nelson B.A., Shuster J.R., Brown S.H., Palmer A.E. and Solomon E.I. (1998). Site-directed mutations in fungal laccase: effect on redox potential, activity and pH profile. *Biochemical Journal* 334(1): 63-70.

Yaropolov A.I., Skorobogatko O.V., Vartanov S.S. and Varfolomev S.D. (1994). Laccase. properties, catalytic mechanism, and applicability. *Applied Biochemistry and Biotechnology* 49: 257-280.

Yuan G.D. (2004). Natural and modified nanomaterials as sorbents of environmental contaminants. *Journal of Environmental Science and Health Part A-Toxic/Hazardous Substances & Environmental Engineering* 39(10): 2661-2670.

Zeddal A., Majchercyzk A. and Huttermann A. (1993). Degradation of polychlorinated biphenyls by white-rot fungi *Pleurotus ostreatus* and *Trametes versicolor* in a solid state system. *Toxicological and Environmental Chemistry* 40: 255-266.

Zielke R.C., Pinnavaia T.J. and Mortland M.M. (1989). Adsorption and reactions of selected organic molecules on clay mineral surfaces. In: Reactions and Movements of Organic Chemicals in Soils. Ed. Kissel D.E., Soil Science Society of America and American Society of Agronomy, Madison, Wisconsin (No. 22), 81-97.

3 Penicillin Biosynthesis and Secretion: Precursor and Enzyme Localization

Carlos García-Estrada[1,2] and Juan Francisco Martín[1,2]

[1]Departamento de Ecología, Genética y Microbiología, Área de Microbiología, Facultad de Ciencias Biológicas y Ambientales, Universidad de León, Campus de Vegazana s/n, 24071, León, Spain. [2]Instituto de Biotecnología de León (INBIOTEC), Parque Científico de la Granja, Av. Real, n°1, 24006, León, Spain.

INTRODUCTION

Since the fortuitous discovery by Sir Alexander Fleming in 1928 of the antimicrobial activity generated by a fungus culture that contaminated a Petri dish with *Staphylococcus sp*., penicillin has been used as a therapeutic compound. Together with Sir Alexander Fleming, the scientists Ernst B. Chain and Howard W. Florey (who developed a massive penicillin production method) were awarded with the Physiology and Medicine Nobel Prize in 1945.

Penicillin was initially isolated from *Penicillium notatum* cultures, though researchers soon found other high producing strain as *P. chrysogenum* NRRL 1951. Only recently the genome regions containing the penicillin gene cluster of these two fungi have been cloned. Genes encoding the penicillin biosynthetic pathways occur also in other filamentous fungi (Laich *et al*., 2003). Works on penicillin biosynthesis soon led to the isolation of penicillin G (benzylpenicillin), a compound obtained when phenylacetic acid was added to the culture medium. Addition of other side chain precursor such as phenoxyacetic acid, allowed researchers to purify a different type of penicillin, penicillin V or phenoxymethylpenicillin.

After the discovery in 1959 of the 6-aminopenicillanic acid (6-APA), which is a penicillin precursor compound (secreted in some strains unable to transform it completely into penicillin) and isopenicillin N (IPN), the synthesis of semisynthetic penicillins started up a new era of chemotherapy. The improvement of the strains using genetic and molecular techniques and optimized culture conditions has given rise to penicillin yields much higher than those generated by the parental *P. chrysogenum* strain NRRL 1951 (Rowlands, 1991). It is estimated that the final penicillin concentration in an industrial fermentation is at least 120 mM, e.g., 40080 mg/liter of culture medium (Crueger and Crueger, 1992).

The structure of penicillin contains the β–lactam ring fused to a sulphur-containing thia-zolidine ring (Aoki and Okuhara, 1980; Demain, 1983; Ingolia and Queener, 1989; Martín *et al*., 1994).

Beside penicillins, the β–lactam ring structure is shared by different compounds such as cephalosporins, monobactams and the carbapenems (Fig. 3.1). These antibiotics affect the assembly of the bacterial cell-wall, interfering with the formation of the peptidoglycan by binding to the DD-transpeptidase and DD-carboxipeptidase, leading to an osmotically sensitive cell wall and therefore to the lysis of the bacterial cell (Giesbrecht *et al.*, 1991; Frère *et al.*, 1993).

In addition to *P. chrysogenum*, *Aspergillus nidulans* and a few other ascomycetes (Aharonowitz *et al.*, 1992; Laich *et al.*, 2003) are able to synthesize penicillins containing an aromatic side chain (e.g., benzylpenicillin or phenoxymethylpenicillin).

In this chapter we would like to present an up-to-date view of the precursor and enzyme localization in relation to penicillin biosynthesis.

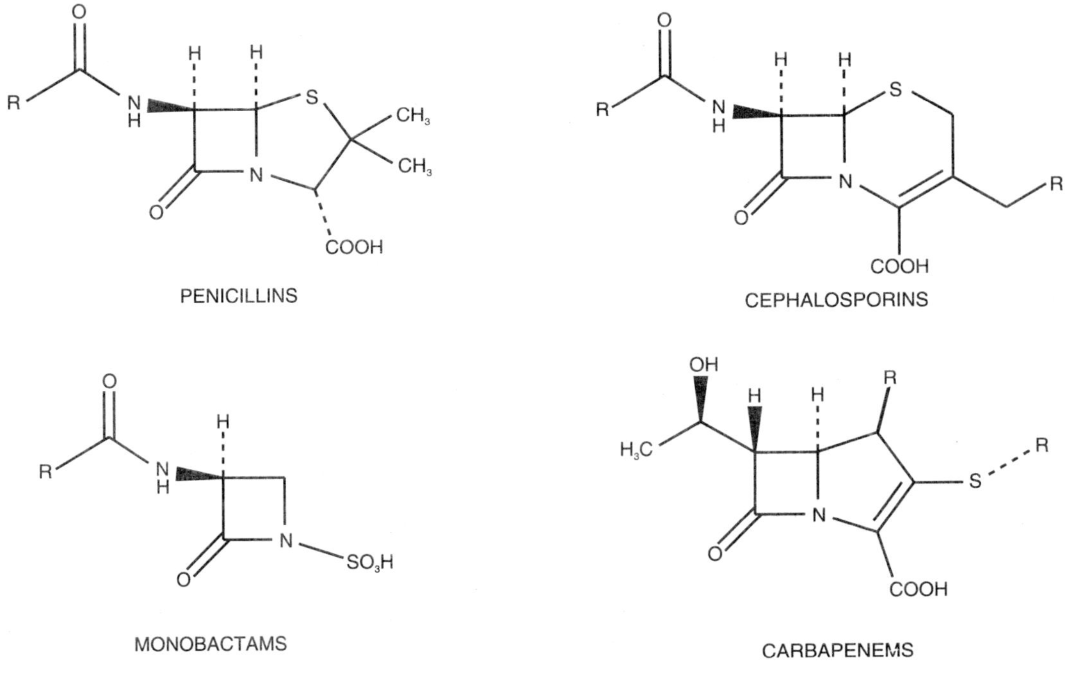

Fig. 3.1. Chemical structures of compounds containing the β-lactam ring.

GENES AND ENZYMES FOR PENICILLIN AND CEPHALOSPORIN BIOSYNTHESIS

The penicillin biosynthetic pathway (Fig. 3.2) starts with the formation of the non-ribosomal tripeptide δ (L-α–aminoadipyl)-L-cysteinyl-D-valine (ACV) formed by the multienzyme δ (α–aminoadipyl)-cysteinyl-valine synthetase (the term synthetase is used for ATP-requiring amino acid condensing enzymes, whereas the term synthase is used for biosynthetic enzymes that do not require ATP), which is encoded by the *pcb*AB gene (Byford *et al.*, 1997; Martín, 2000b). This tripeptide is cyclized to isopenicillin N by the action of isopenicillin N synthase (encoded by the *pcb*C gene). In the last step of penicillin biosynthesis, the isopenicillin N-acyltransferase (IAT; encoded by the *penDE* gene) synthesizes hydrophobic penicillins by substitution of the L-α–aminoadipyl side chain of isopenicillin

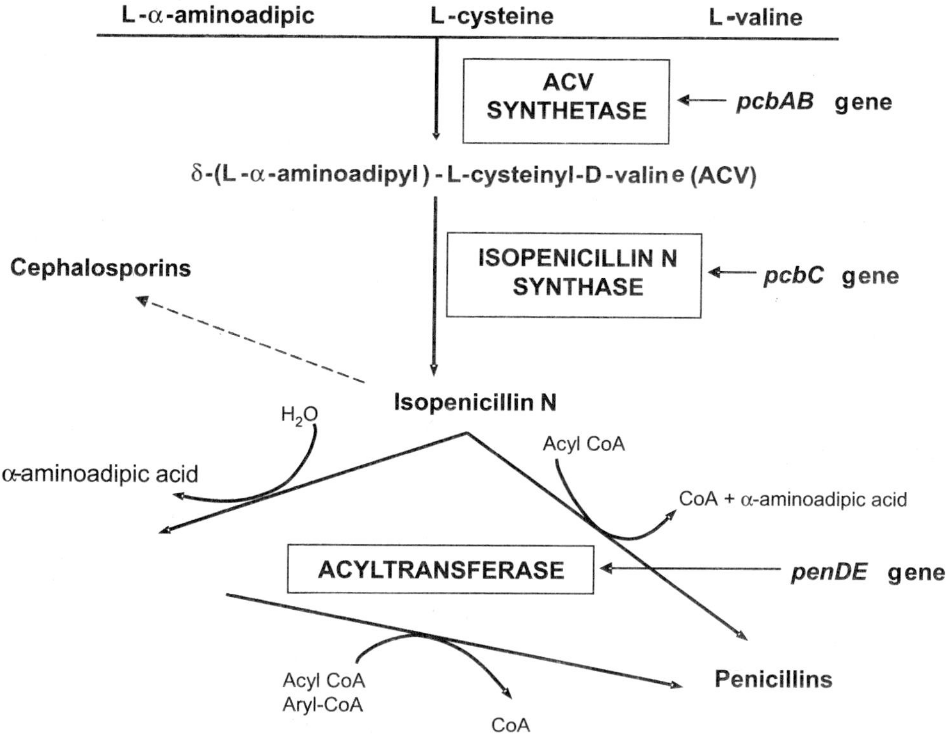

Fig. 3.2. Penicillin biosynthetic pathway.

N by an aromatic acyl side chains (Álvarez *et al.*, 1987; Martín *et al.*, 1990) previous activation of the aromatic acid by a specific aryl-CoA ligase (Lamas-Maceiras *et al.*, 2006).

The genes involved in the biosynthesis of β–lactam antibiotics are arranged in clusters in bacteria as well as in filamentous fungi (Aharanowitz *et al.*, 1992; Martín, 1998) (Fig. 3.3). In *Acremonium chrysogenum*, the genes involved in the biosynthesis of cephalosporin C are arranged in two separate clusters located in different chromosomes (Skatrud and Queener, 1989; Gutiérrez *et al.*, 1999). The "early gene" cluster located in chromosome VII (4.6 Mb) contains the genes *pcbAB* and *pcbC* encoding the enzymes for the first two steps of the pathway, the genes *cefD1* and *cefD2* which are responsible for the epimerization of the isopenicillin N to penicillin N (Ullán *et al.*, 2002a), and the *cefT* gene, encoding a transmembrane protein apparently involved in the secretion of the antibiotic (Ullán *et al.*, 2002b). The "late gene" cluster located in chromosome I (2.2 Mb) contains the genes for the last steps of cephalosporin biosynthesis: *cefEF* and *cefG* (Gutiérrez *et al.*, 1992). In *Penicillium chrysogenum*, the three penicillin biosynthetic genes (*pcbAB*, *pcbC* and *penDE*) are clustered in a 17 kb DNA region (Díez *et al.*, 1990; Smith *et al.*, 1990) located in chromosome I (10.4 Mb) (Fierro *et al.*, 1993). The *pcbAB* gene is 11337 bp long and has no introns. The *pcbC* gene is 996 bp long and has no introns either. The last gene involved in penicillin biosynthesis is the *penDE* gene, which is 1274 bp long and contains three introns. A similar organization of the *pcbAB* (*acvA*), *pcbC* (*ipnA*) and *penDE* (*aat*) has been reported in *Aspergillus nidulans* (MacCabe *et al.*, 1990) but

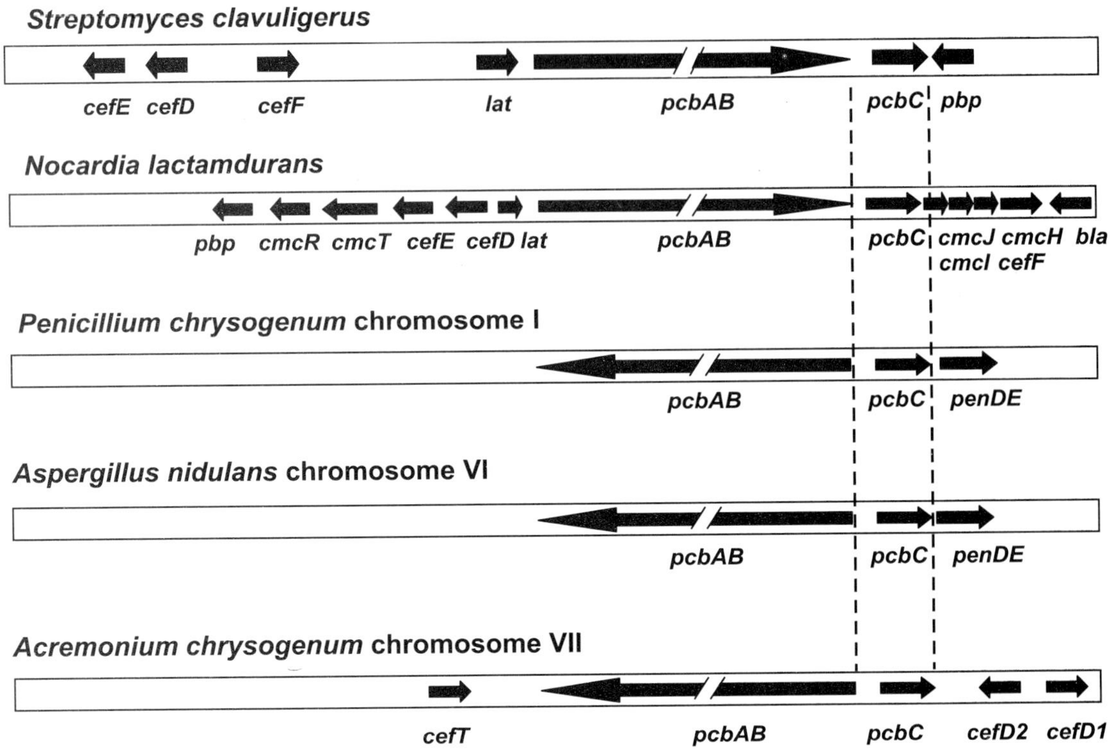

Fig. 3.3. Schematic representation of the β-lactam biosynthetic gene clusters from different microorganisms. All the clusters have been aligned with respect to the *pcbC* gene. Note that the orientation of the *pcbAB* with respect to the *pcbC* gene is opposite in the bacterial clusters (upper part of the Fig.) and in the fungal clusters (lower part).

in this fungus the cluster is present in a single copy in chromosome VI (3.0 Mb) (Montenegro *et al.*, 1992).

It has been suggested that the *pcbAB* and *pcbC* β–lactam genes were horizontally transferred from bacteria to filamentous fungi approximately 370 million years ago (Carr *et al.*, 1986; Ingolia and Queener, 1989; Cooper, 1993; Cohen *et al.*, 1994; Martín *et al.*, 1994). Unlike *pcbAB* and *pcbC* genes, the origin of the fungal *penDE* gene is obscure, since there are no known close eukaryotic or prokaryotic analogous. Several common features of this gene indicate that it was derived by fusion of a eukaryotic with a prokaryotic gene (Martín *et al.*, 1994). The acquisition of these genes by filamentous fungi endowed the latter with important ecological advantages, since secretion of antibiotics reduces the amount of bacterial competitors.

REGULATION OF THE BIOSYNTHESIS OF PENICILLIN

The biosynthesis of hydrophobic penicillins is affected by several factors. Easily utilizable carbon, nitrogen and phosphorous sources have been shown to affect dramatically the production of penicillin. This indicates the existence of complex regulatory processes controlling the synthesis of penicillin

(Sánchez *et al.*, 1981; Abraham, 1986; Chang *et al.*, 1990; Aharonowitz *et al.*, 1992; Brakhage and Turner, 1992; 1995; Brakhage *et al.*, 1992; Feng *et al.*, 1994; Brakhage, 1997, 1998).

CARBON CATABOLITE REGULATION

Carbohydrates exert an important role on regulation of the penicillin biosynthetic genes. Glucose, sucrose, maltose, fructose and galactose (these three to a lesser extent) repress the activity of ACV synthetase and IPN synthase but not (or only to a low extent) acyltransferase in *P. chrysogenum* (Revilla *et al.*, 1984, 1986). Glucose represses the expression of the *pcbAB* and *pcbC* gene promoters (Feng *et al.*, 1994). Sucrose and glucose also repress expression of the *pcbC* gene in *A. nidulans* (Espeso and Peñalva, 1992; Brakhage *et al.*, 1992), although other authors reported that these sugars seem not to have significant effect on *pcbC* or *penDE* genes (Brakhage *et al.*, 1992; Litzka *et al.*, 1995). In this microorganism, carbon repression of penicillin genes occurs via an unknown mechanism distinct from the wide-domain carbon regulators CreA, CreB and CreC (Espeso and Peñalva, 1992).

NITROGEN SOURCE REGULATION

Ammonium has been shown to be an important repressor of both *pcbAB* and *pcbC* gene promoters in *P. chrysogenum* (Feng *et al.*, 1994). When this compound is used as nitrogen source in both *P. chrysogenum* and *A. nidulans*, production of β–lactam antibiotics decreases (Shen *et al.*, 1984; Sánchez *et al.*, 1988). Nitrogen regulation in fungi is mediated by GATA-binding factors (Marzluf, 1997) like AreA in *A. nidulans* (Kudla *et al.*, 1990) and its homologue NRE in *P. chrysogenum* (Haas *et al.*, 1995; Haas and Marzluf, 1995). There are six GATA motifs in the *P. chrysogenum pcbAB-pcbC* intergenic region and only one in the same region of *A. nidulans*; the fact that the introduction of the *P. chrysogenum pcbC* promoter region in *A. nidulans* causes that the expression of the *A. nidulans pcbC* gene becomes sensitive to nitrogen regulation (Kolar *et al.*, 1991), indicates that the same machinery mediating nitrogen repression of penicillin biosynthesis in *P. chrysogenum* is also present in *A. nidulans*.

pH REGULATION

The external pH of the medium has also influence on biosynthesis of hydrophobic penicillins (Shah *et al.*, 1991; Espeso *et al.*, 1993). The transcriptional activator PacC (Espeso *et al.*, 1993; Tilburn *et al.*, 1995; Suárez and Peñalva, 1996), binds to the intergenic region between the *pcbAB* and *pcbC* genes and activates transcription of the *pcbAB* (Then Bergh and Brakhage, 1998) and *pcbC* (Tilburn *et al.*, 1995) genes at alkaline pH. Putative PacC-binding sites have also been identified in the promoter region of the *penDE* gene, which indicates that external pH also controls the expression of this gene.

REGULATION BY AMINO ACIDS

Amino acids are important for biosynthesis of penicillin in two different ways; they are used as precursors and they can also play a regulatory role in the β–lactam biosynthesis. The intracellular concentration of L-α–aminoadipate, unlike that of cysteine or valine seems to be a limiting factor for the synthesis of the tripeptide ACV, which also affects the overall penicillin biosynthesis rate in low-producing strains (Friedrich and Demain, 1978; Honlinger and Kubicek, 1989). It has also been observed that addition of L-lysine to *P. chrysogenum* and *A. nidulans* culture medium gave rise to a decrease in penicillin production (Luengo *et al.*, 1979a; Brakhage and Turner, 1992). Since L-α–aminoadipate is a precursor of L-lysine, it seems that the effect observed by L-lysine is due to a reduction of the L-α–aminoadipate pools. Nevertheless, this is not the only regulatory effect of L-lysine on penicillin biosynthesis, since it has been shown that this amino acid represses the expression of *pcbAB* and *pcbC* genes of *A. nidulans* (Brakhage and Turner, 1992).

The effect of other amino acids on penicillin production has also been studied (Then Bergh and Brakhage, 1998). Some amino acids like L-histidine, L-valine, and L-methionine, in addition to L-lysine, have a negative effect on the expression of *pcbAB* and *pcbC* genes in *A. nidulans* because of acidification of the medium. In case of L-histidine and L-valine the negative effects are probably due to reduced activation by PacC under acidic conditions. Repression by L-methionine of those genes is mediated via a *cis*-regulatory element placed in the intergenic region.

OTHER REGULATORY ELEMENTS

Besides PacC, PENR1 (a CCAAT box-binding protein) represses *pcbAB* and induce *pcbC* and *penDE* genes expression (Brakhage, 1997; 1998). Transcriptional regulation of genes responsible for penicillin biosynthesis has been studied, and *cis* and *trans*-regulatory elements have been identified (Pérez-Esteban *et al.*, 1993; Chu *et al.*, 1995; Feng *et al.*, 1995; Haas and Marzluf, 1995; Litzka *et al.*, 1995; Brakhage, 1997, 1998; Kosalkova *et al.*, 2000). An enhancer region recognized by a transcriptional activator named PTA1 has been characterized (Kosalkova *et al.*, 2000). This transcriptional activator binds to the TTAGTAA heptomer. In addition to transcriptional regulation controlling the synthesis of hydrophobic penicillins, differences in enzyme maturation among different microorganisms and post-translational regulation of such enzymes influence the final production of penicillin.

ANALYSIS OF THE 56.8-KBP REGION AMPLIFIED IN TANDEM REPEATS CONTAINING THE GENES RESPONSIBLE FOR PENICILLIN BIOSYNTHESIS

In *P. chrysogenum* the cluster of the penicillin biosynthesis genes is located in a DNA region that is amplified in tandem repeats bounded by a conserved hexanucleotide $TGTAA^{A}/_{T}$ (e.g., an amplicon) in penicillin overproducing strains (Fierro *et al.*, 1995; Newbert *et al.*, 1997) (Fig. 3.4). Penicillin production is therefore related to the number of amplicons present in different strains. In addition, when the amplified region is deleted (by recombination mechanisms; not fully understood), non-producing mutants like the *P. chrysogenum* npe10 strain are generated. Chromosome analysis has revealed that the DNA fragment which is amplified varies in length depending on the strain (106.5 kbp or 56.8 kbp). At the right end border of the amplified region an internal 3.4 kbp fragment (shift fragment or SF) is able to shift its orientation (Fierro *et al.*, 1995). The SF is part of the amplified region in the industrial strain *P. chrysogenum* E1, but it is outside in strain AS-P-78.

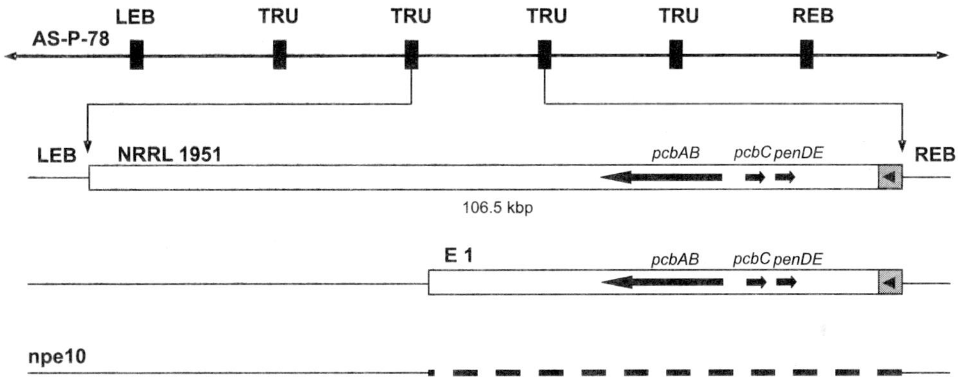

Fig. 3.4. Amplified Region of the penicillin biosynthetic gene cluster of different *P. chrysogenum* strains. The tandem arrangement of repeated units of strain AS-P-78 is indicated at the top. TRU: Tandem Repeat Union; LEB: Left End Border; REB: Right End Border. The amplification units in strains NRRL 1951 and E1 are indicated as white boxes. The DNA region that is deleted in the npe10 mutant is indicated as a dotted line. The shift fragment is shown as a shaded box at the right end of the Amplified Region; the arrow inside the box indicates its different orientation depending on the strain.

The number of tandem repeats depends on the strain (Table 3.1). The wild-type NRRL 1951 and the Wis54-1255 strains have only one copy of the amplified region, the AS-P-78 strain contains approximately 6 copies and the highly producing E1 strain have approximately 14 copies.

Table 3.1. Number of tandem repeats of the Amplified Region containing the three biosynthetic genes responsible for penicillin production in five different *P. chrysogenum* strains.

P. chrysogenum strain	Number of tandem repeats of the Amplified Region
NRRL 1951	1
Wis 54-1255	1 (SF inverted)
npe 10	0 (deleted mutant)
AS-P-78	± 6
E1	±14

The sequence of the 56820 bp amplified region from *P. chrysogenum* AS-P-78 strain has been recently annotated in the databases (GenBank accession number DQ192518). Transcriptional analysis of such region predicted the existence of several transcribed genes and therefore bioinformatic analysis of the whole amplified region was performed. A total of 16 ORFs (including *pcbAB*, *pcbC* and *penDE* genes) were identified (Fierro *et al.*, 2006). Seven ORFs are placed downstream of the *pcbAB* gene and 6 ORFs are located downstream of the *penDE* gene (Fig. 3.5).

Among the 16 ORFs, ORF11, ORF12 and ORF13, which are placed just downstream of the *penDE* gene, are specially remarkable. ORF 11 (1260 pb) contains 1 intron and codes a protein containing a D-amino acid oxidase (deaminating) domain. D-amino acid oxidases of yeasts and fungi are known to convert the D-α–amioadipic chain of cephalosporin C or penicillin N to α–ketoglutarate (Sentheshanmuganathan and Nickerson, 1962; Pollegioni *et al.*, 2004). β–lactam by-products containing α–ketoglutarate instead of α–aminoadipate have been reported (Kitano *et al.*, 1976a, b). These compounds are probably produced by an oxidase-mediated deamination of

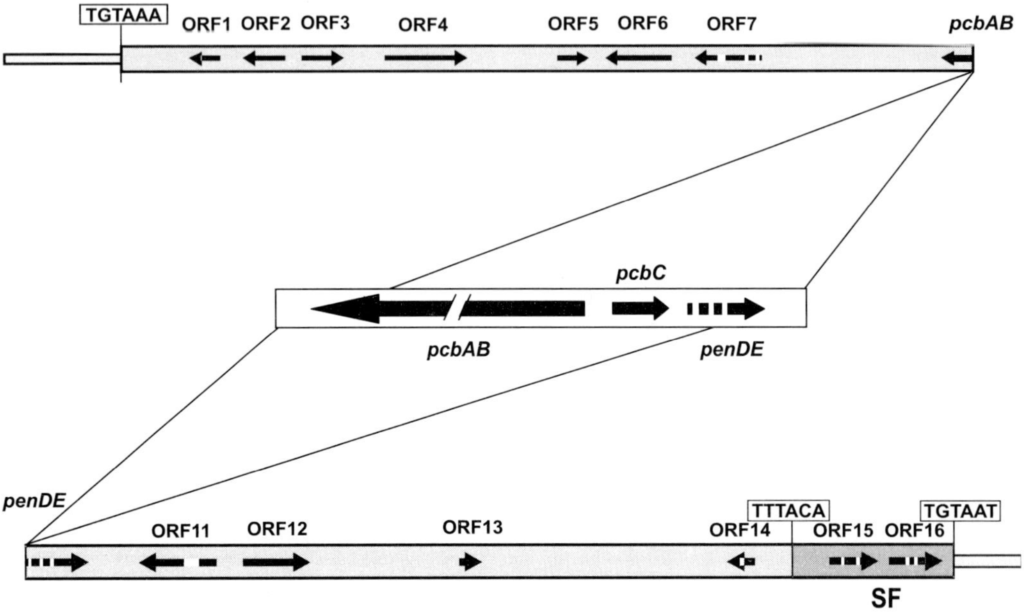

Fig. 3.5. Scheme of the ORFs present in the 56680-bp Amplified Region containing the three structural genes for penicillin biosynthesis in *P. chrysogenum*. ORFs are indicated as black arrows (white boxes represent introns). Hexanucleotide sequences framed by boxes indicate the borders of the Amplified Region and the Shift Fragment (SF), which is highlighted with a darker grey box. The central region of the cluster containing the three penicillin biosynthesis genes is enlarged in the center of the Fig.

a penicillin or cephalosporin intermediate. The ORF11 encoded protein might be involved in such process. The protein encoded by ORF12 has high similarity in its amino terminal half to saccharopine dehydrogenases from different cyanobacteria, especially with a saccharopine dehydrogenase from *Anabaena variabilis*, but it also showed similarity to saccharopine dehydrogenase (LysA) from *Aspergillus nidulans*. The protein encoded by ORF12 showed a good end to end conservation degree (37.7% positives) with *P. chrysogenum lys7* (saccharopine reductase), indicating that both proteins belong to the same family, although they are different from each other. Since the *P. chrysogenum npe10* strain (lacking the whole amplified region including ORF12) is able to grow in minimal medium without the addition of lysine, the protein encoded by ORF12 seems to be a saccharopine dehydrogenase-like isoenzyme different from the primary metabolism saccharopine dehydrogenase (LysA). Therefore, it appears to contribute to lysine biosynthesis through a secondary metabolism pathway in *P. chrysogenum* (Fierro *et al.*, 2006).

ORF13 encodes a hypothetical protein with a conserved Zn (II)2Cys6 fungal-type DNA binding domain, which is similar to that found out in *A. nidulans* transcriptional activator AlcR. Several transcription factors have been reported in *A. nidulans* and *P. chrysogenum* to control expression of the penicillin genes and bind to regulatory boxes in their promoters, including PacC, AreA and CreA (Brakhage, 1998; Martín, 2000a; Fierro *et al.*, 2002). However, no pathway-specific transcriptional regulator for penicillin biosynthesis has been found so far. The protein encoded by ORF13 could be involved in the transcriptional activation of penicillin biosynthesis.

Although the three structural genes responsible for penicillin biosynthesis *pcbAB*, *pcbC* and *penDE* keep a similar arrangement in both *P. chrysogenum* and *A. nidulans*, the regions flanking these genes are not conserved between these two microorganisms (Fierro *et al.*, 2006), indicating that some of the ORFs in *P. chrysogenum* (absent in *A. nidulans*) could be responsible, at least in part, for the difference in penicillin production between these two microorganisms (wild-type strains of *P. chrysogenum* are able to synthesize 30-fold higher levels of penicillin than wild-type *A. nidulans*). It has been shown that in *P. chrysogenum* npe10 strain (which lacks the whole amplified region), the integration of the three genes responsible for penicillin G production (*pcbAB*, *pcbC* and *penDE*) is sufficient to yield penicillin G amounts similar to those found out in the parental strain Wis54-1255.

PENICILLIN BIOSYNTHESIS: COMPARTMENTALIZATION AND ENZYME LOCALIZATION

Penicillin formation comprises three main steps and an additional phenylacetic acid activation reaction, though the biochemical processes involved in such biosynthesis are complex. The penicillin biosynthetic pathway begins with the non-ribosomal condensation of the amino acids L-α–aminoadipate, L-cysteine and L-valine to constitute the tripeptide δ (L-α–aminoadipyl)-L-cysteinyl-D-valine (ACV). This step is catalysed by the 420-kDa ACV synthetase (Martín and Liras, 1989; Zhang and Demain, 1992; Byford *et al.*, 1997; Martín, 2000a) apparently a cytosolic enzyme (Van der Lende *et al.*, 2002) and it is believed that amino acid precursors are stored in vacuoles (Affenzeller and Kubicek, 1991; Lendenfeld *et al.*, 1993). The tripeptide is cyclized to isopenicillin N, which constitutes the β–lactam backbone, by the action of the cytosolic enzyme isopenicillin N synthase (Ramos *et al.*, 1985; Müller *et al.*, 1991). In the last step, the L-α–aminoadipyl side chain of isopenicillin N is exchanged for a more hydrophobic side chain by the isopenicillin N-acyltransferase (IAT) (Álvarez *et al.*, 1987; Martín *et al.*, 1990), which is located inside peroxisomes (microbodies) (Müller *et al.*, 1991, 1992, 1995).

Besides these three main enzymes responsible for penicillin biosynthesis, ancillary enzymes are also required. This is the case of the aryl-CoA ligase, which activates the side-chain aromatic acid (Lamas-Maceiras *et al.*, 2006) and is localized into the microbody (Gledhill *et al.*, 1997; Lamas-Maceiras *et al.*, 2006).

The fact that penicillin biosynthesis is compartmentalized (Fig. 3.6) implies involvement of transport of enzymes, precursors, intermediates and products and indicates that every step has its own optimal environmental conditions. The distinct sub-cellular organization of penicillin biosynthesis enables the spatial separation, regulation and optimisation of the processes involved.

NON-RIBOSOMAL CONDENSATION OF THE TRIPEPTIDE BY THE ACV SYNTHETASE

The ACV synthetase is a big protein with a mass of 405-425 kDa depending on the microorganism (Martín and Liras, 1989; Zhang and Demain, 1992; Aharonowitz *et al.*, 1993; Kleinkauf and von Döhren, 1995). It catalyses the condensation of the precursor amino acids L-α–aminoadipate, L-cysteine and L-valine to constitute the tripeptide ACV (Byford *et al.*, 1997; Martín, 2000b). These precursors are obtained through de novo synthesis, although two major uptake routes for

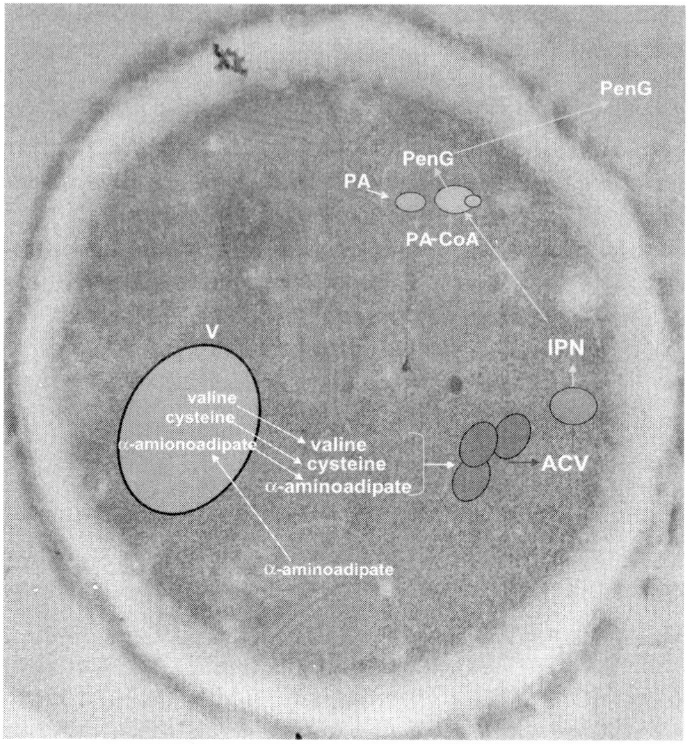

Fig. 3.6. Compartmentalization of the penicillin biosynthesis. Electron microscope photograph of an ultrathin section from a *P. chrysogenum* hyphae by electron microscopy showing a scheme with the organization of the penicillin biosynthetic pathway inside the cell: V: Vacuole; P: Peroxisome (microbody); M; Mitochondria; ACV; α-Aminoadipyl-Cystenyl-Valine; ACVS: ACV synthetase; IPN: isopenicillin N; IPNS: IPN synthase; AT: acyltransferase; CL: CoA Ligase; PA: Phenylacetic acid; PA-CoA: Phenylacetyl-CoA; PenG: Penicillin G.

L-α–aminoadipic acid from the culture medium have been recently described: (i) via the acidic amino acid permease and (ii) via the general amino acid permease (Trip *et al.*, 2004). Unlike L-cysteine and L-valine, which are final products of primary metabolism and common protein constituents, L-α–aminoadipic acid is a key intermediate in the lysine metabolim pathway of fungi (Fig. 3.7). Micro-organisms like yeast and higher fungi, in contrast to animals, are able to synthesize L-lysine *de novo* using α–ketoglutarate and acetyl-CoA as initial substrates *via* the intermediate α–aminoadipic acid (Bhattacharjee, 1985; Bañuelos *et al.,* 1999, 2000; Nishida and Nishiyama 2000; Zabriskie and Jackson, 2000). The condensation of α–ketoglutarate and acetyl-CoA in homocitrate is catalysed by homocitrate synthase. Mutants disrupted in the lys1 gene encoding homocitrate synthase are unable to produce penicillin, as expected (Bañuelos *et al.*, 1999), the latter is transformed in the intermediates homoisocitrate (catalysed by homoaconitase), and α–ketoadipate (catalysed by homoisocitrate dehydrogenase). Then, α–ketoadipate is converted in α–aminoadipate through the aminoadipate aminotransferase. It has been also demonstrated that in *P. chrysogenum*, α–aminoadipate is also obtained from the catabolism of L-lysine by an *omega*-aminotransferase encoded by the *oat1* gene (Martín de Valmaseda *et al.*, 2005).

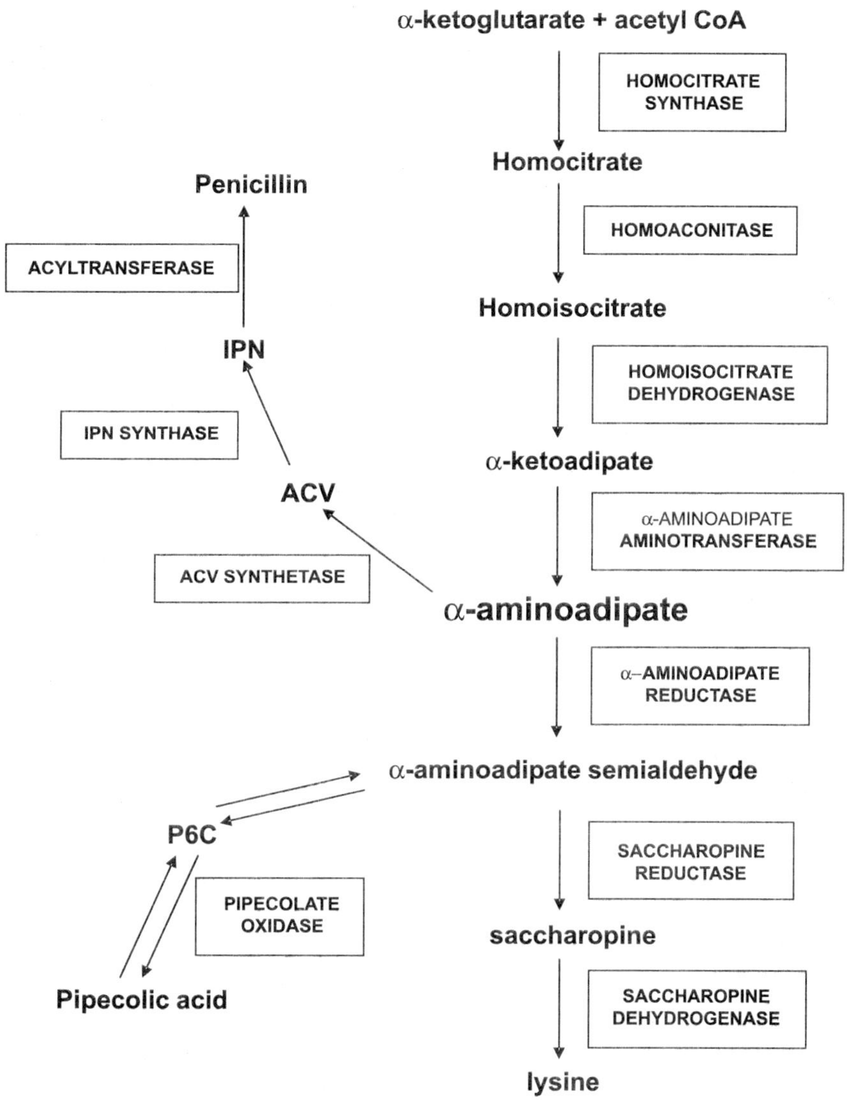

Fig. 3.7. Biosynthesis of lysine and penicillin in *P. chrysogenum.* α-aminoadipate is highlighted as the branching intermediate compound for both pathways.

The ACV synthetase is able to catalyse multiple activities like amino acid adenylation, peptide-bond formation, epimerization and product release by thioesterase (Aharonowitz *et al.*, 1993; Kleinkauf and von Döhren, 1996; Martín, 2000a), since it contains three different modules each of approximately 1000 amino acids, which are relatively conserved among themselves (Kleinkauf and von Döhren, 1996; Zocher and Keller, 1997). The central region of each domain (around 600 amino acids) shows higher conservation degree than the interdomain sequences that comprises about 500 amino acids. The central conserved regions are involved in binding and activation of the precursor amino acids.

Each region contains adenylate-forming (A, for activation), acyl carrier (T, thiolation) and condensation (C) domains consisting of conserved boxes arranged in a characteristic order (Kleinkauf and von Döhren, 1996; Martín, 2000a) (Fig. 3.8). The basic sequence of domains in a non-

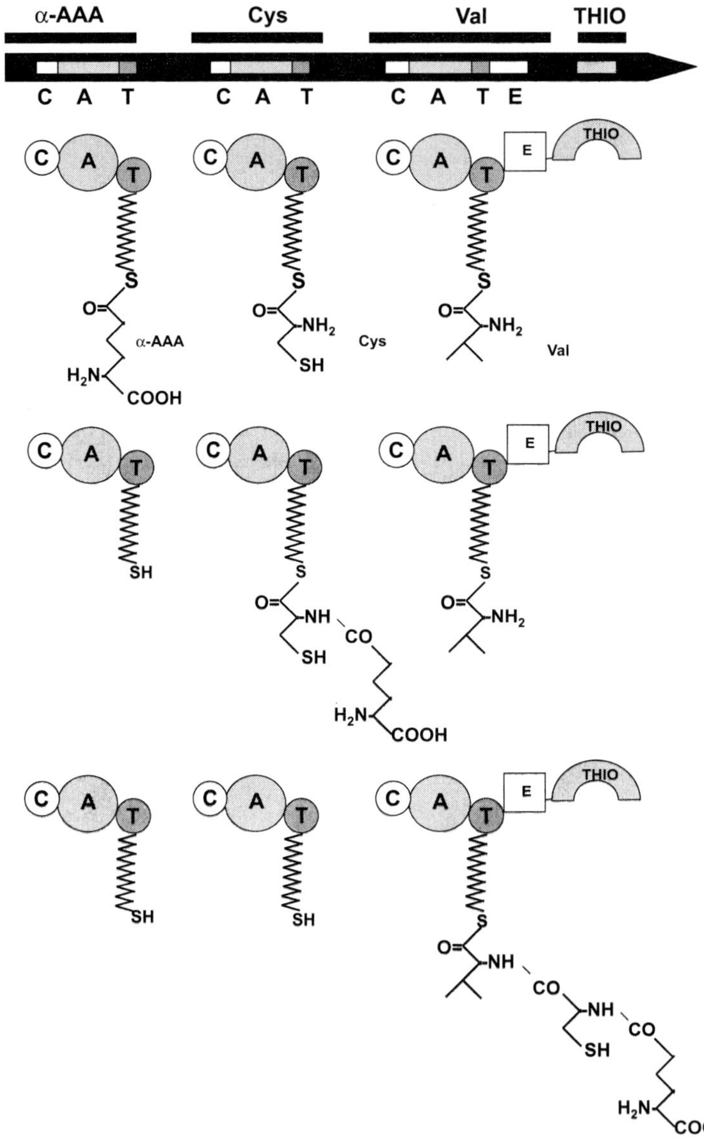

Fig. 8. Organization of domains of the ACV synthetase and a model for peptide chain elongation. Modules for the three functional domains are indicated as a-AAA (corresponding to a-aminoadipic acid), Cys (corresponding to cysteine) and Val (corresponding to valine). THIO represents the thioesterase domain present downstream of the Val module; C: condensation domains; A: activation domains; T: thiolation domains; E: epimerase domain. The phosphopantetheine arm is linked to the T domain and activated amino acids are transferred to the thiol group of phosphopantetheine. In this model, the first condensation step originates the dipeptide L-a-aminoadipyl-L-cysteine and then the tripeptide L-a-aminoadipyl-L-cysteine-L-valine. Therefore, epimerisation of valine occurs in the tripeptide chain attached to the enzyme. Modified from Martín (2000a).

ribosomal peptide synthetase is always CAT CAT CAT, although sometimes there are other interspersed domains (e.g., an epimerization 'E' domain).

The adenylate module is involved in adenylate formation and ATP binding, and contains a region determining the substrate specificity. The acyl carrier module (thiolation) bears a conserved serine residue that binds a thiol-containing phosphopantetheine co-factor that is derived from CoA (Baldwin *et al.*, 1991a; Kleinkauf and von Döhren, 1996) that may in turn form the thioester with the activated amino acid, since pantetheine has been found to be present in ACV synthetase (Baldwin *et al.*, 1991a) and in other peptide synthetases (Ishihara *et al.*, 1975). Phosphopantetheine is added to the apoprotein by an enzyme known as a 4'-phosphopantetheine transferase (PPTase), that derives the phosphopantetheine group from coenzyme A (Lambalot *et al.*, 1996). PPTases involved in secondary metabolism have not been widely characterized in filamentous fungi. The *npgA* gene of *A. nidulans*, encoding a putative 4'-phosphopantetheinyl transferase, has been shown to be necessary for penicillin biosynthesis (Keszenman-Pereyra *et al.*, 2003) and for siderophore biosynthesis (Oberegger *et al.*, 2003) and may be a broad-specificity PPTase required for several steps in primary and secondary metabolism. Recently, the homologue of the *npgA* gene has been also cloned in *P. chrysogenum* (Schoergendorfer and Kuernsteiner, 2005).

The condensation domain of ACV synthetase is responsible for the condensation of two activated amino acids on adjacent modules, catalyses the elongation reaction (de Crécy-Lagard *et al.*, 1995) and it is present between two adjacent modules, i.e., between the thiolation domain of a module and the activation domain of the following one.

An epimerase motif is present at the end of the third module and catalyses the conversion of the precursor amino acid L-valine into D-valine to constitute the final tripeptide δ(L-α–aminoadipyl)-L-cysteinyl-D-valine (Martín, 2000a). The carboxyl terminal region of *P. chrysogenum*, *Aspergillus nidulans* and *Acremonium chrysogenum* ACV synthetases contains a motif (GXSXG) which is homologous to the amino acid sequence present in the active site of the oleoyl-ACP hydrolase (thioesterase I) of vertebrate fatty acid synthetases (THIO box) (Díez *et al.*, 1990; Gutiérrez *et al.*, 1991). This domain seems to be in charge of the hydrolysis of the thioester bond between the ACV tripeptide and the enzyme. However, when the thioesterase domain of *A. nidulans* ACV synthetase was modified (Kallow *et al.*, 2000), the mutant, unexpectedly, showed only a 50% reduction of the ACV peptide formation rate, and the stereoisomer δ(L-α–aminoadipyl)-L-cysteinyl-L-valine was the predominant resulting peptide. Although those results need to be confirmed, it seems that this motif may be involved in the release of the final peptide and in the control of tripeptide epimerization by selection of the isomer to be released.

Several models for peptide chain elongation (L-α–aminoadipyl-L-cysteinyl-D-valine biosynthesis) have been proposed. One of the suggested models describes a thiotemplate mechanism involving ten distinct enzymatic steps (Kleinkauf and von Döhren, 1996). In this model, epimerisation of valine occurs at the peptide stage; the first condensation step would lead to the formation of the dipeptide L-α–aminoadipyl-L-cysteine prior to the formation of the bond between cysteine and valine. However, it has been observed that purified preparations of the ACV synthetase form two diasteroisomeric dipeptides, L-(*O*-methylserinyl)-L-valine and L-(*O*-methylserinyl)-D-valine when the enzyme was incubated with α–aminoadipate, L-valine and the cysteine analogue L-*O*-methylserinyl (instead of α–aminoadipic acid) (Shiau *et al.*, 1995a, b). It has also been reported that the dipeptide L-cysteinyl-D-valine is also formed when ACV synthetase is incubated with L-cysteine and L-valine, suggesting that epimerisation of L-valine takes place at the enzyme-bound dipeptide stage. The peptide bond between α–aminoadipate and cysteine is formed after the L-cysteine-D-valine is synthesized and the thioesterification of valine is not necessary for peptide bond formation (Shiau *et al.*, 1997). In this model, formation of ACV begins with the condensation of L-cysteine and L-valine (Shiau *et al.*, 1995a, b, 1997). In addition to the three amino acids used by ACV synthetase, it has been shown that *A. chrysogenum* ACV synthetase has a relatively broad substrate specificity, being able to form several related tripeptides *in vitro* (Baldwin *et al.*, 1994).

Initial studies on *P. chrysogenum* ACV synthetase associated this protein to membrane structures that were identified as Golgi-like organelles (Kurylowicz *et al.*, 1987). Additional cell fractionation experiments located ACV synthetase attached to or inside vacuoles (Müller *et al.*, 1991; Lendenfeld *et al.*, 1993). However, the biochemical properties of this enzyme like *in vitro* optimal pH 8.4 (which resulted to be well above that of the vacuolar pH), co-factor requirement and protease sensitivity indicated that this enzyme was a cytosolic enzyme. This fact was supported by immunodetection using antibodies generated against ACV synthetase and electron microscopy techniques (Van der Lende *et al.*, 2002). Further work is, however, necessary to confirm the *in vivo* localization of those multifunctional enzymes.

OXIDATIVE CYCLIZATION OF THE TRIPEPTIDE IS MEDIATED BY THE ISOPENICILLIN N SYNTHASE

Isopenicillin N synthase (38 kDa) catalyses the second step in penicillin biosynthesis, which is the cyclisation of the tripeptide ACV to IPN, and it has been widely characterized (Ramos *et al.*, 1985; Carr *et al.*, 1986; Baldwin *et al.*, 1987, 1991b; Cooper, 1993; Durairaj and Jensen, 1996) and crystallized (Roach *et al.*, 1995). Unlike the ACV synthetase, IPN synthase mediates the formation of the bicyclic penicillin nucleus through a single reaction consisting of the transfer of four hydrogen atoms from ACV to molecular oxygen (Baldwin *et al.*, 1991b; Cooper, 1993). Structural analysis of *A. nidulans* IPN synthase (Roach *et al.*, 1995) showed that the active site contains an iron-binding site within an unreactive hydrophobic substrate-binding cavity, structure that allows the reaction to be channelled along a single path. IPN synthase activity strongly depends on the dissolved O_2 concentration and *in vitro* on the presence of Fe^{2+} (Bainbridge *et al.*, 1992; Cooper, 1993), and accepts ACV only in its reduced state (bis-ACV formed by oxidation has to be reduced before it can be converted).

The IPN synthase is co-localized with ACV synthetase in the cytosol (Müller *et al.*, 1991; Van der Lende *et al.*, 2002) and it behaves as a soluble enzyme (Abraham *et al.*, 1981), although its activity in cell free extracts seems to be stimulated by addition of Triton X-100 or sonication (Sawada *et al.*, 1980).

ISOPENICILLIN N-ACYLTRANSFERASE: SUBSTITUTION OF THE α–AMINODIPYL SIDE CHAIN BY AROMATIC ACYL SIDE CHAINS INCREASES HYDROPHOBICITY AND ANTIBIOTIC ACTIVITY

The final step in the biosynthesis of hydrophobic penicillins is mediated by the acyl-CoA:IPN acyltransferase or IAT (Álvarez *et al.*, 1987; Barredo *et al.*, 1989; Tobin *et al.*, 1990; Whiteman *et al.*, 1990; Martín *et al.*, 1994) which replaces the α–aminoadipyl side chain of IPN by aromatic acyl side chain to form hydrophobic penicillins, e.g., penicillin G or V. This exchange originates a significant increase of antimicrobial activity as compared to that of IPN or 6-APA.

The IAT is expressed as a 40-kDa precursor (proacyltransferase, proIAT) which undergoes an autocatalytic self-processing between residues Gly102-Cys103 in *P. chrysogenum*, thus constituting an heterodimer composed of two subunits; α (11 kDa, corresponding to the N-terminal fragment) and β (29 kDa, corresponding to the C-terminal region) (Barredo *et al.*, 1989; Veenstra *et al.*, 1989; Whiteman *et al.*, 1990; Tobin *et al.*, 1990, 1993). Although it was initially believed that the IAT activity was located in the 29-kDa subunit (Álvarez *et al.*, 1987), later it was shown that a higher IAT activity is observed after association of the 29-kDa and 11-kDa subunits (Tobin *et al.*, 1990, 1993), indicating that the most active *P. chrysogenum* IAT form is an α–β heterodimer (Whiteman *et al.*, 1990; Tobin *et al.*, 1993). The IAT obtains the functional quaternary structure via a cooperative folding process involving both subunits (Tobin *et al.*, 1993). More recently it was found that the *A. nidulans* proIAT remains predominantly undissociated as a 40-kDa protein through several purification steps, whereas the *P. chrysogenum* enzyme is efficiently self-processed, rapidly dissociating into the 29-kDa and 11-kDa subunits (Fernández *et al.*, 2003). In fact, only the 29-kDa subunit (but never the 40-kDa form) is isolated from *P. chrysogenum* crude extracts and purified fractions (Álvarez *et al.*, 1987; Fernández *et al.*, 2003). In *A. nidulans* a small amount of both the 29 and 11 kDa subunits is observed and the 40 kDa unprocessed form is always found (Fernández *et al.*, 2003). This difference in IAT processing ability between *P. chrysogenum* and *A. nidulans* might be responsible (at least in part) for the disparity in penicillin biosynthetic ability between these two fungi (wild-type strains of *P. chrysogenum* are able to synthesize 30-fold higher levels of penicillin than wild-type *A. nidulans*).

In vivo, the IAT catalyzes the removal of the L-α–aminoadipyl side chain of isopenicillin N and its replacement with a CoA-derived substitute via an unknown mechanism. In the same process, 6-APA is originated as an intermediate compound and can also be used as substrate for penicillin formation. Since both IPN and 6-APA are able to be used as penicillin precursors it is likely that these exchange reactions performed by IAT are integrated and occur at the same time. IAT catalyses six different reactions related to the cleavage of IPN (releasing 6-APA) and the replacement of the α–aminoadipyl side chain of this intermediate by aromatic acids activated in the form of aryl-CoA derivatives. These six activities are: acyl-CoA:IPN acyltransferase (IAT activity), acyl-CoA:6-APA acyltransferase (AAT activity), IPN amidohydrolase, penicillin transacylase, penicillin amidohydrolase, and acyl-CoA hydrolase (Álvarez *et al.*, 1993). *In vitro*, the AAT activity is six-fold higher than the IAT activity (Álvarez *et al.*, 1993; Luengo, 1995). The IAT is capable of utilizing many different side chains but the presence of an α–methylene group in this substrate seems to be essential for enzyme recognition (Table 3.2). The optimal length of the acyl side-chain is between six and nine and phenylacyl-CoA is always preferentially used (Álvarez *et al.*, 1993).

Table 3.2. Side chain precursors used by the acyltransferase to synthesize different penicillins.

Natural penicillins	Side-chain precursor
Penicillin G (Benzylpenicillin)	$C_6H_5CH_2COOH$
Penicillin F (Pentenylpenicillin)	$CH_3CH_2CH{=}CHCH_2COOH$
Penicillin K (Heptylpenicillin)	$CH_3(CH_2)_6COOH$
Penicillin X (p-Hydroxybenzylpenicillin)	*p*-OH-$C_6H_5CH_2COOH$

P. chrysogenum acyltransferase contains a functional peroxisomal targeting sequence PTS1 (ARL) at the C-terminal end (the three last residues in the protein). The acyltransferase of *A. nidulans*, however, contains a less common peroxisomal targeting sequence (ANI), at the C-terminus. Electron microscopy immunodetection has shown that the acyltransferase of *P. chrysogenum* is located inside peroxisomes (microbodies) (Müller *et al.*, 1991, 1992, 1995). The *A. nidulans* IAT is also supposed to be targeted to the peroxisomes but due to the difference in the PTS1 it might be transported less efficiently.

IN VIVO ROLE OF THE PROIAT: AN UNPROCESSABLE MUTANT FORM OF THE ISOPENICILLIN N ACYLTRANSFERASE (IAT^{C103S}) MODULATES IAT PROCESSING AND PENICILLIN PRODUCTION

As indicated above, the post-translational events acting in the processing of IAT might be responsible for the disparity in penicillin production between *P. chrysogenum* and *A. nidulans*. Many fungal enzymes are formed as proenzymes and the mechanism of self-processing is frequently a step involved in post-translational regulation of enzyme activity. Like other N-terminal nucleophile (NTN) amidohydrolases (Brannigan *et al.*, 1995), the *P. chrysogenum* IAT is a NTN enzyme capable of self-activation. Self-processing may be intramolecular or intermolecular (i.e., one molecule serves as substrate for other identical protein molecule), as reported for the *Thermoplasma acidophilum* 20S proteasome (Seemüller *et al.*, 1996). One of the best known NTN enzymes is the *E. coli* penicillin acylase (Duggleby *et al.*, 1995). Hewitt and co-workers reported evidence indicating that in a slow precursor processing mutant of the *E. coli* penicillin acylase, the linker peptide located between the a and b subunits, blocks the active site cleft of this enzyme (Hewitt *et al.*, 2000).

To test whether the self-processing IAT from *P. chrysogenum* works by a mechanism similar to that of other NTN enzymes and whether the unprocessable proIAT may inhibit the self-processing activity of the IAT and affect the penicillin production, a strain in which the unprocessable IAT mutant is expressed together with the native processable IAT was obtained (García-Estrada *et al.*, 2006). It has been reported (Tobin *et al.*, 1995) that the recombinant mutant IAT^{C103S} is unable to self-process and lacks isopenicillin N acyltransferase activity. These results have been corroborated *in vivo* after expression of such protein mutant in *P. chrysogenum*, indicating that no other proteolytic mechanism different from autocatalysis (absent in the IAT^{C103S} mutant) is involved in the processing of IAT in this fungus (García-Estrada *et al.*, 2006). As indicated before, IAT is located into peroxisomes. However, the rapid self-processing of this enzyme in *P. chrysogenum* makes it impossible until now to ascertain whether this processing into α and β subunits occurs after or before the transport of the proenzyme into microbodies. Subcellular localization of the IAT^{C103S} was determined by immunocytochemistry using electron microscopy techniques. Immunolabelling with antibodies raised against IAT clearly showed that the unprocessable IAT^{C103S} variant is also located on the peroxisomal matrix (García-Estrada *et al.*, 2006), as occurs with the wild type protein (Müller *et al.*, 1991, 1992, 1995). These results indicate that this enzyme is transported into peroxisomes, most likely as the proIAT form; if this is the case, self-processing may occur optimally under the internal peroxisomal ambient conditions. Hence, the high concentration of enzyme in these organelles and the adequate ambient pH value may be essential for efficient self-processing.

Since both the unprocessable IAT^{C103S} mutant and the wild-type IAT are co-localized together, it is tempting to assume that the unprocessable protein is able to exert post-translational regulatory activity on the processing of the IAT. In fact, it was observed that the presence of the unprocessable mutant IAT^{C103S} caused a delay in the increase in the amount of the β subunit generated after wild-type IAT self-processing and a decrease in specific penicillin production (García-Estrada *et al.*, 2006). The increase in the β–subunit amounts was coincident with the gradual decrease of the unprocessable IAT^{C103S}, suggesting that the unprocessable form of the protein exerts a regulatory role on wild-type IAT processing. The decrease in penicillin production observed when the unprocessable protein present supports the hypothesis that the 40-kDa proIAT regulates the processing and activity of IAT. If the wild-type IAT self-processing is inhibited by the unprocessable IAT^{C103S} the decrease in the steady-state levels of IAT C103S would allow a complete self-processing of the wild-type IAT leading to the accumulation of the β subunit.

The effect of the unprocessable $proIAT^{C103S}$ on the wild-type IAT processing was studied in *E. coli*. In this heterologous system the unprocessable protein was able to inhibit almost completely the processing of the wild-type IAT, supporting the conclusion that the excess of the unprocessable IAT^{C103S} affects the processing of the normal IAT (García-Estrada *et al.*, 2006). This inhibition is likely to be exerted by an intermolecular mechanism, as reported for other NTN enzymes. A similar phenomenon was observed in the 20S proteasome, another NTN enzyme (Seemüller *et al.*, 1996). A similar mechanism has been reported for the penicillin acylase from *E. coli* (Böck *et al.*, 1983; Choi *et al.*, 1992), an NTN enzyme that catalyzes hydrolysis of the amide bond of benzylpenicillin releasing 6-APA and phenylacetic acid (Valle *et al.*, 1991). This protein has been shown to be capable of carrying intermolecular autoproteolysis (Kasche *et al.*, 1987). More recently, the same authors have reported that maturation of this enzyme is initiated by intramolecular autoproteolysis (Kasche *et al.*, 1999).

In light of the effects that the unprocessable mutant IAT^{C103S} exerts on IAT processing in *P. chrysogenum*, it seems that the large amounts of pro-IAT that remains unprocessed in *A. nidulans* (Fernández *et al.*, 2003) might be responsible for the slow processing of the IAT in this microorganism, and hence, for the disparity in penicillin production between *P. chrysogenum* and *A. nidulans*.

ARYL-COA LIGASE: ACTIVATION OF THE AROMATIC ACYL SIDE CHAIN

As indicated above, the last reaction of the penicillin biosynthetic pathway involves the removal and replacement of the α-aminoadipyl side chain of IPN by aromatic aryl side chains like phenylacetate or phenoxyacetate. However, previous activation of phenylacetic acid as a thioester with CoA (to form phenylacetyl-CoA) is required (Kogekar and Deshpande, 1982; Álvarez *et al.*, 1993). This reaction is catalysed by phenylacetyl-CoA ligase. This protein belongs to the acyl adenylate protein family (Turgay *et al.*, 1992) which activates the acyl or aryl acids to acyl-AMP or aryl-AMP, respectively, using ATP. After activation, AMP is released and the carboxyl group is transferred to the thiol group of CoA forming a thioester.

The existence of a phenylacetic acid activating activity was reported many years ago (Brunner and Rohr, 1975; Kogekar and Deshpande, 1983), but no significant biochemical characterization of the enzyme could be made. A gene coding an acetyl-CoA synthase (*acuA* or *facA*) has been pointed as candidate for phenylacetyl-CoA ligase gene, but mutation of this gene did not affect penicillin biosynthesis (Martínez-Blanco *et al.*, 1992, 1993; Gouka *et al.*, 1993). Therefore, there must be another gene encoding the real phenylacetyl-CoA ligase.

Recently, the gene *phl*, encoding a phenylacetyl-CoA ligase has been cloned from *P. chrysogenum* (Lamas-Maceiras *et al.*, 2006). Southern hybridizations and analysis of the amplified region containing the three penicillin biosynthetic genes (Fierro *et al.*, 2006) revealed that none of the ORFs predicted was the *phl* gene, which indicates that this gene is not linked to the penicillin gene cluster. This conclusion was confirmed by Southern hybridisation, showing that the *phl* gene is still present in the npe10 mutant (Lamas-Maceiras *et al.*, 2006), which as indicated before, lacks the penicillin gene cluster.

The phenylacetyl-CoA ligase of *P. chrysogenum* lacks a PCP (peptidyl carrier or thiolation) domain (shown as T in Fig. 3.8) and behaves as an aryl-capping enzyme that activates phenylacetic acid and transfers it to the isopenicillin N acyltransferase. This protein showed high identity with 4-coumaroyl-CoA ligases of *Arabidopsis thaliana* and other plants (Lamas-Maceiras *et al.*, 2006). Gene amplification of *phl* gave rise to an eight-fold increase in phenylacetyl-CoA ligase activity and a 35% increase in penicillin production. In addition, resistance to high concentrations of phenylacetic acid was observed. On the other hand, disruption of the *phl* gene resulted in a 40% decrease in penicillin production and a similar reduction of phenylacetyl-CoA ligase activity. Moreover, susceptibility to phenylacetic acid was increased. These results, together with data indicating that complementation of the disrupted mutant with the *phl* gene restored normal levels of penicillin production and resistance to phenylacetic acid (Lamas-Maceiras *et al.*, 2006), indicate that the *phl* gene encodes a phenylacetyl-CoA ligase involved in penicillin production. However, the presence of a second aryl-CoA ligase-like enzyme contributing partially to phenylacetic acid activation was suggested, since disruption of *phl* gene did not completely remove the biosynthesis of benzylpenicillin (Lamas-Maceiras *et al.*, 2006).

Like IAT, phenylacetyl-CoA ligase harbours a peroxisomal type 1 targeting signal (PTS1) (de Hoop and Ab, 1992; McCammon *et al.*, 1994) at the C-terminus and it is localized into microbodies (Gledhill *et al.*, 1997). The peroxisomal colocalization of these two proteins indicates that the last two enzymes of the penicillin pathway form a peroxisomal functional complex.

ORGANELLES INVOLVED IN PENICILLIN BIOSYNTHESIS

As it has been indicated in this chapter, the penicillin biosynthesis process takes place in different cellular compartments, but the microbody (peroxisome) is a key organelle in such pathway (Müller *et al.*, 1991, 1992). Microbodies are organelles ranging in diameter from 200-800 nm surrounded by a single membrane (Müller *et al.*, 1991). They contain enzymes involved in a variety of metabolic pathways like β–oxidation of fatty acids (Kionka and Kunau, 1985; Thieringer and Kunau, 1991a, b; Valenciano *et al.*, 1996, 1998) or karyogamy (Berteaux-Lecellier *et al.*, 1995). Since activation of the aromatic acyl side chain (catalysed by aryl CoA ligase) and its addition to IPN (catalysed by IAT) are performed inside this organelle, transport events through the membrane are expected to be very important. In addition, environmental conditions and concentration of metabolites inside the peroxisome may be very different from those in the cytosol. *P. chrysogenum* microbody luminal pH has been estimated to be slightly alkaline, in the range of 7.0-7.5 (Van der Lende *et al.*, 2002), a pH range that was reported to be optimal for aryl-CoA ligase and IAT (Álvarez *et al.*, 1987, 1993; Gledhill *et al.*, 1997). It cannot be excluded that the pH of the peroxisomal matrix is more alkaline (Van der Lende *et al.*, 2002) as it has been suggested for the lumen of human fibroblast peroxisomes (pH 8.5-9.0) (Dansen *et al.*, 2000). However, these conditions seem not to be similar for all organisms, since in CHO cells the pH of intact peroxisomes was recorded between 6.9 and 7.1 (Jankowski *et al.*, 2001). The involvement of microbodies in penicillin production has been recently demonstrated. Alteration of the number and shape of microbodies in *P. chrysogenum*, as a result of the overexpression of the Pc-Pex11p protein (a peroxin that is involved in microbody abundance), gave rise to an increase in the production of penicillin. Such positive effect on penicillin synthesis is likely to be related to an increased transport of penicillin and/or its precursors across the microbody membrane (Kiel *et al.*, 2005). Since IPN is a hydrophilic compound that is required in the microbody for penicillin formation it is likely that the uptake of this precursor from cytosol is performed through specific carriers. However, the hydrophobic penicillin formed by acyltransferase might be released from the microbody through simple diffusion or by transport using precursor/product antiport. Although 6-APA is a precursor that is originally formed into the microbody by the IAT, it has been shown that this compound can be uptaken from culture media and imported into microbodies to generate penicillin. Due to the hydrophilic nature of 6-APA it is likely that transport of this precursor is performed through a specific carrier or by the same carrier used by IPN.

Although microbodies are very important organelles in penicillin biosynthesis the role of mitochondria is also remarkable. Enzymes like homocitrate synthetase and homoisocitrate dehydrogenase are located within the mitochondrial matrix (Kubicek *et al.*, 1990; Jaklitsch and Kubicek, 1990) and therefore this organelle is essential for the synthesis of the precursor amino acid L-α–aminoadipate. Vacuoles are also important in penicillin biosynthesis since they regulate the levels of L-α–aminoadipate and cysteine, which are toxic at moderate concentrations (Klionsky *et al.*, 1990; Kubicek-Pranz and Kubicek, 1991).

PENICILLIN SECRETION

A wide range of microorganisms synthesize antibiotics, that apart from being useful for these species, are toxic to them in high concentrations. Therefore, it is important for the producer strains to develop mechanisms for protection against their own antibiotics. One of these mechanisms is the membrane permeability barriers coupled with efficient efflux mechanisms for removal and secretion of drug molecules from the cells (Cundliffe, 1991; Martín *et al.*, 2005). Penicillin is not an exception since it was observed that very high concentrations of penicillin G and other hydrophobic penicillins exert toxic effects due to lipophilic interactions with the plasma membrane and intracellular membranes (Sikkema *et al.*, 1995).

Penicillin production has been widely studied, especially from genetic, molecular and biochemical point of views. However, little is known about the mechanism by which penicillin is released into the medium by *P. chrysogenum*. The last step in penicillin biosynthesis consists of the conversion of isopenicillin N and 6-APA into a more hydrophobic penicillin. The penicillins produced by *P. chrysogenum* are amphipatic compounds, moderately hydrophobic and negatively charged at normal cytosolic pH values (Kirschbaum, 1986). Hence, the more plausible mechanism of secretion is by simple diffusion across both the microbody and plasma membrane. In fact, studies with *E. coli* cells and liposomes prepared from *E. coli* lipids, indicated that penicillin G is able to rapidly diffuse across membranes (Yoshimura and Nikaido, 1985; Livermore, 1988). Though these passive efflux mechanisms could be sufficient to release the amounts of penicillin produced by wild-type species, there should be additional transporter mechanisms in highly producing industrial strains which endow them with the ability to protect them against the high levels of penicillin produced. In fact, the high final penicillin titers of at least 120 mM (Crueger and Crueger, 1992) in such strains are indicative of an active secretion process.

Multidrug transporters are membrane proteins present in most living organisms (Marger and Saier Jr., 1993; Dean and Allikmets, 1995; Kuan *et al.*, 1995; Pao *et al.*, 1998) that recognize a wide variety of structurally different compounds and actively export them out from the cytoplasm (Ambudkar *et al.*, 1999; Putman *et al.*, 2000; St Georgiev, 2000; Neyfakh, 2002; Paulsen, 2003). The structure of these transporters is diverse, belonging to one of four different protein superfamilies: ATP-binding cassette (ABC) transporters; major facilitator superfamily (MFS); small multidrug resistance (SMR); and resistance nodulation determinants (RNDs) (Paulsen *et al.*, 1996; Pao *et al.*, 1998; Putman *et al.*, 2000; del Sorbo *et al.*, 2000; Zheleznova *et al.*, 2000). On the basis of bioenergetic criteria, multidrug transporters can be divided into two major classes: the primary active transporters, which include the ABC transporters that require ATP hydrolysis as an energy source, and secondary multidrug transporters (MFS, SMR and RND), which utilize the transmembrane electrochemical gradient of protons or sodium ions to drive the extrusion of drugs from the cell. Whereas the ABC transporters are, in general, multicomponent proteins that are capable of transporting both small molecules and macromolecules in response to ATP hydrolysis, the MFS transporters are single polypeptide secondary carriers that are only capable of transporting small solutes in response to chemiosmotic ion gradients (Martín *et al.*, 2005).

To date, no detailed studies have been carried out that deal with penicillin secretion in *P. chrysogenum* (Martín *et al.*, 2005). However, in *A. nidulans*, it was found that an ABC transporter encoded by the *atrD* gene is involved, somehow, in penicillin secretion (Andrade *et al.*, 2000). Mutants lacking this gene displayed decreased penicillin production, which suggests that the ABC transporter encoded by the *atrD* might be involved, directly or indirectly, in secretion of penicillin within fungal cells. However, the role of the protein coded by the *AtrD* gene on penicillin secretion has not yet been clearly established. The AtrDp could also be part of a signal transduction mechanism or might even be involved in the transport of intermediates to different cellular compartments during penicillin biosynthesis. It is interesting to point out the presence, just downstream of the *penDE* gene in the *A. nidulans* penicillin gene cluster, of two ORFs encoding putative transporter proteins belonging to the MSF transporter (Fierro *et al.*, 2006; *A. nidulans* database, Broad Institute). However, further studies to check the implication of such proteins on penicillin transport have not yet been done.

Irrespective of the mechanism of penicillin secretion in *P. chrysogenum*, the plasma membrane plays a key role in the release of penicillin from the cell. It has been shown that ergosterol has a significant effect on penicillin permeability. Passive diffusion of penicillin across the phospholipid membrane is dramatically reduced by the inclusion of sterols (Hillenga *et al.*, 1994; Hillenga, 1999). Since ergosterol is the predominant sterol in filamentous fungi plasma membranes (*P. chrysogenum* plasma membranes contain 30% ergosterol) (Van der Rest *et al.*, 1995; Hillenga, 1999), the biological membrane could act as a permeability barrier to penicillins. Polyene-resultant mutants of *P. chrysogenum*, apparently containing an abnormal sterol content, were shown to be penicillin oversecretors (Luengo *et al.*, 1979b).

The permeability of the *P. chrysogenum* plasma membrane to penicillin G has been tested *in vivo* by addition of penicillin G to the culture media of the *P. chrysogenum npe10* strain which, as indicated before, lacks the whole penicillin gene cluster and therefore is unable to synthesize penicillin. Analysis of the cell extracts showed that penicillin G was not incorporated into the cell from the culture media, in contrast with 6-APA and IPN, that were able to enter the hyphae. These results together indicate that passive permeation of penicillin G is blocked (perhaps by ergosterol) and therefore passive diffusion seems very unlikely as the mechanism for excretion of β–lactams into the medium.

FUTURE PERSPECTIVES AND CONCLUSIONS

Classical approaches to *P. chrysogenum* and *A. chrysogenum* strain improvement for production of penicillin and cephalosporins have been successful, providing highly producing strains, but the knowledge that we have on the biosynthesis of β–lactam antibiotics is only fragmented. Particularly, there is little information on the global molecular genetics of *P. chrysogenum* and *A. chrysogenum*. Some recent articles on the genome-wide analysis of *P. chrysogenum* catabolite regulation (Castillo *et al.*, 2006) has provided a partial view of a number of genes that are expressed under growth on glucose (penicillin-repressive conditions) or under penicillin-producing conditions on lactose. The 56.8kb region that is amplified in tandem repeats in *P. chrysogenum* chromosome I has been fully sequenced (GenBank accession number DQ192518) and analyzed in detail (Fierro *et al.*, 2006). But these studies are still limited in their scope and a full comprehensive effort is required to understand the genomics, transcriptomics and proteomics of *P. chrysogenum* and *A. chrysogenum*.

A significant advance has been made in the study of the enzymes involved in penicillin and cephalosporin biosynthesis in recent years (Ullán *et al.*, 2002a; Lamas-Maceiras *et al.*, 2006). The enzymes involved in the two last steps of penicillin biosynthesis (phenylacetyl-CoA ligase and isopenicillin N acyltransferase) are located in peroxisomes (microbodies) (Müller *et al.*, 1991, 1992, 1995; Lamas-Maceiras *et al.*, 2006) and this finding raises important questions regarding enzyme targeting and precursor transport. The peroxisomal localization of these penicillin enzymes in the peroxisomal lumen space, where precursors for penicillin biosynthesis are available (e.g., by fatty acid β–oxidation), makes the peroxisome an authentic hydrophobic penicillin cell factory.

The peroxisomal location of the enzymes raises the question of penicillin transport from peroxisomes to the outside of the cell. Although some evidence exists for an ABC transporter apparently involved in penicillin secretion (Andrade *et al.*, 2000) no further conclusive support for its exact role on penicillin secretion has been provided. Similarly, CefT, a MFS protein of 12 transmembrane spanning domain is known to increase cephalosporin secretin in *A. chrysogenum* (Ullán *et al.*, 2002b). However, other cephalosporin transporter must exist in *A. chrysogenum* because the disruption of the *cefT* gene does not prevent cephalosporin secretion.

In summary, knowledge of the molecular biology of those important filamentous fungi is very scarce. It is somehow surprising that in the genome and post-genome era so little economic and scientific effort has been dedicated to advance our understanding of the basic molecular biology of these filamentous fungi. Availability of the genome sequence will provide an important step forward to advance on our scientific knowledge of the biosynthesis of secondary metabolites in these fungi.

REFERENCES

Abraham E.P. (1986). Enzymes involved in penicillin and cephalosporin formation. In: Regulation of secondary metabolite formation. Kleinkauf H., von Döhren H., Dornauer H. and Nesemann G. Eds., *VCH,* Weinheim, FRG., 115-132.

Abraham E.P., Huddlestone J.A., Jayatilake G.S., O'Sullivan J. and White R.L. (1981). Recent Advances in the Chemistry of β–Lactam Antibiotics, In: Gregory G.I. Eds. *Royal Society of Chemistry*, London, United Kingdom, 125-134.

Affenzeller K. and Kubicek C.P. (1991). Evidence for a compartmentation of penicillin biosynthesis in a high- and a low-producing strain of *Penicillium chrysogenum. J. Gen. Microbiol.* 137: 1653-1660.

Aharonowitz Y., Cohen G. and Martín J.F. (1992). Penicillin and cephalosporin biosynthetic genes: structure, regulation, and evolution. *Annu. Rev. Microbiol.* 46: 461-495.

Aharonowitz Y., Bergmeyer J., Cantoral J.M., Cohen G., Demain A.L., Fink U., Kinghorn J., Kleinkauf H., MacCabe A., Palissa H., Pfeifer E., Schwecke T., van Liempt H., von Dohren H., Wolfe S. and Zhang J. (1993). Delta-(L-alpha-aminoadipyl)-L-cysteinyl-D-valine synthetase, the multienzyme integrating the four primary reactions in beta-lactam biosynthesis, as a model peptide synthetase. *Biotechnology* 11: 807-810.

Álvarez E., Cantoral J.M., Barredo J.L., Díez B. and Martín, J.F. (1987). Purification to homogeneity and characterization of the acyl-CoA: 6-APA acyltransferase of *Penicillium chrysogenum. Antimicrob. Agents Chemother.* 31: 1675-1682.

Álvarez E., Meesschaert B., Montenegro E., Gutiérrez S., Díez B., Barredo J.L. and Martín, J.F. (1993). The isopenicillin N acyltransferase of *Penicillium chrysogenum* has isopenicillin N amidohydrolase, 6-aminopenicillanic acid acyltransferase and penicillin amidase activities, all of which are encoded by the single *penDE* gene. *Eur. J. Biochem.* 215: 323-332.

Ambudkar S.V., Dey S., Hrycyna C.A., Ramachandra M., Pastan I. and Gottesman M.M. (1999). Biochemical, cellular, and pharmacological aspects of the multidrug transporter. *Annu. Rev. Pharmacol. Toxicol.* 39: 361-398.

Andrade A.C., Van Nistelrooy J.G., Peery R.B., Skatrud P.L. and De Waard M.A. (2000). The role of ABC transporters from *Aspergillus nidulans* in protection against cytotoxic agents and in antibiotic production. *Mol. Gen. Genet.* 263: 966-977.

Aoki H. and Okuhara M. (1980). Natural β–lactam antibiotics. *Annu. Rev. Microbiol.* 34: 159-181.

Bainbridge Z.A., Scott R.I. and Perry D. (1992). Oxygen utilization by isopenicillin N synthase from *Penicillium chrysogenum. J. Chem. Technol. Biotechnol.* 55: 233-238.

Baldwin J.E., Killin S.J., Pratt A.J., Sutherland J.D., Turner N.J., Crabbe J.C., Abraham E.P. and Willis A.C. (1987). Purification and characterization of cloned isopenicillin N synthetase. *J. Antibiot.* 40: 652-659.

Baldwin J.E., Bird J.W., Field R.A., O'Callaghan N.M., Schofield C.J. and Willis A.C. (1991a). Isolation and partial purification of ACV synthetase from *Cephalosporium acremonium* and *Streptomyces clavuligerus*: evidence for the presence of phosphopantothenate in ACV synthetase. *J. Antibiot.* 44: 241-248.

Baldwin J.E., Lynch G.P. and Schofield C.J. (1991b). Isopenicillin N synthase: A new mode of reactivity. *J. Chem. Soc. Chem. Commun.* 1: 736-738.

Baldwin J.E., Shiau C.Y., Byford M.F. and Schofield C.J. (1994). Substrate specificity of L-delta-(alpha-aminoadipoyl)-L-cysteinyl-D-valine synthetase from *Cephalosporium acremonium*: demonstration of the structure of several unnatural tripeptide products. *Biochem. J.* 301: 367-372.

Bañuelos O., Casqueiro J., Fierro F., Hijarrubia M.J., Gutiérrez S. and Martín J.F. (1999). Characterization and lysine control of expression of the lys1 gene of *Penicillium chrysogenum* encoding homocitrate synthase. *Gene* 226: 51-59.

Bañuelos O., Casqueiro J., Gutiérrez S. and Martín J.F. (2000). Overexpression of the lys1 gene in *Penicillium chrysogenum*: homocitrate synthase levels, alpha-aminoadipic acid pool and penicillin production. *Appl. Microbiol. Biotechnol.* 54: 69-77.

Barredo J.L., van Solingen P., Díez B., Álvarez E., Cantoral J.M., Kattevilder A., Smaal E.B., Groenen M.A.M., Veenstra A.E. and Martín J.F. (1989). Cloning and characterization of the acyl-coenzyme A: 6-aminopenicillanic-acid-acyltransferase gene of *Penicillium chrysogenum. Gene* 83: 291–300.

Berteaux-Lecellier V., Picard M., Thompson-Coffe C., Zickler D., Panvier-Adoutte A. and Simonet J.M. (1995). A nonmammalian homolog of the PAF1 gene (Zellweger syndrome) discovered as a gene involved in caryogamy in the fungus *Podospora anserina. Cell* 81: 1043-1051.

Bhattacharjee JK. (1985). Alpha-Aminoadipate pathway for the biosynthesis of lysine in lower eukaryotes. *Crit. Rev. Microbiol.* 12: 131-151.

Böck A., Wirth R., Schmid G., Schumacher G., Lang G. and Buckel P. (1983). The penicillin acylase from *Escherichia coli* ATCC11105 consists of two dissimilar subunits. *FEMS Microbiol. Lett.* 20: 135-139.

Brakhage A.A. (1997). Molecular regulation of penicillin biosynthesis in *Aspergillus (Emericella) nidulans. FEMS Microbiol. Lett.* 148: 1-10.

Brakhage A.A. (1998). Molecular regulation of beta-lactam biosynthesis in filamentous fungi. *Microbiol. Mol. Biol. Rev.* 62: 547-585.

Brakhage A.A. and Turner G. (1992). L-Lysine repression of penicillin biosynthesis and expression of penicillin biosynthesis genes *acvA* and *ipnA* in *Aspergillus nidulans. FEMS Microbiol. Lett.* 98: 123-128.

Brakhage A.A., Browne P. and Turner G. (1992). Regulation of *Aspergillus nidulans* penicillin biosynthesis and penicillin biosynthesis genes *acvA* and *ipnA* by glucose. *J. Bacteriol.* 174: 3789-3799.

Brakhage A.A. and Turner G. (1995). Biotechnical genetics of antibiotic biosynthesis, In: The Mycota II: Genetics and Biotechnology. Esser K., Lemke P.A. and Kück U. Eds., *Springer-Verlag*, Berlin, 263-285.

Brannigan J.A., Dodson G., Duggleby H.J., Moody P.C.E., Smith J.L., Tomchick D.R. and Murzin A.G. (1995). A protein catalytic framework with an N-terminal nucleophile is capable of self-activation. *Nature* 378: 416-419.

Brunner R. and Rohr M. (1975). Phenacyl:coenzyme A ligase. *Methods Enzymol.* 43: 476-481.

Byford, M.F., Baldwin, J.E., Shiau, C.Y., Schofield, C.J. (1997). The mechanisms of ACV synthetase. *Chem Rev.* 97: 2631-2650.

Carr L.G., Skatrud P.L., Scheetz M.E., Queener S.W. and Ingolia T.D. (1986). Cloning and expression of the isopenicillin N synthetase gene from *Penicillium chrysogenum. Gene* 48: 257- 266.

Castillo N.I., Fierro F., Gutiérrez S. and Martín J.F. (2006). Genome-wide analysis of differentially expressed genes from *Penicillium chrysogenum* grown with a repressing or a non-repressing carbon source. *Curr. Genet.* 49: 85-96.

Chang L.T., McGrory E.L. and Elander R.P. (1990). Penicillin production by glucose-derepressed mutants of *Penicillium chrysogenum. J. Ind. Microbiol.* 6: 165-169.

Choi K.S., Kim J.A. and Kang H.S. (1992). Effects of site-directed mutations on processing and activities of penicillin G acylase from *Escherichia coli* ATCC11105. *J. Bacteriol.* 174: 6270-6276.

Chu Y.W., Renno D. and Saunders G. (1995). Detection of a protein which binds specifically to the upstream region of the *pcbAB* gene in *Penicillium chrysogenum. Curr. Genet.* 28: 184-186.

Cohen G., Argaman A., Schreiber R., Mislovati M. and Aharonowitz Y. (1994). The thioredoxin system of *Penicillium chrysogenum* and its possible role in penicillin biosynthesis. *J. Bacteriol.* 176: 973-984.

Cooper R.D.G. (1993). The enzymes involved in the biosynthesis of penicillin and cephalosporin: their structure and function. *Bioorg. Medic. Chem.* 1: 1-17.

Crueger W. and Crueger A. (1992). Biotechnology, a textbook of industrial microbiology, *Sinauer Associates, Inc.*, Sunderland, USA., 229-270.

Cundliffe E. (1991). Office of Naval Research lecture. Antibiotics and the search for new principles. *J. Ind. Microbiol.* 7: 157-161.

Dansen T.B., Wirtz K.W., Wanders R.J. and Pap E.H. (2000). Peroxisomes in human fibroblasts have a basic pH. *Nat. Cell. Biol.* 2: 51-53.

de Crécy-Lagard V., Marliere P. and Saurin W. (1995). Multienzymatic non ribosomal peptide biosynthesis: identification of the functional domains catalysing peptide elongation and epimerisation. *C. R. Acad. Sci. III* 318: 927-936.

de Hoop M. J. and Ab G. (1992). Import of proteins into peroxisomes and other microbodies. *Biochem. J.* 286: 657-669.

Dean M. and Allikmets R. (1995). Evolution of ATP-binding cassette transporter genes. *Curr. Opin. Genet. Dev.* 5: 779-785.

Del Sorbo G., Schoonbeek H., De Waard M.A. (2000). Fungal transporters involved in efflux of natural toxic compounds and fungicides. *Fungal Genet. Biol.* 30: 1-15.

Demain A.L. (1983). Biosynthesis of β–lactam antibiotics, In: Antibiotics Containing the β–Lactam Structure, Demain A.L. and Solomon N.A. Eds., *Springer-Verlag*, Berlin. 189-228.

Díez B., Gutiérrez S., Barredo J.L., van Solingen P., van der Voort L.H.M. and Martín J.F. (1990). The cluster of penicillin biosynthetic genes. *J. Biol. Chem.* 265: 16358-16365.

Duggleby H.J., Tolley S.P., Hill C.P., Dodson E.J., Dodson G. and Moody P.C. (1995). Penicillin acylase has a single-amino-acid catalytic centre. *Nature* 373: 264-268.

Durairaj M. and Jensen S.E. (1996). Purification and characterization of recombinant *Streptomyces clavuligerus* isopenicillin N stnthase produced in *Escherichia coli. J. Ind. Microbiol.* 16: 197-203.

Espeso E.A. and Peñalva M.A. (1992). Carbon catabolites repression can account for the temporal pattern of expression of a penicillin biosynthetic gene in *Aspergillus nidulans. Molec. Gen. Genet.* 6: 1457-1465.

Espeso E.A., Tilburn J., Arst H.N. Jr. and Peñalva M.A. (1993). pH regulation is a major determinant in expression of a fungal penicillin biosynthetic gene. *EMBO J.* 12: 3947-3956.

Feng B., Friedlin E. and Marzluf G.A. (1994). A reporter gene analysis of penicillin biosynthesis gene expression in *Penicillium chrysogenum* and its regulation by nitrogen and glucose catabolite repression. *Appl. Environ. Microbiol.* 60: 4432-4439.

Feng B., Friedlin E. and Marzluf G.A. (1995). Nuclear DNA-binding proteins which recognize the intergenic control region of penicillin biosynthetic genes. *Curr. Genet.* 27: 351-358.

Fernández F.J., Cardoza R.E., Montenegro E., Velasco J., Gutiérrez S. and Martín, J.F. (2003). The isopenicillin N acyltransferases of *Aspergillus nidulans* and *Penicillium chrysogenum* differ in their ability to maintain the 40-kDa alphabeta heterodimer in an undissociated form. *Eur. J. Biochem.* 270: 1958-1968.

Fierro F., Gutiérrez S., Díez B. and Martín J.F. (1993). Resolution of four large chromosomes in penicillin producing filamentous fungi: The penicillin gene cluster is located on chromosome II (9.6 Mb) in *Penicillium notatum* and chromosome I (10.4 Mb) in *Penicillium chrysogenum. Mol. Gen. Genet.* 241: 573-578.

Fierro F., Barredo J.L., Díez B., Gutiérrez S., Fernández F.J. and Martín J.F. (1995). The penicillin gene cluster is amplified in tandem repeats linked by conserved hexanucleotide sequences. *Proc. Natl. Acad. Sci. USA.* 92: 6200-6204.

Fierro F., Martín J.F. and Kosalkova K. (2002). Sulfur-containing â-lactam antibiotics: enzymes, genes and regulation of the biosynthesis. In: Microbial Secondary Metabolites: Biosynthesis, Genetics and Regulation, Fierro F. and Martín J.F. Eds., *Research Signpost*, Trivandrum, India, 179-210.

Fierro, F., García-Estrada, C., Castillo, I., Rodríguez, R., Velasco-Conde, T., Martín, J.F. (2006). Transcriptional and bioinformatic analysis of the 56.8 kb DNA region amplified in tandem repeats containing the penicillin gene cluster in *Penicillium chrysogenum. Fungal Genet. Biol.* (in press).

Frère J.M, Nguyen-Distéche M., Coyette J. and Joris B. (1993). Mode of action: interaction with the penicillin binding proteins. In: The Chemistry of β–lactams, Page M.I. Eds., *Blackie Academic and Professional.* London, United Kingdom, 148-197.

Friedrich C.G. and Demain A.L. (1978). Uptake and metabolism of alpha-aminoadipic acid by *Penicillium chrysogenum* Wis 54-1255. *Arch. Microbiol.* 119: 43-47.

García-Estrada C., Vaca I., Fierro F., Sjollema K., Veenhuis M. and Martín J.F. (2006). The unprocessable isopenicillin N acyltransferase variant (IAT^{C103S}) of *Penicillium chrysogenum* located into peroxisomes regulates the processing of the wild-type preprotein and decreases the penicillin production. *Biochem. J.* (in press).

Giesbrecht P., Franz M., Krüeger D., Labischinski H. and Wecke J. (1991). Bacteriolysis of Staphylococci is only a side-effect of penicillin-induced death. In: 50 years of penicillin application, history and trends, Kleinkauf H. and von Döhren H. Eds. *PUBLIC Ltd.*, Czech Republic, 353-363.

Gledhill L., Greaves P.A. and Griffin J.P. (1997). Phenylacetyl-CoA ligase from *Penicillium chrysogenum.* Patent IPN WO97/02349, Smithkline Beecham U.K.

Gouka R.J., van Hartingsveldt W., Bovenberg R.A., van Zeijl C.M., van den Hondel C.A., and van Gorcom R.F. (1993). Development of a new transformant selection system for *Penicillium chrysogenum*: isolation and characterization of the *P. chrysogenum* acetyl-coenzyme A synthetase gene (facA) and its use as a homologous selection marker. *Appl. Microbiol. Biotechnol.* 38: 514-519.

Gutiérrez S., Díez B., Montenegro E. and Martín J.F. (1991). Characterization of the *Cephalosporium acremonium pcbAB* gene encoding alpha-aminoadipyl-cysteinyl-valine synthetase, a large multidomain peptide synthetase: linkage to the *pcbC* gene as a cluster of early cephalosporin biosynthetic genes and evidence of multiple functional domains. *J. Bacteriol.* 173: 2354-2365.

Gutiérrez S., Velasco J., Fernández F.J. and Martín J.F. (1992). The *cefG* gene of *Cephalosporium acremonium* is linked to the *cefEF* gene and encodes a deacetylcephalosporin C acetyltransferase closely related to homoserine O-acetyltransferase. *J. Bacteriol.* 174: 3056- 3064.

Gutiérrez S., Fierro F., Casqueiro J. and Martín J.F. (1999). Gene organization and plasticity of the β–lactam genes in different filamentous fungi. *Antonie van Leeuwenhoek* 75: 81-94.

Haas H. and Marzluf G.A. (1995). NRE, the major nitrogen regulatory protein of *Penicillium chrysogenum*, binds specifically to elements in the intergenic promoter regions of nitrate assimilation and penicillin biosynthetic gene clusters. *Curr. Genet.* 28: 177-183.

Haas H., Bauer B., Redl B., Stoffler G. and Marzluf G.A. (1995). Molecular cloning and analysis of *nre*, the major nitrogen regulatory gene of *Penicillium chrysogenum. Curr. Genet.* 27: 150-158.

Hewitt L., Kasche V., Lummer K., Lewis R.J., Murshudov G.N., Verma C.S., Dodson G.G. and Wilson K.S. (2000). Structure of a slow processing precursor penicillin acylase from *Escherichia coli* reveals the linker peptide blocking the active-site cleft. *J. Mol. Biol.* 302: 887-898.

Hillenga D.J. (1999). Transport processes in penicillin biosynthesis. Doctoral Thesis. University of Groningen. http://irs.ub.rug.nl/ppn/181567458.

Hillenga D.J., Versantvoort H.J.M., Driessen A.J.M. and Konings W.N. (1994). Structural and functional properties of plasma membranes from the filamentous fungus *Penicillium chrysogenum. Eur. J. Biochem.* 224: 581-587.

Honlinger C. and Kubicek C.P. (1989). Regulation of delta-(L-alpha-aminoadipyl)-L-cysteinyl-D-valine and isopenicillin N biosynthesis in *Penicillium chrysogenum* by the alpha-aminoadipate pool size. *FEMS Microbiol. Lett.* 53: 71-75.

Ingolia T.D. and Queener S.W. (1989). Beta-lactam biosynthetic genes. *Med. Research Rev.* 9: 245-264.

Ishihara H., Endo Y., Abe S. and Shimura K. (1975). The presence of 4´-phosphopantetheine in the bacitracin synthetase. *FEBS Lett.* 50: 43-46.

Jaklitsch W.M. and Kubicek C.P. (1990). Homocitrate synthase from *Penicillium chrysogenum.* Localization, purification of the cytosolic isoenzyme, and sensitivity to lysine. *Biochem. J.* 269: 247-253.

Jankowski A., Kim J.H., Collins R.F., Daneman R., Walton P. and Grinstein S. (2001). *In situ* measurements of the pH of mammalian peroxisomes using the fluorescent protein pHluorin. *J. Biol. Chem.* 276: 48748-48753.

Kallow W., Kennedy J., Arezi B., Turner G. and von Dohren H. (2000). Thioesterase domain of delta-(l-alpha-Aminoadipyl)-l-cysteinyl-d-valine synthetase: alteration of stereospecificity by site-directed mutagenesis. *J. Mol. Biol.* 297: 395-408.

Kasche V., Haufler U., Markowsky D., Melnyk S., Zeich A. and Galunsky B. (1987). Penicillin amidase from *E. coli. Ann. N. Y. Acad. Sci.* 501: 97-102.

Kasche V., Lummer K., Nurk A., Piotraschke E., Rieks A., Stoeva S. and Voelter W. (1999). Intramolecular autoproteolysis initiates the maturation of penicillin amidase from *Escherichia coli. Biochim. Biophys. Acta* 1433: 76-86.

Keszenman-Pereyra D., Lawrence S., Twfieg M.E., Price J. and Turner G. (2003). The *npgA/ cfwA* gene encodes a putative 4'-phosphopantetheinyl transferase which is essential for penicillin biosynthesis in *Aspergillus nidulans. Curr. Genet.* 43: 186-190.

Kiel J.A, van der Klei I.J., van den Berg M.A., Bovenberg R.A. and Veenhuis M. (2005). Overproduction of a single protein, Pc-Pex11p, results in 2-fold enhanced penicillin production by *Penicillium chrysogenum. Fungal Genet. Biol.* 42: 154-164.

Kionka C. and Kunau W.H. (1985). Inducible beta-oxidation pathway in *Neurospora crassa. J. Bacteriol.* 161: 153-157.

Kirschbaum J. (1986). Penicillin G, potassium. In: Analytical profiles of drug substances Vol. 15, Florey K. Ed., *Academic Press,* New York, 427-509.

Kitano K., Fujisawa Y., Katamoto K., Nara K. and Nakao Y. (1976a). Occurrence of 7b-4-carboxybutanamido)-cephalosporin compounds in the culture broth of some strain of the genus *Cephalosporium. J. Ferment. Technol.* 54: 712-719.

Kitano K., Kintaka K., Suzuki S., Katamoto K., Nara K. and Nakao Y. (1976b). A novel penicillin produced by strains of the genus *Paecillomyces. J. Ferment. Technol.* 54: 705-711.

Kleinkauf H. and von Döhren H. (1995). The nonribosomal peptide biosynthetic system - on the origins of structural diversity of peptides, cyclopeptides and related compounds. *Antonie van Leeuwenhoek* 67: 229-242.

Kleinkauf H. and von Döhren H. (1996). A nonribosomal system of peptide biosynthesis. *Eur. J. Biochem.* 236: 335-351.

Klionsky D.J., Herman P.K. and Emr S.D. (1990). The fungal vacuole: composition, function, and biogenesis. *Microbiol. Rev.* 54: 266-292.

Kogekar R.G. and Deshpande V.N. (1982). Biosynthesis of penicillin *in vitro*: Purification & properties of 'phenyl/ phenoxyacetic acid activating enzyme. *Indian J. Biochem. Biophys.* 19: 257-261.

Kogekar R.G. and Deshpande V.N. (1983). Biosynthesis of penicillin *in vitro*: part II—purification & properties of 6-aminopenicillanic acid—phenylacetyl-CoA/phenoxyacetyl-CoA transferase. *Indian J. Biochem. Biophys.* 20: 208-212.

Kolar M., Holzmann K., Weber G., Leitner E. and Schwab H. (1991). Molecular characterization and functional analysis in *Aspergillus nidulans* of the 5'-region of the *Penicillium chrysogenum* isopenicillin N synthetase gene. *J. Biotechnol.* 17: 67-80.

Kosalkova K., Marcos A.T., Fierro F., Hernando-Rico V., Gutiérrez S. and Martin J.F. (2000). A novel heptameric sequence (TTAGTAA) is the binding site for a protein required for high level expression of *pcbAB*, the first gene of the penicillin biosynthesis in *Penicillium chrysogenum. J. Biol. Chem.* 275: 2423-2430.

Kuan G., Dassa E., Saurin W., Hofnung M. and Saier M.H. Jr. (1995). Phylogenetic analyses of the ATP-binding constituents of bacterial extracytoplasmic receptor-dependent ABC-type nutrient uptake permeases. *Res. Microbiol.* 146: 271-278.

Kubicek C.P., Hönlinger C.H., Jaklitsch W.M., Affenzeller K., Mach R., Gerngross T.-U. and Ying L. (1990). Regulation of lysine biosynthesis in the fungus *Penicillium chrysogenum.* In: Amino Acids: Chemistry, Biology and Medicine, Lubec G. and Rozenthal G.A. Eds., *ESCOM, Science Publishers* B.V., Leiden, The Netherlands.

Kubicek-Pranz E.M. and Kubicek C.P. (1991). Production and biosynthesis of amino acids by fungi. In: Handbook of Applied Mycology Vol. 4: Fungal Biotechnology, Arora D.K., Elander R.P. and Mukerji K.G. Eds., *Marcel Dekker,* Inc. New York, 313-356.

Kudla B., Caddick M.X., Langdon T., Martínez-Rossi N.M., Bennett C.F., Sibley S., Davies R.W. and Arst H.N. Jr. (1990). The regulatory gene areA mediating nitrogen metabolite repression in *Aspergillus nidulans.* Mutations affecting specificity of gene activation alter a loop residue of a putative zinc finger. *EMBO J.* 9: 1355-1364.

Kurylowicz W., Kurzatkowski W. and Kurzatkowski J. (1987). Biosynthesis of benzylpenicillin by *Penicillium chrysogenum* and its Golgi apparatus. *Arch. Immunol. Ther. Exp. (Warsz)* 35: 699-724.

Laich F., Fierro F. and Martín J.F. (2003). Production of penicillin by fungi growing on food products: identification of a complete penicillin gene cluster in *Penicillium griseofulvum* and a truncated cluster in *Penicillium verrucosum. Appl. Environ. Microbiol.* 68: 1211-1219.

Lamas-Maceiras M., Vaca I., Rodríguez E., Casqueiro J. and Martín J.F. (2006). Amplification and disruption of the phenylacetyl-CoA ligase gene of *Penicillium chrysogenum* encoding an aryl-capping enzyme that supplies phenylacetic acid to the isopenicillin N acyltransferase. *Biochem. J.* 395: 147-155.

Lambalot R.H., Gehring A.M., Flugel R.S., Zuber P., LaCelle M., Marahiel M.A., Reid R., Khosla C. and Walsh C.T. (1996). A new enzyme superfamily - the phosphopantetheinyl transferases. *Chem. Biol.* 3: 923-936.

Lendenfeld T., Ghali D., Wolschek M., Kubicek-Pranz E.M. and Kubicek C.P. (1993). Subcellular compartmentation of penicillin biosynthesis in *Penicillium chrysogenum.* The amino acid precursors are derived from the vacuole. *J. Biol. Chem.* 268: 665-671.

Litzka O., Bergh K.T. and Brakhage A.A. (1995). Analysis of the regulation of the *Aspergillus nidulans* penicillin biosynthesis gene aat (*penDE*), which encodes acyl coenzyme A:6-aminopenicillanic acid acyltransferase. *Molec. Gen. Genet.* 249: 557-569.

Livermore D.M. (1988). Permeation of $-lactam antibiotics into *Escherichia coli, Pseudomonas aeruginosa* and other Gram-negative bacteria. *Rev. Infect. Dis.* 10: 691-698.

Luengo J.M. (1995). Enzymatic synthesis of hydrophobic penicillins. *J. Antibiot.* 48: 1195-1212.

Luengo J.M., Revilla G., Villanueva J.R. and Martin J.F. (1979a). Lysine regulation of penicillin biosynthesis in low-producing and industrial strains of *Penicillium chrysogenum. Gen. Microbiol.* 115: 207-211.

Luengo, J.M., G. Revilla, M.J. López, J.R. Villanueva and J.F. Martín (1979b). Penicillin production by mutants of *Penicillium chrysogenum* Wis 54-1255 resistant to polyene-macrolide antibiotics. *Biotechnol. Lett.* 1: 233-238.

MacCabe A.P., Riach M.B.R., Unkles S.E. and Kinghorn J.R. (1990). The *Aspergillus nidulans* npeA locus consists of three contiguous genes required for penicillin biosynthesis. *EMBO J.* 9: 279- 287.

Marger M.D. and Saier M.H. Jr. (1993). A major superfamily of transmembrane facilitators that catalyse uniport, symport and antiport. *Trends Biochem. Sci.* 18: 13-20.

Martín J.F. (1998). New aspects of genes and enzymes for beta-lactam antibiotic biosynthesis. *Appl. Microbiol. Biotechnol.* 50: 1-15.

Martín, J.F. (2000a). α–aminoadipyl-cysteinyl-valine synthetases in β–lactam producing organisms. From Abraham's discoveries to novel concepts of non-ribosomal peptide synthesis. *J. Antibiot.* 53: 1008-1021.

Martín J.F. (2000b). Molecular control of expression of penicillin biosynthesis genes in fungi: Regulatory proteins interact with a bi-directional promoter region. *J. Bacteriol.* 182: 2355-2362.

Martín J.F. and Liras P. (1989). Organization and expression of genes involved in the biosynthesis of antibiotics and other secondary metabolites. *Annu. Rev. Microbiol.* 43: 173-206.

Martín J.F., Ingolia T.D. and Queener S.W. (1990). Molecular genetics of penicillin and cephalosporin antibiotic biosynthesis. In Molecular Industrial Mycology, Leong S.A. & Berka R. Eds., *Marcel Dekker*, New York, USA., 149–195.

Martín J.F., Gutiérrez S., Fernández F.J., Velasco J., Fierro F., Marcos A.T. and Kosalkova K. (1994). Expression of genes and processing of enzymes for the biosynthesis of penicillins and cephalosporins. *Antonie van Leeuwenhoek* 65: 227-243.

Martín, J.F., Casqueiro, J., Liras, P. (2005). Secretion systems for secondary metabolites: how producer cells send out messages of intercellular communication. *Current Opinion in Microbiology* 8: 282-293.

Martín de Valmaseda E.M., Campoy S., Naranjo L., Casqueiro J. and Martín J.F. (2005). Lysine is catabolized to 2-aminoadipic acid in *Penicillium chrysogenum* by an omega-aminotransferase and to saccharopine by a lysine 2-ketoglutarate reductase. Characterization of the omega-aminotransferase. *Mol. Genet. Genomics.* 274: 272-82.

Martínez-Blanco H., Reglero A., Fernández-Valverde M., Ferrero M.A., Moreno M.A. and Luengo J.M. (1992). Isolation and characterization of the acetyl- CoA synthetase from *Penicillium chrysogenum. J. Biol. Chem.* 276: 5475-5481.

Martínez-Blanco H., Orejas M., Reglero A., Luengo J.M., and Peñalva M.A. (1993). Characterisation of the gene encoding acetyl-CoA synthetase in *Penicillium chrysogenum*: conservation of intron position in plectomycetes. *Gene* 130: 265-270.

Marzluf G.A. (1997). Genetic regulation of nitrogen metabolism in the fungi. *Microbiol. Mol. Biol. Rev.* 61: 17-32.

McCammon M. T., McNew J. A., Willy P. J. and Goodman J. M. (1994). An internal region of the peroxisomal membrane protein PMP47 is essential for sorting to peroxisomes. *J. Cell Biol.* 124: 915-925.

Montenegro E., Fierro F., Fernández F.J., Gutiérrez S. and Martín J.F. (1992). Resolution of chromosomes III and VI of *Aspergillus nidulans* by pulsed-field gel electrophoresis shows that the penicillin biosynthetic pathway genes *pcbAB, pcbC*, and *penDE* are clustered on chromosome VI (3.0 megabases). *J. Bacteriol.* 174: 7063-7067.

Müller W.H., van der Krift T.P., Krouwer A.J., Wosten H.A., van der Voort L.H., Smaal E.B. and Verkleij A.J. (1991). Localization of the pathway of the penicillin biosynthesis in *Penicillium chrysogenum. EMBO J.* 10: 489-495.

Müller W.H., Bovenberg R.A., Groothuis M.H., Kattevilder F., Smaal E.B., Van der Voort L.H. and Verkleij A.J. (1992). Involvement of microbodies in penicillin biosynthesis. *Biochim. Biophys. Acta* 1116: 210-213.

Müller W.H., Essers J., Humbel B.M. and Verkleij A.J. (1995). Enrichment of *Penicillium chrysogenum* microbodies by isopycnic centrifugation in nycodenz as visualized with immuno-electron microscopy. *Biochim. Biophys. Acta* 1245: 215-220.

Newbert R.W., Barton B., Greaves P., Harper J. and Turner G. (1997). Analysis of a commercially improved *Penicillium chrysogenum* strain series: involvement of recombinogenic regions in amplification and deletion of the penicillin biosynthesis gene cluster. *J. Ind. Microbiol. Biotechnol.* 19: 18-27.

Neyfakh A.A. (2002). Mystery of multidrug transporters: the answer can be simple. *Mol. Microbiol.* 44: 1123-1130.

Nishida H. and Nishiyama M. (2000). What is characteristic of fungal lysine synthesis through the alpha-aminoadipate pathway? *J. Mol. Evol.* 51: 299-302.

Oberegger H., Eisendle M., Schrettl M., Graessle S. and Haas H. (2003). 4'-phosphopantetheinyl transferase-encoding *npgA* is essential for siderophore biosynthesis in *Aspergillus nidulans*. *Curr. Genet.* 44: 211-215.

Pao S.S., Paulsen I.T. and Saier M.H. Jr. (1998). Major facilitator superfamily. *Microbiol. Mol. Biol. Rev.* 62: 1-34.

Paulsen I.T. (2003). Multidrug efflux pumps and resistance: regulation and evolution. *Curr. Opin. Microbiol.* 6: 446-451.

Paulsen I.T., Skurray R.A., Tam R., Saier M.H. Jr., Turner R.J., Weiner J.H., Goldberg E.B. and Grinius L.L. (1996). The SMR family: a novel family of multidrug efflux proteins involved with the efflux of lipophilic drugs. *Mol. Microbiol.* 19: 1167-1175.

Pérez-Estaban B., Orejas M., Gómez-Pardo E. and Peñalva M.A. (1993). Molecular characterization of a fungal secondary metabolism promoter: transcription of the *Aspergillus nidulans* isopenicillin N synthetase gene is modulated by upstream negative elements. *Mol. Microbiol.* 9: 881-895.

Pollegioni L., Caldinelli L., Molla G., Sacchi S. and Pilone M.S. (2004). Catalytic properties of D-amino acid oxidase in cephalosporin C bioconversion: a comparison between proteins from different sources. *Biotechnol. Prog.* 20: 467-473.

Putman M., van Veen H.W. and Konings W.N. (2000). Molecular properties of bacterial multidrug transporters. *Microbiol. Mol. Biol. Rev.* 64:672-693.

Ramos F.R., López-Nieto M.J. and Martín J.F. (1985). Isopenicillin N synthetase of *Penicillium chrysogenum*, an enzyme that converts delta-(L-alpha-aminoadipyl)-L-cysteinyl-D-valine to isopenicillin N. *Antimicrob. Agents Chemother.* 27: 380-387.

Revilla G., López-Nieto M.J., Luengo J.M. and Martín J.F. (1984). Carbon catabolite repression of penicillin biosynthesis by *Penicillium chrysogenum. J. Antibiot.* 37: 781-789.

Revilla G, Ramos F.R., López-Nieto M.J., Álvarez E. and Martín J.F. (1986). Glucose represses formation of delta-(L-alpha-aminoadipyl)-L-cysteinyl-D-valine and isopenicillin N synthase but not penicillin acyltransferase in *Penicillium chrysogenum. J. Bacteriol.* 168: 947-952.

Roach P.L., Clifton I.J., Fülöp V., Harlos K., Barton G.J., Hajdu J., Andersson I., Schofield C.J. and Baldwin J.E. (1995). Crystal structure of isopenicillin N synthase is the first from a new structural family of enzymes. *Nature* 375: 700-704.

Rowlands R.T. (1991). Industrial improvement of the Panlabs Penicillin Club. In: 50 years of penicillin application, history and trends, Kleinkauf H. and von Döhren H. Eds., *PUBLIC Ltd.*, Czech Republic, 258-264.

Sánchez S., Paniagua L., Mateos R.C., Lara F. and Mora J. (1981). Nitrogen regulation of penicillin G biosynthesis in *Penicillium chrysogenum. Adv. Biotechnol.* [Proc. 6th Int. Ferment. Symp.] 3: 147-154.

Sánchez S., Flores M.E. and Demain A.L. (1988). Nitrogen regulation of penicillin and cephalosporin fermentation. In: Nitrogen source control of microbial processes. Sanchez-Esquivel S. Eds., *CRC Press, Inc.*, Boca Raton, Fla., 121-136.

Sawada Y., Baldwin J.E., Singh P.D., Solomon N.A. and Demain A.L. (1980). Cell-free cyclization of delta-(L-alpha-aminoadipyl)-L-cysteinyl-D-valine to isopenicillin N. *Antimicrob. Agents Chemother.* 18: 465-470.

Schoergendorfer K., and Kuernsteiner H. (2005). Patent WO2005040369.

Seemüller E., Lupas A., and Baumeister W. (1996). Autocatalytic processing of the 20S proteasome. *Nature* 382: 468-470.

Sentheshanmuganathan S., and Nicherson W.J. (1962). Transaminase and D-amino acid oxidase of *Trigonopsis variabilis. J. Gen. Microbiol.* 27: 465-471.

Shah A.J., Tilburn J., Adlard M.W. and Arst H.N. Jr. (1991). pH regulation of penicillin production in *Aspergillus nidulans. FEMS Microbiol. Lett.* 61: 209-212.

Shen Y.Q., Heim J., Solomon N.A., Wolfe S. and Demain A.L. (1984). Repression of beta-lactam production in *Cephalosporium acremonium* by nitrogen sources. *J. Antibiot.* 37: 503-511.

Shiau C.Y., Baldwin J.E., Byford M.F. and Schofield C.J. (1995a) Delta-L-(alpha-aminoadipoyl)-L-cysteinyl-D-valine synthetase: isolation of L-cysteinyl-D-valine, a 'shunt' product, and implications for the order of peptide bond formation. *FEBS Lett.* 16: 373: 303-306.

Shiau C.Y., Baldwin J.E., Byford M.F., Sobey W.J. and Schofield C.J. (1995b). delta-L-(alpha-aminoadipoyl)-L-cysteinyl-D-valine synthetase: the order of peptide bond formation and timing of the epimerisation reaction. *FEBS Lett.* 358: 97-100.

Shiau C.Y., Byford M.F., Aplin R.T., Baldwin J.E. and Schofield C.J. (1997). L-delta-(alpha-Aminoadipoyl)-L-cysteinyl-D-valine synthetase: thioesterification of valine is not obligatory for peptide bond formation. *Biochemistry* 36: 8798-8806.

Sikkema, J., de Bont J.A.M. and Poolman B. (1995). Mechanisms of membrane toxicity of hydrocarbons. *Microbiol. Rev.* 59: 210-222.

Skatrud P.L. and Queener S.W. (1989). An electrophoretic molecular karyotype for an industrial strain of *Cephalosporium acremonium. Gene* 78: 331-338.

Smith D.J., Burnham M.K.R., Bull J.H., Hodgson J.E., Ward J.M., Browne P., Brown J., Barton B., Earl A.J. and Turner G. (1990). β-lactam antibiotic biosynthetic genes have been conserved in clusters in prokaryotes and eukaryotes. *EMBO J.* 9: 741-747.

St Georgiev V. (2000). Membrane transporters and antifungal drug resistance. *Curr. Drug Targets.* 1: 261-284.

Suárez T. and Peñalva M.A. (1996). Characterization of a *Penicillium chrysogenum* gene encoding a PacC transcription factor and its binding sites in the divergent pcbAB-pcbC promoter of the penicillin biosynthetic cluster. *Mol. Microbiol.* 20: 529-540.

Then Bergh K. and Brakhage A.A. (1998). Regulation of the *Aspergillus nidulans* penicillin biosynthesis gene acvA (*pcbAB*) by amino acids: implication for involvement of transcription factor PACC. *Appl. Environ. Microbiol.* 64: 843-849.

Thieringer R. and Kunau W.H. (1991a). Beta-oxidation system of the filamentous fungus *Neurospora crassa*. Structural characterization of the trifunctional protein. *J. Biol. Chem.* 266: 13118-13123.

Thieringer R. and Kunau W.H. (1991b). The beta-oxidation system in catalase-free microbodies of the filamentous fungus *Neurospora crassa*. Purification of a multifunctional protein possessing 2-enoyl-CoA hydratase, L-3-hydroxyacyl-CoA dehydrogenase, and 3-hydroxyacyl-CoA epimerase activities. *J. Biol. Chem.* 266: 13110-13117.

Tilburn J., Sakar S., Widdick D.A., Espeso E.A., Orejas M., Mungroo J., Peñalva M.A. and Arst H.N. Jr. (1995). The Aspergillus PacC zinc finger transcription factor mediates regulation of both acidic- and alkaline-expressed genes by ambient pH. *EMBO. J.* 14: 779-790.

Tobin M.B., Fleming M.D., Skatrud P.L. and Miller J.R. (1990). Molecular characterization of the acyl-coenzyme A: isopenicillin N acyltransferase gene (*penDE*) from *Penicillium chrysogenum* and *Aspergillus nidulans* and activity of recombinant enzyme in *E. coli. J. Bacteriol.* 172: 5908–5914.

Tobin M.B., Baldwin J.E., Cole S.C.J., Miller J.R., Skatrud P.L. and Sutherland J.D. (1993). The requirement for subunit interaction in the production of *Penicillium chrysogenum* acyl-conenzyme A: isopenicillin N acyltransferase in *Escherichia coli. Gene* 132: 199–206.

Tobin M.B., Cole S.C., Miller J.R., Baldwin J.E. and Sutherland J.D. (1995). Amino acid substitutions in the cleavage site of acyl-coenzyme A:isopenicillin N acyltransferase from *Penicillium chrysogenum*: effect on proenzyme cleavage and activity. *Gene* 162: 29-35.

Trip H., Evers M.E., Kiel J.A., Driessen A.J. (2004). Uptake of the beta-lactam precursor alpha-aminoadipic acid in *Penicillium chrysogenum* is mediated by the acidic and the general amino acid permease. *Appl. Environ. Microbiol.* 70: 4775-4783.

Turgay K., Krause M., and Marahiel M.A. (1992). Four homologous domains in the primary structure of GrsB are related to domains in a superfamily of adenylate forming enzymes. *Mol. Microbiol.* 6: 529-546.

Ullán R.V., Casqueiro J., Bañuelos O., Fernández F.J., Gutiérrez S. and Martín J.F. (2002a). A novel epimerization system in fungal secondary metabolism involved in the conversion of isopenicillin N into penicillin N in *Acremonium chrysogenum. J. Biol. Chem.* 277: 46216-46225.

Ullán R.V., Liu G., Casqueiro J., Gutiérrez S., Bañuelos O. and Martín J.F. (2002b). The *cefT* gene of *Acremonium chrysogenum* C10 encodes a putative multidrug efflux pump protein that significantly increases cephalosporin C production. *Mol. Genet. Genomics* 267: 673-683.

Valenciano S., Lucas J.R., Pedregosa A., Monistrol I.F. and Laborda F. (1996). Induction of beta-oxidation enzymes and microbody proliferation in *Aspergillus nidulans. Arch. Microbiol.* 166: 336-341.

Valenciano S., De Lucas J.R., Van der Klei I., Veenhuis M. and Laborda F. (1998). Characterization of *Aspergillus nidulans* peroxisomes by immunoelectron microscopy. *Arch. Microbiol.* 170: 370-376.

Valle F., Balbas P., Merino E. and Bolívar F. (1991). The role of penicillin amidases in nature and in industry. *Trends Biochem. Sci.* 16: 36-40.

Van der Lende T.R., van de Kamp M., Berg M., Sjollema K., Bovenberg R.A., Veenhuis M., Konings W.N. and Driessen A.J. (2002). delta-(L-alpha-Aminoadipyl)-L-cysteinyl-D-valine synthetase, that mediates the first committed step in penicillin biosynthesis, is a cytosolic enzyme. *Fungal Genet. Biol.* 37: 49-55.

Van der Rest M.E., Kamminga A.H., Anraku A.N.Y., Poolman B. and Konings W.N. (1995). The plasma membrane of *Saccharomyces cerevisiae*:structure, function, and biogenesis. *Microbiol. Rev.* 59: 304-322.

Veenstra A.E., van Solingen P., Huininga-Muurling H., Koekman B.P., Groenne M.A.M., Smaal E.B., Kattevilder A., Álvarez E., Barredo J.L. and Martín, J.F. (1989). Cloning of penicillin biosynthesic genes. In: Genetics and Molecular Biology of Industrial Microorganisms, Hershberger C.L., Queener S.W. and Hegeman G., Eds., *American Society for Microbiology*, Washington, DC, USA., 262–269.

Whiteman P.A., Abraham E.P., Baldwin J.E., Fleming M.D., Schofield C.J., Sutherland J.D. and Willis A.C. (1990). Acyl coenzyme A: 6-aminopenicillanic acid acyltransferase from *Penicillium chrysogenum* and *Aspergillus nidulans. FEBS Lett.* 262: 342–344.

Yoshimura F. and Nikaido H. (1985). Diffusion of β–lactam antibiotics through the porin channels of *E.coli* K12. *Antimicrob. Agents. Chemother.* 27: 84-92.

Zabriskie T.M. and Jackson M.D. (2000). Lysine biosynthesis and metabolism in fungi. *Nat. Prod. Rep.* 17: 85-97.

Zhang J. and Demain A.L. (1992). ACV synthetase. *Crit. Rev. Biotechnol.* 12: 245-260.

Zheleznova E.E., Markham P., Edgar R., Bibi E., Neyfakh A.A. and Brennan R.G. (2000). A structure-based mechanism for drug binding by multidrug transporters. *Trends Biochem. Sci.* 25: 39-43.

Zocher R. and Keller U. (1997). Thiol template peptide synthesis systems in bacteria and fungi. *Adv. Microb. Physiol.* 38: 85-131.

4 Current Trends in Molecular Diagnosis of Ochratoxin A Producing Fungi

L. Niessen

Lehrstuhl für Technische Mikrobiologie, Technische Universität München, Weihenstephaner Steig 16, D-85350 Freising, Germany.

INTRODUCTION

Ochratoxin A (OTA, *N*-[(5-chloro-3,4-dihydro-8-hydroxy-3-methyl-1-oxo-1*H*-2-benzopyran-7-yl)-carbonyl]-3-phenyl-L-alanine, CAS No. 303-47-9) is a polyketide mycotoxin first discovered as a metabolite of *Aspergillus ochraceus* (van der Merwe *et al.*, 1965). The compound has been considered growing importance during the last decade because contamination of food and feed poses a serious threat to human and animal health. OTA is nephrotoxic and a possible human carcinogen (Delacruz and Bach, 1990; Huff 1991; IARC 1993; Bennett and Klich, 2003). It has been characterized as one of the most potent toxic substances known from natural sources (Kuiper-Goodman and Scott, 1989). The substance is suspected to be the causal agent of Balkan endemic nephropathy (BEN) and was associated with increased incidence of urinary tract tumors in that region (Petkova-Bocharova *et al.*, 1988). Besides renal toxicity and carcinogenicity, the compound has been observed to cause a wide range of adverse effects in laboratory animals and in cell cultures, e.g., inhibition of protein biosynthesis (Dirheimer and Crepy, 1991), impairment of energetic metabolism (Gekle *et al.*, 2005), induction of oxidative stress and DNA damage (Petrik *et al*, 2003; Kamp *et al.*, 2005), apoptosis (Weber *et al.*, 2003; Assaf *et al.*, 2004; Alvarez *et al.*, 2005), and suppression of the immune system (Alvarez *et al.*, 2004; Surai and Dvorska 2005). OTA is produced in pure cultures of various species in the fungal genus *Aspergillus* (*A. alliaceus, A. carbonarius, A. muricatus, A. niger* and *A. ochraceus*) as well as by *Penicillium verrucosum* and the closely related *P. nordicum* (Larsen *et al.*, 2001; Samson *et al.*, 2004; Seifert and Levésque, 2004). Only recently, Medina *et al.* (2005) demonstrated that also a considerable proportion of strains in *A. tubingensis* from Spanish vine berries were able to produce OTA, a species formerly considered non-toxic. Data were confirmed by Perrone *et al.* (2006), who found an even higher proportion of producers among isolates obtained from Italian grapes. Sources for the toxin to enter the human food chain are cereals (Lund and Frisvad, 2003) and cereal products including beer (Wolf, 2000; Bresch *et al.,* 2000), coffee (Martins *et al.*, 2003; Taniwaki *et al.*, 2003; Pardo *et al.*, 2004), wine (Soleas *et al.*, 2001; Battilani *et al.*, 2003), grape juice (Delage *et al.*, 2003; Drusch and Aumann, 2005), dried vine fruits (Abarca *et al.*, 2003; Tjamos *et al.*, 2004;

Hennel *et al.*, 2005) and other dried fruits (Engel, 2000; Imanaka *et al.*, 2005), cacao and cocoa products (Tafuri *et al.*, 2004; Amezqueta *et al.*, 2005), dry beans (Domijan *et al.*, 2005), and spices (Scheuer and Gareis, 2002). Pork meat and meat products as well as blood products may contain OTA if ingested by animals (Jorgensen and Petersen, 2002; Monaci *et al.*, 2004). This may additionally contribute to the fact that the toxin can be found in human blood sera in considerable concentrations (Rosner *et al.*, 2000).

Although present only in low quantity on cereal grains and other agricultural commodities in the field, *Aspergillus* and *Penicillium* species develop rapidly under typical storage conditions, when the water content of their substrate is most unfavorable to other fungi (Magan *et al.*, 2003). Differences in temperature optimum and a_w requirements parallel with regional differences in the occurrence of both genera (Northolt *et al.*, 1979; Sweeney and Dobson, 1999; Pardo *et al.*, 2006). *A. ochraceus, A. carbonarius* and *A. niger* are regularly isolated from coffee beans, grapes, spices, cocoa, soybeans, groundnuts and rice grown in subtropic to tropic climates (Kuiper-Goodman and Scott, 1989), whereas *P. verrucosum* occurs predominately in stored cereals in temperate regions (Larsen *et al.*, 2001). Isolates resembling *P. verrucosum* but isolated from protein rich substrates, mainly meat, have been summarized under the name "*P. nordicum*". This species contains high OTA producing chemotypes, whereas *P. verrucosum sensu stricto* contains low OTA producing chemotypes. Both taxa have recently been separated using chemo-taxonomical (Larsen *et al.*, 2001) and molecular (Castella *et al.*, 2002) methods.

Lack of knowledge about the genetic background of OTA production as well as missing data on the phylogenetic relationship between relevant species make it difficult to develop assays with high specificity either to species or to toxigenic properties. However, for the food and feed industry, legal actions taken by 37 countries in the world, including the European Union with an obligation to harmonize the regulations in its member countries (van Egmond, 2003), will bear the obligation to ensure no raw materials are processed with OTA levels exceeding the specified maximum residue level (MRL). This can be achieved by setting up concepts based on hazard analysis critical control points (HACCP), which sort out lots with risk of OTA contamination. Concepts according to the HACCP philosophy demand methods which fulfill the following needs:

- The analytical target must be a direct marker for the hazard to be controlled
- The marker must be indicative for the future toxigenic potential of a sample as well as for its mycological history
- Sensitivity of the analysis must be high enough to detect hazardous samples
- Results must enable decisions on whether or not a lot should pass or fail the control point
- Screening of vast numbers of samples must be possible, e.g., the procedure must have an automatable format for high throughput applications
- Results must be available in short time
- Tests must be applicable by low trained personnel
- Tests must be inexpensive and easy to use as must be the equipment used to perform them
- Sample preparation should avoid the use of organic solvents which are hazardous to the environment and operating personnel.

In order to meet the above demands, polymerase chain reaction (PCR) has become a powerful tool in the diagnosis of microbes in food raw materials and in processed food and feed (Schoen *et al.*, 2005). Recently, various species and gene specific PCR primers were developed for detection and identification of OTA producers in conventional and real-time PCR applications. The different strategies selected by authors in order to obtain nucleotide sequence information useful in designing diagnostic PCR assays, the techniques applied and the results obtained will be reviewed in the following chapters. Since the field of PCR detection of OTA producers evolves very rapidly in recent time, information accumulated here can only reflect the state of the art as per finalizing of the current manuscript. Therefore, the current chapter provides an update of a recently published review by Niessen *et al.*, (2005). An overview over the primers discussed in the current paper is given in Table 4.1. The following chapter summarizes the development of PCR as a tool for detection and identification of mycotoxin producers, followed by detailed description of assays, which were particularly developed for producers of OTA.

PCR AS A DIAGNOSTIC TOOL IN MYCOTOXICOLOGY

The first record of a diagnostic PCR assay for a mycotoxigenic fungus dates from 1993, when Tang *et al.* described the use of that technology for the detection of medically relevant fungi in broncheoalveolar lavages (BAL) of patients suffering from invasive pulmonary aspergillosis (IPA). Their assay used the gene coding for alkaline protease as the target nucleotide sequence. Clearly, Tang *et al.*, (1993) set up their assay for the detection of *A. flavus* as a human lung pathogen and not because of its toxigenic properties. Later on, Hayette *et al.*, (2001) developed the earlier assay further by designing a nested PCR, which combined the primers of Tang *et al.*, (1993) with a newly developed pair of internal primers to increase the limit of detection. Assays aimed at detecting fungi in order to evaluate toxigenic potential of food and food raw materials came into existence from 1996 on, when Geisen (1996) and, independently, also Shapira *et al.*, (1996) published the PCR detection of *A. flavus* and *A. parasiticus*. Both fungi are established to be the major sources of aflatoxins, a group of mycotoxins with high acute toxicity and one of the highest carcinogenic potentials observed among naturally occurring compounds (Eaton and Gallagher, 1994). Both authors used genes involved in the biosynthesis of aflatoxins as analytical target to identify the fungi in pure cultures. Others widely used these systems in secondary applications to trace the fungi in naturally contaminated materials, such as, figs (Färber *et al.*, 1997), cereal grain (Shapira *et al.*, 1996; Geisen *et al.*, 1998), peanut kernels (Chen *et al.*, 2002), and in mixed feeds (Zachova *et al.*, 2003). Recently, Mayer *et al.*, (2003) used primers specific for the nor-1 gene in *A. flavus* in a real-time PCR application to establish a positive correlation between quantitative PCR signal and number of colony forming units (cfu) in foods.

In the period between 1996 and 1999 several PCR based assays for the diagnosis of toxigenic *Fusarium*-species were developed. Studies focused mainly on producers of trichothecenes and fumonisins, which are the most hazardous secondary metabolites produced by species within the genus (Bennett and Klich, 2003). Most of the assays published during that period used random amplification of polymorphic DNA (RAPD) to generate genomic markers with specificity for the respective fungal species as sequence target for their primers. In none of the assays marker sequences could be linked to a certain function in the fungal genome (anonymous marker). Schilling *et al.* (1996) developed assays for the detection of *F. avenaceum, F. culmorum*, and *F. graminearum*, the latter two of which are the major pathogens in *Fusarium* Head Blight (FHB) of cereals and the typical producers of hazardous mycotoxins like deoxynivalenol (DON) and nivalenol (NIV). Similarly, Yoder and Christianson (1998) developed diagnostic PCR-based assays for *F. crookwellense* (Syn.

Table 1.1. Primers used for detection and identification of ochratoxin-A producing fungi

Primer name	*Primer sequences 5´- 3´*	*Target species*	*DNA target*	*Product size (bp)*	*Reference*
CARBO1	AAGCGAATCGATAGTCCACAAAGAATAC	*A. carbonarius*	*cmdA* gene	371	Perrone *et al.*, 2004
CARBO2	TCTGGCAGAAGTTAATATCCCGGTT				
A1B_fw	GAATTCACCACACATCATAGC		AFLP marker (AJ516957)	189	Schmidt *et al.*, 2004a
A1B_rv	TTAACTAGGATTTGGCATTGAAC				
C1B_fw	GAATTCACGGTGCTCGACCC		AFLP marker (AJ516956)	351	Schmidt *et al.*, 2004a
C1B_rv	TTAACTGTCGGCGGAACAGGC				
OPX7F$_{809}$	AGGCTAATGTTGATAACGGATGAT		RAPD marker OPX7$_{809}$	809	Fungaro *et al.*, 2004
OPX7R$_{809}$	GCTGTCAGTATTGGACCTTAGAG				
CAR1	GCATCTCTGCCCCTCGG		rRNA gene ITS1-ITS2 region	420	Patino *et al.*, 2005
CAR2	GGTTGGAGTTGTCGGCAG				
ITS1	TCCGTAGGTGAACCTGCG	*A. niger/ A. tubingensis*	rRNA gene ITS1-ITS2 region	420	González-Salgado *et al.*, 2005
NIG	CCGGAGAGAGGGGACGGC				
ANDF78	AACATTCAAGGCCGACTTTCCC		topoisomerase II gene	241	Kanbe *et al.*, 2002
ANDR102	GAGGAGAGCAAGTGATGATTGG				
ASPU	ACTACCGATTGAATGGCTCG		rRNA gene ITS1 region	not specified	Sugita *et al.*, 2004
Ni1r	ACGCTTTCAGACAGTGTTCG				
OPX7f$_{372}$	CAGTCGTCCAGTACCCTAAC		RAPD marker OPX7$_{372}$	372	Sartori *et al.*, 2006 (in press)
OPX7R$_{372}$	GAGCGAGGCTGATCTAAGTG				
OCA-V	ATACCACCGGGTCTAATGCA	*A. ochraceus*	AFLP marker (AJ511647)	260	Schmidt *et al.*, 2003
OCA-R	TGCCGACAGACCGAGTGGATT				
3for	GGGCTGTCCTGACCGATG		expressed marker for OTA production	470	Mühlencoert, 2004
3rev2	ACGGGAGGCGGAGGAAGA				
OCRA1	CTTCCTTAGGGGTGGCACAGC		rRNA gene ITS1-ITS2 region	400	Patino *et al.*, 2005
OCRA2	GTTGCTTTTCAGCGTCGGCC				
AoOTA-L	CATCCTGCCGCAACGCTCTATCTTTC		*pks* gene, KS domain	690	Dao *et al.*, 2005
AoOTA-R	CAATCACCCGAGGTCCAAGAGCCTCG				
S3	ACGGTGAATGCCGCGTCGTAGT		719 bp unassigned DNA sequence from A. ochraceus	*353*	*Lebrihi et al., 2003*
S4	GCATCGTGATACCCATGCTGCG				
AoLC35-12L	GCCAGACCATCGACACTGCATGCTC	OTA/citrinin producers	*pks* gene, KS domain	520	Dao *et al.*, 2005
AoLC35-12R	CGACTGGCGTTCCAGTACCATGAGCC				
S5	GGACGAGATCCAAGGCTCGCAG		484 bp unassigned DNA sequence from A. ochraceus	296	*Lebrihi et al., 2003*
S6	TGGGCACCCATGTGGAAACAGA				
4for	GACGACGGTGTTCTCGGA	*A. ochraceus/ A. carbonarius*	expressed marker for OTA production	410/350	Mühlencoert, 2004
4rev2	CCGGGCGAGAGTTCAATC				
otapks1	TACGGCCATCTTGAGCAACGGCACTGCC	*P. nordicum*	*otapks*PN gene	500	Geisen *et al.*, 2004
otapks2	ATGCCTTTCTGGGTCCGATA				
otapkstaq1	CACGGTTTGGAACACCACAAT		*otapks*PN gene	not specified	Geisen *et al.*, 2004
otapkstaq2	TGAAGATCTCCCCCGCCT				
otapksPNprobe	FAM-CGTACCAATCCCCATCCAGGGCTC-TAMRA				
otanps_for	AGTCTTCGCTGGGTGCTTCC	*P. nordicum/P.verrucosum*, OTA producing isolates	*otanps*PN gene	750	Bogs *et al.*, 2006 (in press)
otanps_rev	CAGCACTTTTCCCTCCATCTATCC				

F. cerealis), and the closely related species *F. sambucinum, F. torulosum,* and *F. venenatum,* which are potential producers of trichothecene mycotoxins like T-2 and HT-2 toxins, diacetoxyscirpenol, and NIV (Moss and Thrane, 2004). Other assays for the detection of *Fusarium* spp. targeted well characterized genes, e.g., the cutinase gene in *F. solani* (Alexandrakis *et al.*, 1998), the gaoA gene coding for galactose oxidase in *F. graminearum* (Niessen and Vogel, 1997), the gene coding for elongation factor 1 α (EF1α) (Aoki and O'Donnell, 1999). The internal transcribed spacers 1 and 2 (ITS1, ITS2) of the fungal ribosomal RNA gene were used by Grimm and Geisen (1998) to set up a quantitative microplate colorimetric PCR assay for the group of *Fusarium* spp. known to produce fumonisins. The first assay using a gene from the trichothecene biosynthetic pathway for the detection of toxigenic *Fusarium* spp. was published by Niessen and Vogel (1998) who used the trichodiene synthase gene sequence (tri5) as target for the amplification of a group specific DNA fragment in a PCR assay applied in quality control of cereals and malt in the brewing industry.

It has to be noted that the assays mentioned above are the ones, which describe the first use of PCR for the detection of the respective species and groups of species as mentioned in literature. Many more such assays were set up in subsequent studies using either the same or different target sequences or PCR technology. In particular, the use of various trichothecene biosynthetic genes became popular since the availability of sequences and functional linkage of these genes to physiological properties. This enabled use of certain genes for differentiation of strains into chemotypes according to mycotoxin production (Lee *et al.*, 2001; Bluhm *et al.*, 2002; Chandler *et al.*, 2003; Jennings *et al.*, 2004; Nicholson *et al.*, 2004; Tan *et al.*, 2004; Li *et al.*, 2005). Since trichothecenes are one of the biggest group of fungal secondary metabolites, similar compounds are also produced by species other than *Fusarium* spp. Peltola *et al.* (2002) demonstrated specific detection of the highly toxic *Stachybotrys chartarum*, a fungus most typical for damp housing and water damage. They used the tri5 gene from the satratoxin biosynthetic pathway of *S. chartarum* as target for PCR amplification to differentiate toxigenic and non-toxigenic isolates within the species.

Apart from the trichothecenes, fumonisins are another group of hazardous *Fusarium* mycotoxins with a high toxicological potential (Dutton, 2004). Like in many fungal secondary metabolites, genes involved in their biosynthesis, regulation and transport are organized in a cluster in physical vicinity (Plattner *et al.*, 1996; Proctor *et al.*, 1999a; Proctor *et al.*, 1999b; Seo *et al.*, 2001; Desjardins *et al.*, 2002; Proctor *et al.*, 2003). The fum5 gene, which codes for a polyketide synthase in *Gibberella fujikuroi* was used by Bluhm *et al.*, (2002) as target for a pair of primers which gave a PCR product with typical fumonisin producing fungi in a group specific manner. Recently, Sanchez-Rangel *et al.*, (2005) accomplished a similar detection system with primers targeted to the fum1 gene.

Besides the typical producers of *Fusarium* mycotoxins, attempts to detect fungi from other taxonomical groups via genes involved in the biosynthesis of the toxins were successful. Paterson *et al.* (2000) developed primers for the isoepoxydon dehydrogenase (IDH) gene which detected typical producers of patulin, e.g., *P. expansum, P. roquefortii,* and *A. clavatus*. Finally, the cyclodepsipeptide mycotoxin enniatin is synthesized via extra-ribosomal peptide synthesis by the multienzyme complex enniatin synthase. The enzyme is coded on the ennsyn gene, which was used by Nicholson *et al.,* (2004) as target for a pair of group specific PCR primers for the detection of typical enniatin producers, e.g., *F. avenaceum, F. acuminatum, F. chlamydosporum, F. lateritium,* and *F. tricinctum.*

PCR DIAGNOSIS OF OCHRATOXIN A—PRODUCING FUNGI

The following paragraphs of this chapter deal with PCR-based methods which were developed during recent years in order to provide analytical tools for detection and identification of those fungal organisms which were described as producers of ochratoxin A. Most of the systems were developed to analyze pure cultures or contaminated commodities to estimate toxicological hazards associated with a fungal strain. However, it is worthwhile to note that some of the diagnostic tools described in the following chapters were rather aimed at the organisms in their role as human or animal pathogens. This is especially true for *A. niger*, a fungus formerly considered a GRAS organism (generally regard as safe) and used widely in the industrial production of (food) enzymes and citric acid or even in the desulfurization of coal (Raman *et al.,* 1995). Recent findings which show that between 3 and 10 % of isolates in *A. niger* are able to produce ochratoxin A—under favorable conditions might put some constraints to safety considerations when using that organism. However, Schuster *et al.,* (2002) concluded that isolates can still be used in safe biotechnological applications provided that non-toxigenic properties have properly been tested. The PCR-based methods now available may provide useful tools to analyze either the taxonomic classification or even the mycotoxigenic potential of a given isolate. They will thus make both products and processes safer.

PRIMERS TARGETED TO ANONYMOUS GENOMIC MARKERS

Anonymous genomic markers are sections within the genome which are specifically linked to a taxonomical unit, a certain phenotype or physiological property without being assigned to a defined genetic function. Typically they are generated either by treatment of the genomic DNA with restriction endonucleases, amplification of DNA regions flanked by microsatellites or other repetitive sequence elements, or amplification of DNA using short randomly priming oligonucleotides under low stringency conditions. All the different techniques available have in common that polymorphic banding patterns are generated which may be employed to distinguish taxonomical groups. Sequence information obtained from anonymous markers have widely been used to set up PCR primers for specific detection and identification of fungal organisms.

The following sections provide information on how these techniques have been employed to set up species specific detection systems for potentially OTA producing fungi.

AFLP MARKER BASED PRIMERS FOR *A. OCHRACEUS* AND *A. CARBONARIUS*

Amplified fragment length polymorphism (AFLP) was introduced by Vos *et al.,* (1995) as a technique for genetic mapping of plants. Majer *et al.,* (1996) appear to be the first who used the technique to detect genetic variation in fungi (*Cladosporium fulvum* and *Pyrenopaziza brassicae*). For a recent review of the AFLP technique see Bensch and Akesson (2005). In AFLP, genomic DNA of the organism under study is fragmented by restriction endonucleases (mostly *Eco*RI together with *Mse*I) as a first step. In the second step, adapter oligonucleotides are ligated to the sticky ends of the resulting restriction fragments. Oligonucleotide primers incorporating the adapter sequence and one of either restriction sites are used to amplify all restriction fragments in a preamplification step.

Preamplified product is used as template in a second round PCR using oligonucleotides similar to the first round primers but with a 3´-end extention of one to three nucleotides which lead to selective amplification of only a fraction of the total preamplified fragments. Typically, about 100 fragments are amplified which are size separated by polyacrylamide gel electrophoresis (PAGE) and detected by either silver staining or a fluorescent dye attached to one of the primers. Resulting banding patterns can be normalized using internal and external size markers and compared using appropriate software packages. Data are used to set up phylogenetic trees, for taxonomical identification down to the level of strains or individuals, to study population structures, gene mapping and linkage studies or marker analysis in all groups of organisms. Since AFLP fingerprints are highly reproducible they can be stored in reference data bases and used for species identification of unknown samples circumventing cumbersome multi locus sequencing.

Schmidt *et al.,* (2003, 2004a) used AFLP to detect specific markers for *A. ochraceus* and *A. carbonarius*. Both fungal species were found to be the predominant producers of OTA in green coffee (Taniwaki *et al.*, 2003; Martins *et al.*, 2003; Pardo *et al.*, 2004). Moreover, the latter species has been identified as the major producer of that toxin in grapes (Battilani *et al.*, 2003; Serra *et al.*, 2003; Serra *et al.*, 2006) and raisins (Abarca *et al.*, 2003; Tjamos *et al.*, 2004). Schmidt and coworkers compared strains of both species isolated from coffee related sources as well as closely related taxa and demonstrated that the resulting AFLP fingerprints were well distinct at the species level in both taxa. Several of the amplified fragments were found to be characteristic to either of the species and could be used as specific AFLP markers for *A. ochraceus* (Schmidt *et al.*, 2003) and *A. carbonarius* (Schmidt *et al.*, 2004a), respectively. The marker fragments were cloned and sequenced and primers designed from the sequenced markers enabled detection of both species. PCR with primer pair OCA-V/OCA-R designed to detect *A. ochraceus* (Schmidt *et al.*, 2003, Table 4.1) as well as *A. carbonarius* specific primers A1B_fw/A1B_rv (189 bp product) and C1B_fw/C1B_rv (351 bp product) (Schmidt *et al.*, 2004a, Table 4.1) were optimized for sensitivity by variation of buffer composition and pH as well as optimization of the thermal cycling protocol. Specificity of the primers was tested with DNA of several different target strains as well as closely related *Aspergillus* and *Penicillium* spp. and DNA isolated from non-infected green coffee. Based on the *A. ochraceus* specific primer pair, Schmidt *et al.* (2004b) set up a real-time PCR assay for quantitative estimation of *A. ochraceus* DNA. The authors used a LightCycler™ (Roche diagnostics) system with SYBR Green I as intercalating fluorescent dye to quantify the concentration of DNA of the fungus in 30 samples of naturally infected green coffee from various regions. The assay had a detection limit of 4.7 pg per reaction compared to a detection limit of approximately 400 pg per reaction in conventional PCR with detection in ethidium bromide stained agarose gels. Correlation between DNA concentration found with the assay and OTA content of the respective samples was low ($r = 0.55$) but statistically significant. Recently, Perrone *et al.* (2006) used AFLP to distinguish isolates of closely related *Aspergillus* section *Nigri* species derived from grapes and to compare the results with OTA producing capacities of the isolates under study. They found that isolates assigned to *A. carbonarius, A. niger*, and the morphologically indistinct *A. tubingensis* were clearly distinct by their AFLP fingerprint. Moreover, the authors demonstrated that 25 % of the *A. tubingensis* strains analyzed were able to produce ochratoxin A in pure cultures with a proportion of toxin producers similar to that found in *A. niger*. From that result it was concluded, that *A. tubingensis* strains may well account for OTA concentrations found in samples of Italian wine (Otteneder and Majerus, 2000; Pietri *et al.*, 2001; Brera *et al.*, 2005). Similar observations were made by

Medina *et al.*, (2005) who analyzed *Aspergillus* section *Nigri* isolates from Spanish grapes for their ability to produce OTA. In their study 14% of the *A. tubingensis* isolates produced the toxin.

DETECTION OF *A. OCHRACEUS* AND *A. CARBONARIUS* WITH cDNA AFLP BASED PRIMERS

cDNA AFLP is a variation of the original protocol of Vos *et al.* (1995) in which expressed genes are analyzed rather than the complete genome of an organism (Bachem *et al.*, 1996, Bachem *et al.*, 1998). Messenger RNA is isolated and used as template for the preparation of cDNA by reverse transcription which then undergoes the regular AFLP procedure. In an attempt to spot genes differentially expressed under conditions promoting OTA production in *A. ochraceus*, Mühlencoert (2004) used this technique to find expressed markers for genes involved in biosynthesis of the toxin. Most interestingly, Mühlencoert *et al.*, (2004) reported that OTA production in *A. ochraceus* NRRL 3174 was depend on the initial pH of the culture broth (AM medium, Adye and Mateles, 1964) rather than on a specific combination of carbon and nitrogen source as is the case in *P. nordicum* (Färber and Geisen, 2004). OTA production in *A. ochraceus* NRRL 3174 grown at starting pH 5.0 was induced by shifting the medium to pH 6.5 after 80 h of cultivation. AFLP of cDNA produced from mycelia at different growth stages of cultures grown under the respective starting pH conditions resulted in four fragments which were present only under OTA permissive conditions. Two of the fragments could be cloned successfully and sequenced. Pairs of PCR primers were designed from the nucleotide sequences of the 550 bp fragment 3 and the 470 bp fragment 4. PCR primers designed from the longer fragment 3 showed high specificity to DNA isolated from various strains of *A. ochraceus*, whereas primers specific to fragment 4 revealed two products of different size with DNA of *A. ochraceus* (410 bp) and *A. carbonarius* (350 bp), respectively. Although Mühlencoert (2004) demonstrated by reverse transcription quantitative real-time PCR analysis that both primer pairs amplified a specific product only from cDNA present under OTA permissive conditions, neither fragment could be clearly assigned to a gene potentially involved in the biosynthesis of the toxin. The 3′-end of the sequence of fragment 3 shared limited similarity with some zinc finger proteins in *Mus musculus*. The only similarity to a fungal sequence was with the *Nectria haematococca* kinesin related protein 1 which is involved in chromosome movement during the mitotic cell cycle.

RAPD MARKER BASED PRIMERS FOR *A. CARBONARIUS* AND *A. NIGER*

Randomly amplified polymorphic DNA (RAPD) is a technique first proposed by Williams *et al.* (1990) which uses short decamer oligonucleotides under low stringency conditions to amplify fragments from the genomic DNA of the organisms under study. Usually, sets of primers containing all possible and sensible combinations of ten nucleotides are tested with the genomic DNA of an array of tester strains resulting in a pattern of 5-15 bands after ethidium bromide staining in agarose gels. If successful, banding patterns are polymorphic and can be used to distinguish groups of organisms. Polymorphic bands can be cut out from the gel, cloned and sequenced and the sequence information used to develop a set of specific PCR primers. Typically, the DNA sequence of the decamer random primer is elongated by another ten to twelve nucleotides according to the sequence obtained of the 5′- and 3′-end to design a specific primer pair for the fragment.

Fungaro *et al.,* (2004) described the development of primers with high specificity to *A. carbonarius*, one of the major OTA producers in coffee (Joosten *et al.*, 2001; Buecheli and Taniwaki, 2002; Taniwaki *et al.*, 2003; Magnani *et al.*, 2005). Fungaro *et al.,* (2004) subjected strains of *A. carbonarius*, *A. tubingensis*, and *A. niger* to RAPD analysis. Primer OPX7 revealed informative patterns which resolved the three taxa. A 809 bp fragment differentiated *A. carbonarius* from the other species. Primers designed from the sequence of fragment $OPX7_{809}$ resulted in amplification of a 809 bp PCR product exclusively from *A. carbonarius* DNA (Table 4.1). However, primers could not distinguish between toxigenic and non-toxigenic isolates of that species.

Based on the same methodology, Sartori *et al.,* (2006, in press) recently developed a pair of primers which leads to amplification of a 372 bp PCR product specifically with isolates of *A. niger* (Table 4.1). In a multiplex PCR application, the authors combined three primer pairs for the simultaneous detection of *A. ochraceus* (primers OCA-V/R, Schmidt *et al.*, 2003), *A. carbonarius* (primers $OPX7F_{809}$/$OPX7R_{809}$, Fungaro *et al.*, 2004), and *A. niger* (Sartori *et al.*, 2006, in press) in DNA mixtures as well as in artificially and naturally infected green coffee beans using a CTAB extraction protocol originally described for leaf tissue (Doyle and Doyle, 1987). Sartori *et al.,* (2006, in press) stated that their assay was the first to detect *A. niger* in a species specific assay. However, reviewing the literature revealed that three different pairs of primers have already been described for that purpose in earlier publications. The earliest was by Kanbe *et al.* (2002) who used the topoisomerase II gene as their sequence source for primer design. Sugita *et al.,* (2004) described successful use of a primer pair designed upon the ITS1 sequence of *A. niger* to amplify a PCR product of non-specified length in a species specific assay. The former two primer pairs were developed for medical applications, whereas Gonzalez-Salgado *et al.,* (2005) focused on detection of *A. niger* in vine berries. However, none of the earlier publications gave proof of applicability of the respective assays to detect the target fungus in contaminated materials.

PRIMERS TARGETED TO GENETICALLY DEFINED SEQUENCES

Nucleotide sequences of a great variety of fungal genes can be readily retrieved from sequence data bases, e.g., GenBank, RefSeq or PDB, opening the opportunity to develop and even test PCR primers *in silico* by searching for sequence similarities with the meanwhile 89 billion bases (as of June 2005) accessible with the Entrez Nucleotide database and the BLAST tool (http://www.ncbi.nlm.nih.gov/entrez/query.fcgi?db=Nucleotide). The hits produced by this *in silico* hybridization already gave a good estimate of the range of taxa covered by the primers under development. Of course, this kind of approach does not prevent from testing the newly developed primers for specificity with genomic DNA isolated from several strains of the taxon under study as well as from closely related taxa. Also testing of the primers with DNA isolated from the matrix should be exercised when the assay will subsequently be applied in contaminated sample materials.

In the following section of this chapter, several PCR assays are reviewed which use the nucleotide sequence of well defined fungal genes in order to set up primers for detection and identification of potential OTA producing fungi.

RIBOSOMAL RNA GENE BASED APPROACHES FOR THE PCR DIAGNOSIS OF OTA PRODUCING FUNGI

Genes coding for 18S, 5.8S, and 28S ribosomal RNA are divided by two intergenic transcribed spacer (ITS) regions, ITS1 and ITS2, which are highly valuable tools in the molecular taxonomic distinction of fungal species (Iwen *et al.*, 2002). Ribosomal RNA genes are multicopy genes, varying in copy number from 50 to 300 between species, however, there is also considerable variation between isolates in certain fungal species (Howlett *et al.*, 1997). Because of the higher copy numbers, assays based on rRNA genes tend to be more sensitive as compared to single copy genes. Presence of two distinct types of ITS2 sequences was described in isolates belonging to the *Gibberella fujikuroi* species complex (O´Donnell *et al.*, 1998). Similar observations were made for *Ascochyta* spp., where Fatehi and Bridge (1998) demonstrated the presence of multiple PCR fragments of identical length but with significant sequence differences. This might pose problems when analyzing results of a PCR in which one or both primers bind to that part of the rDNA. However, it is unknown to date if this is a general phenomenon or only restricted to the group where it was first described.

The first PCR-based assay for detection of *A. niger* used sequences from the 18S ribosomal RNA gene as target for primer binding (Jimenez *et al.*, 1999). The authors used primers originally published by Melchers *et al.*, (1994), which they applied to analyze presence of the fungus in artificially contaminated cosmetical and pharmaceutical raw materials and products after an enrichment step. It has to be noted that the primer pair used in the studies was not designed for specific detection of *A. niger* but rather is genus specific for *Aspergillus*. Primers were not included in Table 4.1 of the chapter review since their application to naturally contaminated samples will show a high degree of false positive results from the presence of fungi other than *A. niger*.

Specific variation within the ITS-region of the rRNA gene was recently used to set up species specific detection systems for species in *Aspergillus* section *Nigri*. In an attempt to develop a generic PCR assay for the recognition of the seven most frequently encountered opportunistic pathogens among *Aspergillus* spp., Sugita *et al.*, (2004) successfully used ITS sequences as the source for primer development. During the study, species specific assays were developed for detection of *A. fumigatus, A. flavus,* and *A. niger* as the most frequently occurring fungi in human pulmonary infections. In order to optimize the sensitivity of the assays developed, primers were used in a nested PCR with a pair of *Aspergillus* specific primers for the first round and respective species specific oligonucleotides as nested primers (Table 4.1). However, the authors did not clearly tend to detect the latter fungus because of its toxigenic potential but rather they were interested in its role as inducer of pulmonary infections in humans. Aimed at developing molecular biological tools to facilitate discrimination of species within the *Aspergillus* section *Nigri* ("black aspergilli"), González-Salgado *et al.* (2005) combined a generic forward primer hybridizing in the ITS1 region of the rRNA gene (ITS1, White *et al.*, 1990) with various reverse primers to design PCR assays detecting and identifying of *A. niger, A. japonicus, A. heteromorphus,* and *A. ellipticus,* respectively. Testing of the primers with a wide range of black aspergilli from different hosts and regions revealed specific reaction with strains of the respective species. However, the assay could not distinguish strains of *A. niger* from *A. tubingensis*, a feature which would be a great surplus because according to the authors only the former species produces OTA but is morphologically indiscernible from the latter. Digestion of the resulting PCR product with the restriction endonuclease *Rsa*I, however, revealed that the 420 bp *A. niger* fragment was restricted into two portions of 345 and 76 bp while the *A. tubingensis* product remained uncut.

Two ITS based assays specific for *A. carbonarius* and *A. ochraceus*, respectively, were developed by Patino *et al.*, (2005). Both species are considered the most important producers of OTA in coffee (Taniwaki *et al.*, 2003) and grapes (Cabenes *et al.*, 2002) in warmer climates. Primer pairs were designed to bind to the ITS1 and ITS2 regions flanking the 5.8S rRNA gene in fungal genomes. The authors selected those parts of the aligned sequences for primer design, which displayed the maximum number of bias between *A. carbonarius* and *A. ochraceus*. Although this strategy promises a high degree of specificity of the primers selected, it has the disadvantage of producing products of equal length in both PCR assays, which renders impossible their use in a multiplex assay for both species. However, multiplexing several primer pairs in one assay is highly recommended if PCR is going to be used as a screening tool in quality control because the number of tests run and the time needed for their performance must be kept as low as possible. The assays developed proofed to be highly specific when tested with DNA of a wide range of nontarget organisms. However, no test was run using DNA isolated from non-infected matrix material, e.g., coffee beans and grapes. The detection limit of 1-10 pg of purified DNA per assay demonstrated high sensitivity of the method developed.

CALMODULIN GENE TARGETED PRIMERS FOR DETECTION OF *A. CARBONARIUS* AND *A. JAPONICUS*

Calmodulin is a CALcium MODULated proteIN (Cheung, 1970; Kakiuchi and Yamazaki, 1970) abundantly present in the cytoplasm of all eukaryotic cells. Its structure has been highly conserved during evolution. It confers calcium sensitivity on a number of different enzymes including Ca^{2+}/calmodulin dependent protein kinases involved in the regulation of the mycotoxin aflatoxin in *A. parasiticus* (Jayashree *et al.*, 2000). Similar to ribosomal RNA genes, β-tubulin (*ben*A) or elongation factor 1α (EF-1α), the calmodulin (*cmd*A) gene provides highly conserved as well as variable sequence regions suitable for the development of species specific PCR primers. Positions of the variable regions within the gene are highly stable in eukaryotes. Based on the alignment of a partial sequence of representative strains of *A. carbonarius, A. japonicus*, and *A. aculeatus,* Perrone *et al.* (2004) detected a high degree of homology between strains assigned to *A. carbonarius* (99.98%) and *A. japonicus*/*A. aculeatus* (99.40%), respectively. This fact provides further evidence for identity of the latter two species. Apart from highly conserved regions the authors identified three variable regions suitable for design of specific PCR primers. Sequences of *cmd*A within the *A. niger* group isolates showed only 89% homology between isolates and no sequences were found which distinguished all *A. niger* group isolates from other OTA producers analyzed during the study. Primer pairs designed for detection of *A. carbonarius* (CARBO1/CARBO2, Table 4.1) and for *A. japonicus*/*A. aculeatus* were found to be highly specific for their respective target when tested with DNA from closely related *Aspergillus* spp. of sections *Nigri* and *Circumdati*. Detection of *A. carbonarius* DNA was demonstrated to be highly sensitive in an optimized PCR assay with a detection limit of 12.5 pg of DNA per reaction. The assay was demonstrated to be useful in screening vast numbers of isolates of black aspergilli from grapes in order to evaluate the toxigenic potential connected with this commodity. However, it was not used by the authors to detect the target organisms directly from sample materials which would be of great importance in quality control and HACCP applications.

PRIMERS FOR OTA BIOSYNTHETIC PATHWAY GENES

Following the biosynthetic pathway proposed by Huff and Hamilton (1979), presence of various enzymes catalyzing key reactions in the formation of OTA can be anticipated (Niessen *et al.*, 2005). A polyketide synthase can be postulated to link together the building block C2-bodies of the isocumarine portion of the toxin. Also some kind of cyclase will be necessary to build a closed ring system from a polyketide chain. An enzymatic activity can be anticipated to be present, which leads to chlorination of the isocumarine body to result in OTα, the actual precursor of OTA, and finally an enzyme with peptide synthase activity must be postulated to link together OTα and phenylalanine to result in OTA as the toxic end product. As with several other mycotoxins, e.g., trichothecenes, aflatoxins, fumonisins, it might further be anticipated that the genes coding for these enzymes as well as genes coding for possible regulators and transporters will form a cluster of genes in close physical vicinity within the fungal genome (for review see Keller and Hohn, 1997).

Based on the aforementioned assumptions, Geisen *et al.* (2004) applied a reverse genetical approach to detect and characterize a polyketide synthase gene from OTA producing *Penicillium* spp. They used generic primers LC3 and LC5 to amplify a 750 bp portion of a fungal polyketide synthase of the MSAS type (Bingle *et al.*, 1999) from *P. nordicum* BFE487. The primers gave rise to a 750 bp fragment with DNA isolated from *P. nordicum* strains but not with the closely related *P. verrucosum*. The fragment was sequenced (GenBank accession number AY196315) and compared to published sequences. It displayed homology to the polyketide synthase gene *pksL2* from *Aspergillus flavus* (Feng and Leonhard, 1998). After transferring the nucleotide sequence (the fragment contained one continuous open reading frame) to an amino acid sequence, an even more stringent homology with different fungal polyketide synthases over the whole fragment length was observed. A nested specific primer pair (otapks 1 and otapks 2, Table 4.1) was deduced from the identified nucleotide sequence and tested for specificity with DNA isolated from several food borne fungi as well as closely related *P. verrucosum* strains. Primers yielded a 500 bp PCR fragment exclusively with DNA from *P. nordicum* strains, whereas all other analyzed food related fungi yielded negative results. TaqMan® quantitative real-time PCR with a specific primer pair and a fluorogenic probe (Table 4.1) was applied to monitor the growth kinetics of *P. nordicum* BFE487 in artificially inoculated wheat grains (Geisen *et al.*, 2004). Results of that experiment showed congruence between growth kinetics established by counts of colony forming units (cfu) and quantitative real-time PCR (qRT-PCR). The same primers were applied by the authors to quantify the level of expression of the *otapks*PN gene during growth of *P. nordicum* BFE487 in artificially inoculated wheat (Geisen *et al.*, 2004). Data were compared to concentrations of OTA found in the respective samples. It was demonstrated that OTA concentrations became detectable at day four of the experiment, when expression of the *otapks*PN gene was first detected by qPCR with cDNA. Levels of gene expression declined after day eight, when OTA concentrations were found to remain at a constantly high level. Using the qRT-PCR system described above, Geisen (2004) demonstrated that environmental parameters prevailing under food production conditions had a substantial influence on the expression of the *otapks*PN gene in *P. nordicum* and concluded that expression analysis of OTA biosynthetic key genes might be useful as critical control point in HACCP concepts for the food industry.

In order to isolate a genomic clone containing the complete *otapks*PN gene and additional adjacent genomic regions, a lambda phage genomic DNA gene bank was constructed (Karolewiez and Geisen, 2005). It was anticipated that the genes of the ochratoxin A biosynthetic pathway displaying similarly clustered organization as can be found in other secondary metabolite biosynthetic pathway genes, e.g., aflatoxins (Yu *et al.*, 2004). The 500 bp otapks 1/2 PCR product was digoxigenin labeled and used as a probe to screen the library. Five different phage clones were found to yield a hybridization signal with the probe, indicating that the *otapks*PN gene or at least a part of it was present in the fungal insert DNA. Excision of the inserted fragment from the multiple cloning site revealed a 10 kbp genomic DNA fragment of *P. nordicum* BFE487, which hybridized with the *otapks*PN gene probe. The fragment was completely sequenced. Karolewiez and Geisen (2005) detected four genes located on the fragment: *asph*PN (alkaline serine protease homologue), *asp*PN (alkaline serine protease), *nps*PN (non-ribosomal peptide synthase) and *otapks*PN (ochratoxin polyketide synthase). Two of these genes had homologies to genes which are to be anticipated to be present in the ochratoxin biosynthetic pathway as described above. Involvement of the *otapks*PN gene in the biosynthesis of OTA was demonstrated by knocking the gene out. The resulting transformant was unable to produce the toxin under conditions under which the parent wild type produced the metabolite.

PCR primers were designed from each of the three open reading frames present on the 10 kbp genomic fragment. Primers designed from the sequence of a putative *nps*PN gene (GeneBank accession number AY534879, coding for a non ribosomal peptide synthase) yielded a 750 bp PCR product from strains of both *P. nordicum* and *P. verrucosum*, whereas the primer pair derived from the *otapks*PN gene (GeneBank accession number AY199315) yielded a 500 bp product exclusively with DNA isolated from strains of *P. nordicum* (Bogs *et al.*, 2006). The latter two primer pairs were useful in differentiating *P. verrucosum* from *P. nordicum* by analyzing isolates with both PCR assays. Specificity of the primers designed was fully verified when target DNA isolated from various *Penicillium* spp. and other fungal contaminants typically occurring in food and feed related sources were used as template in a PCR with *otapks*PN and the *otanps*PN specific primer pairs. When OTA producing capabilities of *P. nordicum* and *P. verrucosum* strains isolated from cured meat and from meat production environments were compared to the results obtained with the *otapks*PN and *otanps*PN specific primers, the latter PCR displayed positive reaction exclusively with strains capable of producing detectable amounts of OTA in pure culture (HPLC analysis).

It must be noted here that none of the sequences described for OTA biosynthetic pathway genes in *Penicillium nordicum* have homologies with genes found in OTA producing *Aspergillus* spp. In particular, no homologies were found between sequences of the *otapks*PN gene and a polyketide synthase potentially involved in OTA biosynthesis in *A. ochraceus* as demonstrated by knockout experiments (O'Callaghan *et al.*, 2003). Also, neither of the gene specific primers described above for *P. nordicum* showed any cross reactivity when tested with *A. ochraceus* or other OTA producing *Aspergillus* spp., even under low stringency conditions (R. Geisen, personal communication). These results indicate that OTA biosynthesis in *A. ochraceus*, but also in other OTA producing *Aspergillus* spp., might follow a different route as compared to *Penicillium*.

During their studies on OTA biosynthetic pathway genes, O´Callaghan *et al.,* (2003) did not only clone a polyketide synthase from *A. ochraceus* but cloned and sequenced several more differentially expressed cDNA fragments of yet unassigned origin from typical producers of OTA. Based on that work, University College Cork (Ireland) filed world wide (WO 2004/072224 A2) as well as European (EP 1 592 705 A2) patent applications based on an Irish priority application dated 12.03.2003 (IE 20030095). Claims cover the use of those sequences for the purpose of detecting OTA producers in pure cultures and sample materials as well as the use of those sequences for primer walking to uncover sequence beyond those already cloned.

Finally, Dao *et al.,* (2005) used a strategy similar to that described by Geisen (2004) to sequence the ketosynthase domain of a polyketide synthase gene in *A. ochraceus* NRRL 3174. According to the authors, part of their sequence (not accessible in GenBank at the time of writing this review) was identical with the sequence published by O'Callaghan *et al.* (2003) (accession no. AY272043). From the sequences obtained during their study, Dao *et al.* (2005) designed a PCR assay which specifically detected *A. ochraceus* (primers AoOTA-L/AoOTA-R, Table 4.1) and another one for detection of fungi potentially able to produce OTA and the mycotoxin citrinin in a group specific manner (primers AoLC35-12L/AoLC35-12R, Table 4.1). The spectrum of species detected by the latter assay comprised *A. carbonarius, A. melleus, A. ochraceus, P. citrinum* and *P. verrucosum*, but also *Monascus ruber* as a typical producer of citrinin in food and feed, among other *Monascus* spp. (Wang *et al.*, 2005). Detecting producers of both toxins may be an advantage because they can be found to be co-occurring in contaminated cereals (Vrabcheva *et al.*, 2000). Primer pairs designed from the sequences of two unassigned DNA fragments cloned from *A. ochraceus* (presumably strain NRRL 3174 or M333) were patented by Evialis, a French based company specialized in animal nutrition products. Patent application EP 1 329 521 A1 from 20.01.2003 was based on a French priority application (FR 0200682) of 21.01.2002 (Librihi *et al.*, 2003). The claim covers the use of unassigned DNA sequences cloned from *A. ochraceus* and from *P. viridicatum* for detection of fungi producing OTA or citrinin. Nucleotide-nucleotide BLAST search against the EMBL nucleotide database revealed weak similarities of the sequences with fungal polyketide synthases. Specificity of the primers pairs S3/S4 and S5/S6 were identical with the primers described by Dao *et al.* (2005), with the former primers being specific for *A. ochraceus* and the latter group specific for producers of OTA and citrinin. However, primers described in the two publications are not identical nor were the primers of Dao *et al.,* (2005) based on the patented sequences, although the senior author of that publication is listed as inventor in EP 1 329 521 A1 (Librihi, 2003).

FUTURE PERSPECTIVES

Systems for the PCR-based detection and identification of nearly all of the fungal species known to produce OTA have been set up over the last few years. According to Frisvad and Samson (2000), *A. muricatus* and *A. alliaceus* are producers of OTA within section Circumdati of *Aspergillus*, which also contains *A. ochraceus* as one of the most predominant producers of the toxin in tropical and subtropical regions. No specific PCR assays are available yet for their detection in commodities used in food and feed production. However, such systems could be useful in order to establish information about the epidemiology and ecology of these species. Also in section *Nigri* of *Aspergillus* several new species have recently been described, among which *A. lacticoffeatus* and *A. sclerotiniger* were found to be producers of OTA in pure culture (Samson *et al.*, 2004). Both new species were isolated from coffee beans and might therefore be of relevance to OTA contamination of that commodity. As with the new species in section *Circumdati*, specific detection by PCR would help to study the epidemiology of those species. Very recently it was found that about 25% of grape isolates assigned to *A. tubingensis* were able to produce OTA in pure culture (Perrone *et al.*, 2006). Since the species is morphologically indistinguishable from *A. niger*, a PCR based system which differentiates both species would be of great advantage to study the distribution of both species in the environment. Future developments might bring up more PCR based diagnostics for the species assigned above.

Review of the current literature dealing with molecular biological methods for detection and identification of fungi with a potential to produce the mycotoxin OTA revealed a number of assays developed for that purpose within the last three to four years. In all cases authors were able to show high specificity of the respective primers developed, irrespective of the kind of sequence used to design them. However, with only a few exceptions the assays were only demonstrated to work with target DNA isolated from pure fungal cultures. Most authors did not attempt to detect their target organisms in contaminated sample materials, which is an essential if assays are to be used in practical applications. It is, therefore, necessary to further develop the PCR-based assays described in this chapter towards complete quality control and hazard analysis applications, including extensive testing for specificity, optimization of sensitivity in sample materials, rapidness and ease of use for trained and untrained personnel. Also sampling, sample preparation and DNA extraction must further be optimized and tested with high numbers of samples. Assays applied should allow for quantitative estimation of the contamination at hand in order to rate the hazard associated with a sample under study. In this respect, work done by R. Geisen and his group at Federal Research Centre for Nutrition and Food (Karlsruhe, Germany) for *P. nordicum* as well as by the groups lead by A. Dobson (University College Cork, Ireland) and by A. Lebrihi (Ecole Nationale Supèrieure Agrognomique de Toulouse (ENSAT), France) on detection of OTA biosynthetic pathway genes may lead the way here because they are directly connected to the hazard to be analyzed. To date, although still incomplete, the most profound knowledge about genetics of the OTA biosynthetic pathway is available for *P. nordicum*, the principle producer of that toxin within the genus *Penicillium*. Usefulness of the *otapks*PN gene as a critical control point in the evaluation of environmental conditions during food processing steps has recently been demonstrated (Geisen, 2004) and may be applied as a model for further developments. These might include microarray DNA chips comprising genes involved in biosynthesis and regulation of OTA. With the capacities available in microarray technology, probes for OTA biosynthetic genes might be combined with probes for genes involved in the production of other relevant mycotoxins, e.g., aflatoxins, trichothecenes, fumonisins, aimed at optimizing food production processes for minimized risk of mycotoxin production. In analogy to computer aided manufacturing (CAM) this would be addressed as genome aided processing (GAP) in food production. Still, enormous work has to be done in order to accomplish that state. However, keeping in mind that most of the developments described in this chapter were done in very recent years, GAP based systems may become reality within short turn and will be applied for the welfare of consumers.

CONCLUSIONS

Detection of ochratoxin A producing moulds is routinely accomplished by plating serial dilutions of ground commodity or by direct plating of samples on plates of suitable medium, e.g., DRBC (King *et al.*, 1979) or DG18 (Hocking and Pitt, 1980). Preparation of samples is cumbersome and time consuming and specialization is needed for species identification. Although PCR-based assays may not entirely replace the classical methods, they display many advantages which make them suitable for quality control, e.g., rapidness of analysis, high specificity and sensitivity, applicability to detect life and dead biomass, no expertise in fungal taxonomy necessary, applicable to genomic and expressed targets. The most advanced systems are based on genes involved in the biosynthesis of mycotoxins, which enable a direct correlation between PCR signal and mycotoxins produced. Usefulness of such assays was demonstrated for aflatoxin producers (Shapira *et al.*, 1996; Geisen, 1996), trichothecene producers (Niessen and Vogel, 1998; Edwards *et al.*, 2001; Lee *et al.*, 2001; Bluhm *et al.*, 2002), fumonisin producers (Bluhm *et al.*, 2002; González-Jaén *et al.*, 2004), producers of ergot alkaloids (Boichenko *et al.*, 2001) and patulin (Paterson *et al.*, 2000). More systems may follow in that line in the near future.

Assays which detect potential producers of OTA have been shown to work well with pure fungal cultures. However, applicability and usefulness in the routine analysis of commodities remains to be demonstrated for most of them. Setting up rapid and robust assay systems in a user-friendly format which can be applied in routine analysis with a wide acceptance in science and industry must be developed as a follow up of the systems available to date.

REFERENCES

Abarca M.L., Accensi F., Bragulat M.R., Castellá G., and Cabañes F.J. (2003). *Aspergillus carbonarius* as the main source of ochratoxin A contamination in dried vine fruits from the Spanish market. *Journal of Food Protection* 66: 504-506.

Adye J. and Mateles R.I. (1964). Incorporation of labelled compounds in aflatoxins. *Biochimica et Biophysica Acta* 86: 418-420.

Alexandrakis G., Sears M., and Gloor P. (1998). Postmortem diagnosis of *Fusarium* panophthalmitis by the polymerase chain reaction. *American Journal of Ophthalmology* 121(2): 221-223.

Alvarez L., Gil A.G., Ezpeleta O., Garcia-Jalon J.A., and Lopez de Cerain A. (2004). Immunotoxic effects of Ochratoxin A in wistar rats after oral administration. *Food and Chemical Toxicology* 42: 825-834.

Alvarez L., Leache C., Gonzalez-Penas E., and Lopez-de-Cerain A. (2005). Alterations induced *in vitro* by ochratoxin A in rat lymphoid cells. *Human and Experimental Toxicology* 24: 459-466.

Amezqueta S., Gonzalez-Penas E., Murillo M., and Lopez de Cerain A. (2005). Occurrence of ochratoxin A in cocoa beans:effect of shelling, *Food Additives and Contaminants* 22: 590-596.

Aoki T. and O´Donnell K. (1999). Morphological and molecular characterization of *Fusarium pseudograminearum* sp. nov., formerly recognized as the group 1 population of *F. graminearum. Mycologia* 91: 597-609.

Assaf H., Azouri H., and Pallardy M. (2004). Ochratoxin A induces apoptosis in human lymphocytes through down regulation of Bcl-x_L. *Toxicological Sciences* 79: 335-344.

Bachem C.W.B, van der Hoeven R.S., de Bruijn S.M., Vreugdenhil D., Zabeau M., and Visser R.G.F. (1996). Visualization of differential gene expression using a novel method of RNA fingerprinting based on AFLP: Analysis of gene expression during potato tuber development. *Plant Journal* 9(5): 745-753.

Bachem C.W.B., Oomen R.J.F.J., and Visser R.G.F (1998). Transcript imaging with cDNA-AFLP: a step-by-step protocol. *Plant Molecular Biology Reporter* 16: 157-173.

Battilani P., Pietri A., Bertuzzi T., Languisco L., Giorni P., and Kozakiewicz Z. (2003). Occurrence of ochratoxin A-producing fungi in grapes grown in Italy. *Journal of Food Protection* 66: 633-636.

Bennett J.W. and Klich M. (2003). Mycotoxins. *Clinical Microbiology Reviews* 16(3): 497-516.

Bensch S. and Akesson M. (2005). Ten years of AFLP in ecology and evolution: why so few animals? *Molecular Ecology* 14: 2899-2914.

Bingle L.E.H., Simpson T.J., andLazarus C.M. (1999). Ketosynthase domain probes identify two subclasses of fungal polyketide synthase genes. *Fungal Genetics and Biology* 26: 209-223.

Bluhm B.H., Flaherty J.E., Cousin M.A., and Woloshuk C.P. (2002). Multiplex polymerase chain reaction assay for the differential detection of trichothecene- and fumonisin-producing species of *Fusarium* in cornmeal. *Journal of Food Protection* 65: 1955-1961.

Bogs C., Battilani P., and Geisen R. (2006). Development of a molecular detection and differentiation system for ochratoxin A producing *Penicillium* species and its application to analyse the occurrence of *P. nordicum* in fermented meats. *International Journal of Food Microbiology* 107: 39-47.

Boichenko L.V., Boichenko D.M., Vinokurova N.G., Reshetilova T.A., and Arinbasarov M.U. (2001). Use of polymerase chain reaction for searching for producers of ergot alkaloids from among microscopic fungi. *Mikrobiologiia* 70: 360-364.

Brera C., Soriano J.M., Debegnach F., and Miraglia M. (2005). Exposure assessment to ochratoxin A from the consumption of Italian and Hungarian wines. *Microchemical Journal* 79(1-2): 109-113.

Bresch H., Urbanek M. and Hell K. (2000). Ochratoxin A in coffee, tea and beer. *Archiv fuer Lebensmittelhygiene* 51(4-5): 89-94.

Buecheli P. and Taniwaki M.H. (2002). Research on the origin, and on the impact of post-harvest handling and manufacturing on the presence of ochratoxin A in coffee. *Food Additives and Contaminants* 19: 655-665.

Cabenes F.J., Accensi F., Bragulat M.L., Abarca M.L., Castella G., Minguez S., and Pons A. (2002). What is the source of ochratoxin A in wine? *International Journal of Food Microbiology* 79: 213-215.

Castella G., Larsen T.O., Cabañes F.J., Schmidt H., Alboresi A., Niessen L., Färber P., and Geisen R. (2002). Molecular characterization of ochratoxin A producing strains of the genus *Penicillium. Systematic and Applied Microbiology* 25: 74-83.

Chandler E.A., Simpson D.R., Thomsett M.A., and Nicholson P. (2003). Development of a PCR assay to Tri7 and Tri13 trichothecene biosynthetic genes, and characterisation of chemotypes of *Fusarium graminearum, Fusarium culmorum* and *Fusarium cerealis. Physiological and Molecular Plant Pathology* 62: 355-367.

Chen R.S., Tsay J.G., Huang Y.F., and Chiou R.Y.Y. (2002). Polymerase chain reaction-mediated characterization of molds belonging to the *Aspergillus flavus* group and detection of *Aspergillus parasiticus* in peanut kernels by a multiplex polymerase chain reaction. *Journal of Food Protection* 65(5): 840-844.

Cheung W.Y. (1970). Cyclic 3',5'-nucleotide phosphodiesterase : Demonstration of an activator. *Biochemical and Biophysical Research Communications* 38: 533-538.

Dao H.P., Mathieu, F., and Lebrihi A. (2005). Two primer pairs to detect OTA producers by PCR method. *International Journal of Food Microbiology* 104: 61-67.

Desjardins A.E., Munkvold G.P., Plattner R.D., and Proctor R.H. (2002). FUM1 - a gene required for fumonisin biosynthesis but not for maize ear rot and ear infection by *Gibberella moniliformis* in field tests. *Molecular Plant-Microbe Interactions* 15: 1157-1164.

Delacruz L. and Bach P.H. (1990). The role of ochratoxin A metabolism and biochemistry in animal and human nephropathy. *Journal of Biopharmaceutical Science* 1: 277-304.

Delage N., d'Harlingue A., Ceccaldi B.C., and Bompeix G. (2003). Occurrence of mycotoxins in fruit juices and wine. *Food Control* 14: 225-227.

Dirheimer G. and Crepy E.E. (1991). Mechanism of action of ochratoxin A. *IARC Science Publications* 115: 171-186.

Domijan A.M., Peraica M., Zlender V., Cvjetkovic B., JurjevicZ., Topolovec-Pintaric S., and Ivic D. (2005). Seed-borne fungi and ochratoxin A contamination of dry beans (*Phaseolus vulgaris* L.) in the Republic of Croatia. *Food and Chemical Toxicology* 433: 427-432.

Doyle J.J. and Doyle J.L. (1987). A rapid DNA isolation procedure for small quantities of fresh leaf tissue. *Phytochemical Bulletin* 19: 11-15.

Drusch S. and Aumann J. (2005). Mycotoxins in fruits: microbiology, occurrence, and changes during fruit processing. *Advances in Food and Nutrition Research* 50: 33-78.

Dutton M.F. (2004). Fumonisin B1 in animal and human health, In: Kushwaha R.K.S. (ed.) *Fungi in Human and Animal Health*, Scientific Publishers, Jodhpur, Rajasthan (India). 317-353.

Eaton D.L. and Gallagher E.P. (1994). Mechanism of aflatoxin carcinogenesis. *Annual Review of Pharmacology and Toxicology* 34: 135-172.

Edwards S.G., Pirgozliev S.R., Hare M.C., and Jenkinson P. (2001). Quantification of trichothecene-producing *Fusarium* species in harvested grain by competitive PCR to determine efficacies of fungicides against *Fusarium* head blight of winter wheat. *Applied and Environmental Microbiology* 67: 1575-1580.

Engel G. (2000). Ochratoxin A in sweets, oil seeds and dairy products. *Archiv fuer Lebensmittelhygiene* 51: 98-101.

Färber P., Geisen R., and Holzapfel W.H. (1997). Detection of aflatoxigenic fungi in figs by a PCR reaction. *International Journal of Food Microbiology* 36: 215-220.

Färber P. and Geisen R. (2004). Analysis of differentially-expressed ochratoxin A biosynthesis genes in *Penicillium nordicum. European Journal of Plant Pathology* 110: 661-669.

Fatehi J. and Bridge P. (1998). Detection of multiple rRNA-ITS regions in isolates of *Ascochyta. Mycological Research* 102: 762-766.

Feng G. H. and Leonard T.J. (1998). Culture conditions control expression of the genes for aflatoxin and sterigmatocystin biosynthesis in *Aspergillus parasiticus* and *A. nidulans. Applied and Environmental Microbiology* 64: 2275-2277.

Frisvad J.C. and Samson R.A. (2000). *Neopetromyces* gen. Nov. and an overview of teleomorphs of *Aspergillus* subgenus *Circumdati. Studies in Mycology* 45: 201-207.

Fungaro M.H.P., Vissoto P.C., Sartori D., Vilas-Boas L.A., Furlaneto M.C., and Taniwaki M.H. (2004). A molecular method for detection of *Aspergillus carbonarius* in coffee beans. *Current Microbiology* 49: 123-127.

Geisen R. (1996). A multiplex PCR reaction for the detection of aflatoxin and sterigmatocystin producing fungi. *Systematic and Applied Microbiology* 19: 388-392.

Geisen R., Mulfinger S., andNiessen L. (1998). Detection of *Aspergillus flavus* in wheat by PCR. *Journal of Food Mycology* 1(4): 211-218.

Geisen R., Mayer Z., Karolewiez A., and Färber P. (2004). Development of a real-time PCR system for detection of *Penicillium nordicum* and for monitoring ochratoxin A production in foods by targeting the ochratoxin polyketide synthase gene. *Systematic and Applied Microbiology* 27: 501-507.

Geisen R. (2004). Molecular monitoring of environmental conditions influencing the induction of ochratoxin A biosynthesis genes in *Penicillium nordicum. Molecular Nutrition and Food Research* 48: 532-540.

Gekle M., Sauvant C., and Schwerdt G. (2005).Ochratoxin A at nanomolar concentrations: A signal modulator in renal cells. *Molecular Nutrition and Food Research* 49: 118-130.

González-Jaén M.T., Mirete S., Patino B., López-Errasquín E., and Vázquez C. (2004). Genetic markers for the analysis of variability and for production of specific diagnostic sequences in fumonisin-producing strains of *Fusarium verticillioides. European Journal of Plant Pathology* 110: 525-532.

González-Salgado A., Patino B., Vázquez C., and González-Jaén M.T. (2005). Discrimination of *Aspergillus niger* and other *Aspergillus* species belonging to section *Nigri* by PCR assays. *FEMS Microbiology Letters* 245: 353-361.

Grimm C. and Geisen R. (1998). A PCR-ELISA for the detection of potential fumonisin producing *Fusarium* species. *Letters in Applied Microbiology* 26(6): 456-462.

Hayette M.P., Vaira D., Susin F., Boland P., Christiaens G., Melin P., and de Mol P. (2001). Detection of *Aspergillus* species DNA by PCR in broncheoalveolar lavage fluid. *Journal of Clinical Microbiology* 39(6): 2338-2340.

Hennel M., Beinling B., Eppert D., Meyer D., and Schoning G. (2005). Ochratoxin A in sultanas. *Mycotoxin Research* 21(1): 43-45.

Hocking A.D. and Pitt J.I. (1980). Dichloran-glycerol medium for enumeration of xerophilic fungi from low moisture foods. *Applied and Environmental Microbiology* 39: 488-492.

Howlett B.J., Rolls B.D., and Cozijnsen A.J. (1997). Organisation of ribosomal DNA in the ascomycete *Leptosphaeria maculans. Microbiological Research* 152(3): 261-267.

Huff J.E. (1991). Carcinogenicity of ochratoxin A in experimental animals. In: *Mycotoxins, Endemic Nephropathy and Urinary Tract Tumors*, Castegnaro M., Plestina R., Dirheimer G., Chernozemsky N., Bartsch H.. (eds)., IARC, Lyon, 229-244.

Huff W.E. and Hamilton P.B. (1979). Mycotoxins - their biosynthesis in fungi: Ochratoxins - metabolites of combined pathways. *Journal of Food Protection* 42: 815-820.

IARC (International Agency for Research on Cancer) (1993). *Some naturally occurring substances: food items and constituents, heterocyclic aromatic amines, and Mycotoxins*, IARC Monograph on the Evaluation of Carcinogenic Risk of Chemical to Humans, vol. 56, International Agency for Research on Cancer Lyon, France, 571 pp.

Imanaka B.T., Taniwaki M.H., Menezes H.C., Vicente E. and Fungaro M.H.P. (2005). Incidence of toxigenic fungi and ochratoxin A in dried fruits sold in Brazil. *Food Additives and Contaminants* 22: 1258-1263.

Iwen P.C., Hinrichs S.H.and Rupp M.E. (2002). Utilization of the internal transcribed spacer regions as molecular targets to detect and identify human fungal pathogens. *Medical Mycology* 40: 87-109.

Jayashree T., Praveen-Rao J., andSubramanyam C. (2000). Regulation of aflatoxin production by Ca2+/calmodulin-dependent protein phosphorylation and dephosphorylation. *FEMS Microbiology Letters* 83(2): 215-219.

Jennings P., Coates M.E., Turner J.A., Chandler E.A., and Nicholson P. (2004). Determination of deoxynivalenol and nivalenol chemotypes of *Fusarium culmorum* isolates from England and Wales by PCR assay. *Plant Pathology* 53: 182-190.

Jimenez L., Bosko Y., Smalls S., Ignar R., and English D. (1999). Molecular detection and identification of *Aspergillus niger* contamination in cosmetic/pharmaceutical raw materials and finished products. *Journal of Rapid Methods and Automation in Microbiology* 7(1): 39-46.

Joosten H.M.L.J., Goetz J., Pittet A., Schellenberg M., and Bucheli P. (2001). Production of ochratoxin A by *Aspergillus* carbonarius on coffee cherries. *International Journal of Food Microbiology* 65(1/2): 39-44.

Jorgensen K. and Petersen A. (2002). Content of ochratoxin A in paired kidney and meat samples from healthy Danish slaughter pigs. *Food Additives and Contaminants* 19(6): 562-567.

Kakiuchi S. and Yamazaki R. (1970). Calcium dependent phosphodiesterase activity and its activating factor (PAF) from brain: Studies on cyclic 3',5'-nucleotide phosphodiesterase (III). *Biochemical and Biophysical Research Communications* 41(5): 1104-1110.

Kamp H.G., Eisenbrand G., Schlatter J., Wurth K., and Janzowski C. (2005). Ochratoxin A: induction of (oxidative) DNA damage, cytotoxicity and apoptosis in mammalian cell lines and primary cells. *Toxicology* 206: 413-425.

Kanbe T., Yamaki K., and Kikuchi A. (2002). Identification of the pathogenic *Aspergillus* species by nested PCR using a mixture of specific primers to DNA topoisomerase II. *Microbiology and Immunology* 46(12): 841-848.

Karolewiez A.and Geisen R. (2005). Cloning a part of the ochratoxin A biosynthetic gene cluster of *Penicillium nordicum* and characterization of the ochratoxin polyketide synthase gene. *Systematic and Applied Microbiology* 28: 588-595.

Keller N. and Hohn T.M. (1997). Metabolic pathway gene clusters in filamentous fungi. *Fungal Genetics and Biology* 21: 17-29.

King A.D., Hocking A.D., and Pitt J.I .(1979). Dichloran-rose bengal medium for enumeration of molds from foods. *Applied and Environmental Microbiology* 37: 959-964.

Kuiper-Goodman T., and Scott P.M. (1989). Risk assessment of the mycotoxin ochratoxin A. *Biomedical and Environmental Sciences* 2: 179-243.

Larsen T.O., Svendsen A. and Smedsgaard J. (2001). Biochemical characterization of ochratoxin A-producing strains of the genus *Penicillium. Applied and Environmental Microbiology* 67: 3630-3635.

Lee T., Oh D.W., Kim H.S., Lee J., Kim Y.H., Yun S.H,.and Lee Y.W. (2001). Identification of deoxynivalenol- and nivalenol-producing chemotypes of *Gibberella zeae* by using PCR. *Applied and Environmental Microbiology* 67: 2966-2972.

Li H.P., Wu A.B., Zhao C.S., Scholten O., Löffler H., and Liao Y.C. (2005). Development of a generic PCR detection of deoxynivalenol- and nivalenol-chemotypes of *Fusarium graminearum. FEMS Microbiology Letters* 243: 505-511.

Librihi, A. (2003). Procédé de détection de champignons producteurs d´ochratoxine A ou de citrinine. EP 1 329 521 A1, 23.07.2003, 35 pp.

Lund F. and Frisvad J.C. (2003). *Penicillium verrucosum* in wheat and barley indicates presence of ochratoxin A. *Journal of Applied Microbiology* 95: 1117-1123.

Magan N., Hope R., Cairns V., and Aldred D. (2003). Post-harvest fungal ecology: Impact of fungal growth and mycotoxin accumulation in stored grain. *European Journal of Plant Pathology* 109: 723-730.

Magnani M., Fernandes T., Prete C.E.C., Homechim M., Ono E.Y.S., Vilas-Boas L.A., Sartori D., Furlaneto M.C., and Fungaro M.H.P. (2005). Molecular identification of *Aspergillus* spp. isolated from coffee beans. *Scientia Agricola* 62(1): 45-49.

Majer D., Mithen R., Lewis B.G., Vos P., and Oliver R.P. (1996). The use of AFLP fingerprinting for the detection of genetic variation in fungi. *Mycological Research* 100(9): 1107-1111.

Martins M.L., Martins H.M., and Gimeno A. (2003). Incidence of microflora and of ochratoxin A in green coffee beans (*Coffea arabica*). *Food Additives and Contaminants* 20: 1127-1131.

Mayer Z., Bagnara A., Faerber P., and Geisen R. (2003). Quantification of the copy number of nor-1, a gene of the aflatoxin biosynthetic pathway by real-time PCR, and its correlation to the cfu of *Aspergillus flavus* in foods. *International Journal of Food Microbiology* 82: 143-151.

Medina A., Mateo R., Lopez-Ocana L., Valle-Algarra F.M., and Jimenez M. (2005). Study of Spanish grape mycobiota and ochratoxin A production by isolates of *Aspergillus tubingensis* and other members of *Aspergillus* section *Nigri. Applied and Environmental Microbiology* 71(8): 4696-4702.

Melchers W.J.G., Verweil P.E., van den Hurk P., van Belkum A., de Pauw B.E., Hoogkamp-Korstanje J.A.A. and Meis J.F.G.M. (1994). General primer-mediated PCR for detection of *Aspergillus* species. *Journal of Clinical Microbiology* 32(7): 1710-1717.

Monaci L., Tantillo G. and Palmisano F. (2004). Determination of ochratoxin A in pig tissues by liquid-liquid extraction and clean-up and high-performance liquid chromatography. *Analytical and Bioanalytical Chemistry* 378(7): 1777-1782.

Moss M.O. and Thrane U. (2004). *Fusarium* taxonomy with relation to trichothecene formation. *Toxicology Letters* 153: 23-28.

Mühlencoert E. (2004). Ochratoxin A production by *Aspergillus ochraceus*, thesis, Technische Universität München, Munic, Germany, 124 pp. (http://mediatum.ub.tum.de/mediatum/servlets/MCRFileNodeServlet/mediaTUM_derivate_000000000001191/mediaTUM_derivate_000000000001191.pdf).

Mühlencoert E., Mayer I., Zapf M.W., Vogel R.F., and Niessen L. (2004). Production of ochratoxin A by *Aspergillus ochraceus. European Journal of Plant Pathology* 110: 651-659.

Nicholson P., Simpson D.R., Wilson,A.H. Chandler, E., and Thomsett, M. (2004). Detection and differentiation of trichothecene and enniatin-producing *Fusarium* species on small-grain cereals. *European Journal of Plant Pathology* 110: 503-514.

Niessen M.L. and Vogel R.F. (1997). Specific identification of *Fusarium graminearum* by PCR with *gaoA* targeted primers. *Systematic and Applied Microbiology* 20(1): 111-123.

Niessen L. and Vogel R.F. (1998). Group specific PCR-detection of potential trichothecene-producing *Fusarium* species in pure cultures and cereal samples. *Systematic and Applied Microbiology* 21: 618-631.

Niessen L., Schmidt H., Mühlencoert E., Färber P., Karolewiez A., and Geisen R. (2005). Advances in the molecular diagnosis of ochratoxin A-producing fungi. *Food Additives and Contaminants* 22: 324-334.

Northolt M.D., van Egmont H.P., and Paulsch W.E. (1979). Ochratoxin A production by some fungal species in relation to water activity and temperature. *Journal of Food Protection* 42: 485-490.

O'Callaghan J.O., Caddick M.X. and Dobson A.D.W. (2003). A polyketide synthase gene required for ochratoxin A biosynthesis in *Aspergillus ochraceus*. *Microbiology* 149: 3485-3491.

O'Donnell K., Cigelnik Eand Nirenberg H.I. (1998). Molecular systematics and phylogeography of the *Gibberella fujikuroi* species complex. *Mycologia* 90(3): 465-493.

Otteneder H. and Majerus P. (2000). Occurrence of ochratoxin A (OTA) in wines: influence of the type of wine and its geographical origin *Food Additives and Contaminants* 17(9): 793-798.

Pardo E., Marín S., Ramos A.J. and Sanchis V. (2004). Occurrence of ochratoxigenic fungi and ochratoxin A in green coffee from different origins. *Food Science and Technology* 10: 45-50.

Pardo E., Malet M., Marin S., Sanchis V. and Ramos A.J. (2006). Effects of water activity and temperature on germination and growth profiles of ochratoxigenic *Penicillium verrucosum* isolates on barley meal extract agar. *International Journal of Food Microbiology* 106: 25-31.

Paterson R.R.M., Archer S., Kozakiewicz Z., Lea A., Locke T. and O´Grady E. (2000). A gene probe for the patulin metabolic pathway with potential for use in patulin and novel disease control. *Biocontrol Science and Technology* 10: 509-512.

Patino B., González-Salgado A., González-Jaén M.T., and Vázquez C. (2005). PCR detection assays for the ochratoxin-producing *Aspergillus carbonarius* and *Aspergillus ochraceus* species. *International Journal of Food Microbiology* 104: 207-214.

Peltola S., Niessen L., Nielsen K.F., Jarvis B.B., Andersen B., Salkinoja-Salonen M., and Möller, E.M. (2002). Toxigenic diversity of two different RAPD groups of *Stachybotrys chartarum* isolates analyzed by potential for trichothecene production and for boar sperm cell motility inhibition. *Canadian Journal of Microbiology* 48: 1017-1029.

Petkova-Bocharova T., Chernozymsky I.N., and Castegnaro M. (1988). Ochratoxin A in human blood in relation to Balkan endemic nephropathy and urinary system tumors in Bulgaria. *Food Additives and Contaminants* 5: 299-301.

Perrone G., Susca A., Stea G., and Mulè G. (2004). PCR assay for identification of *Aspergillus carbonarius* and *Aspergillus japonicus*. *European Journal of Plant Pathology* 110: 641-649.

Perrone G., Mulè G., Susca A., Battilani P., Pietri A., and Logrieco A. (2006). Ochratoxin A production and amplified fragment length polymorphism analysis of *Aspergillus carbonarius*, *Aspergillus tubingensis*, and *Aspergillus niger* strains isolated from grapes in Italy. *Applied and Environmental Microbiology* 72(1): 680-685.

Petrik J., Zanic-Grubisic T., Barisic K., Pepeljnjak S., Radic B., Ferenicic Z., and Cepelak I. (2003). Apoptosis and oxidative stress induced by ochratoxin A in rat kidney. *Archives of Toxicology* 77: 685-693.

Pietri A., Bertuzzi T., Pallanori L., and Piva G. (2001). Occurrence of ochratoxin A in Italian wines. *Food Additives and Contaminants* 18(7): 647-654.

Plattner R.D., Desjardins A.E., Leslie J.F., and Nelson P.E. (1996). Identification and characterization of strains of *Gibberella fujikuroi* mating population A with rare fumonisin production phenotypes. *Mycologia* 88: 416-424.

Proctor R.H., Desjardins A.E. and Plattner R.D. (1999a). Biosynthetic and genetic relationships of B-series fumonisins produced by *Gibberella fujikuroi* mating population A. *Natural Toxins* 7: 251-258.

Proctor R.H., Desjardins A.E., Plattner R.D., and Hohn T.M. (1999b). A polyketide synthase gene required for biosynthesis of fumonisin mycotoxins in *Gibberella fujikuroi* mating population A. *Fungal Genetics and Biology* 27: 100-112.

Proctor R.H., Brown D.W., Plattner R.D., and Desjardins A.E. (2003). Co-expression of 15 contiguous genes delineates a fumonisin biosynthetic gene cluster in *Gibberella moniliformis*. *Fungal Genetics and Biology* 38: 237-249.

Raman V.K., Pandey R.A., and Bal A.S. (1995). Reactor systems for microbial desulfurization of coal: an overview. *Critical Reviews in Environmental Science and Technology* 25(3): 291-312.

Rosner H., Rohrmann B. and Peiker G. (2000). Ochratoxin A in human serum. *Archiv fuer Lebensmittelhygiene* 51: 104-107.

Samson R.A., Houbraken J.A.M.P., Kuijpers A.F.A., Frank M., and Frisvad J.C. (2004). New ochratoxin A or sclerotium producing species in *Aspergillus* section *Nigri*. *Studies in Mycology* 50: 45-61.

Sanchez-Rangel D., SanJuan-Badillo A., and Plasencia, J. (2005). Fumonisin production by *Fusarium verticillioides* strains isolated from maize in Mexico and development of a polymerase chain reaction to detect potential toxigenic strains in grains. *Agricultural and Food Chemistry* 53: 8565-8571.

Sartori D., Furlanto M.C., Martins M.K., Fereira de Paula M.R., Pizzirani-Kleiner A.A., Taniwaki M.H.and Pelegrinelli Fungaro M.H. (2006). PCR method for the detection of potential ochratoxin-producing *Aspergillus* species in coffee beans. *Research in Microbiology*, in press (doi: 10.1016/j.resmic.2005.09.008).

Scheuer R. and Gareis M. (2002). Occurrence of ochratoxin A and B in spices. *Mycotoxin Research* 18(2): 62-66.

Schilling A.G., Moeller E.M., and Geiger H.H. (1996). Polymerase chain reaction-based assays for species-specific detection of *Fusarium culmorum*, *F. graminearum* and *F. avenaceum*. *Phytopathology* 86(5): 515-522.

Schmidt H., Ehrmann M., Vogel R.F., Taniwaki M.H., and Niessen L. (2003). Molecular typing of *Aspergillus ochraceus* and construction of species specific SCAR-primers based on AFLP. *Systematic and Applied Microbiology* 26: 434-438.

Schmidt H., Taniwaki M.H., Vogel R.F., and Niessen L. (2004a). Utilization of AFLP markers for PCR-based identification of *Aspergillus carbonarius* and indication of its presence in green coffee samples. *Journal of Applied Microbiology* 97: 899-909.

Schmidt H., Bannier M., Vogel R.F. and Niessen L. (2004b). Detection and quantification of *Aspergillus ochraceus* in green coffee by PCR. *Letters in Applied Microbiology* 38: 464-469.

Schoen C.D., Szemes M., Bonants P.J.M., Speksnijder A., Klerks M.M., van den Boogert P.H.J.F., Waalwijk C., van der Wolf J.M., and Zijlstra C. (2005). Novel molecular and biochemical techniques for quality control and monitoring in the agrofood production chain, In: *Rapid Methods for Biological and Chemical Contaminants in Food and Feed*, van Amerongen A., Barug D., and Lauwaars M. (eds.) Wageningen, Netherlands, Wageningen Academic Publishers, 151-176.

Schuster E., Dunn-Coleman N., Frisvad J.C., and van Dijck P.W.M. (2002). On the safety of *Aspergillus niger* - a review. *Applied Microbiology and Biotechnology* 59(4-5): 426-435.

Seifert K.A. and Lévesque C.A. (2004). Phylogeny and molecular diagnosis of mycotoxigenic fungi. *European Journal of Plant Pathology* 110: 449-471.

Seo J.A., Proctor R.H., and Plattner R.D. (2001). Characterization of four clustered and coregulated genes associated with fumonisin biosynthesis in *Fusarium verticillioides. Fungal Genetics and Biology* 34: 155-165.

Serra R., Abrunhosa L., Kozakievicz Z., and Venancio A. (2003). Black *Aspergillus* species as ochratoxin A producers in Portuguese wine grapes. *International Journal of Food Microbiology* 88: 63-68.

Serra R., Mendoca C., and Venancio A. (2006). Fungi and ochratoxin A detected in healthy grapes and wine production. *Letters in Applied Microbiology* 42: 42-47.

Shapira R., Paster N., Eyal O., Menasherov M., Mett A., and Salomon R. (1996), Detection of aflatoxinogenic molds in grains by PCR. *Applied and Environmental Microbiology* 62: 3270-3273.

Soleas G.J., Yan J., and Goldberg D.M. (2001). Assay of ochratoxin A in wine and beer by high pressure liquid chromatography photodiode array and gas chromatography mass selective detection. *Journal of Agricultural Chemistry* 49: 2733-2740.

Sugita T., Kamimura K., Uchida K., Yamaguchi H., and Nagai A. (2004). PCR identification system for the genus *Aspergillus* and three major pathogenic species: *Aspergillus fumigatus, Aspergillus flavus* and *Aspergillus niger. Medical Mycology* 42: 433-437.

Surai P.F. and Dvorska J.E. (2005). Effects of mycotoxins on antioxidant status and immunity, In: *The Mycotoxin Blue Book*, Diaz D.E. (ed.),. Nottingham, UK, Nottingham University Press, 93-137.

Sweeney M. and Dobson A.D.W. (1999). Molecular biology of mycotoxin biosynthesis. *FEMS Microbiology Letters* 175: 149-163.

Tafuri A., Ferracane R., and Ritieni A. (2004). Ochratoxin A in Italian marketed cocoa products. *Food Chemistry* 88(4): 487-494.

Tan M.K., Simpfendorfe, S., Backhouse D., and Murray G.M. (2004). Occurrence of *Fusarium* head blight (FHB) in southern NSW in 2000: Identification of causal fungi and determination of putative chemotype of *Fusarium graminearum* isolates by PCR. *Australasian Plant Pathology* 33: 385-392.

Tang C.M., Holden D.W. Aufauvre-Brown A. and Cohen J. (1993). The detection of *Aspergillus* spp. by the polymerase chain reaction and its evaluation in bronchoalveolar lavage fluid. *American Review of Respiratory Disease* 148(5): 1313-1317.

Taniwaki M.H., Pitt J.I., Teixeira A.A. and Imanaka B.T. (2003). The source of ochratoxin A in Brazilian coffee and its formation in relation to processing methods. *International Journal of Food Microbiology* 82: 173-179.

Tjamos S.E., Antoniou P.P., Kazantzidou A., Antonopoulos D.F., Papageorgiou I., and Tjamos E.C. (2004). *Aspergillus niger* and *Aspergillus carbonarius* in Corinth raisin and wine-producing vineyards in Greece: Population composition, Ochratoxin A production and chemical control. *Journal of Phytopathology* (Berlin) 152: 250-255.

van der Merwe K.J., Steyn P.S., Fourie L.F., Scott D.B., and Theron J.J. (1965). Ochratoxin A, a toxic metabolite produced by *Aspergillus ochraceus* Wilh. *Nature* 205: 1112-1113.

Van Egmond H.P. (2003) Mycotoxins and Regulations: an update. In: *Proceedings of the 2nd World Mycotoxin Forum*, Nordwijk, The Netherlands, February 2003.

Vos P., Hogers R., Bleeker M., Reijans M., van de Lee T., Hornes M., Frijters A., Pot J., Peleman J., Kuiper M., and Zabeau M. (1995). AFLP: a new technique for DNA fingerprinting. *Nucleic Acids Research* 23: 4407-4414.

Vrabcheva T., Usleber E., Dietrich R., and Märtlbauer E. (2000). Co-occurrence of ochratoxin A and citrinin in cereals from Bulgarian villages with a history of Balkan endemic nephropathy. *Journal of Agricultural and Food Chemistry* 48: 2483-2488.

Wang Y.Z., Ju X.L., and Zhou Y.G. (2005). The variability of citrinin production in Monascus type cultures. *Food Microbiology* 22: 145-148.

Weber F., Schwerdt G., Freudinger R., and Gekle M. (2003). Programmed cell death by interaction of ochratoxin A with other nephrotoxins. *Mycotoxin Research* 19(1): 20-23.

White T.J., Burns T., Lee S., and Taylor J.W. (1990) Amplification and direct sequencing of fungal ribosomal RNA genes for phylogenetics. In: *PCR Protocols: a Guide to Methods and Applications*, Inis M.A., Gelfand D.H., Sninsky J.J., White T.J. (eds.), Academic Press, New York, 315-322.

Williams J.G.K., Kubelik A.R., Livak K.J., Rafalski J.A., and Tingey S.V. (1990). DNA polymorphisms amplified by arbitrary primers are useful as genetic markers. *Nucleic Acids Research* 18(22): 6531-6536.

Wolf J. (2000). Ochratoxin A in cereals and cereal products. *Archiv fuer Lebensmittelhygiene* 51(4-5): 85-88.

Yoder W.T. and Christianson L.M. (1998). Species-specific primers resolve members of *Fusarium* section *Fusarium*: Taxonomic status of the edible "Quorn" fungus reevaluated. *Fungal Genetics and Biology* 23(1): 68-80.

Yu J., Chang P.K., Ehrlich K.C., Cary J.W., Bhatnagar D., Cleveland T.E., Payne G.A., Linz J.E., Woloshuk C.P., and Bennett J.W. (2004). Clustered pathway genes in aflatoxin biosynthesis. *Applied and Environmental Microbiology* 70(3): 1253-1262.

Zachova I., Vytrasova J., Pejchalova M., Cervenka L., and Tavcar-Kalcher G. (2003). Detection of aflatoxigenic fungi in feed using the PCR Method. *Folia Microbiologica* 48(6): 817-821.

5 Fungal Chitinases - An Overview on Regulation and Cloning

Neetu Dahiya

Department of Biotechnology, Panjab University, Chandigarh-160014, India

INTRODUCTION

The fungal cell wall is a complex structure composed typically of chitin, 1, 3 β and 1, 6 β glucan, mannan and proteins. There is evidence of extensive cross-linking between chitin, glucan and other cell wall components (Cabib *et al.*, 2001). The wall is a highly dynamic structure subject to constant change, for example, during cell expansion and division in yeasts, and during spore germination, hyphal branching and septum formation in filamentous fungi. The maintenance of this plasticity in situations of morphogenetic changes occur in dimorphic fungi depends upon the activities of various enzymes that are associated with the cell wall. Chitinase is one of the most important enzymes involved in maintaining cell wall structure in fungi. Most of the fungal chitinases characterized to date have exochitinase, endochitinase or N-acetyl hexosaminidase activity and a number of these enzymes also exhibit transglycosylase activity. They may therefore contribute to breakage and reforming of bonds within and between polymers, leading to re-modelling of the cell wall during growth and morphogenesis.

Ghormade *et al.,* (2000) detected chitinase and N-acetylglucosaminidase activities in cell wall-bound and free fractions in the dimorphic fungus *Benjaminiella poitraisii.* These enzymes were found to be involved in the yeast-to-mycelium transition. Chitinases are also associated with the biology of insect mycopathogens. Fungal chitinases can disrupt the cuticle barrier, providing access to nutrients (Wattanalai *et al.*, 2004). At a late stage of infection, internal fungal cells must emerge from the insect to produce conidiophores. At this stage the insect endocuticle is digested, suggesting that extracellular chitinases play a major role in infection. Chitinases can also inhibit the development of other microbial competitors. Lorito *et al.,* (1998) showed that certain fungal endochitinases, such as the one isolated from *Trichoderma harzianum,* could act as potent anti-fungal enzymes.

Chitinases (EXC 3.2.1.14) catalyses the hydrolytic cleavage of the β–1,4 glycosidic bonds present in biopolymers of N-acetylglucosamine mainly in chitin. A number of proteins demonstrating chitinolytic activity were identified in fungi (Robert *et al.*, 1999; Mercedes *et al.*, 2001; ElKatatny *et al.*, 2001; Li *et al.*, 2002). These proteins present a large and diverse groups of enzymes, they differ not only in spatial and temporal localization but also in their molecular structure and substrate specificity.

STRUCTURE OF CHITIN

Chitin is a polymer of N-acetyl D-glucosamine units linked by β–1-4 bonds (Fig. 5.1). When the number of N-acetyl D-glucosamine units is higher than 50%, the biopolymer is termed chitin. The X-ray diffraction studies revealed that it occurs in three polymorphic forms, i.e., α-, β- and γ-chitins, which differ in the arrangement of molecular chains within the crystal cell (Fig. 5.2). In α-chitin, chains are arranged in anti-parallel fashion whereas in β-chitin chains are parallel. In γ-chitin chains are in mixed form (Peberdy, 1985; Deshpande *et al.*, 1987). Among these α-chitin is the most abundant form of chitin.

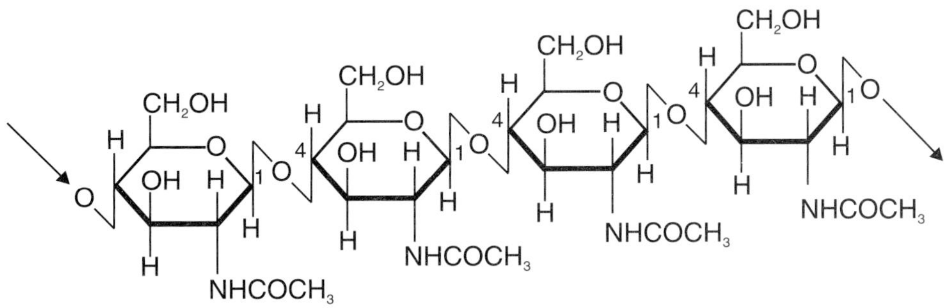

Fig. 5.1. Structure of chitin

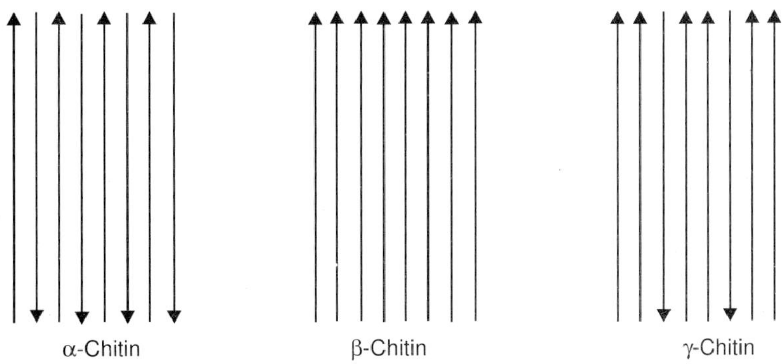

Fig. 5.2. Diagramatic representation of unit cells of three different forms of chitin. (http://vefur.rf.is/TAFT2003/PPtskjol/Fyrirlestrar/L70%20-20Kristberg.pdf.)

CHITIN DEGRADING ENZYMES

Chitinases can be classified into two major categories. Endochitinases (EC 3.2.1.14) cleave chitin randomly at internal sites, generating low molecular mass multimers of N-acetyl D-glucosamine, such as chitotetraose, chitotriose and diacetylchitobiose. Exochitinases can be divided into two subcategories: chitobiosidases (EC 3.2.1.29) which catalyze the progressive release of diacetylchitobiose starting at the nonreducing end of chitin microfibril; and 1,4-β N-acetyl glucosaminidases (EC 3.2.1.30) which cleave the oligomeric products of endochitinases and chitobiosidases generating the monomer, i.e., N-acetyl D-glucosamine (Sahai and Manocha, 1993). An alternative pathway involves the deacetylation of chitin to chitosan, which is finally converted to glucosamine residues by the action of chitosanase (EC 3.2.1.99). Fig. 5.3 shows pathway of microbial degradation of chitin and transport of hydrolyzed products across the membrane (Howard *et al.*, 2003).

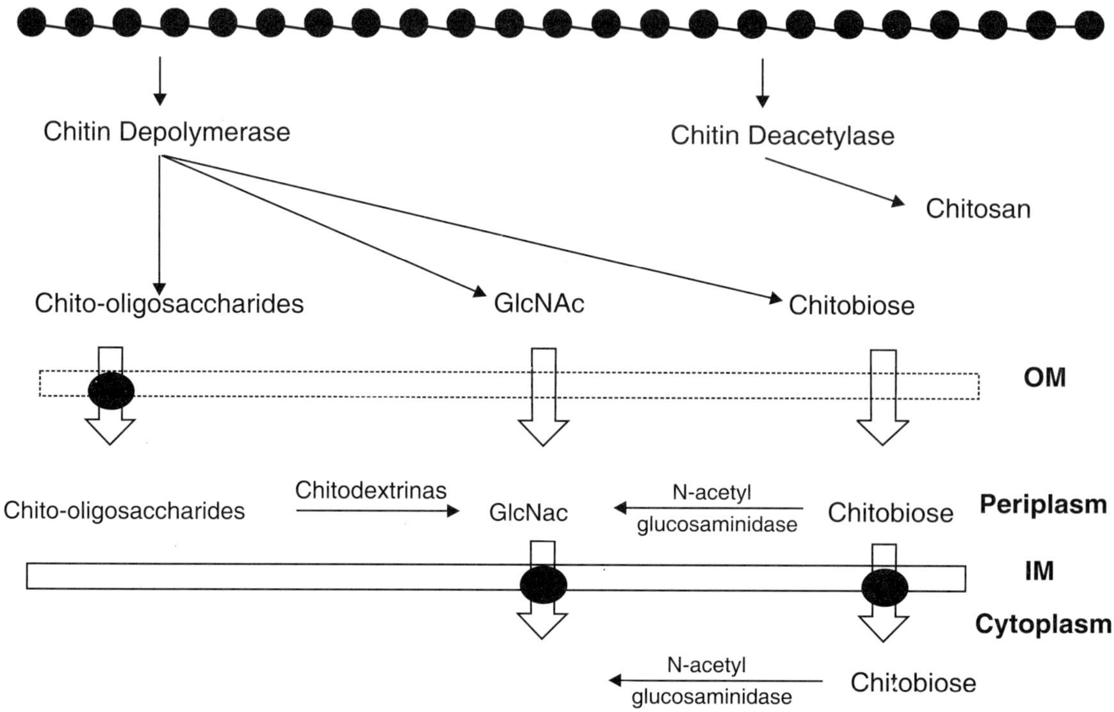

Fig. 5.3. Pathway of microbial degradation of chitin (Howard *et al.*, 2003). Multiple enzymes are required for degradation of chitin. The chitin polymer is degraded to chito-oligosaccharides, N-acetyl D-glucosamine (GlcNAc), and chitobiose by the activity of chitin depolymerases. Alternatively, chitin can be deacetylated by chitin deacetylase to form chitosan. Chito-oligosaccharides are then transported to the periplasm, perhaps via a specific outermembrane porin, where they are degraded by chitodextrinases to form GlcNAc. Chitobiose, which enters the periplasm through non-specific porins, is degraded to GlcNAc by the activity of periplasmic N-acetylglucosaminidases, though some is transported to cytoplasm. GlcNAc is subsequently transported to the cytoplasm where it is further metabolized or shunted to cell wall biogenesis. Cytoplasmic chitobiose is eventually degraded by cytoplasmic N-acetylglucosaminidases to form additional GlcNAc. *Shaded ovals* Transporters, *OM* outer membrane, *IM* inner membrane.

MODE OF ACTION

There are two general mechanistic pathways for acid catalyzed glycosyl hydrolysis, which results in the following:

(i) Retention of the stereochemistry of the anomeric oxygen at C1 relative to the initial configuration (scheme 1) or
(ii) Inversion of the stereochemistry (scheme 2) (Tews *et al.*, 1997)

An example of retaining mechanism is hen egg white lysozyme (HEWL), which has been reported to require two acidic residues. One of which is protonated. The mechanism is believed to proceed as follows: The β–(1,4) glycosidic oxygen is first protonated (leading to an oxocarbenium ion intermediate) which is stabilized by a second carboxylate (either through covalent or electrostatic interactions). Nucleophilic attack by water yields the hydrolysis products, which necessarily retains the initial anomeric configuration. This is commonly referred as double displacement mechanism of hydrolysis proposed for family 18 chitinases (Fig. 5.4a). Although the X-ray crystal structure of family 19 chitinase reveals a lysozyme, like fold (suggesting a double displacement mechanism), the hydrolysis products for two, family 19 chitinases show inversion of anomeric configration. This leads to the second commonly discussed hydrolysis mechanism a concerted single displacement reaction in which a bound water molecule acts as the nucleophile (Fig. 5.4b). All of the family 18 chitinases reported to date yield a β-anomer hydrolysis product (retaining mechanism) whereas family 19 chitinase result in the α-anomer (inverting mechanism).

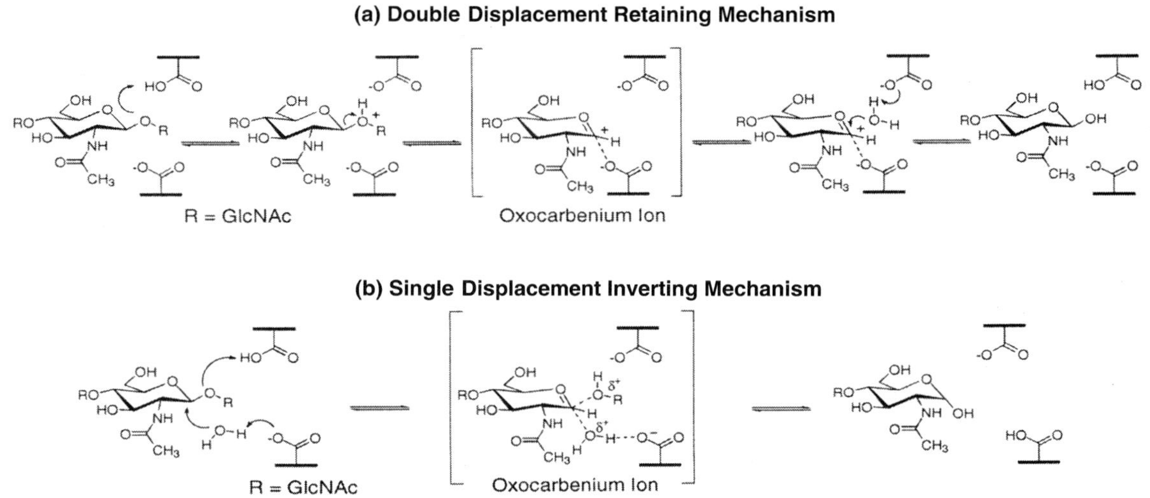

Fig. 5.4 (a). The double-displacement hydrolysis mechanism proposed for family 18 chitinases. Protonation of a GlcNAc residue in a boat conformation leads to an oxazoline intermediate, which may be hydrolyzed to form a product with retention of the anomeric configuration. **(b)** The single displacement hydrolysis mechanism proposed for family 19 chitinases. Two acidic residues are required in the active site and the hydrolysis product has inversion of the anomeric configuration (Brameld and Goddard, 1998).

Assay of chitinolytic activity

Using solid media

In many chitinolytic systems, chitin is hydrolyzed by secreted chitin depolymerases. Because these enzymes are able to diffuse through agar assays to identify chitinolytic bacteria or genomic clones encoding chitin-degrading enzymes can be performed by monitoring the degradation of polymeric chitin incorporated into an agar medium. Though these assays have limited sensitivity, they represent a simple and inexpensive method to identify chitinolytic microorganisms. Initial screens for chitinase activity should include both α and β chitin as different forms of chitin may induce the expression of specific chitinases (Svitil *et al.*, 1997). Alternative substrates such as chitin azure may be included in agar media to facilitate the visualization of chitin depolymerase activity. Ethylene glycol chitin can be incorporated with Trypan Blue in molten agar. After culturing, clear halos will appear around colonies of chitinolytic bacteria.

Using analogs

Chitin agar screens identify only chitin depolymerase enzymes and not other chitin or chito-oligosaccharide modifying proteins. Chitin analogs have been well documented as a valuable tool for initial identification of chitinolytic enzyme producing strains and clones (Morimoto *et al.*, 1997; Svitil *et al.*, 1997; Techkarnjanaruk and Goodman, 1999; Cottrell *et al.*, 2000; Lonhienne *et al.*, 2001). 4-Methylumbelliferyl β-D-N-N′ diacetyl chitobioside and 4-methylumbelliferyl β-D-N, N′, N″ triacetyl chitotrioside (Fig. 5.5) are prepared by linking a 4MUF moiety to the reducing end of chitobiose or chitotriose, respectively (O'Brien and Colwell, 1987). When cleaved by a chitin degrading enzyme, methylumbelliferon is released, which can be detected by its bright blue fluorescence under UV light (O'Brien and Colwell, 1987). A variety of enzymes can be detected with these analogs including some chitin depolymerase, chitodextrinases and N-acetylglucosaminidases (chitobiases).

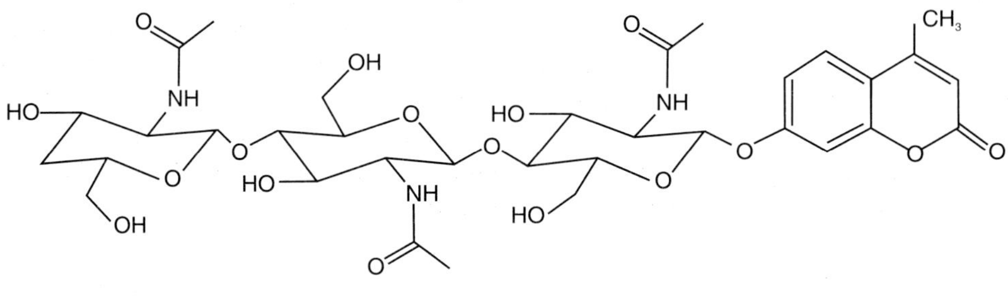

Fig. 5.5. Structure of 4-Methylumbelliferyl β–D-N, N′, N″ triacetyl chitotrioside

In addition to MUF derivatives, paranitrophenol (PNP) chitin analogs 4-nitro-phenyl N-acetyl β–D-glucosaminide, 4-nitro-phenyl N, N′-diacetylchitobioside and 4-nitro-phenyl N, N′, N″-triacetyl chitotrioside can be used for analysis of chitinolytic activity. When PNP is liberated from the carbohydrate polymer it can be detected spectrophotometrically by monitoring the absorbance at 410 nm (Sakai *et al.*, 1994; Matsuo *et al.*, 1999).

DETECTION OF CHITINASES BY THE RELEASE OF REDUCING SUGARS

Chito-oligomers, chitobiose and GlcNAc all of which have a reducing end released as chitin is degraded. This increase in reducing activity can be detected and quantified using a variety of methods. The earliest practical method for the determination of hexosamine was developed by Elson and Morgan (1933). Commercially available kits for monitoring reducing sugar levels can be used but an easy and inexpensive method described by Gracia *et al.,* (1993) makes use of bicinchoninic acid.

An alternative to the above protocol is the dinitrosalicylic acid reducing sugar test optimized by Miller (1959) and estimation of reducing sugars by the method of Reissig *et al.,* (1955). Rondle and Morgan (1955) described another method for the determination of glucosamine and galactosamine. Dubois *et al.,* (1956) developed phenol-sulphuric acid reaction for determination of reducing sugars and related substances. In 1966, Jeuniaux described viscometric methods and nephlometric (turbidometric) method for determining the chitinase activities using acid and alkali treated shrimp shells as the substrate.

Molano *et al.,* (1977) developed a rapid and extremely sensitive assay for chitinase using radioactive chitin as the substrate. A reliable colorimetric microassay adapted to microtitre plate was established by Wirth and Wolf (1990), which was based on the precipitability of carboxymethylchitin remazol brilliant violet 5R from buffered solutions with hydrochloric acid. In this method a carboxy methyl substituted soluble chitin and reprecipitated colloidal chitin derivatives were covalently linked with remazol brilliant violet 5R. These dye labeled substrates were used for qualitative as well as quantitative assay of chitinase.

Using Zymograms

Tronsmo and Harman (1993) came forward with another method of detection and quantification of chitinase in solutions and on gels. An alternative method for producing zymograms has been described that detected not only the presence of some chitin depolymerases but also chitodextrinase and N-acetylglucosaminidase activity using MUF analog of choice (Li *et al.*, 2002). The enzymes were visualized as fluorescent bands by using an agarose overlay containing 4-methyumbelliferyl derivatives of N-acetyl-β-D-glucosaminide, β-D-N, N′-diacetyl chitobioside or β-D-N, N′, N″-triacetyl chitotriose for N-acetyl β-glucosaminidase, chitobioside or endochitinase, respectively.

PREPARATION OF CHITIN

Preparation of swollen chitin

Swollen chitin was prepared by the method of Monreal and Reese (1969). It was carried out by stirring 10g of flake chitin into 100ml of 85% phosphoric acid at 4°C for 48h and reprecipitating it by pouring the gelatinous mixture into an excess of ice cold water. Excess water was removed and precipitate was washed with distilled water till pH becomes neutral.

Preparation of colloidal chitin

Colloidal chitin was prepared by the method described by Jeuniaux (1966). One gram flake chitin in 250ml distilled water was mixed with 320ml of cold concentrated H_2SO_4 (below 10°C). The viscous acid solution (100ml) was filtered through glass wool in a funnel into 900ml of distilled water under continuous stirring. The chitin precipitate was washed with distilled water up to neutral pH.

Preparation of glycol chitin

Glycol chitin was prepared by the method of Yamada and Imoto (1981). 4.0g flake chitin was suspended in 100ml of 42% (w/w) NaOH solution and kept at room temperature for 4h with occasional swirling. The alkaline chitin thus obtained was filtered off, washed with 42% (w/w) NaOH solution and pressed well to remove most of excess NaOH solution. The alkaline cake was transferred to a beaker and vigorously mixed with finely crushed ice (~50-70g, pre-cooled in a freezer before use). After mixing for 20-30min a highly viscous alkaline solution was diluted to 250ml with NaOH solution so as to make final concentration of NaOH 14% (w/w).

The alkaline chitin solution was cooled in an ice bath and 10ml of 2-chloroethanol was added drop wise using an autopipette with mixing during 30min and the mixture was left overnight at room temperature. The mixture was re-cooled in an ice bath and 10ml acetic anhydride was added dropwise using an autopipette with stirring. After stirring for 30min, the mixture in an ice bath was made neutral with acetic acid and dialyzed against running water for 3days.

Preparation of regenerated chitin

Regenerated chitin was prepared by the method of Molano *et al.*, (1977). One gram chitosan was mixed with 20ml of 10% acetic acid while grinding and left overnight at room temperature. 90ml methanol was added slowly with mixing and the cloudy solution was filtered through glass wool. The filtrate was placed on a magnetic stirrer and 1.5ml of acetic anhydride was added. After about a minute the mixture gels and the magnet comes to standstill. The gel was allowed to stand for 30min and then cut into small pieces with a spatula. The gel fragments were transferred into a homogenizer, covered with methanol and homogenized for 1min at maximum speed. The chitin was filtered through sintered glass funnel and washed with water to neutrality.

Preparation of fungal cell wall

The fungi were grown in potato dextrose broth, incubated at 30°C, and then harvested after 72h growth. The mycelium thus obtained was homogenized in a glass homogenizer and used as crude fungal cell wall.

Chitinases of yeast

Dickinson *et al.*, (1989) reported an extracellular chitinase from *Candida albicans*. The enzyme was optimally active at pH 6.5 and temperature 45°C and the K_m with chitin was 3.9mg/ml. Jackson *et al.*, (1996) purified three chitinases of molecular weights 33kDa, 58kDa and 79kDa from *Candida albicans*. They were purified by anion exchange chromatography. Niimi *et al.*, (1997) purified N-acetylglucosaminidase from *Candida albicans*, which was optimally active at pH 4.0-4.5.

Chitinases of fungi

An extracellular chitinase (molecular weight 40kDa) was purified from *Trichoderma harzianum* by precipitation with ammonium sulphate followed by ion exchange chromatography on Q-Sepharose, gel filtration on Sephadex G-100 and hydrophobic interaction on Phenyl-sepharose CL-4B. The pH and temperature optima of enzyme were 4.0-4.5 and 40°C, respectively. The chitinase hydrolyzed swollen chitin more readily than glycol chitin and powdered chitin, but did not attack N, N′-diacetylchitobiose. Zn^{2+} and Cu^{2+} inhibited chitinase activity by 57% and 22%, respectively (Ulhoa and Peberdy, 1992).

Lorito *et al.,* (1994) purified an N-acetyl β-glucosaminidase from culture filtrate of *Trichoderma harzianum* strain P1. The molecular weight and pI were 72kDa and 4.6, respectively.

Sakurda *et al.,* (1996) purified a 42kDa chitinase from *Piromyces communis* OTS1 by affinity chromatography, gel filtration and chromatofocusing. The chitinase was optimally active at pH 6.2 and 60°C. It was inhibited by Ag^{+}, Hg^{+} and allosamidin at 1mM concentration. A β-N-acetyl hexosaminidase, molecular weight 190kDa and pI 4.3 was purified from *Aspergillus nidulans* (Reyes *et al.*, 1989). It was optimally active at pH 5.0. Hg^{2+}, Ag^{+} and acetate inhibited hexosaminidase activity. It was an exoenzyme, which degraded chitin oligomers with release of N-acetyl D-glucosamine.

Pera *et al.,* (1997) purified β-N-acetyl D-glucosaminidase from *Aspergillus niger* using polypropylene glycol 1025, DEAE-Sephacel and Sephadex G-150 chromatography. Pinto *et al.,* (1997) purified chitinase from *Metarhizium anisopliae*. The purified chitinase had a molecular weight of 30kDa and was optically active at pH 4.5-5.0 and temperature 40-45°C, respectively.

A 43kDa chitinase was purified from *Trichoderma harzianum* Rifai T24 using ammonium sulphate precipitation followed by hydrophobic interaction chromatography (Phenyl Sepharose) and gel filtration chromatography. The K_m and V_{max} values for the chitinase were 3.8g/l and 49.3pkat/ml, respectively. Chitinase was stable at 30°C. Its half-life at 60°C was 15min (El-Katatny *et al.*, 2001).

Two chitinases P-1 and P-2 were purified from *Isaria japonica* by chromatography on DEAE-Bio gel chromatography and gel filtration with Superdex 75PG. The molecular weights of enzymes P1 and P2 were 43.273kDa and 31.134kDa, respectively. The optimum pH and temperature were 3.5-4.0 and 50°C for P-1 and 4.0-4.5 and 40°C for P-2. The products from chitin hexamer obtained with P-1 were almost all dimers with only small amount of trimer whereas those with P-2 were mainly trimers with some dimer and tetramer (Kawachi *et al.*, 2001).

Souza *et al.,* (2003) purified 43kDa endochitinase from *Colletotrichum gloeosporioides*. The pH and temperature optima were 7.0 and 50°C, respectively. The enzyme showed high stability at 50°C and pH 7.0. Two isozymes I and II (molecular weight 67kDa) of N-acetyl β-D-glucosaminidases were purified from *Fusarium oxysporum* F3 using ammonium sulphate precipitation followed by anion exchange chromatography (DEAE-Sepharose CL6B), gel filtration (Sephacryl S-100) and hydrophobic interaction chromatography (t-butyl column). The optimum pH of isozymes I and II were 5.0 and 6.0, respectively whereas maximum activity of both the isozymes was obtained at 40°C. The K_m of isozymes I and II was 49.6 and 48.6 mM and the V_{max} 1.24 and 0.26 μM/mg/min, respectively on p-nitrophenyl N-acetyl-β-D-glucosaminide as substrate (Gkargkas *et al.*, 2003).

Multiple chitinases have been found in hyphae of filamentous fungi, which are presumed to have various functions during hyphal growth. An intracellular chitinase named chitinase III from

Rhizopus oligosporus hydrolysed chromogenic derivatives of chitobiose, but had no N–acetylglucosaminidase activity.

Chit 33 endochitinase of *Trichoderma harzianum* CECT2413 is induced either by carbon or nitrogen source starvation or to a low degree also under conditions of temperature stress (Mercedes *et al.*, 2001).

An endochitinase of *Aspergillus fumigatus* NCPF2140 was induced by chitin (as sole carbon source) and repressed by GlcNAc indicating regulation by a negative feed back mechanism (Escott *et al.*, 1998).

Regulation of fungal chitinase genes

Chitinase gene expression has been reported to be controlled by a repressor/inducer system in which chitin or other products of degradation act as inducers. High levels of chitinase were found in cultures of *Metarhizium anisopilae* supplied with chitin but not with other polysaccharides such as pectin, xylan or cellulose. Studies with restricted cultures to prevent catabolite repression revealed that the most efficient inducers of chitinase and chitobiase were N-acetylglucosamine and glucosamine, respectively (St Leger *et al.*, 1986). Another chitinase gene *chi*3 encoding CHIT30 from *Metarhizium anisoplae* is upregulated by chitin, tick culture and low concentrations of N-acetylglucosamine (0.25%) and is down regulated by high N-acetylglucosamine (1.0%) and glucose (1%) concentrations (da Silva *et al.*, 2005).

Vasseur *et al.*, (1990) found that chitinase gene expression in a chitinase overexpressing mutant *A. album* E3 was induced by N-acetyl glucosamine while it was repressed by glucose. Chitin or other oligomers did not enhance chitinase expression in *A. album* E3. On the other hand, in gene regulation of *T. harzianum* high chitinase activity was found only in cultures supplied with chitin but not with cellulose, chitosan or chitobiose. In *T. harzianum* N-acetylglucosamine did not enhance the chitinase production but instead repressed its synthesis. Basal and constitutive levels of chitinase were reported in cultures of *A. album* E3 and *T. harzianum* even under starvation conditions. These low levels of chitinase may be sufficient to initiate chitin degradation and to release soluble oligomers, which in turn induce chitinase synthesis (Vasseur *et al.,* 1990; Ulhoa and Peberdy, 1991).

Repression of chitinase synthesis in almost all organisms indicates that catabolite repression may be involved in the regulation of microbial chitinase genes. The gene coding for a 33 kDa chitinase (CHIT33) in *T. harzianum* CECT 2413 was strongly repressed by glucose and de-repressed under starvation conditions or in presence of fungal cell wall containing chitin. The CHIT33 gene was not induced by chitin or its oligomers (Mercedes *et al.*, 2001).

Mach *et al.,* (1999) reported regulation of two major chitinase genes *ech*42 (encoding the CHIT42 endochitinase) and *nag*I (encoding CHIT 73 N-acetyl-beta-D-glucosamine). N-acetylglucosamine, di-N-acetylchitobiose or tri-N-acetylchitotriose *nagI* gene express when added to mycelia pregrown on different carbon sources. *Ech*42 expression was also observed during growth on fungal cell walls but in contrast was not triggered by addition of chito-oligomers to pregrown mycelia. Significant *ech*42 expression was observed after prolonged carbon stravation. In addition *ech*42 gene transcription was triggered by physiological stress such as low temperature, high osmotic pressure or the addition of ethanol. Viterbo *et al.,* (2002) reported induction of chit36Y (coding for CHIT36) from *Trichoderma harzianum* by stress conditions, colloidal chitin and N-acetyl D-glucosamine.

Cloning of fungal chitinase genes

Pishko *et al.*, (1995) reported cloning and sequencing of two chitinase (CTS)-encoding genes (*cts*) from *Coccidioides immitis* (Ci), a respiratory fungal pathogen of humans. They reporetd that *cts*1 gene contains five introns and a 1281-bp ORF which translates a 427-amino-acid (aa) protein of 47.4 kDa and *cts*2 gene contains two introns and a 2580-bp ORF which translates a 860-aa protein of 91.4 kDa. The deduced CTS1 protein showed highest homology to the *Aphanocladium album* and *Trichoderma harzianum* CTS (74% and 76%, respectively), while CTS2 showed highest homology to the CTS of *Saccharomyces cerevisiae* (Sc) and *Candida albicans* (47% and 51%, respectively). The putative N-terminal sequence of the mature CTS1 protein also showed 89% homology to the reported N-terminal sequence of a 48-kDa complement fixation antigen (CF-Ag) of Ci, which has demonstrated chitinase activity. The CF-Ag is a clinically important antigen used in serodiagnosis of this fungal disease. CTS2 showed several of the conserved features of the Sc CTS, including putative catalytic and Ser/Thr-rich domains, and a C-terminal Cys-rich region.

Takaya *et al.*, (1998) reported cloning of chitinase-encoding gene from *Aspergillus nidulans* by polymerase chain reaction using degenerated oligonucleotide primers designed from the conserved amino acid sequences among chitinases from yeasts and *Rhizopus* spp. The cloned gene, named *chi*A, encoded a polypeptide consisting of 660 amino acids. Disruption of *chi*A had no effect on hyphal or conidiophore morphology, but germination frequency and hyphal growth rate decreased substantially. They studied expression of *chi*A gene using *Escherichia coli* beta-galactosidase as a reporter enzyme and reported that the beta-galactosidase activity was present during hyphal growth and increased twice as the conidiophores developed. Using *in situ* staining they found high expression of beta-galactosidase in metulae, phialides, and conidia during conidiophore development, indicating developmentally regulated expression of *chi*A.

Boga *et al.*, (1998) reported the complete nucleotide sequence and analysis of the chromosomal and full-length cDNA copies of the regulated gene (*chit*1) coding one of the chitinases produced by the biocontrol agent *Metarhizium anisopliae*. They used degenerated primers, encompassing conserved regions of other fungal chitinases, to amplify a 650-bp DNA fragment, which was used to isolate genomic and cDNA clones from *M. anisopliae*. Albeit at least two different chitinases are characterized in this fungus, only one *chit* gene was isolated. The *chit*1 gene is interrupted by three short typical fungal introns and has a 1,521-bp ORF, which encodes a protein of 423 amino acids with a stretch of 35 amino acid residues displaying characteristics of signal peptide. The deduced sequence of the mature protein predicts a 42-kDa protein with pI of 5.8. Southern analysis of genomic DNA indicates a single copy of *chit*1 in the *M. anisopliae* genome.

Another chitinase gene was reported from *Metarhizium anisopliae* grown in a medium containing chitin as the sole carbon source (Kang *et al.*, 1998). Comparisons of nucleotide sequence of the isolated gene with those of other fungal chitinase genes showed low sequence identity (24.4-36.4%) except for the active site of chitinase. In addition, molecular mass determination of the fused gene product separated on a gel showed that the fused chitinase seems to be about 70 kDa, while the molecular mass calculated from the deduced amino acid sequence can be about 58 kDa. These molecular masses were different from values of 33 kDa for an endochitinase and 110 kDa for an exochitinase (N-acetylglucosaminidase) from *M. anisopliae* published previously.

Takaya *et al.*, (1998) reported the primary structure of an intracellular chitinase (chitinase III) from *Rhizopus oligosporus*, a zygomycete filamentous fungus. Chitinase III was purified to homogeneity from actively growing mycelia of *R. oligosporus* using three steps of column chromatography. Its molecular mass was 43.5 kDa and the pH optimum was 6.0 when p-nitrophenyl N, N′, N″-beta-D-triacetylchitotrioside was used as a substrate. Chitinase III also hydrolysed chromogenic derivatives of chitobiose, but had no N-acetylglucosaminidase activity. The gene encoding chitinase III (*chi*3) was cloned using PCR with degenerate oligonucleotide primers from the partial amino acid sequence of the enzyme. The deduced amino acid sequence of *Chi*3 was similar to that of bacterial chitinases and chitinases from mycoparasitic fungi, such as *Aphanocladium album* and *Trichoderma harzianum*, but it had no potential secretory signal sequence in its amino terminus. Northern blot analysis showed that *chi*3 was transcribed during hyphal growth. These results suggest that chitinase III may function during morphogenesis in *R. oligosporus*.

Kim *et al.,* (2002) reported the characterization of a diverse set of chitinase genes from *T. virens*. The full-length genomic clones were isolated and characterized, while sequencing of the corresponding cDNA clones and manual annotation provided a basis for establishing gene structure. Based on homology of the deduced amino acid sequences, they identified three members of the 42Kd endochitinase gene family, two 33Kd exochitinases and two exochitinases with homology to N-acetylglucosaminidases. Using northern blot analysis and enzymatic assays they reported interdependent regulation of this important gene family in *T. virens*.

Expression of the endochitinase encoding *ech*42 gene of the mycoparasite *Trichoderma atroviride* is reported to be derepressed by carbon starvation (Brunner *et al.*, 2003). For identifying promoter areas involved in control by this condition, they prepared fusions of several mutant forms of the *ech*42 promoter to the *Aspergillus niger gox*A gene as a reporter. They reported that removal of a 130-bp fragment comprising a binding site for the carbon catabolite repressor Cre1, an AGGGG element and three separate binding sites identical and highly similar, respectively, to those for the *Aspergillus nidulans* regulator of conidiation *Brl*A resulted in a three-fold increase in derepression under carbon starvation. A truncation of the promoter to 196 bp, which removed all of the observed DNA binding motifs, resulted in five-fold derepression. Using protein-DNA binding analyses they reported that only the *Brl*A-like sites, but neither the AGGGG element nor the *Cre*1 binding site, bound proteins from cell-free extracts from carbon-starved mycelia of *T. atroviride* thus they identified a new regulator of chitinase gene expression in *Trichoderma*.

A chitinase gene encoding an inducible 45 kDa chitinase of *Aspergillus fumigatus* was cloned by Jaques *et al.,* (2003). Based on deduced amino acid sequence, it was identified as a chitinase of the fungal/bacterial class, which was, designated ChiB1. Recombinant ChiB1, expressed in *Pichia pastoris*, was shown to function by a retaining mechanism of action. Cleavage patterns with the N-acetylglucosamine (GlcNAc) oligosaccharide substrates GlcNAc(4), GlcNAc(5) and GlcNAc(6) indicated that the predominant reaction involved hydrolysis of GlcNAc(2) from the non-reducing end of each substrate. Products of transglycosylation were also identified in each incubation. Following disruption of chiB1 by gene replacement, growth and morphology of disruptant and of the wild-type strain were essentially identical. However, during the autolytic phase of batch cultures the level of chitinase activity in culture filtrate from a disruptant was much lower than the activity from the wild type.

Fang *et al.,* (2005) cloned a chitinase gene coding for an endochitinase (Bbchit1) from *Beauveria bassiana*. They reported that Bbchit1 was intronless, and there was a single copy in *B. bassiana*. Its regulatory sequence contained putative CreA/Crel carbon catabolic repressor binding domains, which was consistent with glucose suppression of Bbchit1. At the amino acid level, Bbchit1 showed significant similarity to a *Streptomyces avermitilis* putative endochitinase, a *Streptomyces coelicolor* putative chitinase, and *Trichoderma harzianum* endochitinase Chit36Y. However, Bbchit1 had very low levels of identity to other chitinase genes previously isolated from entomopathogenic fungi, indicating that Bbchit1 was a novel chitinase gene from an insect-pathogenic fungus. A gpd-Bbchit1 construct, in which the *Aspergillus nidulans* constitutive promoter drove Bbchit1, was transformed into the genome of *B. bassiana*, and three transformants that overproduced Bbchit1 were obtained.

FUTURE PERSPECTIVES AND CONCLUSIONS

The success in using chitinases for different applications depends on the supply of highly active preparation at a reasonable cost. Most of the suppliers use either natural microbial biodiversity or genetically engineered chitinase overproducing microbial strains to obtain efficient preparations. The present research into the cloning of chitinase genes is directed mainly towards the study of gene regulation and catalytic activity. Especially in fungi, fusion techniques (both chemical and electrical) could be tried for enhanced production of chitinase in fungi and yeast.

The use of chitinases for production of chito-oligosaccharides and N-acetyl D-glucosamine is one of the major applications in human health care sector. There are few reports on the industrially available production of GlcNAc from chitin by using chitinolytic enzymes since the enzyme reaction on chitin progress very slowly and also its yield is low from the standpoint of practical use. Chitinases can also be employed for the biocontrol of plant pathogens and for developing transgenic plants. Although biocontrol of phytopathogenic fungi using chitinase producing micro-organisms and plants has been studied extensively but more work is required for studying microbe-microbe, plant-microbe and chitinase-microbe interaction which would be of great help in using chitinase as biocontrol agent. Developing a bioprocess for production of N-acetyl D-glucosamine from chitin will be a novel approach.

REFERENCES

Bogo M.R., Rota C.A., Pinto H.Jr., Ocampos M., Correa C.T., Vainstein M.H. and Schrank A. (1998). A chitinase encoding gene (chit1 gene) from the entomopathogen *Metarhizium anisopliae*: isolation and characterization of genomic and full-length cDNA. *Current Microbiology* 37(4): 221-5.

Brameld K.A. and Goddard III W.A. (1998). The role of enzyme distortion in the single displacement mechanism of family 19 chitinases. *Proceedings National Acadamy of Science, USA* 120: 4276-4281.

Brunner K., Montero M., Mach R.L., Peterbauer C.K. and Kubicek C.P. (2003). Expression of the *ech42* (endochitinase) gene of *Trichoderma atroviride* under carbon starvation is antagonized via a BrlA-like cis-acting element. *FEMS Microbiology Letters* 218(2): 259-64.

Cabib E., Dong-Hyun R., Schmidt M., Crotti L.B. and Varma A. (2001). The yeast cell wall and septum as paradigms of cell growth and morphogenesis. *The Journal of Biological Chemistry* 276: 19679–19682.

Cottrell M., Wood D., Yu Y. and Kirchman D. (2000). Selected chitinase genes in cultured and uncultured marine bacteria in the alpha- and gamma-subclasses of the proteobacteria. *Applied Environmental Microbiology* 66: 1195-1201.

da Silva M.V., Santi L., Staats C.C., da Costa A.M., Colodel E.M., Driemeier D., Vainstein M.H. and Schrank A. (2005). Cuticle-induced endo/exoacting chitinase CHIT30 from *Metarhizium anisopliae* is encoded by an ortholog of the *chi3* gene. Research in *Microbiology* 156(3): 382-92.

Deshpande M.V., Balkrishanan H., Ranjekar P.K. and Shanker, V. (1987). Isolation and immobilization of *Sclerotium rolfsii* protoplasts. *Biotechnology Techniques* 9: 49-52.

Dickinson K., Keer V., Hitchcock C.A. and Adams D.J. (1989). Chitinase activity from *Candida albicans* and its inhibition by allosamidin. *Journal of General Microbiology* 135: 1417-1421.

Dubois M., Gillcs K.A., Hamilton J.K., Rebers P.A. and Smith F. (1956). Colorimetric method for determination of sugars and related substances. *Analytical Chemistry* 28: 350-356.

El-Katatny M.H., Gudelj M., Robra K.H., Elnaghy M.A. and Gubitz G.M. (2001). Characterizatioin of a chitinase and endo β-1,3 glucanase from *Trichoderma harzianum* Rifai T24 involved in control of the phytopathogen *Sclerotium rolfsii. Applied Microbiology and Biotechnology* 56: 137-143.

Elson L. A. and Morgan W.T.J. (1933). A colorimetric method for the determination of glucosamine and chondrosamine. *Biochemical Journal* 27: 1823-1828.

Escott G.M., Hearn V.M. and Adams D.J. (1998). Inducible chitinolytic system of *Aspergillus fumigatus. Microbiology* 144: 1575-1581.

Fang W., Leng B., Xiao Y., Jin K., Ma J., Fan Y., Feng J., Yang X., Zhang Y. and Pei Y. (2005). Cloning of *Beauveria bassiana* chitinase gene *Bbchit1* and its application to improve fungal strain virulence. *Applied Environmental Microbiology* 71(1): 363-370.

Ghormade V.S., Lachke S.A. and Deshpande M.V. (2000). Dimorphism in *Benjaminiella poitrasii*: involvement of intracellular endochitinase and N-acetylglucosaminidase activities in the yeast-mycelium transition. *Folia Microbiology* 45: 231-238.

Gkargkas K., Mamma D., Nedey G., Topakas E., Christakopoulos P., Kekos D. and Macris B.J. (2003). Studies on a N-acetyl-β-D-glucosaminidase produced by *Fusarium oxysporum* F3 grown in solid-state fermentation. *Proceedings Biochemistry* 39: 1599-1605.

Gracia E. (1993). Assessment of endo-1, 4-beta-D-glucanase activity by a rapid colorimetric assay using disodium-2-2'-bi-cinchoninate. *Journal of Food Biochemistry* 17: 135-145.

Howard M.B., Ekborg N.A., Weiner R.M. and Hutcheson S.W. (2003). Detection and characterization of chitinases and other chitin modifying enzymes. *Journal of Industrial Microbiology and Biotechnology* 30: 627-635. http://vefur.rf.is/TAFT2003/PPtskjol/Fyrirlestrar/L70%20-20Kristberg.pdf.

Jackson A.J., Saunders V.A., Gooday G.W. and Humphreys A.M. (1996). Chitinase activities from yeast and hyphal cells of *Candida albicans. Mycological Research* 100: 321-327.

Jaques A.K., Fukamizo T., Hall D., Barton R.C., Escott G.M., Parkinson T., Hitchcock C.A. and Adams D.J. (2003). Disruption of the gene encoding the ChiB1 chitinase of *Aspergillus fumigatus* and characterization of a recombinant gene product. *Microbiology* 149: 2931-2939.

Jeuniaux, C. (1966). Chitinases. *Methods In Enzymology* 8: 644-650.

Kang S.C., Park S. and Lee D.G. (1998). Isolation and characterization of a chitinase cDNA from the entomopathogenic fungus, *Metarhizium anisopliae. FEMS Microbiology Letters* 65(2): 267-71.

Kawachi I., Fujieda T., Ujita M., Ishii Y., Yamagishi K., Sato H., Funaguma T. and Hara A. (2001). Purification and properties of extracellular chitinases from the parasitic fungus *Isaria japonica. Journal of Bioscience and Bioengineering* 92: 544-549.

Kim D.J., Baek J.M., Uribe P., Kenerley C.M. and Cook D.R. (2002). Cloning and characterization of multiple glycosyl hydrolase genes from *Trichoderma virens. Current Genetics* 40(6): 374-84.

Li H., Morimoto K., Katagiri N., Kimura T., Sakka K., Lun S. and Ohmiya K. (2002). A novel β-N-acetylglucosaminidase of *Clostridium paraputrificum* M-21 with high activity on chitobiose. *Applied Microbiology and Biotechnology* 60: 420-427.

Lonhienne T., Mavromatis K., Vorgias C., Buchon L., Gerday C. and Bouriotis V. (2001). Cloning, sequencing and characterization of two chitinase genes from the Antarctic *Arthrobacter* sp. strain TAD20: Isolation and partial characterization of the enzymes. *Journal of Bacteriology* 183: 1773-1779.

Lorito M., Hayes C.K., Di Pietro A., Woo S.L. and Harman G.E. (1994). Purification, characterization and synergistic activity of a glucan 1,3-β-glucosidase and N-acetyl D-glucosaminidase from *Trichoderma harzianum. Phytopathology* 84: 398-405.

Lorito M., Woo S.L., Garcia I., Colucci G., Harman G.E., Pintor-Toro J.A., Filippone E., Muccifora S., Lawrence C.B., Zoina A., Tuzun S., Scala F. and Fernandez I.G. (1998). Genes from mycoparasitic fungi as a source for improving plant resistance to fungal pathogens. *Proceedings National Academy of Science, USA* 95: 7860-7865.

Mach R.L., Peterbauer C.K., Payer K., Jaksits S., Woo S.L., Zeilinger S., Kullnig C.M., Lorito M. and Kubicek C.P. (1999). Expression of two major chitinase genes of *Trichoderma atroviride* (*T. harzianum* P1) is triggered by different regulatory signals. *Applied Environmental Microbiology* 65(5): 1858-63.

Matsuo Y., Kurita M., Park J.K., Tanaka K., Nakagawa T., Kawamukai M. and Matsuda, H. (1999). Purification, characterization and gene analysis of N-acetylglucosaminidase from *Enterobacter* sp. G-1. *Bioscience Biotechnology and Biochemistry* 63: 1261-1268.

Mercedes Dana M., Limón M.C., Mejías R., Robert L.M., Tahía B., José A. P.T and Kubicek C.P. (2001). *Regulation of chitinase 33 (chit33) gene expression in Trichoderma harzianum. Current Genetics* 38(6): 335-42.

Miller G. (1959). Use of dinitrosalicylic acid reagent for detection of reducing sugars. *Analytical Chemistry* 31: 426-428.

Molano J., Duran A. and Cabib E. (1977). A rapid and sensitive assay for chitinase activity using tritiated chitin. *Analytical Biochemistry* 83: 648-656.

Monreal J. and Reese E.T. (1969). The chitinase of *Serratia marcescens*. *Canadian Journal of Microbiology* 15: 689-696.

Morimoto K., Karita S., Kimura T., Sakka K. and Ohmiya K. (1997). Cloning, sequencing and expression of the gene encoding *Clostridium paraputrificum* chitinase ChiB and analysis of the functions of a novel cadherin-like domains and a chitin binding domain. *Journal of Bacteriology* 179: 7306-7314.

Niimi K., Niimi M., Shepherd M.G. and Cannon R.D. (1997). Regulation of N-acetylglucosaminidase production in *Candida albicans*. *Archives of Microbiology* 168: 464-472.

O'Brien M. and Colwell R.R. (1987). A rapid test for chitinase activity that uses 4-methylumbelliferyl-N-acetyl-β-D-glucosaminide. *Applied Environmental Microbiology* 53: 1718-1720.

Peberdy J.F. (1985). Mycolytic enzymes. In: Fungal protoplasts-applications in biochemistry and genetics, Peberdy, J.F. and Ferenczy, L. Eds., New York: Marcel Dekker. 31-44.

Pera L.M., Baigori M.D. and Callieri D. (1997). Purification of β-N-acetyl-D-glucosaminidase from *Aspergillus niger* using thermostabilization and heat as the initial step. *Biotechnology Techniques* 11: 35-37.

Pinto A. deS., Barreto C.C., Schrank A., Ulhao C.J. and Vainstein M.H. (1997). Purification and characterization of an extracellular chitinase from the entomopathogen *Metarhizium anisopliae*. *Canadian Journal of Microbiology* 43: 322-327.

Pishko E.J., Kirkland T.N., Cole G.T. (1995). Isolation and characterization of two chitinase-encoding genes (cts1, cts2) from the fungus *Coccidioides immitis*, *Gene* 167(1-2): 173-7.

Reissig J.L., Strominger J.L. and Leloir L.F. (1955). A modified colorimetric method for the estimation of N-acetylamino sugars. *The Journal of Biological Chemistry* 217: 959-966.

Reyes F., Calatayud J., Vazquez C. and Martinez M.J. (1989). Beta-N-acetylglucosaminidase from *Aspergillus nidulans* which degrades chitin oligomers during autolysis. *FEMS Microbiology Letters* 53: 83-87.

Robert L.M., Clemens K.P., Kathrin P., Sylvia J., Sheridan L.W., Susanne Z., Cornelia M.K., Matteo L., and Christian P.K. (1999). Expression of Two Major Chitinase Genes of *Trichoderma atroviride* (*T. harzianum* P1) Is Triggered by Different Regulatory Signals. *Applied Environmental Microbiology* 65: 1858-1863.

Rondle C.J.M. and Morgan W.T.J. (1955). The determination of glucosamine and galactosamine. *Biochemical Journal* 61: 586-589.

Sahai A.S. and Manocha M.S. (1993) Chitinases of fungi and plants: their involvement in morphogenesis and host-parasite interaction. *FEMS Microbiology Reviews* 11: 317-338.

Sakai K., Narihara M., Kasama Y., Wakayama M. and Moriguchi M. (1994). Purification and characterization of thermostable N-acetylhexosaminidase of *Bacillus stearothermophilus* CH-4 isolated from chitin containing compost. *Applied Environmental Microbiology* 60: 2911-2915.

Sakurda M., Morgavi D.P., Komatani K., Tomita Y. and Onodera R. (1996). Purification and characterization of a cytosolic chitinase from *Piromyces communis* OTS1. *FEMS Microbiology Letters* 137: 75-78.

Souza R.F., Gomes R.C., Coelho R.R., Alviano C.S. and Soares R.M. (2003). Purification and characterization of an endochitinase produced by *Colletotrichum gloeosporioides*. *FEMS Microbiology Letters* 222: 45-50.

St Leger R.J., Cooper R.M. and Charnley A.K. (1986). Cuticle-degrading enzymes of entomopathogenic fungi: Cuticle degradation *in vitro* by enzymes from entomopathogens. *Journal of Invertebrate Pathology* 47: 167-177.

Svitil A., Chadhain S., Moore J. and Kirchman D. (1997). Chitin degradation proteins produced by the marine bacterium *Vibrio harveyii* growing on different forms of chitin. *Applied Environmental Microbiology* 63: 408-413.

Takaya N., Yamazaki D., Horiuchi H., Ohta A. and Takagi M. (1998). Intracellular chitinase gene from *Rhizopus oligosporus*: molecular cloning and characterization. *Microbiology* 144: 2647-2654.

Takaya N., Yamazaki D., Horiuchi H., Ohta A. and Takagi M. (1998) Cloning and characterization of a chitinase-encoding gene (*chiA*) from *Aspergillus nidulans*, disruption of which decreases germination frequency and hyphal growth. *Bioscience Biotechnology and Biochemistry* 62(1): 60-65.

Techkarnjanaruk S. and Goodman A. (1999). Multiple genes involved in chitin degradation from the marine bacterium *Pseudoalteromonas* sp. strain S91. *Microbiology* 145: 925-934.

Tews I., Scheltinga A.C., Terwisscha V., Perrakis A., Wilson K.S. and Dijkstra B.W. (1997). Substrate assisted catalysis unifies two families of chitinolytic enzymes. *Journal of American Chemical Society* 119: 7954-7959.

Tronsmo A. and Harman G.E. (1993). Detection and quantification of N-acetyl-β-D-glucosaminidase, chitobiosidase and endochitinase in solutions and on gels. *Analytical Biochemistry* 208: 74-79.

Ulhoa C.J. and Peberdy J.F. (1991). Regulation of chitinase synthesis in *Trichoderma harzianum*. *Journal of general Microbiolgy* 137: 2193-2169.

Ulhoa C.J. and Peberdy J.F. (1992). Purification and some properties of the extracellular chitinase produced by *Trichoderma harzianum*. *Enzyme and Microbial Technology* 14: 236-240.

Vasseur V., Arigoni F., Andersen H., Defago G., Bompeix G. and Seng J.M. (1990). Isolation and characterization of *Aphanocladium album* chitinase overproducing mutants. *Journal of General Microbiology* 136: 2561-2567.

Viterbo A, Montero M, Ramot O, Friesem D, Monte E, Llobell A, and Chet I. (2002). Expression regulation of the endochitinase chit36 from *Trichoderma asperellum* (*T. harzianum* T-203). *Current Genetics* 42(2): 114-122.

Wattanalai R., Wiwat C., Boucias D.G. and Tartar A. (2004). Chitinase gene of the dimorphic mycopathogen *Nomuraea rileyi*. *Journal of Invertebrate Pathology 85*: 54-57.

Wirth S.J. and Wolf G.A. (1990). Dye-labelled substances for the assay and detection of chitinase and lysozyme activity. *Journal of Microbiology Methods* 12: 197-205.

Yamada H. and Imoto T. (1981). A convenient synthesis of glycol chitin, a substrate of lysozyme. *Carbohydrate Research* 92: 160-162.

6 Arbuscular Mycorrhizal Fungi in Agricultural Ecosystem

Cesar Arriagada Escamilla[1] and Mahendra Rai[2]

[1]Estación Experimental del Zaidín, Departamento de Microbiología del Suelo, Profesor Albareda 1. CP-18008, Granada-España.
[2]Department of Biotechnology, SGB Amravati University, Amravati-444 602, Maharashtra, India.

INTRODUCTION

The interest of scientists in manipulation of soil microorganisms and their metabolic processes arose from the numerous studies showing that these microorganisms have a great potential for improvement of soil quality, degradation or immobilization of xenobiotic compounds, plant growth and plant disease resistance.

Arbuscular mycorrhizal (AM) fungi develop a symbiotic relationship with higher plants. This symbiosis is biotrophic and normally mutualistic. The long-term compatible interaction being based on bi-directional nutrient transfer between the symbionts, normally benefit plant growth, particularly through enhanced nitrogen and phosphorus uptake (Pearson and Jackobsen, 1993), and have been reported to enhance the plant growth by supplying phosphate (Stoppler *et al.*, 1990; Baon *et al.*, 1992), nitrogen (Cliquet and Stewart, 1993) and other nutrients from the soil and translocating them to the host plant (Smith *et al.*, 1986), improving soil aggregation (Schreiner and Bethenfalvay, 1995). AM fungi protect plants against the toxic effects of excessive concentrations of heavy metals (Fabig, 1982; Gildon and Tinker, 1983; Heggo *et al.*, 1990; Haselwandter and Berreck, 1994; Rivera-Becerril *et al.*, 2002). However, previous studies have shown conflicting results as to whether AM fungi increase or decrease Cd uptake by plants (Fabig 1982; Heggo *et al.*, 1990; Karagiannidis and Nickolaou, 2000). Cd accumulation in AM plants seems to depend on Cd soil concentration and on the kind of AM fungi (Karagiannidis and Nickolaou, 2000; Arriagada *et al.*, 2004). AMF also increases resistance to drought and salinity and acts as bioprotectors (Benhamou *et al.*, 1994). A survey of literature reveals that a number of investigators have significantly contributed in this field (Gianinazzi *et al.*, 1990; Berta *et al.*, 1990; Guillemin *et al.*, 1992; Schubert *et al.*, 1992; Azcon-Aguilar *et al.*, 1992; Rapparini *et al.*, 1996; and Morte *et al.*, 1996).

In vitro culture induces anatomical and physiological modifications that render plants sensitive to water stress. In fact, micropropagated plantlets are exposed to different conditions, notably a weak vapors pressor deficit (VPD), reduced light and presence of sucrose in the growth media. These conditions induce a number of adaptations characterized by non-functional stomata and less developed cuticle.

Acclimatization of micropropagated plantlets is a crucial stage for crop and tree production of economic importance. The relevance of mycorrhizas for the acclimatization success of micropropagated plants has gained importance in the recent past. Mycorrhization provide better adaptation in water stress in plantlets. Mycorrhizas seems to act in two ways: i) due to the symbiosis the plant can reach its best performance, ii) mycorrhizas are able to buffer the action of stress during acclimatization. Mycorrhiza formation appears to play a key role in favouring *ex vitro* development of micropropagated plantlets.

Inspite of such complexities, the mycorrhizal fungi have dominant importance in the root region by virtue of their close relationship with cells and tissues of their hosts, on the one hand, and with soil particles and soil solutions on the other hand. The carbon compounds derived from photosynthesis are utilized by both the host and the fungus. The latter absorbs soil-borne nutrients, part of which is translocated to the host. The system, therefore, consists of hyphae ramifying in the soil which connect with the root tissue or even penetrate the root cells. Nutrients synthesized or absorbed are, shared between the fungus and host. In these plants the fungus supplies the whole symbiotic system with carbon and inorganic nutrients. In these special cases, the plant is often considered to be dependent on its fungal partner and indeed the type of fungus involved, one capable of obtaining carbon compounds from the soil or from living plants, is different. The general plan of functioning of mycorrhizal plants is well substantiated, what is needed and being actively sought is a greater knowledge of the physiological and biochemical processes and mechanism of absorption, translocation and release of substances by the partners and the ecological input of the symbiosis, and hence to sustain the ecosystem itself.

MAINTAINING FITNESS OF HOST-PLANT

Arbuscular mycorrhizal fungi maintain healthy growth of host-plants. They

- increase nutrient and water absorption through improved absorptive area
- increase nutrient mobilization through biological weathering - increase in feeder root longevity by provision of a biological deterrent to root infection by soil-borne pathogens
- accumulate nitrogen, phosphorus, potassium, calcium, magnesium and zinc and translocate them to the host
- permeate deep to the forest floor, and thus, minerals mobilized in this zone can be absorbed before they reach the sub-soil system
- can degrade complex minerals and organic substrates and make these essential elements available to the plants
- protection of delicate root-system from attack by pathogenic fungi through the following strategies: a) use of surplus carbohydrates b) provision of physical barrier c) secretion of antibiotics d) favouring protective rhizosphere microbes.
- provide host plant with growth hormones including auxins, cytokinis, gibberellins, and growth regulating B-vitamin important contributors to organic matter turnover and nutrient cycling in forest ecosystem
- AM fungal biomass can account for as much as 15% of the net primary production and a majority of it enters the soil organic matter annually
- mycorrhization of micropropagated plant increases the rate of out planting performance of these plant almost 90-100%.

As a biohardening tool

The technique of micropropagation has been applied recently for clonal selection and mass production of arbuscular mycorrhiza in woody plants (Salamanca *et al.*, 1992). Menge (1977) stated that inoculation of AMF to nursery plants has been proven both necessary and feasible, and it has been extended to micropropagated plants (Kiernan *et al.*, 1984; Ravolanirina *et al.*, 1989; Gianinazzi *et al.*, 1990; Guillemin *et al.*, 1992; Varma and Schuepp, 1995). Inoculants have already been commercialized (Sieverding, 1991). The available information mainly concerned tree species and reforestation programmes in mediterranean ecosystems (Morte *et al.*, 1996). Salamanca *et al.*, (1992) studied mycorrhizal inoculation of micropropagated woody legumes used in revegetation programmes for desertified mediterranean ecosystems. Vestberg (1992) reported that the general use of micropropagation for production of elite stock of strawberry has opened up even more possibilities (Niemi and Vestberg, 1992). Varma and Schuepp (1995) and Lovato *et al.*, (1996) stated that inoculation of micropropagated plantlet with active cultures of AMF appears to be critical for their survival and growth. This avoids "Transient transplant shock" and stunted growth on transfer in the field. Endomycorrhization can modify root architecture to give a root system which is better adapted for uptake of mineral nutrient and water (Berta *et al.*, 1990), as well as increasing hormone production (Allen, 1985) and resistance to pesticides or root pathogens.

Micropropagated plantlets inoculated with AM spores increases the survival rate and growth in potted condition. *Vitis vinifera* plantlets inoculated with *Glomus epigeous* significantly increased percentage infection, biomass, photosynthesis of host and increased resistance to downy mildew disease, compared with control.

As Bioprotectors

A review of literature indicates that a lot of work has been done on this aspect (Schenck, 1981; Caron, 1989; Jalali and Jalali, 1991; Sharma *et al.*, 1992; Norman *et al.*, 1996). According to recent view there is no single response and degree of protection is offered by the symbiosis, at all. It varies with plant species. However, in majority of the cases, a reduction of disease is reported.

The recent developments in research about mechanisms involved in the resistance of mycorrhizal plants to biotic or abiotic stresses are dedicated to the study of a single mechanism (chitinase, phenol, proteins) or to the comparison of the infection process of a pathogen and a mycorrhizal fungus (Dehne and Schonbeck, 1978). High levels of amino acids, especially arginine, which occur in mycorrhizal plants, have been found to suppress sporulation in *Thielaviopsis* (Baltruschat and Schonbeck, 1975). Increase in peroxidases and phytoalexins have also been reported for AMF plants and it has been suggested that this confers resistance. In addition, it is known that there are proteins that are synthesized only in roots colonized by AMF (Schellenbaum *et al.*, 1992).

Interactions between AMF and plant pathogens can also result in the development of the pathogens as a site for mycorrhizae establishment or increase the disease incidence (Varma and Shankar, 1994). The evaluation of AM mycorrhizae as biocontrol agents remains one of the challenging areas. If prior root colonization by AMF can reduce root disease, then this would be of vital interest for many tropical crops that are grown in nurseris before transplantation. This dimension of mycorrhizae in inducing suppression of soil-borne disease holds great promise.

As bioremediation alternative for contaminated soil

Contamination of the biosphere with heavy metals due to man-made activities generates a great environmental and human health problem. Higher concentration of heavy metals in soils select plant population with a high resistance to these elements but with a low diversity of species in different trophic levels (Ernst, 1990). Pollutant can remain in soil for a long time, for that the use of tolerant plants to accumulate metal for subsequent processing is both technically and economically attractive.

The entry of contaminants into the environment results from either natural processes or human activity. Natural contamination originates from either excessive withering of mineral and metal ions from rocks or from displacement of certain contaminants from the groundwater or subsurface layer of the soil. The most common human-assisted routes for entry of inorganic contaminants, and heavy metal in particular, into agricultural and non-agricultural lands are via disposal of industrial effluents, sewage sludges, deposition of air borne, industrial wastes, military operations, combustion of fossil fuel, petrochemical processing, mining and smelting, pesticides, herbicides, fertilizers and any other process.

The presence of high concentrations of toxic metals in sludge poses a constraint on the application of these sludges on agricultural land (US EPA, 1993). A number of studies have been carried out in the last few years on the risks associated with the application of metal-charged sludge on agricultural land (Brookes and Verstraete, 1989). Metal accumulation in plants following sludge application on agricultural land was demonstrated for antimony, arsenic, cadmium, copper, iron, mercury, molybdenum, nickel, lead, selenium and zinc. Metals can also make it to the food chain by sticking to the surface of vegetation when metal containing sludge is spread on agricultural land (Chaney *et al.*, 1998).

Application of sewage sludge to agricultural land is a practice highly developed. The sewage improves the soil physical properties and nutrients status (Sauerbeck, 1987). These sludge can, nevertheless, contain considerable concentrations of heavy metals persist long time in soil after application (Juste and Mench, 1992). At present in the European Union, among the 50 and 70% of sewage sludge are disposed of on hand. This represents an average application of 50-100 Kg of dry matter ha^{-1} of agricultural soil y^{-1} and the trend is rising (Brookes and Verstraete, 1989). Therefore, the contribution of sludge to the overall input of metals to the environment is considerable. Land application of municipal sewage sludge is recognized as a valuable resource that can be recycled as fertilizer, source of organic matter, improving the structure and also contains phosphorus and nitrogen, essential nutrients of plants, is a major concern for the bioaccumulation of cadmium and other metals in the food chain (Alloway, 1995). Current European Union regulations for the maximum metal levels in agricultural soils (CEC, 1986), and the maximum admissible metal concentration of food (Ewers, 1991) are based on numerous studies related with phytotoxicity and zootoxicity.

Pollution caused by heavy metals restrict the establishment of plants at the same time the number and activity of soil microorganisms and microbial processes, also reduced. The high toxicity of heavy metals to soil microbes and microbiological processes, associated to the long term effects in the soil, are recognized as important facts at the moment although there is little information.

All organisms, including microorganisms show resistance to heavy metals by "tolerance" when the organism survives in the presence of high internal metal concentrations, or by "avoidance" when the organism is able to restrict metal uptake (Baker 1987; Turnau *et al.,* 1996). The concept of using plants to accumulate metal for subsequent processing is both technically and economically attractive. The use of plants to remove toxic metals from soils (phytoremediation) is emerging as a potential strategy for cost-effective and environmentally sound remediation of contaminated soils.

The generic term "phytoremediation" consist of the Greek prefix phyto (plant) attached to the Latin root remedium (to correct or remove an evil). Phytoremediation can be defined as the process of utilizing plants to absorb, accumulate, and detoxify contaminants in the growth substrate (soil, water and air) through physical, chemical or biological processes. To sum up, phytoremediation is the use of plants to remediate contamination of soil with organic wastes (Cunningham and Berti, 2000).

The use of plants for remediating heavy metal contaminated soils has multifold advantages as follows:

- Large scale application
- A major landscape valuation of contaminated sites
- Potential economic benefit of an agronomy/forestry based technology, and some plant species can have potential economic returns
- Plants concentrate the contaminants within their tissues
- Concentrated hazardous waste requires smaller reclamation facilities for extracting the heavy metals.

Apart from direct advantages, plants provide indirect benefits to the contaminated sites such as:

- Increased aeration of the soil which in term enables microbial degradation of organic contaminants and microbe-assisted uptake of metal contaminants
- Reduced eroded soil
- Increase in soil microbiota maintaining a healthy ecosystem
- Improve the chemical, physical and biological properties of the soil.

As soil metals cannot be biodegraded, remediation of soil contaminated with heavy metals, mainly limited to immobilization or extraction/concentration processes. The phytoremediation option has appeared in these last years as a valid tool.

Phytoremediation is the use of vegetation for *in situ* treatment of soil pollution, sediments, and water. Many researchers have described general processes by which plants can be used to remediate metal-contaminated soils (Adriano *et al.*, 1995; Cunningham *et al.*, 1995; Salt *et al.*, 1995; Saxena *et al.*, 1999; Gert, 2000).

In much polluted soils mycorrhizal diversity appears to be reduced or sometimes eliminated. AM fungi can improve heavy metal tolerance of their host plants and this should not always be attributed to an improved nutrition alone (Colpaert, 1998).

Mycorrhizal colonization of plant roots can reduce translocation of heavy metals to shoots by binding of the heavy metals to the cell walls of the fungal hyphae in roots (Dehn and Schüepp, 1989). It is possible that the plants and not the fungus were tolerant to metals (Leyval *et al.*, 1997) and that the benefit could have been due to mycorrhizal effects other than alleviation of metal toxicity. However, differences in efficiency of mycorrhizal fungi on plant metal uptake have been reported (Medeiros *et al.*, 1994) suggesting, as observed for ectomycorrhizae, that the association between the plant and the fungus is crucial to the mycorrhizal benefit in metal-polluted soils. In this way, mycorrhiza can help higher plants to adapt and survive in contaminated habitats (Wilkinson and Dickinson, 1995).

Plants can be used in the establishment of contaminated soils with heavy metals. In fact, plants have mechanisms for accumulation, tolerating or alleviation of high level of heavy metals in contaminated soil (Khan *et al.*, 2000). Many of the accumulative plants used belong to the family *Brassicaceae* which do not form AM symbiosis. Some varieties of *Thlaspi* and ecotypes of *Silene vulgaris* have been found to be Cd accumulators (Brown *et al.*, 1994; Ernst *et al.*, 2000). However, these plants produce little biomass being more interesting plant with higher productivity such as trees (Greger and Landberg, 1997). *Eucalyptus globulus* is a tree species with a wide plasticity to grow in impoverished or marginal soils. Studies of AM fungi symbiosis in trees are scarce (Wilkinson and Dickinson, 1995). The *Eucalyptus* species are able to develop AM and Ectomycorrhizal symbiosis and show great tolerance to heavy metals Cu, Zn and Cd (Pereira, 1998; Arriagada *et al.*, 2004). Arriagada *et al.,* (2004) found that AM fungi *Glomus mosseae* and *G. deserticola* increased the dry weight, root length, total N, P and K concentration and the quantity of chlorophyll of *E. globulus* shoot. The protection for this plant by AM fungi against the action of the heavy metals was more evident when this plant was grown as intercrop with *Glycine max* than as monoculture.

On the other hand, it is known that soil microorganisms influenced the development of the AM symbiosis. The saprobe fungi are important and common components of rhizosphere soil where they obtain greater nutritional benefit from organic and inorganic compounds released from living roots together with sloughed cells (Dix and Webster, 1995), and are used to degrade toxic substances (Wainwright, 1992; Madrid *et al.*, 1996). These facts indicate that saprobe and AM fungi can be important in the recovery or restoration of contaminated soils with heavy metals and these problems reduce the numbers of soil microorganisms and their activities (Baath, 1989). The existence of synergistic effects of saprobe fungi such as, *Fusarium concolor* and *Trichoderma koningii* on plant root colonization by AM fungi and on the effectiveness of AM fungi on plant resistance to heavy metals in soil has been proved (Fracchia *et al.*, 2000; Srinath *et al.*, 2003; Arriagada *et al.*, 2004).

AS NUTRITION SUPPLIER

One of the major limitations to the wide spread use of AMF is the lack of reliable inoculant in suitable forms and amounts economically adapted to plant protection (Gianinazzi and Schüepp, 1994; Varma and Hock, 1995).

The most important benefits of mycorrhiza is the increase in the phosphorous uptake by the plant. After its absorption P is allocated directly for various structural and physiological function of the plant or stored in vesicles of AMF for later utilization. The mycorrhizal effect generally decreases with an increase in phosphorous level in soil. The enzymes of polyphosphate metabolism are present in AMF.

AM fungi help plants in the uptake of micronutrients, such as, copper, zinc, magnesium, manganese and cobalt. Plants absorb and store these elements so as to prevent their concentration to reach toxic levels. Concentration of these elements have been significantly correlated with the developmental stages of the plant. Iron has also been implicated in ion uptake and metabolism in the *Rhizobium*-legume symbiosis (Koide and Schreiner, 1992).

Nitrogen is needed for the formation of amino acids, purines, pyrimidines and thus indirectly involved in protein and nucleic acid synthesis. Mycorrhizae affect nitrogen cycle in several ways. Mycorrhizal fungal hypae have the ability to extract nitrogen and transport it from soil to plant because of the enhance absorption surface extended by them. A mycorrhizal fungus contains enzymes which break down the organic nitrogen reductase for altering the forms of nitrogen in soil. Nodulation by the endogenous is greatly improved by AM fungi (Haymen, 1983).

Hormones accumulation in the host tissue is affected by mycorrhizal colonization with changes in the levels of cytokinins, abscissic acid, gibberelins like substances. Alteration in the biomass partitioning between shoot and roots as well as morphological effect may be harmonal mediated. The effect of AM fungi on photosynthesis and host morphology could also be hormonal. *Glomus mosseae* has been shown to synthesise phytohormones. Some ectomycorrhizal fungi produce auxin, cytokinins and gibberellins.

In addition to the nutrient uptake, mycorrhizae also enhance the transport of water from soil to plant. Since mycorrhizal plants have higher transpiration rates, the water uptake per unit root length and thus the mass flow of soil solution to the root surface is twice as higher as in non-mycorrhizal plants.

ARBUSCULAR MYCORRHIZAL IN FRUIT TREES

AM fungi are a major component of the soil microbial biomass in most ecosystems (Allen, 1991). Arbuscular mycorrhizal ubiquitous symbioses between fungi of the order *Glomales* and roots of higher plants constitute a crucial link at the root-soil interface. The mineral nutritional aspects of AM symbiosis are widely appreciated (Smith and Read, 1997).

Citrus roots normally form symbiotic associations with AM fungi (Camprubi *et al.*, 1995) and apple plants were benefited by improving growth and nutrition (Geddeda *et al.*, 1984). In mycorrhizal associations, plants supply labile photosynthates to fungi, while fungi aid in the uptake of nutrients, especially P.

Benson and Covey (1976) reported that inoculation of apple seedlings with *G. fasciculatum* in a fumigated soil caused greater growth stimulation than that of *G. mosseae*. Gunze and Hennessy (1980) found an increased number of AM structures in inoculated *Vigna unguiculata* (L.) Walp. plants treated with indolebutyric acid (IBA).

Gnekow and Marschner (1989) concluded that apple cuttings (*Malus pumila* var. *domestica* cv. M26) rely on AM P-uptake at least in low P soils. In high P soils, apple cuttings may profit predominantly from the uptake of Zn and Cu by the fungal symbionts.

Yano-Melo *et al.,* (1999) reported that *Glomus clarum* and *G. etunicatum* enhanced growth, photosynthesis and transpiration rates of *Musa*, but these effects were probably caused by improved host P nutrition. AM fungi increased root length of *Citrus jambhiri* L. and increased whole plant transpiration when soil was moist. Schubert and Lubraco (2000) demonstrated an increase in growth by inoculation with *G. mosseae* in apple (*Malus pumila* L.) plants of the rootstock clone MM106.

Fidelibus *et al.,* (2000) observed four AM fungus isolates of the genus *Glomus* from areas with different edaphic and climatic conditions stimulated root growth and P uptake of 'Volkamer' lemon seedlings under conditions of high P. Some studies between AM-hoist fruit trees and herbicides, show that AM fungi are beneficial to apple tree growth (Plenchette *et al.,* 1981). High levels of herbicide adversely affected the population of mycorrhizae which in turn could reduce tree vigour (Granger *et al.*, 1995).

Growth promotion in *Spilanthes calva* and *Withania somnifera* Dunal

Spilanthes calva and *Withania somnifera* are important tropical medicinal plants commonly used in ayurveda (Indian natural therapy) and other traditional systems of medicine (Bhargava and Singh 1978; Dey 1980; Sankaran 1984; Sahni and Srivastava 1993; Tripathi *et al.,* 1996). *W. somnifera* is also known as Indian Ginseng and belongs to the family Solanaceae. There are more than ninety-one pharmaceutical products produced from the roots of this plant. *S. calva*, a member of Asteraceae, is another medicinal herb known as "toothache plant" or "virus blocker" and is well known for enhancing immunity. Because of high medicinal value, there is an increasing demand for these plants in national and international market. Therefore, there is a greater need to enhance the growth and bio-active agents of these plants. This can be achieved by inoculation of microorganisms like arbuscular mycorrhizal fungi or other growth promoting endophytes.

Enhancement of growth of *Spilanthes calva* and *Withania somnifera* Dunal : A case study

A pronounced growth response following inoculation with *Piriformospora indica* (=arbuscular mycorrhiza-like) was observed in *Spilanthes calva* and *Withania somnifera.* Data revealed that the plants treated with *P. indica* were superior in development to control plants (Tables 6.1, 6.2 and 6.3). A significant increase in the shoot length was observed in the inoculated plants. The microscopic examination of stained root samples revealed a high colonization of *S. calva* and *W. somnifera* by *P. indica* in 62 and 73 per cent root length, respectively.

Table 6.1. Influence of *Piriformospora indica* on shoot and root length and on per cent root colonization of *S. calva* and *W. somnifera* in a field trial. The control plants were treated with an equal amount of autoclaved mycelium

Treatment	Shoot length (cm)		Root length (cm)		Percent colonization	
	S. calva	***W. somnifera***	***S. calva***	***W. somnifera***	***S. calva***	***W. somnifera***
+ *P. indica*	28.76 (±0.68)	67.3 (±0.26)	14.46 (±0.45)	16.53 (±0.25)	62.33 (±2.08)	73 (±2.64)
Control	25.16 (±0.20)	42.23 (±0.2)	12.13 (±0.15)	13.03 (±0.15)	Nil	Nil

All values are Mean ± S.D.; differences between inoculated and control plants are significant at P<0.05

Table 6.2. Influence of *P. indica* on morphology and growth of host plants 90 days after inoculation in a field trial. The control plants were treated with equal amount of autoclaved mycelium

Hosts	Treatment	Basal area (cm^2)	Leaf area (cm^2)	Length of Head (cm^2)	Diameter of Inflorescences (cm)	No. of flower/ Inflorescences	No. of seeds/ fruit
Spilanthes calva	***+P. indica***	7.06 (±0.47)	37.67 (±2.28)	2.49 (±0.09)	5.06±0.32	48.57 (±0.4)	1006 (±7.63)
	Control	4.11 (±0.57)	26.00 (±2.87)	1.48 (±0.03)	4.16±0.76	11.50 (±3.6)	716 (±0.36)
Withania somnifera	***+P. indica***	11.40 (±0.61)	45.59 (±0.34)	Nil	Nil	307.40 (±0.53)	46.33 (±5.77)
	Control	5.57 (±0.49)	13.08 (±0.73)	Nil	Nil	81.80 (±1.57)	35.33 (±4.93)

All values are Mean ± S.D.; differences between inoculated and control plants are significant at P<0.05

Table 6.3. Influence of *P. indica* on biomass of field grown *S. calva* and *W. somnifera*. The control plants were treated with equal amount of sterile fungus biomass

Hosts	Treatment	Fresh weight (gm)		Dry weight (gm)		NPP gm/	ED plant/day
		AGP	UGP	AGP	UGP		
Spilanthes calva	**+P. indica**	74.74 (±0.65)	9.26 (±0.15)	14.76 (±0.11)	2.13 (±0.23)	0.06	211.13
	Control	8.74 (±0.55)	6.26 (±0.35)	6.54 (±0.06)	1.46 (±0.06)	0.02	
Withania somnifera	**+P.indica**	152.53 (±0.76)	10.70 (±0.26)	63.03 (±0.15)	4.63 (±0.15)	0.23	671.90
	Control	19.07 (±1.1)	3.77 (±0.25)	8.67 (±0.35)	1.40 (±0.36)	0.12	

Note: NPP, Net primary productivity; AGP, Aboveground parts; UGP, Underground parts; ED, Endophyte dependency
All values are Mean ± S.D.; differences between inoculated and control plants are significant at $P<0.05$

The basal and leaf area of treated plants was also high (Table 6.2). Interestingly, in inoculated *S. calva* plants, some large, kidney-shaped heads were observed among the normal round heads. These kidney-shaped heads were never observed in control plants. The length of the inflorescence and the number of flowers in inoculated *S. calva* were also increased (Table 6.2). Similarly, in the inoculated plants of *W. somnifera,* the number of flowers was higher (Table 6.2) as compared to controls. In both medicinal plants, seed count was higher as compared to controls.

The overall root biomass in the inoculated plants was higher than in the corresponding controls. The fresh and dry-weight of under- and aboveground parts of inoculated plants of *S. calva* and *W. somnifera* plants was greater than that of control (Table 6.3). The Net Primary Productivity (NPP) of inoculated *S. calva* and *W. somnifera* plants was 0.06 and 0.23 g/plant/day, respectively. These values were higher than for control plants (0.02 and 0.12 g/plant/day, respectively).

MD is used as an index to compare receptivity of different plant species to AM fungi (Gerdemann, 1975). This can also be used for other endophytes, like *P. indica.* In the present study, the term ED (Endophyte dependency) is used instead of MD as the test organism does not develop a typical mycorrhizal association. The ED value for *S. calva* was 211.13 and that for *W. somnifera* was 671.90. The data suggests that the later plant is more dependent than the former on *P. indica* for growth.

MD of a host plant can be altered by a number of factors such as soil-type, soil P content, mycorrhizal species, etc. (Menge *et al.*, 1978; Azcon and Ocampo, 1981). Amongst reasons proposed for differences in MD in different plants or varieties of the same species, Baylis (1975) reported that the root-hair length and root thickness can determine the MD level of a plant species. Rajapakse and Miller (1988) observed that average length of fine roots was negatively correlated with MD in cowpea. The value of ED in *W. somnifera* was greater than that of *S. calva.* The more intense root proliferation in treated plants could be due to synthesis of unidentified extracellular phytohormones by *P. indica* (Varma *et al.*, 2001).

In a nutschell, there was a significant increase in growth and yield of both the plants, viz., *S. calva* and *W. somnifera* inoculated with *P. indica* as compared to uninoculated controls. The growth differences observed between inoculated and control may have been caused by a greater absorption of water and mineral nutrients by extensive colonization of *P. indica* in roots. The ability of *P. indica* to continue improving growth of *S. calva* and *W. somnifera* even during the hot March-

June summer season (the day temperature raises above 40°C) suggests that the fungus may improve drought tolerance of the plants. However, this should be tested elsewhere in other field and soil conditions. The increase in fresh and dry-weight in treated *S. calva* and *W. somnifera* could be related to increased root P uptake. This assumption is based on the earlier report of Sudha *et al.* (1998), where they had found an active translocation of phosphate (P^{32}) in rice and transformed carrot roots on inoculation with *P. indica* (Varma *et al.*, 2001). Our present findings widen the potential for using *P. indica* for large scale cultivation of *S. calva* and *W. somnifera*.

Bioremediation of Cadmium in soils by Eucalyptus, Arbuscular Mycorrhizal and Saprobe fungi: A case study

The experiments were carried out using Eucalyptus (*Eucalyptus globulus* Labill.). Plants were grown in a greenhouse and watered every week with 10 mL of a nutrient solution with 50 mgL^{-1} of P (Hewitt, 1952). The AM fungi *Glomus deserticola* (Trappe, Bloss and Menge) and saprobe fungi *Trichoderma koningii* were used. Plants were inoculated at the time of transplantation, after 3 weeks of growth. The saprobe fungi were inoculated at the same time as *G. deserticola*. Five replicate pots per treatment were used. Cd was applied to eucalyptus pots at the concentration of 0, 50 and 100 mg Cd kg^{-1} of substrate pot. Our work was to evaluate the influence of the inoculation of arbuscular mycorrhizal fungi and several saprobe fungi on development of *E. globulus* in contaminated soils with Cadmium (Fig. 6.1).

The sensibility of plants to the phtotoxicity of Cd was diverse in the different plant species (Huang and Cunningham, 1996). The application of 50 mg kg^{-1} of Cd to soil decreased the shoot dry weight of eucalyptus. The percentage of AM root length colonization of eucalyptus also decreased by these doses of Cd. In our experiments we observed that *T. koningii* was able to increase the shoot dry weight of eucalyptus inoculated with *G. deserticola* as well as the percentage of AM root length colonization (Fig. 6.1). The plant protection by saprobe and AM fungi to toxicity of Cd was dependent on the type of plant, microorganism and Cd concentration (Arriagada *et al.*, 2004).

The presence of *T. koningii* increased the protective action of *G. deserticola* in eucalyptus when 50 mg kg^{-1} of Cd was applied. Our results show that about 80 % of Cd was accumulated in the stem of *Eucalyptus* especially when 50 mg kg^{-1} of Cd were applied. The higher accumulation of Cd in less metabolically active part of the plant indicates that the damage caused on the physiology of the plant will be minor as compared to the metabolically active parts of the plant (Leep and Dickinson, 1998).

The doses of 100 mgL^{-1} of Cd decreased the AM root length colonization and metabolic activity of *G. deserticola* and also eliminate the synergistic effect of *T. koningii* on AM root colonization by *G. deserticola*.

A close relationship between *Chlorophyll* determinate after acetone extraction and SPAD values, using a *Chlorophyll* meter Minolta 502, was obtained. This *Chlorophyll*-meter used was valuable to measure *Chlorophyll*, taken into account the inherent limitations of the method (Monge and Bugbee, 1992). The Relationship between chlorophyll content and SPAD values in *E. globulus* with Cd-treatment were analysed (Fig. 6.3) and found that a perfect correlation in two methods to determine

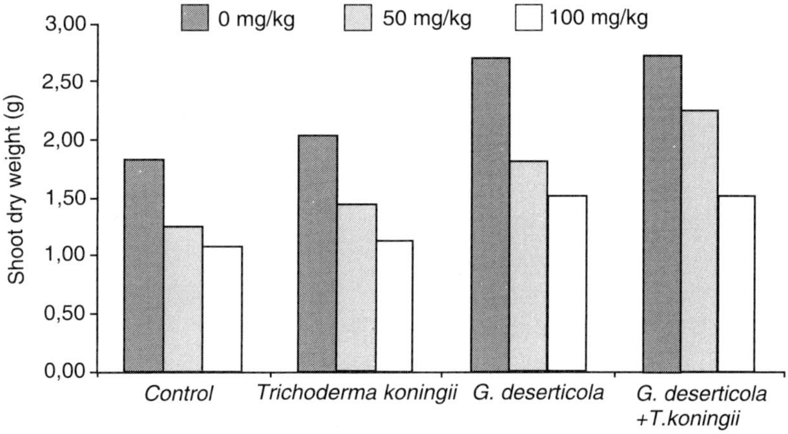

Fig. 6.1. Shoot dry weight of *Eucalyptus* colonised with AM and saprobe fungi plants cultivated with 0, 50 and 100 mg kg^{-1}Cd.

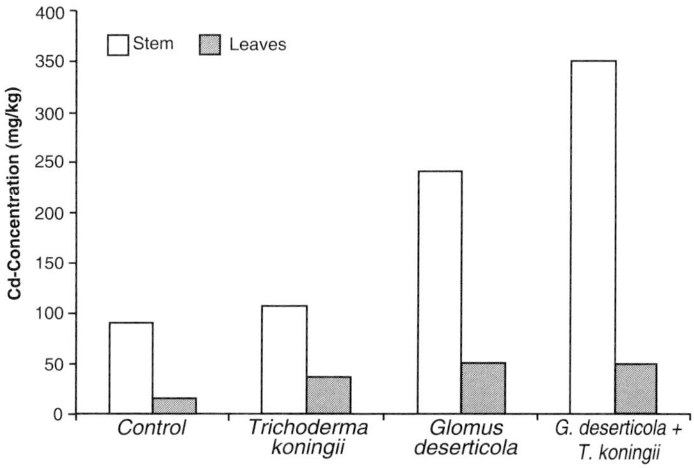

Fig. 6.2. Cd concentration in *Eucalyptus* plants colonised with AM and saprobe fungi at 50 mg/kg of Cd in soil.

Chlorophyll content. Heavy metals usually decreased the total chlorophyll in higher plants. In fact, the heavy metals are reported to inhibit *Chlorophyll* biosynthesis, principally by inhibiting d-amino-levulinic acid dehydrogenase and protochlorophyllide reductase (Ouzounidou, 1995). Exposure of whole plants to Cd affects *Chlorophyll* and *Chloroplast* development in young leaves. Fifty and 100 mg kg^{-1} of Cd applied with essential nutrients or well competing for the central ion isolated *Chlorophyll* was substituted by heavy metals ions. It was discovered that the substitution of the central Mg in *Chlorophyll* by heavy metals *in vivo* is an important type of damage in metal stressed plants (Prasad and Strzalka, 1995). This effect is more notorious in non-inoculated plants with the arbuscular mycorrhizal fungi, since upon increasing the Cd concentration, essential nutrients of the plants decreases. It has been observed that an increment in Cd concentration decreased Fe-Mg, Ca and K disponibility (Ouzounidou *et al.*, 1997). Arbuscular mycorrhizal colonized plants produce the major total *Chlorophyll* with or without Cadmium (Table 6.4, Fig. 6.3).

Table 6.4. Total Chlorophyll content and Chlorophyll relative (SPAD values) in *E. globulus* with Cd-treatment and inoculated with *G. deserticola* and saprophytic fungi. Control (C), *Fusarium concolor* (Fc), *Pleurotus ostreatus* (Po), *Trichoderma sp* (Tsp), *Glomus deserticola* (Gd). Values in the same column followed by an identical letter are not significantly different (P£0.05, Tukey-test).

Treatment	*Total Chlorophyll mg/g F.W.*		*Relative Chlorophyll SPAD*	
0 ppm Cd				
C	0,573	ab	28,2	a
Fc	0.535	ab	28,0	a
Po	0,464	a	27,8	a
Tsp	0,562	ab	28,1	a
Gd	0,916	d	37,6	b
Gd+Fc	0,926	d	38,0	b
Gd+Po	0,898	d	37,3	b
Gd+Tsp	0,915	d	37,6	b
50 ppm Cd				
C	0,371	a	26,3	a
Fc	0,477	a	27,3	a
Po	0,364	a	25,1	a
Tsp	0,465	a	27,9	a
Gd	0,827	d	38,1	b
Gd+Fc	0,888	d	36,9	b
Gd+Po	0,843	d	37,6	b
Gd+Tsp	0,904	d	38,1	b
100 ppm Cd				
C	0,284	a	22,4	a
Fc	0,211	a	21,1	a
Po	0,229	a	22,5	a
Tsp	0,241	a	25,3	a
Gd	0,674	c	34,6	b
Gd+Fc	0,711	c	34,8	b
Gd+Po	0,676	c	35,0	b
Gd+Tsp	0,698	c	34,3	b

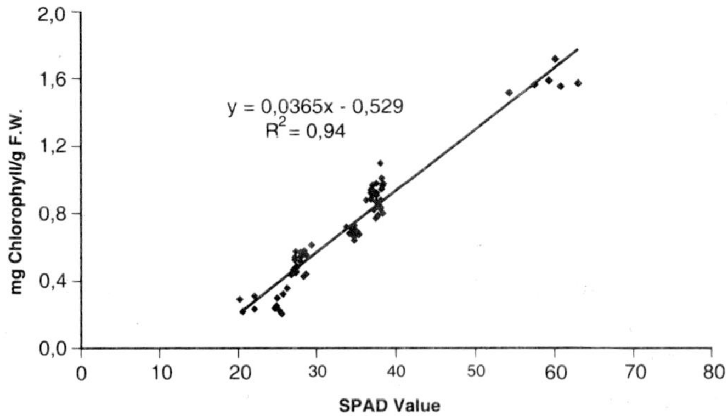

Fig. 6.3. Relationship between chlorophyll content and SPAD values in *E. globulus* with Cd-treatment and inoculated with *G. deserticola* and saprophytic fungi.

The saprophytic and AM fungi are most important because they represent a substantial proportion of microbial biomass in soil. Furthermore, the heavy metals degrades all levels to the ecosystem, reducing the size of the microbial population, and under these unfavorable environmental conditions, the role of AM and saprophytic fungi in phytoremediation of these soils may be very important. These microorganisms have an enormous potential for environment of soil quality, degradation or immobilization of xenobiotic compounds, plant growth and plant disease resistance. In conclusion the physical-chemical properties of soil, the host plant, the AM fungi, the concentration of metals in soil and the interaction of other microorganisms such as saprobe fungi, can improve the soil bioremediation incorporating these soils to the productive system.

CONCLUSIONS

The arbuscular mycorrhizal fungi represent as biofertilizer, biopesticides, and bio-hardning agents for agriculture crop, tissue-cultured-raised plantlets as well as forest biomass. AM fungi converts arid soil into a fertile and productive one. Chemical pesticides cause a serious problem of pollution and disturb the ecosystem. When the chemical fertilizers are used in the field about 75% of the phosphorus applied to the crops is not used during the first year by the crop and reverts to form unavailable to the plant. In the soil of high pH, aluminum and calcium levels almost 100% of the chemical phosphate becomes immobilized during chemical reaction. AM fungi increase the uptake of phosphorus, mineral, water and the phytohormone concentration in the plant. Although the exact biochemical relationship of these symbionts and the mechanism of the absorption of these minerals is not known clearly. But the mycorrhizal technology will be a useful tool in future for improving the crop production in sustainable agriculture.

REFERENCES

Adriano D.A., Chlopecka D. , Kaplan H. ,Clijsters and Vangronsveld J. (1995). Soil contamination and remediation: philosophy, science and technology. In: Proceedings of Third International Conference On The Biogeochemistry of Trace Elements. Paris (France) 15-19 May 1995, pp 465-504.

Allen M.F. (1985). Phytohormone action: an integrative approach to understanding diverse mycorrhizal responses. In: Proc 6th N Am Conf Mycorrhizae (R Molina, ed) For Scri Lab, Corvallis, OR, 158-160.

Allen M.F. (1991). The ecology of mycorrhizae. Cambridge University Press. Cambridge. pp.184

Alloway B.J. (1995). Cadmium. In: Heavy metal in soils. Alloway, B.J. (ed.). pp: 122-151. London. Blackie, Cop.

Arriagada C.A., Herrera M.A., Garcia-Romera I. and Ocampo J.A (2004). Tolerance to Cd of soybean (*Glycine max*) and eucalyptus (*Eucalyptus globulus*) inoculated with arbuscular mycorrhizal and saprobe fungi. *Symbiosis* 36: 285-299.

Azcon-Aguilar C., Barcelo A.,Vidal M.T.,de la Vina G. (1992). Further studies on the growth and development of micropropagated Avocado plants. *Agronomie* 12: 837-840.

Azcon R. and Ocampo J.A. (1981). Factors affecting the vesicular-arbuscular mycorrhizal infection and mycorrhizal dependency of thirteen wheat cultivars. *New Phytol* 87: 677-685.

Baath E. (1989). Effects of heavy metals in soil on microbial processes and populations. *Water, Air Soil Poll.* 47: 335-379.

Baker A.J.M. (1987). Metal tolerance. *New Phytologist* 106: 93-111.

Baltruschat H. and Schonbeck F. (1975). The influence of endotropic mycorrhiza on the infestation of tobacco by *Thielaviopsis basicola. Phytopathologische Zeitschrift* 84: 172-188.

Baon J.B., Smith S.E., Alston A.M., and Wheeler R.D. (1992). Phosphorus efficiency of three cereals as related to indigenous mycorrhizal infection. *Australian J. of Agric Res.* 43: 479-491.

Baylis G.T.S. (1975). The magnoloid mycorrhiza and mycotrophy in root systems derived from it. In: Sanders FE *et al.* (eds) Endomycorrhizas, Academic Press, London.

Benhamou N., Fortin J.A., Hamel C., St.-Arnaud M., Shatilla A. (1994). Resistance responses of mycorrhizal Ri T-DNA-transformed carrot roots to infection by *Fusarium oxysporum* f. sp. *chrysanthemi*. *Phytopath.* 84: 958-968.

Benson N.R. and Covey R.P. (1976). Response of apple seedlings to zinc fertilization and mycorrhizal inoculation. *HortScience* 11: 252-253.

Berta G, Fusconi A., Trotta A., Scannerini S. (1990). Morphogenetic modifications induced by the mycorrhizal fungus *Glomus* strain E3 in the root system of *Allium porum* L. *New Phytol* 114: 207-215.

Bhargava N.C., and singh O.P. (1978). Fortege: an indigenous drug in common sexual disorders in males. *Mediscope* 21: 140-141.

Brookes P.C. and Verstraete W. (1989). The functioning of soil as an ecosystem. In: Soil Quality Assessment: State of the Art Report on Soil Quality. Final Report to the Commission of the European Communities, Directorate-General XII, Contract EV4A/0008/NL, Agricultural University Wageningen.

Brown S., Chaney R., Angle J., Baker A. (1994). Phytoremediation potential of *Thlaspi caerulescens* and bladder campion for zinc- and cadmium-contaminated soil. *Journal of Environmental Quality* 23: 1151-1157.

Camprubi A., Calvet C. and V. Estaun (1995). Growth enhancement of Citrus reshni after inoculation with *Glomus intraradices* and *Trichoderma auroviride* and associated effects on microbial populations and enzyme activity in potting mixes. *Plant Soil* 173: 233-238.

Caron M. (1989). Potential use of mycorrhizae in control of soilborne diseases. *Can. J. Plant Pathol.* 11: 177-179.

CEC (Commission of the European Community). (1986). Council directive on the protection of the environment and in particular of the soil, when sewage sludge is used in agriculture. Official Journal European Communities. 181: 6-12.

Chaney R, Brown S, Angle J. (1998). Soil–root interface: ecosystem health and human food-chain protection. In: Huang P, ed .*Soil chemistry and ecosystem health.* Madison, Winsconsin: Soil Science Society of America, 279-311.

Cliquet J.B. and Stewart G.R. (1993). Ammonia assimilation in *Zea mays* L. infected with a vesicular-arbuscular mycorrhizal fungus *Glomus fasciculatum. Plant Physiology*, 101(3): 865-871.

Colpaert J.V. (1998). Biological Interactions: The significance of root microbial symbioses for phytorestoration of metal-contaminated soils. In: Metal Contaminated Soils: *In situ* Inactivation and Phytorestoration. Vangronsveld, J and Cunningham S.D (eds.) pp: 75-91.

Cunningham S.D. and Berti W.R. (2000). Phytoextraction and Phytostabilization: Technical, Economic, and regulatory considerations of the soil-lead issue. In: Phytoremediation of contaminated soils and water. Terry, N. and Bañuelos, G. (eds.), CRC Press, Boca Raton, Florida, pp: 359-376.

Cunningham S.D., Berti W.R. and Huang J.W.W. (1995). Phytoremediation of contaminated soils. *Trends Biotechnol* 13: 393-397.

Dehn B. and Schüepp H. (1989). Influence of VA mycorrhizae on the uptake and distribution of heavy metals in plants. *Agricultural, Ecosystem and Environment.* 29: 79-83.

Dehne H.W. and Schonbeck F. (1978). The influence of endotrophic mycorrhiza on plant diseases. 3. Chitinase activity and ornithine-cycle. *Z. Planzenkr. Pflanzzenschutz.* 82: 666-678.

Dey A.C. (1980). Indian medicinal plants used in ayurvedic preparations. Bishen Singh Mahendra Pal Singh, Dehra Dun.

Dix N.J., and Webster J. (1995). *Fungal Ecology*. Chapman and Hall, England.

Ernst W.H.O. (1990). Mine vegetation in Europe. In, Heavy Metal Tolerance in Plants: Evolutionary Aspects, A.J.Shaw (ed.), CRC Press, CRC Press, pp. 21-37.

Ewers U. (1991). Standards, guidelines, and legislative regulations concerning metals and their compounds. In: Metals and their Compounds in the Environment: Ocurrence, Analysis and Biological Relevance. Merian, E. (ed.), VCH, Weinheim, pp. 687-711.

Fabig B. (1982). Einflub von Al und den Schwermetallen Fe, Mn, Zn, Cu, Pb und Cd auf die efficienz der VA-Mykorrhiza bei tropischen und subtropischen Pflanzen. Ph.D. thesis, University of Göttingen, Göttingen, Germany.

Fidelibus M.W., Martin C.A, Wright G.C. and Stutz J.C. (2000). Effect of arbuscular mycorrhizal (AM) fungal communities on growth of 'Volkamer' lemon in continually moist or periodically dry soil. *Scientia Horticulturae* 84: 127-140.

Fracchia S., García-Romera I., Godeas A., and Ocampo J.A. (2000). Effect of the saprophytic fungus *Fusarium oxysporum* on arbuscular mycorrhizal colonization and growth of plants in greenhouse and field trials. *Plant and Soil* 223: 175-184.

Geddeda Y.I., Trappe J.M. and Stebbins R.L. (1984). Effects of vesicular-arbuscular mycorrhizae and phosphorus on apple seedlings. *Journal American Society Horticultural Science.* 109: 24-27.

Gerdemann J.W. (1975). Vesicular-arbuscular mycorrhizae. In: *The development and function of roots* (Eds. Torrey J.G. and Clarkson, D.T). Academic press, London, pp 618.

Gert A. (2000). Phytoremediation of soil and sludge with special examination of heavy metal contamination. In: Bioremediation of Contaminated Soils. Wise, D *et al.* (eds.) pp: 787-809. Dekker, Inc. New York.

Gianinazzi S. and Scheupp H. (1994). *Arbuscular Mycorrhizas on sustainable agriculture and natural ecosystem* (Eds. Gianinazzi, S. and Scheupp, H.) Birkhauser-Verlag, Switzerland.

Gianinazzi S., Trouvelot A., Gianinazzi-Pearson V. (1990). Role and use of mycorrhizas in horticultural crop production. *Adv. Hortic Sci.* 4: 25-30.

Gildon A. and Tinker P.B. (1983). Interactions of vesicular-arbuscular mycorrhizal infection and heavy metals in plants. I. The effects of heavy metals on the development of vesicular-arbuscular mycorrhizas. *New Phytologist* 95: 247-261.

Gnekow M. A. and Marschner H. (1989). Role of VA mycorrhiza in growth and mineral nutrition of apple (*Malus pumila* var *domestica*) rootstock cuttings. *Plant Soil* 119(2): 285-293.

Granger R.L., Khanizadeh S. , Meheriuk M. , Berard L.S. and Dalpe Y. (1995). Effects of simazine on the mycorrhizal population in soil beneath an apple tree canopy. *Fruit Varieties Journal* 49(2): 90-93.

Greger M. and Landberg T. (1997). Use of willow clones with high Cd accumulating properties in phytoremediation of agricultural soils with elevated Cd levels. in R.Prost (ed.): Proceeding of 3rd Int. congress on the Biogeochemistry of trace elements. NRA Editions, pp. 505-511.

Guillemin J.P., Gianinazzi S. and Trouvelot A. (1992). Screening of arbuscular mycorrhizal fungi for establishment of micropropagated pineapple plants. *Agronomie.* 12: 831-836.

Gunze C. M. B. and Hennessy C. M. (1980). Effect of host applied aouxing on development of endomycorrhiza in cowpeas. *Trans. Br. mycol. Soc* 74: 247-251.

Haselwandter K. and Berreck, M. (1994). Accumulation of radionuclides in fungi. In: Metal Ions in Fungi. Winkelmann, G. and Winge, D.R. (eds)., Marcel Dekker, New York. pp. 259-277.

Hayman D.S. (1983). The physiology of vesicular-arbuscular endomycorrhizal symbiosis. *Can. J. Bot.* 61: 944-963.

Heggo A., Angle J.S., and Chaney R.L. (1990). Effects of vesicular-arbuscular mycorrhizal fungi on heavy metal uptake by soybeans. *Soil Biology and Biochemistry* 22: 865-869.

Hewitt E.J. (1952). Sand water culture methods used in the study of plant nutrition. Commonwealth Agricultural Bureau, Technical Communication No. 22.

Huang J.W. and Cunningham S.D. (1996). Lead phytoextraction: species variation in lead uptake and translocation. *New Phytologist* 134: 75-84.

Jalali B.L. and Jalali I. (1991). Mycorrhiza in plant disease control, In: *Handbook of Applied Mycology, Soil and Plants,* Vol. I. (Eds. Arora, K., Rai, B., Mukerji, K.G. and Knudsen, G.R.), Marcel Dekker, New York, pp 131-154.

Juste C. and Mench M. (1992). Long-term application of sewage sludge and its effects on metal uptake by crops. In: Biogeochemistry of Trace Metals: Advances in Trace Substances Research. Adriano, D.C. (ed.) pp. 159-193. Lewis, Boca Raton, Florida.

Karagiannidis N. and Nikolaou N. (2000). Influence of arbuscular mycorrhizae on heavy metal (Pb an Cd) uptake, growth, and chemical composition of *Vitis vinifera* L. (cv. Razaki). *American Journal of Enology and Viticulture.* 51: 269-275.

Khan G., Kuek C., Chaudhry T.M., Khoo C.S. and Hayes W.J. (2000). Role of plants, mycorrhizae and phytochelators in heavy metal contaminated land remediation. *Chemosphere* 41: 197-207.

Kiernam J.M., Hendrix J.W., Soltz L.P., Maronek D.M. (1984). Characterization of strawberry plants produced by tissue culture and infected with specific mycorrhizal fungi. *Hortic. Sci.* 19: 883-885.

Koide R.T. and Schreiner R.P. (1992). Regulation of the vesicular-arbuscular mycorrhizal symbiosis. *Annu. Rev. Plant Physiol. Plant Molec. Biol.* 43: 557-581.

Leep N.W. and Dickinson N.M. (1998). Biological Interactions: The role of woody plants in phytorestoration. In: *Metal Contaminated Soils. In situ Inactivation and Phytorestoration.* J. Vangronsveld and S. Cunningham, eds. Springer Verlag, Berlin. pp. 67-73.

Leyval C., Turnau K. and Haselwandter K. (1997). Effect of heavy metal pollution on mycorrhizal colonization and function: physiological, ecological and applied aspects. *Mycorrhiza* 7(3): 139-153.

Lovato A., Guillemin J.P. and Gianinazzi S. (1992). Application of commercial arbuscular endomycorrhizal fungal inoculants to the establishment of micropropagated grapevine rootstock and pineapple plants. *Agronomie* 12: 873-880.

Lovato A, Schuepp H, Trouvelot A and Gianinazzi S. (1996). Application of arbuscular mycorrhizal fungus (AMF) in orchard and ornamental plants. In 'Mycorrhiza: structure, function, molecular biology and biotechnology' (eds. Varma A and Hock B), Springer-Verlag, Berlin, Heidelberg, New York, pp. 443-468.

Madrid F., De La Rubia T. and Martinez J. (1996). Effect of *Phanerochaete flavido-alba* on aromatic acids in olive oil mill waste waters. *Technological Environmental Chemistry* 51: 161-168.

Medeiros C.A.B., Clark R.B. and Ellis J.R. (1994). Effects of excess manganese on mineral uptake in mycorrhizal sorghum. *Journal of Plant Nutrition* 17: 2203-2219.

Menge J.A. (1977). Utilization of mycorrhizal fungi in *Citrus* nurseries. *Proc. Int. Soc. Citricult.* 1: 129-132.

Menge J.A., Davis R.M., Johnson E.L.V. and Zentmyer G.A. (1978). Mycorrhizal fungi increase growth and reduce transplant injury in avocado. *Calif Agric.* 32: 6.

Monge O.A. and Bugbee B. (1992). Inherent limitations of non-destructive chlorophyll meters: a comparasion of two types of meters. *HortScience* 27: 69-71.

Morte M.A., Diaz G., Honrubia M. (1996). Effect of arbuscular mycorrhizal inoculation on micropropagated *Tetraclinis articulata* growth and survival. *Agronomie*. 16: 633-637.

Niemi M. and Vestber M. (1992). Inoculation of commercially grown strawberry with VA mycorrhizal fungi. *Plant Soil* 95: 133-142.

Norman J.R., Atkinson D. and Hooker J.E. (1996). Arbuscular mycorrhizal fungal-induced alteration to root architecture in strawberry and induced resistance to the root pathogen *Phytophthora fragariae*. *Plant and Soil*, 185: 191-198.

Ouzounidou G. (1995). Cu-ions mediated changes in growth, chlorophyll and other ion contents in a Cu-tolerant Koeleria splendens. *Biol. Plant.*, 37: 71-79.

Ouzounidou G., Moustakas M. and Eleftheriou E.P. (1997). Physiological and ultrastructural effects of cadmium on wheat (*Triticum aestivum* L.) leaves. *Arch. Environ. Contam. Toxicol.* 32: 154-160.

Pearson J.N. and Jakobsen I. (1993). The relative contribution of hyphae and roots to phosphorus uptake by arbuscular mycorrhizal plants, measured by dual labelling with ^{32}P and ^{33}P. *The New Phytologist* 124: 489-494.

Pereira G.E. (1998). Efecto de las micorrizas vesículo arbusculares en plántulas de *Eucalyptus globulus* (Labill.) y *E. camadulensis* (Dehnh.) en relación a la tolerancia de sustancias fitotóxicas. Tesis Doctoral, Universidad de Córdoba, Spain. *Plant Pathol.*, 11: 177-179.

Plenchette C., Furlan V. and Fortin J.A. (1981). Growth stimulation of apple trees in unsterilized soil under field conditions with VA mycorrhiza inoculation. *Canadian Journal of Botany* 59: 2003-2008.

Prasad M.N.V. and Strzalka J. (1995). Impact of heavy metals on photosynthesis. In: Heavy Metal Stress in Plants. From molecules to ecosystem. Prasad and Hagemeyer (eds.), Springer – Verlag Berlin, pp 117-138.

Rajapakse S. and Miller (Jr.) J.C. (1988). Relationship between cowpea root systems and mycorrhizal dependency. *Hort. Sci.* 23: 568-570.

Rapparini F., Bertazza G. and Baraldi R. (1996). Growth and carbohydrate status of *Pyrus communis* L. plantlets inoculated with *Glomus* sp. *Agronomie* 16: 653-661.

Ravolanirina F., Gianinazzi S., Trouvelot A. and Carre M. (1989). Production of endomycorrhizal explants of micropropagated grapevine rootstocks. *Agriculture, Ecosystem and Environment*. 29: 323-327.

Rivera-Becerril F., Calantzis C., Turnau K., Caussanel J.P., Belimov A.A., Gianinazzi S., Strasser R.J. and Gianinazzi-Pearson V. (2002). Cadmium accumulation and buffering of cadmium-induced stress by arbuscular mycorrhiza in three *Pisum sativum* L. genotypes. *Journal of Experimental Botany* 53: 1177-1185

Sahni Y.P. and Srivastava D.N. (1993). *Withania somnifera*: an indigenous anti-ulcergenic drug. *Indian J. Ind. Med.* 10: 53-56.

Salamanca C.P., Herrea M.A., and Barea J.M. (1992). Mycorrhizal inoculation of micropropagated woody legumes used in revegetation programmes for desertified Mediterranean ecosystems. *Agronomie*. 12: 869-872.

Salt D.E., Blaylock M., Kumar N.P.B.A., Dushenkov V., Ensley B.D., Chet I., and Raskin I. (1995). Phytoremediation: a novel strategy for the removal of toxic metals from the environment using plants. *Biotechnology* 13: 468-474.

Sankaran J.R. (1984). Problem of male virility - an oriental therapy. *J. Natl. Med. Assoc.* 26: 315-317.

Sauerbeck D.R. (1987). Effect of agricultural practices on the physical, chemical and biological properties of soils: Part II. Use of sewage sludge and agricultural wastes. In: Scientific Basis for Soil Protection in the European Community. Barth, H., L'Hermite, P (eds.), Elsevier Applied Science, London, pp. 181-210.

Saxena P., Krishnaraj S., Dan T., Perras M., and Vettakkorumakankav N. (1999). Phytoremediation of heavy metal contaminated and polluted soils. In: Heavy Metal Stress in Plants. From molecules to ecosystem. Prasad and Hagemeyer (eds.), Springer – Verlag Berlin, pp 305-329.

Schellenbaum L, Gianinazzi S. and Gianinazzi-Pearson V. (1992). Comparision of acid soluble protein synthesis in roots of endomycorrhizal wild type *Pisum sativum* and corresponding isogenic mutants. *Journal of Physiology,* 141: 2-6.

Schenck N.C. and Kellam M.K. (1978). The influence of vesicular-arbuscular mycorrhizae on disease development. *Agriculture Experimental Station Technical Bulletin,* Florida, 798.

Schenck N. C. (1981). Can mycorrhizae control root disease? *Plant Disease* 65: 230-234.

Schreiner R.P. and Bethlenfalvay G.J. (1995). Mycorrhizal interactions in sustainable agriculture. *Critical Reviews in Biotechnology* 15: 3-4 271-285.

Schubert A., Bodrino C. and Gribaudo I. (1992). Vesicular arbuscular mycorrhizal inoculation of Kiwifruit (*Actinidia deliciosa*) micropropagated plants. *Agronomie* 12: 847-850.

Schubert A. and Lubraco G. (2000). Mycorrhizal inoculation enhances growth and nutrient uptake of micropropagated apple rootstocks during weaning in commercial substrates of high nutrient availability. *Applied Soil Ecology* 15: 113-118.

Sharma A.K., Johri B.N. and Gianinazzi S. (1992). Vesicular arbuscular mycorrhizae in relation to plant disease. *World Journal of Microbiology and Biotechnology*. 8: 559-563.

Sieverding E. (1989). Should VAM inocula contain single or several fungal species? *Agric. Ecosystems Environ.* 29: 369-390.

Smith S.E. and Read D.J. (1997). Mycorrhizal symbiosis, 2nd edn. Academic, San Diego.

Smith G.S. and Roncadori R.W. and Hussey R.S. (1986). Interaction of Endo-mycorrhizal fungi, superphosphate and *Meloidogyne incognita* on cotton in microplot and field studies. *Journal of Nematology*, 18: 206-216.

Srinath J., Bagyaraj D.J., and Satyanarayana B.N. (2003). Enhanced growth and nutrition of micropropagated *Ficus benjamina* to *Glomus mosseae* co-inoculated with *Trichoderma harzianum* and *Bacillus coagulans. World J Microbiol Biotechnol* 19, 69-72.

Stoppler H., Kolsch E. and Vogtmann H. (1990). *Biological agriculture and Horticulture*, 7: 191-199.

Tripathi A.K., Shukla Y.N., Kumar S. (1996). Ashwagandha (*Withania somnifera*): a status report. *J. Med. Aroma Plant Sci.* 18: 46-62.

Turnau K., Kottke I., and Dexheimer J. (1996). Toxic element filtering in *Rhizopogon roseolus / Pinus sylvestris* mycorrhizas collected from calamine dumps. *Mycological Research* 100: 16-22.

Tyler G., Balsberg Phalsson A-M., Bengtsson G., Baath E., and Tranvik L. (1989). Heavy-metal ecology of terrestrial plants, microorganisms and invertebrates. *Water Air Soil Pollution* 47: 189-215.

US EPA. (1993). Standards for the use or disposal of sewage sludge. *Fed. Regist.* 58: 9248-9415.

Varma A.K. and Hock B. (1995). Mycorrhiza: Structure, function, molecular biology and biotechnology. Springer-Verlag, Berlin.

Varma A.K. and Schuepp H. (1995). Mycorrhizae, their application in micropropagated plantlets. *Critical Reviews in Biotechnology* 15(3/4): 313-328.

Varma A. and Shankar A. (1994). Mycorrhizae. In: *Botany in India, History and Progress*, John, B.N., Ed., Oxford & IBH Publishing House, New Delhi, 345-360.

Varma A., Singh A., Sudha Sahay N.S., Sharma J., Roy A., Kumari M., Rana D., Thakaran S., Deka D., Bharti K., Hurek T., Blechert O., K-H Rexer, Kost G., Hahn A., Maier W., Walter M., Strack D. and Kranner I. (2001). *Piriformospora indica*: an axenically culturable mycorrhiza-like endosymbiotic fungus. In: Hock B. (ed.), *Mycota* IX. Springer, Berlin, Heidelberg, New York pp 123-150.

Vestberg M. (1992). The effect of vesicular arbuscular mycorrhizal inoculation on the growth and root colonization of ten strawberry cultivars. *Agricultural Science in Finland* 1(5): 527-535.

Wainwright M. (1992).The impact of fungi on environmental biogeochemistry. in G.C.Carrol, and D.T.Wicklow (ed.): The fungal community. Marcel Dekker, pp. 601-618.

Wilkinson D.M. and Dickinson N.M. (1995). Metal Resistance in Trees: The Rolc of Mycorrhizae. *Oikos* 72, 298-300.

Yano-Melo A.M., O.J. J.M. Saggin, Lima-Filho, Melo N.F., and Maia L.C. (1999). Effect of arbuscular mycorrhizal fungi on the acclimatization of micropropagated banana plantlets. *Mycorrhiza* 9: 119-123.

7 Strategies for Control of Disease Caused by Different Species of *Phytophthora* Using Arbuscular Mycorrhizal Fungi

Mahendra Rai[1], Adriana P.D. Silveira[2], and Soraya C. Franca[2]

[1]Department of Biotechnology, SGB Amravati University, Amravati-444 602, Maharashtra, India
[2]Department of Microbiology, Institute of Agronomy (IAC), Campinas, Av. Barao de Itapura 1.480, C.P. 28-13001-970, State of São Paulo, Brazil

INTRODUCTION

In the present agriculture system, the high input of chemicals to control plant diseases has generated many environmental problems. Alternatively, there is a great need for biological control of soil-borne pathogens, which would be good for sustainable agriculture and would be environmentally friendly. The replacement of pesticides by the controlled use of microorganisms to protect plants against pathogens can be a key component in a sustainable agriculture (Barea *et al.,* 1996). The process of biological control is economically profitable with minimized environmental pollution (Bethlenfalway and Linderman, 1992). Biological control of soil-borne diseases is complex because these diseases occur in rhizosphere soil, which is defined as the soil surrounding a root that is affected by it. The rhizosphere is characterized by rapid change, intense microbial activity, and high populations of bacteria compared with non-rhizosphere soil (Handelsman and Stabb, 1996). Weller (1988) emphasized that microbes growing in the rhizosphere are excellent agents for biocontrol of soil-borne diseases as they provide front-line defense against root-pathogens.

Phytophthora species pose severe problems as causal pathogens responsible for root rot of many economic plants. There are more than 80 species of *Phytophthora* of which nine are saprophytes, whereas most species are plant pathogens and distributed ubiquitously (Tucker, 1931; Waterhouse, 1970; Ho, 1981; Zentmyer, 1983; Ho and Jong, 1987; Farr *et al.*, 1989; Gerrettson-Cornell, 1989; Holiday, 1989). Some important species of *Phytophthora* include *P. arecae, P. botryosa, P. cactorum, P. cajani, P. cambivora, P. capsici, P. cinnamomi, P. citricola, P. citrophthora, P. clandestina, P. colocasiae, P. fragariae, P. palmivora, P. phaseoli, P. infestans, P. inflata, P. iranica, P. insolita, P. clandestina, P. colocasiae, P. nicotianae, P. quercina,* and *P. sinensis*. The most common diseases caused by *Phytophthora* are root and tuber-rot, damping-off of seedlings, collar and trunk canker, fruit-rot, wilt and leaf-blight (Farr *et al.*, 1989; Holiday, 1989). Although, there are chemical treatments for control of *Phytophthora,* there is a need to develop proper, effective, economically viable, and environment friendly sustainable strategies for biological control of *Phytophthora.*

Arbuscular mycorrhizal fungi (AMF) form symbiotic associations with more than 90% of plant taxa. Fossils and molecular studies indicate that AMF existed over >400 million years ago (Pirozynski and Dalpe, 1989; Remy *et al.*, 1994). In addition to improving P nutrition, these fungi reduce the severity of disease caused by soil-borne pathogens, and thus their potential can be used in biocontrol (Baltruschat and Schonbeck, 1975; Schenck *et al.*, 1975; Dehne and Schonbeck, 1975, 1979; Roncadori and Hussy, 1977; Schonbeck and Dehne, 1977; Davis *et al.*, 1978; Bartschi *et al.*, 1981; Dehne, 1982; Krishna and Bagyaraj, 1983; Rosendahl, 1985; Caron *et al.*, 1986; Graham, 1986; Caron, 1989; Perrin, 1990; Jalali and Jalali, 1991; Paulitz and Linderman, 1991; Benthlenfalvay and Linderman, 1992; Hwang *et al.*, 1992; Guillemin *et al.*, 1994; Hooker *et al.*, 1994; Linderman, 1994, 2001; Liu, 1995; Barea and Jeffries, 1995; Calvet *et al.*, 1995; Azcon-Aguilar and Barea, 1996; Trotta *et al.*, 1996; Kjoller and Rosendahl, 1996; Bodker *et al.*, 1998; Branzanti *et al.*, 1998; Cordier *et al.*, 1998, 1999; Borowicz, 2001; Larsen and Bodker, 2001; Greipsson and El-Mayas, 2002; Singh *et al.*, 2003). However, other AMF did not affect disease (Ross, 1972; Daft and Okusanya, 1973; Ramirez *et al.*, 1975; Matare and Hattingh, 1977; Davis *et al.*, 1978; Davis, 1979; Davis and Menge, 1980; Baath and Hayman, 1983, 1984; Greipsson and El-Mayas, 2002; Jeffries *et al.*, 2003). On the contrary, in some cases AMF promoted growth of pathogens and not the host plant (Vaast *et al.*, 1998).

There are excellent reviews pertaining to the role of AMF in control of plant diseases (Schonbeck, 1979; Dehne, 1982; Bagyaraj, 1984; Schenck, 1987; Smith, 1987; Zambolin, 1987; Caron, 1989; Jalali and Jalali, 1991; Paulitz and Linderman, 1991; Sharma *et al.*, 1992; Hooker *et al.*, 1994; Linderman, 1994, 2001; and Azcon-Aguilar and Barea, 1996, 1997).

The present chapter is focused at the strategic control of *Phytophthora* diseases by AMF, including selection of effective AMF, mechanisms of interaction, use of mixed microbial cultures, *Piriformospora indica* as a new bioprotector, and future strategies for control of various diseases caused by *Phytophthora.*

MECHANISM OF PATHOGEN SUPPRESSION AND BIOTIZATION OF HOSTS

Many protection mechanisms for interaction between AMF and soil-borne pathogens were proposed by Perrin (1990). But still we need a deeper understanding of the interaction among AMF, the host plant, and other microbes for biological control processes. Some important points for the explanations (Azcon-Aguilar and Barea, 1996 and Trotta *et al.*, 1996) of protection-mechanisms of host-plant against disease are as follows:

- Improvement of host plant nutrient
- Competition with the pathogen for carbon
- Competition for infection/ colonization sites
- Morphological and anatomical changes in root system
- Microbial changes in the mycorrhizosphere
- Changes in active compounds related to plant resistance response

- Alleviation of physical stress
- Biotization of host plant

Ross (1972) reported that the susceptibility of soybean to *Phytophthora sojae* increases after inoculation with AMF. Davis and Menge (1980) stated that the influence of a AMF fungus on disease may be due to the associated increase in mineral absorption rather than to a direct influence of the mycorrhizal fungus itself. Most mechanisms involve the accumulation of plant-derived compounds in response to the mycorrhizal infection like arginine, phenolics and isoflavonoids (Baltruschat and Schonbeck, 1975; Dehne and Schonbeck, 1978; Morandi *et al.*, 1984; Morandi, 1996) that have antimicrobial properties. Moreover, phenolics are also involved in signaling or controlling plant-microbe interactions (Morandi, 1996). Morandi (1996) further explained that phenolic metabolism in plant roots could be involved in biocontrol. According to him, "Even low accumulation of antimicrobial phenolics or their precursors potentially protects the mycorrhizal roots to some extent from pathogens, and better prepares them to react more quickly by releasing defense phenolics". AMF cause the accumulation of phenolics, in particular phytoalexins and associated flavonoids and isoflavonoids in the roots of the hosts. In addition to these phenolics, mycorrhizal roots contain other phenolics in higher percentage, which are responsible for antimicrobial properties. This must be one of the reasons that root pathogens do not find mycorrhizal roots to be suitable host as compared to non-mycorrhizal ones. Morandi (1996) further suggested that conjugate isoflavonoids that are not toxic and accumulate at higher concentration in mycorrhizal roots serve as storage form of phytoalexins. In case of attack of pathogen, these compounds are readily hydrolyzed. Borowicz (2001) found that overall inoculation with AMF had a large indirect negative effect on pathogens. AMF plants may even sustain greater attack by pathogens yet grow better than non-mycorrhizal counterparts. A survey of literature indicates that AMF enhance plant resistance and biotize the plant (Dehne and Schonbeck, 1979; Davis and Menge, 1981; Borowicz, 2001). The term 'Biotization' was proposed by Nowak (1998) for induced resistance response caused by inoculants. Larsen and Bodker (2001) pointed out that there is a keen competition between the symbiotic fungus and pathogen.

As a general response to attack by pathogens, the major inducible defense mechanisms triggered in plants include the following as suggested by many investigators (Bowles, 1990; Collinge *et al.*, 1994; Ryals *et al.*, 1996; Gianinazzi-Pearson, 1996; Dassi *et al.*, 1998, 1999; Dixon and Harrison, 1990):

- Cell-wall modifications
- Enhancement of secondary metabolism
- Accumulation of pathogenesis-related (PR) proteins

Gianinazzi-Pearson (1996) opined that the above mentioned responses are usually observed during arbuscular mycorrhizal symbiosis. For instance, in mycorrhized tomato roots, it has been found that pathogenesis-related (PR) proteins play a significant role in bioprotection of hosts from the soil-borne pathogen *Phytophthora parasitica.* PR proteins involved in the defense response of plants include acidic or basic chitinases and β–1-3-glucanases, which are antimicrobial hydrolases (Collinge *et al.*, 1994). Interestingly, Pozo *et al.* (1999) analyzed different polypeptide profiles from tomato roots after interaction between *Glomus mosseae* and *Phytophthora parasitica.*

The question can be raised as to whether mycorrhizas interact directly or indirectly ? Graham (2001) addressed this question and stated that AMF interact directly with root pathogens that have similar nutrient requirements. There is a keen competition between AMF and pathogen for food. Consequently, they suppress each others growth and colonization. Larsen and Bodker (2001) also reported that there are direct trophic interactions based on competition for carbon and possibly other nutrients between the symbiont and other organisms colonizing the root cortex and mycorrhizosphere. Unfortunately, only limited evidence has emerged that the AMF compete directly with other biotrophic organisms for the same cortical space and resources (Graham, 2001). Borowicz (2001) stated that reduced pathogen growth in mycorrhizal plants does not mean increased resistance. There is much competition between AMF and pathogens like *Phytophthora* for infection sites and also for metabolites/ food produced by the plants. This might be a reason for poor pathogenicity (Smith, 1988; Muchovej *et al.*, 1991; Traquair, 1995). The infection by pathogens in mycorrhizal plants is generally considered as increasing resistance in host plants (Graham and Menge, 1982; Strobel *et al.*, 1982; Smith *et al.*, 1986). But its not a justified conclusion as if percentage or density increases, it may alter the resistance. Thus, more research in this direction is required in order to reach a definite conclusion.

The host-AMF-pathogen system is a complex system, and may be affected by many environmental factors. Variations are reported among and within plant-AMF-pathogen systems (Dehne, 1982; Hussey and Ronacadori, 1982; Ingham, 1988; Roncadori, 1997; Borowicz, 2001) due to host genotype, inoculum density of pathogenic species (=*Phytophthora*) and AMF species/ strain. There may be a great influence of these factors on the direction and magnitude of AMF interaction (Davis and Menge, 1981; Strobel *et al.*, 1982; Kaye *et al.*, 1984; Diederichs, 1987; Thomas *et al.*, 1994; Price *et al.*, 1995; Norman *et al.*, 1996; Vaast *et al.*, 1998). Moreover, there are some other factors, likc timing of AM-fungus colonization, presence of number and type of soil organisms and abiotic factors, which directly affects the interaction (Siddique and Mahmood, 1995; Trotta *et al.*, 1996; Vaast *et al.*, 1998).

More recent studies provide convincing evidence that intraradical hyphae of AMF efficiently absorb sugar from the roots, and the host regulates colonization by providing resources to the root cells and rhizospheres (Graham, 2000). Doud *et al.*, (2000) found that AMF hyphae rapidly assimilate hexoses, but extramatrical hyphae are incapable of taking up sugars.

There is a need to know more about cellular, molecular and physiological reaction of this tripartite host-symbiont-pathogen combination. Molecular studies may shed light on the expression of plant resistance due to interaction of AMF and plant pathogens. Recent research provides evidence that there are some defense-related genes, responsible for bioprotective effects in AM plant (Gianinazzi-Pearson *et al.*, 1992, 1996; Harrison and Dixon, 1994; Lambais and Mehdy, 1995). The molecular mechanisms involved in bioprotection by AMF are poorly understood (Hooker *et al.*, 1994; St. Arnaud *et al.*, 1995; Azcon-Aguilar and Barea, 1996; Linderman, 2001). The induction of additional chitinase and chitosanase isoforms were reported pertaining to molecular changes caused by AMF symbiosis (Dumas-Gaudot *et al.*, 1992, 1994; Pozo *et al.*, 1996, 1998; Slezack *et al.*, 1996).

HOST- AMF-PHYTOPHTHORA INTERACTION

A survey of literature indicates that AMF are used against root-rot disease caused by different species of *Phytophthora* (Table 7.1). Matare and Hattingh (1977) reported that mycorrhizal fungus *Glomus fasciculatus* stimulated growth of avocado seedlings in the presence or absence of

Table 7.1. Influence of AMF on different diseases caused by species of *Phytophthora*

S.No.	*AMF*	*Host & Disease*	*Phytophthora sp.*	*Reference*
1.	*Endogone sp.*	*Root-* rot of Soybean (*Glycine max* (L.) Merr.)	*P. megasperma*	Ross (1972)
	Endogone mosseae	*Root*-rot of Soybean (*Glycine max* (L.) Merr.)	P. *megasperma* var. *sojae*	Chou and Schmitthenner (1974)
2.	*G. fasciculatus*	Root-rot of Avocado (*Persea americana* Mill.)	*P. cinnamomi*	Matare and Hattingh (1977)
3.	*G. fasciculatus*	Root-rot of Citrus sweet orange (*C. sinensis* L.)	*P. parasitica*	Davis *et al.*, (1978)
		Root-rot of Avocado (*Persea americana* Mill.)	*P. cinnamomi*	
		Root-rot of Alfalfa (*Medicago sativa* L.)	*P. megasperma*	
4.	*G. fasciculatus*	Root-rot of Citrus sweet orange (*C. sinensis* L.)	*P. parasitica*	Davis and Menge (1980)
5.	*G. fasciculatus, G. constrictus*	Root-rot of sweet orange *(Citrus sinensis)*	*P. parasitica*	Davis and Menge (1981)
6.	*G. mosseae, Gi. margarita*	Root-rot of sweet orange *(Citrus sinensis)*	*P. parasitica*	Davis and Menge (1981)
7.	*Sclerocytis sinuosa*	Root-rot of sweet orange *(Citrus sinensis)*	*P. parasitica*	Davis and Menge (1981)
8.	*G. fasciculatus, G. constrictus*	Troyer citrange (*Poncirus trifoliata x Citrus sinensis)*	*P. parasitica*	Davis and Menge (1981)
9.	*Sclerocytis sinuosa*	Troyer citrange (*Poncirus trifoliata x Citrus sinensis*	*P. parasitica*	Davis and Menge (1981)
10.	*G. mosseae*	Port-Orford-cedar (*Chamaecyparis lawsoniana* (A. Murr.) Parl)	*P. cinnamomi*	Bartschi *et al.*, (1981)
11.	*G. mosseae G. caledonium*	Alpine strawberry *(Frageria vesca* Linn.*)*	*P. fragariae*	Baath and Hayman (1984)
12.	*Glomus* Spp.	Pineapple (*Ananas comosus* (L.) Merr.)	*P. cinnamomi*	Guillemin *et. al.*, (1994)
13.	*G. mosseae* V57, *G. hoi* V98 and	Crown rot of strawberry *(Frageria vesca* Linn.*)* *G. fistulosum* V128	*P. cactorum*	Vestberg *et al.*, (1994)
14.	*G. mosseae*	Tomato (*Lycopersicum esculentus)*	*P. nicotianae* var. *parasitica*	Pozo *et al.*, (1996)
15.	*G. mosseae*	Tomato (*Lycopersicum esculentus)*	*P. nicotianae* var. *parasitica*	Cordier *et al.*, (1996*a)*
16.	*G. mosseae*	Wild cherry (*Prunus avium)*	*P. cinnamomi*	Cordier *et al.*, (1996*b)*
17.	*G. mosseae*	Tomato (*Lycopersicum esculentus)*	*P. nicotianae* var. *parasitica*	Trotta *et al.*, (1996)
18.	*G. mosseae*	Tomato (*Lycopersicum esculentus)*	*P. nicotianae* var. *parasitica*	Gnavi *et al.*, (1996)
19.	*G. fasciculatum G. etunicatum*	*(*Strawberry) *Frageria x ananassa*	*P. fragariae*	Norman *et al.*, (1996)
20.	*G. intraradices*	Root-rot of sweet orange *(Citrus sinensis*	*P. parasitica*	Nemec *et al.*, (1996)
21.	*G. fistulosum*	Wild strawberry *(Fragaria vesca)*	*P. fragariae*	Mark and Cassells (1996)
22.	*G. deserticola*	Root stocks of apple	P. *cactorum*	Sainz *et al.*, (1998)
23.	*G. mosseae*	Tomato (*Lycopersicum esculentus)*	*P. nicotianae* var. *parasitica*	Cordier *et al.*, (1998)
24.	*G. mosseae*	Tomato (*Lycopersicum esculentus)*	*P. nicotianae* var. *parasitica*	Fusconi *et al.*, (1999)
25.	*Acaulospora scrobiculata G. deserticola*	*Taxus baccata*	*P. cinnamomi.*	Sainz *et al.*, (2000)
26.	*G. mosseae* G. intraradices	Tomato *(Lycopersicum esculentus)*	*P. parasitica*	Pozo *et al.*, (2001)
27.	*G. intraradices*	*Citrus limonia*	*P. parasitica*	Agnani (2002)

Phytophthora cinnamomi without affecting *P. cinnamomi*. Davis *et al.*, (1978) studied influence of AMF on *Phytophthora* root-rot of *Citrus sinensis*, *Persea americana* and *Medicago sativa* caused by *P. parasitica*, *P. cinnamomi* and *P. megasperma*, respectively. They reported little or no differences in response to the *Phytophthora* species, mycorrhizal and non-mycorrhizal citrus or alfalafa seedlings, but mycorrhizal avocado seedlings were affected more severely by *P. cinnamomi* than were non-mycorrhizal seedlings. In citrus and avocado experiments, the growth stimulus caused by mycorrhizal fungus was eliminated when *Phytophthora* was present. Sporulation by *G. fasciculatus* on citrus was affected adversely by *P. parasitica*. Davis and Menge (1980) studied interaction between *G. fasciculatus* and *Phytophthora parasitica* on citrus. They reported that in soil fertilized with less than 15 μg P/g soil, roots of sweet orange seedlings infected with *P. parasitica* and the mycorrhizal fungus *G. fasciculatus* were healthier and weighed more than roots infected with *P. parasitica* alone. Further, they also found that in leaves of mycorrhizal plants P content was more as compared to non-mycorrhizal. Trotta *et al.*, (1996) also studied interactions between *Phytophthora nicotianae* var. *parasitica* and *Glomus mosseae* in tomato plants. Genotype dependence in the interaction between *Glomus fistulosum*, *Phytophthora fragariae* and the wild strawberry (*Fragaria vesca)* was studied by Mark and Cassells (1996). Pozo *et al.*, (2001) reported that *Glomus mosseae* was effective in reducing disease symptoms produced by *P. parasitica* infection, and evidence points to a combination of local and systemic mechanisms being responsible for this bioprotection effect. Usually only selected species of *Glomus* is applied for biological control, except one example of *Endogone,* which was used for disease control in soybean (Ross and Harper, 1970; Ross, 1972).

The molecular studies concerning host-symbiont-pathogen relationship is receiving attention. Encouragingly, Boehm *et al.*, (1999) used real-time quantitative PCR: DNA calculation of spores isolated from the mycorrhizal fungus *Glomus mosseae* and monitoring of *Phytophthora* pathogens in host plant extracts. It seems that in future more studies in this direction will give new insights into the process of interaction.

SCREENING DIFFERENT AMF FOR SELECTION OF POTENTIAL CANDIDATES

Due to the complexity of the interactions involved in biocontrol and influence of various environmental factors, only one strain may not suppress the growth of pathogen. It is evident from recent studies that all isolates of same species are not similar. Molecular studies have proved that multiple genomes can co-exist within individual (Kuhn *et al.*, 2001; Rodriguez *et al.*, 2001). Kerr and Htay (1974) stated that the genetic diversity of microbes with disease suppressive potential remains a powerful yet largely untapped resource for biocontrol of plant disease. Therefore, there is a need to seek new biocontrol strains, particularly those adapted to the soil/environmental conditions, where they will be used (Stirling *et al.*, 1992; Cook, 1993; Stabb *et al.*, 1994). Stirling *et al.*, (1992) screened biological control potential of bacteria isolated from soil suppressive to *Phytophthora cinnamomi*. Azcon-Aguilar and Barea (1997) opined that selection of efficient strains of AMF and host/fungus/substrate combination will give the maximum benefits. Borowicz (2001) also remarked that in most of the experiments, single species is selected as inoculant. These studies may not be justified towards finding a positive effect on plant-pathogen relations if the investigator selects the species of *Glomus* that are excellent growth boosters. Single species may not give desired results in green-house experiments as in natural conditions more than one species colonize the plants. Hussey and Roncadori

(1982) found that highly artificial greenhouse and growth-chamber conditions used in most studies may not capture essential environmental factors influencing plant-pathogen interactions.

Selection of candidate arbuscular mycorrhizal fungus is necessary for better bioprotection effect. One species of AMF may be effective for one pathogen, while the same may not produce the desired effect on other pathogen. For selecting effective strains of AMF, its better to isolate the locally available candidate AMF from the same environmental/soil conditions in which they are to be used. In order to save the time and expenses, initially AMF can be applied in pot-experiments before going to the fields. But, one may or may not expect the same results as field conditions are sometimes completely different from the green house conditions.

Root-colonization is considered to be an important phenomenon for biological control (Weller, 1983; de Weger *et al.*, 1987; Parke, 1990). The percentage of root colonization can also be regarded as an alternative approach without concern for their biocontrol activity. It is assumed that biocontrol agents of root pathogens should be good root colonists (Weller, 1988). Therefore, a "superior colonist"/ "Super strain" should be selected for effective biological control.

Bodker *et al.,* (2002) were probably the first to screen species of *Glomus, viz., G. mosseae, G. geosporum, G. claroideum* and *G. caledonium* against *Aphanomyces euteiches.* However, these isolates did not demonstrate any significant effect on disease development. Such screening of AMF is needed for other pathogens, like *Phytophthora parasitica* or other species of *Phytophthora* causing diseases on several plants of economic importance.

PIRIFORMOSPORA INDICA AS A NEW CANDIDATE FOR BIOPROTECTION

P. indica is a new and emerging fungal endosymbiont discovered by Verma *et al.* (1998) from sandy desert soil of Rajasthan in North-West India. It belongs to Basidiomycotina and has great potential for colonization and growth promotion of plants. The molecular studies of the fungus revealed that it is closely related to *Rhizoctonia.* The typical pear-shaped chlamydospores were found to be efficient in successfully colonizing plants. The important feature of *P. indica* is that it can be cultured in agar-medium unlike the other arbuscular mycorrhizal fungi and resemble AM fungi in other characteristics (Singh *et al.*, 2000, 2003; Varma *et al.*, 2001). The fungus grows on wide range of synthetic and complex media. *P. indica* does not colonize the members of cruciferae like other AMF. There are more than 90% of the micropropagated plantlets, treated with the fungus, survive transfer from laboratory to farms (Sahay and Varma, 2000). The fungus has tremendous growth promotion effect on economic plants like *Withania somnifera,* and *Spilanthes calva* (Rai *et al.*, 2001). The fungus shows biocontrol activity against some root pathogenic fungi (Varma *et al.*, 2001). It is an efficient phosphate solubilizer and transporter (Sudha *et al.*, 1999; Varma *et al.*, 2001).

Tapping the potential of other effective bio-controllers

Generally, AMF alone may not prove to be effective biocontrol agent. These fungi needs help of other beneficial microorganisms, viz., Plant Growth Promoting Rhizobacteria (PGPR,s) and *Trichoderma* as later are effective bicontrol agents (Amorim *et al.,* 2002) and often act synergistically in combination with AMF (Jeffries *et al.*, 2003). Carruthers *et al.,* (1995) demonstrated the significance of antibiotic production by *Pseudomonas aureofaciens* PA147-2 for biological control of *Phytophthora megasperma* root rot of asparagus. Brewster *et al.* (1997) used *Enterobacter aerogenes* for the control of *Phytophthora cactorum in vitro.* Azcon-Aguilar and Barea (1997) suggested that the interactions between arbuscular mycorrhizal fungi and Plant Growth Promoting Rhizobacteria could be a biotechnological tool essential for the growth promotion of plants. Smith *et al.,* (1990) demonstrated the potential for biological control of *Phytophthora* root and crown rots of apple by *Trichoderma* and *Gliocladium* spp.

Mcleod *et al.,* (1995) studied *Trichoderma* for biological control of avocado root rot in bark medium artificially infested with *Phytophthora cinnamomi*. Soler *et al.* (1998) selected *Trichoderma* as antagonist for *Phytophthora cinnamomi* and *Rosellinia necatrix.*

Okamoto *et al.,* (1998) reported biocontrol of *Phytophthora capsici* by *Serratia marcescens* F-1-1. Menge *et al.,* (1999) found *Myrothecium roridum, M. verrucaria, Trichoderma harzianum, Epiccocum purpurascens, Catenaria anguillae, Humicola fuscoatra, Anguillospora pseudolongissma, Hypochytrium catenoides, Streptomyces griseoalbus, S. violascens Micromonospora carbonacea,* and *Ceraceomyces tessulatus* as inhibitory to *P. cinnamomi* via competition, antibiosis or parasitism. Now-a-days, there are several commercial biocontrol products available in the market for control of *Phytophthora* spp. (Table 7.2). Godfrey *et al.,* (2000) used *Pseudomonas aureofaciens* for the control of *Phytophthora megasperma* var. *sojae* which causes root-rot of asparagus. López-Herrera (1999) studied the biocontrol of *Phytophthora cinnamomi* by using *Trichoderma* spp. *Myrothecium roridum* was also found as a potencial biocontrol agent against *Phytophthora cinnamomi* (Gees and y Coffey, 1989). Xiao *et al.,* (2002) screened 53 antibiotic-producing *Streptomyces* isolated from soils from Minnesota, Nebraska, and Washington for their ability to inhibit plant pathogenic *Phytophthora medicaginis* and *Phytophthora sojae in vitro.* Out of these, eight isolates were found to possess pathogen-inhibitory capabilities. They were subsequently tested for their ability to control *Phytophthora* root rots on alfalfa and soybean in sterilized vermiculite and naturally infested field soil. The *Streptomyces* isolates tested significantly reduced root rot severity in alfalfa and soybean caused by *P. medicaginis* and *P. sojae*, respectively.

Table 7.2. Some commercial biocontrol agent other than AMF available in the market for control of *Phytophthora* species

Product name	*Biocontrol agent for Phytophthora spp.*	*Manufacturer/ Distributor*
Bio-Fungus, Supervit	*Trichoderma* spp.	www.bioplant.dk and dcm@dcmpronatura.com
Companion	*Bacillus subtilis* GB03; *B. subtilis, B. lichenformis, B. megaterium*	www.growthproducts.com
Mycostop	*Streptomyces griseoviridis* strain K61	www.agbio-inc.com
Polyversum (formerly Polygandron	*Pythium oligandrum*	biopreparaty@mbox.vol.cz
Trichopel	*Trichoderma harzianum* and *T. viride*	j.hunt@agrimm.co.nz

FUTURE DIRECTIONS AND CONCLUSIONS

The biological control techniques hold much promise as tool for sustainable plant production as high input of chemical has generated several acute environmental problems. To avoid these deleterious effects, it is necessary to use biological methods in general, and natural microbial enemies in particular.

Arbuscular mycorrhizal fungi possess enormous potential for the management of root-borne-fungal diseases. The need of the hour is to screen AM fungi in order to select powerful strains for effective control of diseases caused by *Phytophthora.* Another area of research, which needs more emphasis in order to search for effective biocontrol agent is improvement of strains by using molecular genetics as biological control involves complex interactions among the host, pathogen, antagonist and environment. In addition, physical and chemical factors of the soil that influence both colonization and the expression of traits important to antagonism in rhizosphere deserve necessary attention. Not only the arbuscular mycorrhizal fungi but more research is needed to formulate potential combinations of microbes for the control of disease caused by *Phytophthora.*

The mechanisms concerning suppression of *Phtophthora* by arbuscualr mycorrhizal fungi are still to be discovered and needs more research. In general, there are several protection mechanisms that have been proposed for interaction between AMF and soil borne pathogens. But still we need to have a deeper understanding of the interaction of *Phytophthora.* The mechanism as to how AMF attenuates the potential of *Phytophthora* by their presence and activities, is to be unraveled. In fact, the molecular studies are required in order to open up new vistas in unzipping the mysteries of complex host-symbiont-pathogen-interaction.

In fact, there is a pressing need to screen AMF for the effective control of *Phytophthora*. Another future strategy can be to evaluate other endophytes like *Piriformospora indica* and Dark septate endophytes (DSE) for growth promotion of host plants and also for combating the disease caused by *Phytophthora.*

The technology developed in the laboratory should be economically viable and eco-friendly so that it can be used by marginal farmers. But, it's a major challenge in case of arbuscular-mycorrhizal-technology as these are obligate symbionts. Most experiments carried in laboratory of green house conditions fail to justify the natural environmental conditions. The examples of successful practical application of this technology for plant protection are rare (Hooker *et al.*, 1994 and Linderman, 1994). Therefore, for better understanding of the plant-AMF-*Phytophthora* interaction, it is necessary to plan experiments in natural systems so that it is exposed to diverse environmental factors.

REFERENCES

Agnani D.R.G. (2002). Interactions of arbuscular mycorrhizal fungi, biological control and *Phytophthora parasitica* in *Citrus limonia,* M.S. Thesis, Institute of Agronomy, University of São Paulo, pp 71.

Amorim E.P. da R. and de Melo I.S. (2002). Antagonism of rhizobacteria to *Phytophthora parasitica* and *Phytophthora citrophthora* and the effect on development of citrus seedlings. *Revista-Brasileira-de-Fruticultura* 24(2): 565-568.

Azcon-Aguilar C. and Barea J.M. (1996). Arbuscular mycorrhizas and biological control of soil-borne plant pathogens—an overview of the mechanisms involved. *Mycorrhiza* 6: 457-464.

Azcon-Aguilar C. and Barea J.M. (1997). Applying mycorrhiza biotechnology to horticulture: significance and potentials. *Scientia Horticulturae* 68: 1-24.

Bååth, E. and Hayman D.S. (1983). Plant growth responses to vesicular-arbuscular mycorrhiza. XIV. Interactions with *Verticillium* wilt on tomato plants. *New Phytol.* 95: 419-426.

Bååth E. and Hayman D.S. (1984). No effect of VA mycorrhiza on red core disease of strawberry. *Trans. Br. Mycol. Soc.* 82: 534-536.

Bagyaraj D.J. (1984). Biological interactions with VA mycorrhizal fungi. In: C.I. Powell and D.J. Bagyaraj (eds). *VA Mycorrhiza.* CRC, Bocoa Raton, Fla. pp. 131-153.

Baltruschat H. and Schonbeck F. (1975). The influence of endotropic mycorrhiza on the infestation of tobacco by *Thielaviopsis basicola. Phytopathologische Zeitschrift* 84: 172-188.

Barea J.M. and Jeffries P. (1995). Arbuscular mycorrhizas in sustainable soil plant systems. In: B. Hock and A. Varma (eds), *Mycorrhizae Structure and Function, Molecular Biology and Biotechnology.* Springer-verlag, Berlin. pp. 521-559

Barea J.M., Calvet C., Estaun V. and Camprubi A. (1996). Biological control as a key component in sustainable agriculture. *Plant and Soil* 185: 171-172.

Bärtschi H., Gianinazzi-Pearson V. and Vegh I. (1981). Vesicular-arbuscular mycorrhiza formation and root rot disease (*Phytophthora cinnamomi*) development in *Chamaecyparis lawsoniana. Phytopathol. Z.* 102: 213-218.

Bethlenfalway G.J. and Linderman R.G. (1992). *Mycorrhiza for sustainable agriculture.* ASA special Publication Madison, W1 pp.124.

Bodker L., Kjoller R. and Rosendahl S. (1998). Effect of phosphorus and the arbuscular mycorrhizal fungus *Glomus intraradices* on disease severity of root-rot of peas (*Pisum sativum*) caused by *Aphanomyces euteiches. Mycorrhiza* 8: 169-174.

Bodker L., Kjoller R., Kristensen K. and Rosendahl S. (2002). Interactions between indigenous arbuscular mycorrhizal fungi and *Aphanomyces euteiches* in field grown pea. *Mycorrhiza* 12: 7-12.

Böhm, J., Hahn A., Schubert S., Bahnweg G., Adler N., Nechwatal J., Oehlmann R. and Oswald W. (1999). Real-time quantitative PCR: DNA calculation of spores isolated from the mycorrhizal fungus *Glomus mosseae* and monitoring of *Phytophthora* pathogens in host plant extracts. *J. Phytopathol.* 147: 409-416.

Borowicz V.A. (2001). Do arbuscular mycorrhizal fungi alter plant-pathogen relations? *Ecology* 82 (11): 3057-3068.

Bowles D.J. (1990). Defence-related proteins in higher plants. *Annu. Ver. Biochem.* 59: 873-907.

Branzanti M.B., Gentili M., Neri D. (1998). Control of crown rot in strawberry by arbuscular mycorrhizal fungi [Fragaria]. *Rivista-di-Frutticoltura-e-di-Ortofloricoltura,* 60(12): 77-80.

Brewster D.T., Spiers A.G., Hopcroft D.H. (1997). Biocontrol of *Phytophthora cactorum in vitro* with *Enterobacter aerogenes. New Zealand Journal of Crop and Horticultural Science 25*: 9-18.

Calvet C., Pinochet J., Camprubi A. and Fernadez C. (1995). Increased tolerance to the root lesion nematode *Pratylenchus vulnus* in mycorrhizal micropropagated BA-29 quince rootstock. *Mycorrhiza* 5: 253-258.

Caron M. (1989). Potential use of mycorrhizae in control of soil-borne diseases. *Can. J. Plant Pathol.* 11: 177-179.

Caron M., Fortin J.A. and Richard C. (1986). Effect of inoculation sequence on the interaction between *Glomus intraradices* and *Fusarium oxysporum* f. sp. *radicis-lycopersici* in tomatoes. *Can. J. Plant Pathol.* 8: 12-16.

Carruthers F.L., Shum-Thomas T., Conner A.J. and Mahanty H.K. (1995). The significance of antibiotic production by *Pseudomonas aureofaciens* PA147-2 for biological control of *Phytophthora megasperma* root rot of asparagus. *Plant and Soil* 170: 339-344.

Chou G.L. and Schmitthenner A.F. (1974). Effect of *Rhizobium japonicum* and *Endogone mosseae* on soybean root rot caused by *Pythium ultimum* ar *Phytophthora megasperma* var. *sojae. Plant Dis. Rep.* 58: 221-225.

Collinge D.B., Gregersen P.L. and Thordal-Christensen H. (1994). The induction of gene expression in response to pathogenic microbes. In: A.S. Basra (ed.) *Mechanisms of Plant Growth and Improved Productivity: Modern Approaches and Perspectives*, New York: Marcel Dekker, pp 391-433.

Cook R.J. (1993). Making greater use of introduced microorganisms for biological control of plant pathogens. *Phytopathol.* 31: 53-80.

Cordier D., Gianinazzi S and Gianinazzi-Pearson V. (1996a). Colonization patterns of root tissues by *Phytophthora nicotianae* var. *parasitica* related to reduced disease in mycorrhizal tomato. *Plant and Soil* 185: 223-232.

Cordier D., Trouvelot A., Gianinazzi S. and Gianinazzi-Pearson V. (1996*b).* Arbuscular mycorrhiza technology applied to micropropagated *Prunus avium* and to protection against *Phytophthora cinnamomi. Agronomie* 16: 679-688.

Cordier C., Ferrol N., J.M. Barea, S. Gianinazzi and Gianinazzi-Pearson, V. (1998). Cellular and molecular events underlying the induction of resistance against *Phytophthora parasitica* in mycorrhizal tomato plants, 2nd International Conference on Mycorrhiza, Uppsala, Sweden, 5-10 July 1998 (abstract).

Cordier C., Pozo M.J., Barea J.M., Gianinazzi S., Gianinazzi-pearson V. (1999). Cell defense responses associated with localized and systemic resistance to *Phytophthora parasitica* induced in tomato by an arbuscular mycorrhizal fungus. *Molecular-plant-microbe-interactions* 11(10): 1017-1028.

Daft M.J. and Okusanya B.O. (1973). Effect of *Endogone* mycorrhiza on plant growth.V. Influence of infection on the multiplication of viruses in tomato, Petunia, and Strawberry. *New Phytol.* 72: 975-983.

Dassi B., Dumas-Gaudot E. and Gianinazzi S. (1998). Do pathogenesis-related (PR) proteins play a role in bioprotection of mycorrhizal tomato roots towards *Phytophthora parasitica* ? *Physiological and Molecular Plant Pathology* 52: 167-183.

Dassi B., Samra A., Dumas-Gaudot E. and Gianinazzi S. (1999). Different polypeptide profiles from tomato Roots following interactions with arbuscular mycorrhizal *(Glomus mosseae)* or pathogenic *(Phytophthora parasitica)* fungi. *Symbiosis* 26: 65-77.

Davis R.M. (1979). Influence of vesicular-arbuscular mycorrhizae on *Phytophthora* root-rot of citrsus, Ph.D. Thesis, University of California, Riverside, pp. 102.

Davis R.M. and Menge J.A. (1980). Influence of *Glomus fasciculatus* and soil phosphorus on *Phytophthora* root rot of Citrus. *Ecology and Epidemiology* 70 (5): 447-452.

Davis R.M. and Menge J.A. (1981). *Phytophthora parasitica* inoculation and intensity of vesicular-arbuscular mycorrhizae in citrus. *New Phytol.* 87: 705-715.

Davis R.M., Menge J.A. and Zentmyer G.A. (1978). Influence of vesicular-arbuscular mycorrhizae on *Phytophthora* root-rot of three crop plants. *Ecology and Epidemiology* 68: 1614-1617.

Dehne H.W. (1982). Interaction between vesicular-arbuscular mycorrhizal fungi and plant pathogens. *Phytopath.* 72: 1115-1119.

Dehne H.W. and Schonbeck F. (1975). The influence of endotrophic mycorrhiza on the fusarial wilt of tomato. *Z. Planzenkr. Pflanzzenschutz.* 82: 630-632.

Dehne H.W. and Schonbeck F. (1978). The influence of endotrophic mycorrhiza on plant diseases. 3. Chitinase activity and ornithine-cycle. *Z. Planzenkr. Pflanzzenschutz.* 82: 666-678.

Dehne H.W. and Schoenbeck F. (1979). The influence of endotrophic mycorrhiza on plant diseases. II. Phenol metabolism and lignification. *Phytopathol. Z.* 95: 210-216.

de Weger L.A, C.i.M van der Vlugt, A H.M. Wijfjes, P.AH.M Bakker B., Schippers and Lugtenberg B. (1987). Flagella of a plant-growth–stimulating *Pseudomonas fluorescens* strain are required for colonization of potato roots. *J. Bacteriol.* 169: 2769-2773.

Diederichs C. (1987). Interactions between five endomycorrhizal fungi and the root-knot nematode *Meloidogyne javanica* on chickpea under tropical conditions. *Tropical Agriculture* (Trinidad) 64: 353-355.

Dixon R.A and Harrison M.J. (1990). Activation, structure and organization of genes involved In microbial defence in plants. *Adv. Genet.* 28: 165-234.

Douds Jr D.D., Pfeffer P.E. and Shachar-Hill Y. (2000). Application of *in vitro* methods to study carbon uptake and transport by AM fungi. *Plant and Soil* 226: 255-261.

Dumas-Gaudot E., Furlan V., Grenier J. and Asselin E. (1992). New acidic chitinase isoforms induced in tobacco roots by vesicular arbuscular mycorrhizal fungi. *Mycorrhiza* 1: 133-136.

Dumas-Gaudot E., Asselin E. , Gianinazzi-Pearson V. Gollotte A. and Gianinazzi S. (1994). Chitinase isoforms in roots of various pea genotypes infected with arbuscular mycorrhizal fungi. *Plant Sci.* 99: 27-37.

Farr D.F., Bills G.F., Chamuris C.P. and Rossman A.Y. (1989). *Fungi in plants and plant products in the United States*, American Phytopathological Society Press, St. Paul, MN, pp. 1252.

Fusconi A, Gnavi E., Trotta A. and Berta G. (1999). Apical meristems of tomato roots and their modifications induced by arbuscular mycorrhizal and soilborne pathogenic fungi. *New Phytol.* 142: 505-516.

Gees R. and Coffey M.D. (1989). Evaluation of strain of *Myrothecium roridum* as a potential biocontrol agent against *Phytophthora cinnamomi. Phytopathology* 79(10): 1079-1084.

Gerrettson-Cornell L. (1989). A compendium and classification of the species of the genus *Phytophthora* de Bary by the canon of the traditional taxonomy. *Forest Commission*, New South Wales, Sydney, pp. 103.

Gianinazzi-Pearson V., Tahiri-Alaoui A., Antoniw J.F., Gianinazzi S. and Dumas-Gaudot E. (1992). Weak expression of the pathogenesis related PR-bl gene and localization of related protein during symbiotic ectomycorrhizal interactions in tobacco roots. *Endocytobiosis Cell Res* 8: 177-185.

Gianinazzi-Pearson V., Dumas-Gaudot E., Gollotte A., Tahiri-Alaoui A. and Gianinazzi S. (1996). Cellular and molecular defence-related root responses to invasion by arbuscular mycorrhizal fungi. *New Phytol.* 133: 45-58.

Gnavi E., Fusconi A., Cordier C., Gianinazzi-Pearson V. and Betra G. (1996). Possible involvement of root apical meristems of mycorrhizal tomato plants in resistance to the pathogenic fungus. *Phytophthora nicotianae* var. *parasitica. Giornale Botanico Italianao* 130: 934-937.

Godfrey S.A.C., Silby M.W., Falloon P.G. and Mahanty H.K. (2000). Biological control of *Phytophthora megasperma* var. *sojae*, causal agent of *Phytophthora* rot of asparagus, by *Pseudomonas aureofaciens* PA147-2: a preliminary field trial. *New Zealand Journal of Crop and Horticultural Science* 28(2): 97-103.

Graham J.H. (1986). Citrus mycorrhizae: Potential benefits and interactions with pathogens. *Hort. Sci.* 2: 1302-1306.

Graham J.H. (2000). Assessing costs of arbuscular mycorrhizal symbiosis in agroecosystems. In: G.K.Podila and D.D. Douds Jr (eds), *Current advances in mycorrhizae research*. St. Paul, MN, USA: The American Phytopathological Society Press, pp. 127-140.

Graham J.H. (2001). What do root pathogens see in mycorrhizas ? *New Phytol.* 149: 357-359.

Graham J.H. and Menge J.A. (1982). Influence of vesicular-arbuscular mycorrhizae and soil phosphorus on take-all disease of wheat. *Phytopathol.* 72: 95-98.

Greipsson S. and El-Mayas H. (2002). Synergistic effect of soil pathogenic fungi and nematodes reducing bioprotection of arbuscular mycorrhizal fungi on the grass *Leymus arenarius* 47(6): 715-727

Guillemin J.P., Gianinazzi S., Gianinazzi-Pearson V. and Marchal J. (1994). Contribution of arbuscular mycorrhizas to biological protection of micropropagated pineapple (*Ananas comosus* (L.) Merr.) against *P. cinnamomi* rands. *Agric. Sci. Finland.* 3: 241-251.

Handelsman J. and Stabb E.V. (1996). Biocontrol of soilborne plant pathogens, *The Plant Cell* 8: 1855-1869.

Harrison M.J. and Dixon R.A. (1994). Spatial patterns of expression of flavonoid/isoflavonoid pathway genes during interactions between roots of *Medicago truncatula* and the mycorrhizal fungus *Glomus versiforme. Plant J.* 6: 9-20.

Ho H.H. (1981). Synoptic keys to the species of *Phytophthora, Mycologia* 73: 705-714.

Ho H.H. and Jong S.C. (1987). Critical problems in the taxonomy of *Phytophthora* species in culture. *Mycotaxon* 29: 207-232.

Holiday P. (1989). *A Dictionary of Plant Pathology*. Cambridge University Press, New York, NY, pp. 369.

Hooker J.E., Jaizme-Veja M. and Atkinson D. (1994). Management of positive interactions of arbuscular mycorrhizal fungi with essential groups of soil microorganisms. In: S. Gianinazzi and H. Schüepp (eds), *Impact of arbuscular mycorrhizas on sustainable agriculture and natural ecosystems.* Birkhäuser, Basel, pp. 191-200.

Hussey R.S. and Roncadori R.W. (1982). Vesicular-Arbuscular Mycorrhizae may limit nematode activity and improve plant growth. *Plant Disease* 66: 9-14.

Hwang S.F., Chang K.F. and Chakravarty P. (1992). Effects of vesicular-arbuscular mycorrhizal fungi on the development of *Verticillium* and *Fusarium* wilts of alfalafa. *Plant Dis.* 76: 239-243.

Ingham R.E. (1988). Interactions between nematodes and vesicular-arbuscular mycorrhizae. *Agriculture, Ecosystems and Environment* 24: 169-182.

Jalali B.L. and Jalali I. (1991). Mycorrhiza in plant disease control. In: D.K.Arora, B. Rai, K.G. Mukerji, G.R. Knudsen, G.R. (eds), *Handbook of applied mycology.* Dekker, New York, pp. 131-154.

Jeffries P., Gianinazzi S., Perotto S., Turnau K. and Barea J.M. (2003). The contribution of arbuscular mycorrhizal fungi in sustainable maintenance of plant health and soil fertility. *Biol. Fertil. Soils.* 37: 1-16.

Kaye J.W., Pfleger F.L. and Stewart E.L. (1984). Interaction of *Glomus fasciculatum* and *Pythium ultimum* on green house-grown poinsettia. *Canadian J. of Botany* 62: 1575-1579.

Kerr A. and Htay K. (1974). Biological control of crown gall through bacteriocin production. *Physiol. Plant Pathol.* 4: 37-44.

Kjoller R. and Rosendahl S. (1996). The presence of the arbuscular mycorrhizal fungus *Glomus intraradices* influences enzymatic activities of the root pathogen *Aphanomyces euteiches* in pea roots. *Mycorrhiza* 6: 487-491.

Krishna K.R. and Bagyaraj D.J. (1983). Interaction between *Glomus fasciculatum* and *Sclerotium rolfsii* in peanut. *Can. J. Bot.* 61: 2349-2351.

Kuhn G., Hijri M. and Sanders I.R. (2001). Evidence for the evolution of multiple genomes in arbuscular mycorrhizal fungi. *Nature* 414: 745-748.

Lambais M.R. and Mehdy M.C. (1995). Differential expression of defense-related genes in arbuscular mycorrhiza. *Can. J. Bot.* 73 suppl. (1): S533-S540.

Larsen J. and Bodker L. (2001). Interactions between pea root-inhabiting fungi examined using signature fatty acids. *New Phytol.* 149: 487-493.

Linderman R.G. (1994). Role of VAM fungi in biocontrol. In: F.L. Pfleger and R.G. Linderman (eds), *Mycorrhizae and plant health.* APS, St Paul, pp. 1-26.

Liu R.J. (1995). Effect of vesicular arbuscular mycorrhizal fungi on *Verticillium* wilt of cotton, *Mycorrhiza* 5: 293-297.

López-Herrera C.J., Pérez-Jiménez R.M., Llobel A., Monte-Vázquez E. and Zea-Bonilla T. (1999). Estudios *in vivo* de *Trichoderma* como agente de biocontrol contra *Phytophthora cinnamomi* Y *Rosellinia* necatrix en aguacate. *Revista Chapingo Serie Horticultura* 5: 261-265.

Mark G.L. and Cassells A.C. (1996). Genotype-dependence in the infection between *Glomus fistulosum*, *Phytophthora fragariae* and the wild strawberry (*Fragaria vesca*). *Plant and Soil* 185: 233-239.

Matare R. and Hattingh M.J. (1977). Effect of mycorrhizal status of avocado seedlings on root rot caused by *Phytophthora cinnamomi. Plant and Soil* 49: 433-435.

Mcleod A., Labuschagne N., and Kotzé J.M. (1995). Evaluation of *Trichoderma* for biological control of avocado root rot in bark medium artificially infested with *Phytophthora cinnamomi. South African Avocado Growers´ Association Yearbook* 18: 32-37.

Menge J.A., Peggy A.M. and George Z. (1999). Control of *Phytophthora cinnamomi* root rot of avocado. In: M.L. Arpaia and R. Hofshi (eds), *Proceedings of Avocado Brainstorming '99.* pp. 133-138.

Morandi D. (1996). Occurrence of phytoalexins and phenolic compounds in endomycorrhizal interactions, and their potential role in biological control, *Plant and Soil* 185: 241-251.

Morandi D., Bailey J.A. and Gianinazzi-Pearson V. (1984). Isoflavonoid accumulation in soybean roots infected with vesicular-arbuscular mycorrhizal fungi. *Physiol. Plant Pathol.* 24: 357-364.

Muchovej J.J., Muchovej R.M.C. and Goncalves E.J. (1991). Effect of kind and method of fungicidal treatment of bean seed on infections by the VA mycorrhizal fungus *Glomus macrocarpum* and by the pathogenic fungus *Fusarium solani, Palnt and Soil* 132: 47-51.

Nemec S., Datnoff L.E. and Strandberg J. (1996). Efficacy of biocontrol agents in planting mixes to colonize plant roots and control root diseases of vegetables and citrus. *Crop Protection* 15: 735-742.

Norman J.R., Atkinson D. and Hooker J.E. (1996). Arbuscular mycorrhizal fungal-induced alteration to root architecture in strawberry and induced resistance to the root pathogen *Phytophthora fragariae. Plant and Soil.* 185: 191-198.

Nowak J. (1998). Benefits of *in vitro* biotization of plant tissue cultures with microbial inoculants. *In vitro Cell Dev. Biol. Plant.* 34: 122-130.

Okamoto H., Sato M., Sato Z. and Isaka M. (1998). Biocontrol of *Phytophthora capsici* by *Serratia marcescens* F-1-1 and analysis of biocontrol mechanisms using transposon-insertion mutants. *Annals of the Phytopathol. Soc. of Japan.* 64: 287-293.

Parke J.L. (1990). Population dynamics of *Pseudomonas cepacia* in the pea spermosphere in relation to biocontrol of *Pythium. Phytopathol.* 80: 1307-1311.

Paulitz T.C. and Linderman R.G. (1991). Mycorrhizal interactions with soil organisms. In D.K. Arora, B. Rai, K.G. Mukerji and G. Knudsen (eds). *Handbook of applied mycology.* Dekker, New York. pp. 77-129.

Perrin R. (1990). Interactions between mycorrhizae and diseases caused by soil-borne fungi. *Soil Use Manage* 6: 190-194.

Pirozynski K.A and Dalpe Y. (1989). Geological history of the Glomaceae, with particular reference to mycorrhizal symbiosis. *Symbiosis* 7: 1-36.

Pozo M.J., Dumas-Gaudot E., Slezack S., Cordier C., Asselin A., Gianinazzi, V. Gianinazzi-Pearson S., Azcon-Aguilar C. and Barea J.M. (1996). Induction of new chitinase isoforms in tomato roots during interactions with *Glomus mosseae* and/or *Phytophthora nicotianae* var. *parasitica. Agronomie* 16: 689-697

Pozo M.J., Azcon-Aguilar C., Dumas-Gaudot E. and Barea J.M. (1998). Chitosanase and chitinase activities in tomato roots during interactions with arbuscular mycorrhizal fungi or *Phytophthora parasitica. Journal of Experimental Botany* 49: 1729-1739.

Pozo M.J., Azcon-Aguilar C., Dumas-Gaudot E. and Barea J.M. (1999). Beta-1,3-glucanase activities in tomato roots inoculated with arbuscular mycorrhizal fungi and/or *Phytophthora parasitica* and their possible involvement in bioprotection. *Plant-Science* (Netherlands). 141(2): 149-157.

Pozo M.J., Cordier C., Dumas-Gaudot E., Gianinazzi S., Barea J.M. and Azcon-Aguilar C. (2001). Localized versus systemic effect of arbuscular mycorrhizal fungi on defence responses to *Phytophthora* infection in tomato plants. *J Exp Bot.* 53(368): 525-534.

Price N.S., Roncadori R.W. and Hussey R.S. (1995). The growth of nematode "tolerant" and "intolerant" soyabeans as affected by phosphorus, *Glomus intraradices* and light, *Plant Pathol.* 44: 597-603.

Rai M.K., Acharya D., Singh A. and Varma A. (2001). Positive growth responses of the medicinal plants *Spilanthes calva* and *Withania somnifera* in inoculation by *Piriformospora indica* in a field trial. *Mycorrhiza* 11: 123-128.

Ramirez B.N., Mitchell D.J. and Schenck N.C. (1975). Establishment and growth effects of three vesicular-arbuscular mycorrhizal fungi on papaya. *Mycologia* 67: 1039-1041.

Remy W., Taylor T.N., Hass H. and Kerp H. (1994). Four hundred-million year old vesicular arbuscular mycorrhizae. *Proc. Natl. Acad. Sci.* USA 91: 11841-11843.

Rodriguez A., Dougall T., Dodd J.C. and Clapp J.P. (2001). The large subunit ribosomal RNA genes of *Entrophosphora infrequens* comprise sequences related to two different glomalean families. *New Phytol.* 152: 159-167.

Roncadori R.W. (1997). Interactions between arbuscular mycorrhizas and plant parasitic nematodes in agro-ecosystems. In: A C. Gange and V.K. Brown (eds). *Multitrophic interactions in terrestrial systems*-the 36th Symposium of the British Ecological Society, Blackwell Science, Oxford, U.K. pp.101-113.

Roncadori R.W. and Hussy R.S. (1977). Interaction of endomycorrhizal fungus *Gigaspora margarita* and root-knot nematode on cotton. *Phytopathol.* 67: 1039-1041.

Rosendahl S. (1985). Interactions between the vesicular arbuscular mycorrhizal fungus *Glomus fasciculatum* and *Aphanomyces euteiches* root-rot of peas. *Phytopathol. Z.* 114: 31-40.

Ross J.P. (1972). Influence of *Endogone* mycorrhiza on *Phytophthora* rot of soybean. *Phytopathol.* 62: 896-897.

Ross J.P. and Harper J.A. (1970). Effect of *Endogone* mycorrhiza on soybean yields. *Phytopathol.* 60: 1552-1556.

Ryals J.A., Neuenschwander U.H., Willits M.G., Molina A., Steiner H.Y. and Hunt M.D (1996). *Systemic acquired resistance. Plant Cell* 8: 1809-1819.

Sahay N.S. and Varma A. (1999). *Piriformospora indica* : A new biological hardening tool for micropropagated plants. *FEMS Microbiol. Lett.* 181: 297-302.

Sahay N.S. and Varma A. (2000). Biological approach towards increasing the survival rates of micropropagated plants. *Curr. Sci.* 78: 126-129.

Sáinz M.J., Vilariño A., García-Berrios J., Pintos and Pedro Mansilla C., J. (1998). Interaction between *Glomus deserticola* and *Phytophthora cactorum* on the growth of two apple rootstocks, 2nd International Conference on Mycorrhiza, Uppsala, Sweden, 5-10 July 1998 (abstract 2).

Sainz M.J., Iglesias I., Pintos C., Mansilla J.P. and Vilariño A. (2000). Improved production of nursery stock of *Taxus baccata* l. through management of the arbuscular mycorrhizal symbiosis. *Acta Hort.* (ISHS) 536: 379-384.

Schenck N.C. (1987). Vesicular-arbuscular mycorrhizal fungi and the control of fungal root disease. In: I. Chet (ed) *Innovative approaches to plant disease control.* Wiley, New York, pp. 179-191.

Schenck N.C., Kinloch R.A. and Dickson D.W. (1975). Interaction of endomycorrhizal fungi and root-knot nematode on soybean. In: F.E. Saunders, Barbara Mosse and P.B. Tinker (eds.), *Endomycorrhizas.* Academic press, New York and London, pp. 605-617.

Schonbeck F. (1979). Endomycorrhiza in relation to plant diseases. In: B. Schippers and W. Games (eds), *Soil-borne plant pathogens*. Academic Press, New York. pp. 271-280.

Schonbeck F. and Dehne H.W. (1977). Damage to mycorrhizal and non-mycorrhizal cotton seedlings by *Thielaviopsis basicola. Plant Disease Reporter* 61: 266-267.

Sharma A.K., Johri B.N. and Gianinazzi S. (1992). Vesicular-arbuscular mycorrhizae in relation to plant disease. *World J. Microbiol. Biotechnol.* 8: 550-563.

Siddiqui Z.A and Mahmood I. (1995). Some observations on the management of the wilt disease complex of pigeon pea by treatment with a vesicular arbuscular fungus and biocontrol agents for nematodes. *Bioresource Technology* 54: 227-230.

Singh A., Sharma J., Rexer K-H. and Varma A (2000). Plant productivity determinants beyond minerals, water and light, *Piriformospora indica*-A revolutionary plant promoting fungus, *Curr. Sci.* 79: 101-106.

Singh Anjana; Singh A., Meera Kumari, Rai M. and Varma A. (2003). Biotechnological importance of *Piriformospora indica* Verma *et al.*- A novel symbiotic mycorrhiza-like fungus: an overview. *J. of Biotechnology* 2: 65-75.

Slezack D., Dassi B. and Dumas-Gaudot E. (1996). Arbuscular mycorrhiza-induced chitinase isoforms. In: R.A. Muzzarelli (ed.), *Chitin Enzymology*, Grottamare, Italy:Atec Edizioni, pp. 339-347.

Smith G.S. (1987). Interactions of nematodes with mycorrhizal fungi. In: J.A.Veech and D.W. Dickson (eds), *Vistas on Nematology*. Society of nematology, Hyattsville, Md. pp. 292-300.

Smith G.S. (1988). The role of phosphorus nutrition in interactions of vesicular arbuscular mycorrhizal fungi with soil-borne nematodes and fungi. *Phytopathol.* 78: 371-374.

Smith G.S., Roncadori R.W. and Hussey R.S. (1986). Interaction of endomycorrhizal fungi, superphosphate, and *Meloidogyne incognita* on cotton in microplot and field studies, *J. of Nematology* 18: 208-216.

Smith V.L., Wilcox W.F. and Harman G.E.Y. (1990). Potential for biological control of *Phytophthora* root and crown rots of apple by *Trichoderma* and *Gliocladium* spp. *Phytopathol.* 80: 880-885.

Soler A., Grondona I., Montero M., López H., Llobell A. and Monte E. (1998). Selección de cepas de *Trichoderma antagonistas* de *Phytophthora cinnamomi* y *Rosellinia necatrix* para su aplicación en cultivos de aguacate. *Programa y resúmenes del IX Congreso de la Sociedad Española de Fitopatología. Salamanca, España.* p. 285.

Stabb E.V., Jacobson L.M. and Handelsman J. (1994). Zwittermicin-A-producing strains of *Bacillus cereus* from diverse soils. *Appl. Environ. Microbiol.* 60: 4404-4412.

St. Arnaud M., Hamel C., Caron M. and Fortin J.A. (1986). Endomycorhizes VA et sensibilite des plantes aux maladies: synthese de la litterature et des mechanisms dínteraction potentiels. In: J.A Fortin, C. Charest and Y. Piche (eds), *La symbiose mycorhizienne: etat des connaissances.*ORTIS, Frelisghsburg, Quebec, Canada. pp. 51-87.

Stirling A.M., Hayward A.C. and Pegg K.G. (1992). Evaluation of the biological control potential of bacteria isolated from soil supressive to *Phytophthora cinnamomi. Australian Plant Pathol.* 21(4): 133-142.

Strobel N.E., Hussey R.S. and Roncadori R.W. (1982). Interactions of vesicular arbuscular mycorrhizal fungi, *Meloidogyne incognita*, and soil fertility on peach. *Phytopathol.* 72: 690-694.

Sudha, Kaur H., Narula A., Kumar S., Srivastava P.S. and Varma A. (1999). Mycorrhiza-aided biological hardening of *in vitro* raised plantlets. In: J.P. Tiwari, T.N. Lakhanpal, J. Singh, V.P. Chamola and R. Gupta (eds), *Advances in Biotechnology*, APS Publishing Corporation, New Delhi. pp. 469-485.

Thomas L., Mallesha B.C. and Bagyaraj D.J. (1994). Biological control of damping-off of cardamom by the VA mycorrhizal fungus, *Glomus fasciculatum. Microbiol. Res.* 149: 413-417.

Traquair J.A. (1995). Fungal biocontrol of root diseases: endomycorrhizal suppression of *Cylindrocarpon* root-rot. *Can. J. Bot.* 73: S89-S95.

Trotta A., Varese G.C., Gnavi E., Fusconi A., Sampo S. and Berta G. (1996). Interactions between the soilborne root pathogen *Phytophthora nicotianae* var. *parasitica* and the arbuscular mycorrhizal fungus *Glomus mosseae* in tomato plants. *Plant and Soil*, 185: 199-209.

Tucker C.M. (1931). Taxonomy of the genus *Phytophthora* de Bary. Mycol. Pap 92. Commonwealth Mycological Institute, Kew, Surrey, England, pp. 9.

Vaast P., Caswell-Chen E.P. and Zasoki R.J (1998). Influences of a root-lesion nematode, *Pratylenchus coffeae*, and two arbuscular mycorrhizal fungi, *Acaulospora mellea* and *Glomus clarum*, on coffee (*Coffea arabica* L.). *Biology and Fertility of Soils* 26: 130-135.

Varma A., Singh A. Sudha, Sahay N.S., Sharma J., Roy A., Kumari M., Rana D., Thakran S., Deka D., Bharti K., Hurek T., Blechert O, Rexer K.-H., Kost G., Hahn A., Maier W., Walter M., Strack D. and Kranner I. (2001). *Piriformospora indica*: an axenically culturable mycorrhiza-like endosymbiotic fungus. In: B. Hock (ed), *The Mycota IX, Fungal Associations,* Springer-Verlag, Heidelberg. pp. 125-150.

Verma S., Varma A., Rexer K-H., Hassel A., Kost G., Sarbhoy A., Bisen P., Buetehorn B. and Franken P. (1998). *Piriformospora indica*, gen. et sp. nov., a new root-colonizing fungus. *Mycologia* 90: 896-903.

Vestberg M., Palmujoki H., Parikka P. and Uosukainen M. (1994). Effects of arbuscular mycorrhiza on crown rot (*Phytophthora cactorum*) in micropropagated strawberry plants. *Agric. Sci. Finl.* 3: 289-295.

Waterhouse G.M. (1970). *The genus Phytophthora* de Bary. 2nd ed., Mycol papers 122. CMI, Kew, Surrey, England, pp. 59.

Weller D.M. (1983). Colonization of wheat roots by a fluorescent pseudomonad suppressive to take-all. *Phytopathol.* 73: 1548-1553.

Weller D.M. (1988). Biological control of soilborne plant pathogens in the rhizosphere with bacteria. *Ann. Rev. Phytopathol.* 26: 379-407.

Zambolin L. (1987). *Tolerancia de plantas micorrizadas a fitone-matoides.* II. Reuniao Brasileira sobre Micorrizas, São Paulo, pp.103-125.

Zentmyer G.A. (1983). The world of *Phytophthora.* In: D.C. Erwin, S. Bartnicki-Garcia and P.H. Tsao (eds), *Phytophthora: its Biology, Taxonomy, Ecology and Pathology.* American Phytopathological Society, St. Paul, MN. pp. 1-7.

Xiao K., Linda L.K. and Deborah A.S. (2002). Biological control of *Phytophthora* Root rots on alfalfa and soybean with *Streptomyces, Biological Control* 23(3): 285-295.

8 Fungal Culture Collections in the Mycological Work

López, Clara E[1*]. and Zacchino, Susana A.[2]

[1]CEREMIC (Mycological Work Reference Center), [2]Pharmacognosy, School of Biochemical and Pharmaceutical Sciences, Suipacha 531- (2000) Rosario, Argentina

INTRODUCTION

It is a recognized scientific principle that all research must be reproducible. Reproducibility in mycological work means that any research performed with a fungal strain can be conducted using the same strain, used in the original study. To accomplish with this requirement, it is generally accepted that *voucher specimens* should be deposited in publicly accessible reference Fungal Culture Collections (FCC) (Ammirati, 1979 and Watanabe *et al.,* 2004). In FCC, the cultures are retained either as dry materials or as permanent preserved living cultures, where they are safely stored and may be revived at any time.

The unequivocal citation of the allocated accession number of *voucher specimens* and cultures in each publication (Agerer *et al.,* 1994) is generally sufficient to ensure reproducibility of scientific results and confirm the identity of the species studied (Badalyan, 2001). Voucher numbers are necessary not only when new fungi are described but they should be cited during any type of scientific research, whether it is conducted by taxonomists, systematists, physiologists, chemists, molecular biologists, pathologists, ecologists, clinicians – in fact, by anyone studying an organism scientifically.

The aim of this chapter is to remark the importance of the maintenance and preservation of fungal cultures for the realization of fundamental and applied mycological work and the role played by Fungal Culture Collections in the reception of fungal cultures, morphological and genetic identification, short and long term *ex situ* preservation, creation of Information Databases, publication of FCC catalogues, and distribution of strains.

Several methods of preservation are described, indicating the advantages and limitations of each one for maintaining the morphological, physiological and genetic integrity of the cultures. In addition, the relation between costs and term of preservation and the requirements for the shipment of fungal strains are also discussed in this chapter.

HISTORY OF FUNGAL CULTURE COLLECTIONS

Microbial Culture Collections have been the consequence of the work by Robert Koch, who first employed solid media for obtaining pure cultures (Uruburu, 1991). This technique allowed microbiologists to establish collections of different microbial strains, but faced the problem of maintaining them in order to be useful for future applications.

Prof. Frantisek Kral's collection of Praga (Uruburu, 1993) constituted the first collection of fungal cultures and gave place to the first published catalogue of available fungal strains. Kral's collection then moved to Vienna, where most cultures were lost during the Second World War (Uruburu, 1994).

At present, there are about 500 culture collections in 65 countries registered in the World Directory of Culture Collections (WDCC) (http://wdcm.nig.ac.jp/statistics.html). This directory supplies addresses of recognized FCC, listing the accepted acronyms that are used to identify the collections in scientific publications. Among the CC listed in WDCC, 165 are supported by government; 33 are semi-governmental; 140 are supported by universities and 7 by industries and 175 CC produce catalogues of holdings.

There are a total of 1,152,175 microbes in the above culture collections, 381,519 of them are fungi.

FUNCTIONS OF FUNGAL CULTURE COLLECTIONS

The main functions of Fungal Culture Collections are:

- **To be a Public Deposit.** This is the most important function of these institutions which comprises the collection, characterization, maintenance, preservation, and distribution of fungal strains, helping to promote scientific comparability of results between researchers. These activities will be further amplified in this chapter.
- **To offer Secure Deposits.** In this type of activity, the distribution of the strains is restricted, in accordance to the instructions of the depositor.
- **To offer a Deposit of Microorganisms for the Purpose of Patent Procedures.** For developing this activity, it is necessary to have reached the status of "Depositary Authority" under the Budapest Treaty on the International Recognition of the Deposit of Microorganisms for the purpose of Patent procedures. For details, see the Document "Industrial Property" of the World Intellectual Property Organisation (www.wipo.int/portal/index.html.es).
- **To offer Training Courses.** To develop researches dealing with new techniques for preservation, package, maintenance or taxonomy, among others.

  Although not all CC provide all services listed above, all of them constitute very important support to the scientific work dealing with fungi.

COLLECTION

CC are open to receive strains from individuals or institutions wishing to deposit fungi. They must write to the CC's Mycology Department enclosing as much information as possible related to the strain. A form including a series of data (history of the strain, source of isolation including geographical site, available references, recommended medium and growth conditions, among others) must be completed and submitted by the contributor for review, before a strain can be accepted.

The expression, "Type culture collection" is reserved to the first strain of a species deposited in a CC.

Fungi to be deposited in CC should have been processed soon after collection, since it is well-known that fungal strains can be deteriorated when separated from their habitat. The procedures adopted to manage fungi must aim at guaranteeing their identity including the transportation to the CC Institutions. It must be considered that mite infestation is a chronic problem in laboratories where large number of fungi are cultivated (Okuda *et al.,* 2005). So, it is advisable to take in account the guidelines for the control of mites recommended by Stevens *et al.,* (1981) and Gams (1998).

CHARACTERIZATION

The correct classification of a strain is always critical to allow the re-identification of the organism at a later time and opens the possibility of detecting contaminations after a long period of time (Stadler and Hellwig, 2005). Many collections possess qualified technicians which identify the strains and re-identify them in case the species concept of an organism changes.

The conservation of dried fruitbodies from which cultures have been obtained is indispensable for gaining information about the anatomical and morphological features of a strain, not reproducible in cultures.

In recent years, molecular biology added valuable alternatives for the unequivocal classification of cultures such as PCR fingerprinting and DNA sequencing of a strain, these techniques complements the morphological data and help to the establishment of phylogenetic relationships.

A necessary concourse of methodologies including classical morphology, chemotaxonomy, biochemistry and molecular methods will aid to solve the problems arising from the misidentification of fungal cultures.

MAINTENANCE OF FUNGAL STRAINS

In contrast with bacteria, the maintenance of a fungal collection is difficult due to their eukaryotic nature; fungi are often subjected to genetic variability and diminution of viability.

The *voucher* culture deposited in the FCC is assigned with a number following the acronym of the CC (for example, *Candida albicans* ATCC10231) (ATCC: American Type Culture Collection P.O. Box 1549 Manassas, Virginia 20108, USA). The maintenance of this original fungal culture requires succesive new cultures from time to time, which are known as "sub-cultures". So, each microbial strain of a collection is the result of succesive sub-cultures which sometimes can lead to pleomorphyzation.

The different methods used to preserve fungal cells performed by FCC have the objective to minimize the changes produced by sub-culturing, highly insuring a high survival and genetic stability (Fries, 1985 and Burgess, 1995).

The following factors influence the survival of fungal cells:

Selection of proper culture media (Hunter-Cervera and Belt, 1991)

It is taken as a general rule that the selection of media with low O_2 and limited substrates will lead to more resistant cells. Nevertheless, media should possess accessible nutrients too, in order to facilitate the nutrition when it must re-develop from a latent phase. In addition to these general requirements, each fungal spp. requires a proper medium for its development.

Age and status of the cells

The cells in stationary phase are more resistant than young cells due to their bigger wall. Cultures with resistant forms, such as spores, are preferred for assuring a better survival.

PRESERVATION TECHNIQUES

Freezing

Fungal cultures grown on solid media and placed in a freezer at -20°C to -80°C have demonstrated a long viability particularly in the presence of cryoprotectants. The freezing has the advantage that cells are maintained in latent life avoiding genetic changes of fungal strains. It has been shown that the failure rate of medically important fungi, stored at -70°C for 13 years was 2.3% (Pasarell and Mc Ginnis, 1992).

The method can be applied to sporulated as well as non-sporulated cultures although the sporulated cultures survive the freezing process better than the non-sporulated cultures.

Cryoprotectants

They are substances added to a cellular suspension which aid to reduce the injuries during the freezing and thawing.

Cryoprotectants are classified into two types according to their capacity of penetrating the fungal membrane:

- agents that can penetrate the fungal membrane such as defatted milk, glycerol and dimethyl sulfoxide (DMSO), protecting the fungal cell from inside (Nakasone *et al.,* 2004).
- agents that cannot penetrate the fungal membrane such as sugars, dextran, sodium glutamate, honey, peptone, polyvinyl-pyrrolidone, hydroxyethyl starch among others, which protect the fungal cell from outside.

Among the possible cryoprotectants defatted milk 10%, glycerol 5, 10, 15 % and DMSO 10% are the most successfully used (Umino and Manfio, 2004).

Rate of freezing

There are three tested protocols:

- 4°C (1 h) + freezing at -80°C.

- 4°C (1h) + (-20°C) 1 h + freezing at -80°C.
- 4°C (1h) + plunged into liquid N_2 at -196°C for 30 sec + freezing at -80°C.

The pre-freezing at 4°C allows the cryoprotectant to better penetrate the cell membrane, leading to a better survival of the cells (Kirsop and Doyle, 1991). The most convenient rate of freezing is 1°C/min ((Hunter-Cervera and Belt, 1996).

The methodology used in the Mycological Work Reference Center [Centro de Referencia en Micología, Universidad Nacional de Rosario (CEREMIC)] is the following:

I. Cells are suspended into 10% glycerol.
II. The suspension is pre-frozen at 4°C for 1h.
III. Cells are frozen at -196°C by plunging into liquid N_2 in a sealed glass ampoule or a thin polypropylene tube.
IV. Tubes or ampoules are moved to a freezer at -80°C.

RECOVERY OF THE FROZEN CULTURE

Different methodologies are employed for the thawing of fungal cultures:

- Quick defrost in a water bath.
- Defrost in steps: 1h at -20°C, 1 h at 4°C and 30 min. at room temperature.
- To leave the strain at room temperature up to total defrost.

The best results are obtained with the quick defrost (Experience of the Mycological Reference Center of National University of Rosario, Suipacha 531 (2000) Rosario):

The ampoule containing the cells are plunged into a water-bath at 35-37°C. The tube must be taken out immediately after it is defrost.

LYOPHILIZATION

This methodology consists the dehydration of fungal cells by freeze-drying based on the sublimation of ice under vacuum. This is considered the most useful technique for preserving fungal cells because of its following advantages:

- Cells are maintained in latent life that avoids genetic changes in the fungal strains
- The delivery of strains is easy since it is not necessary to send them with special cares. In addition, the addressee has no hurry of reviving the fungal strain since lyophilized cultures can survive more than 30 years.
- Easy storage.

The drawback of this methodology is that not all microorganisms resist lyophilization. This methodology can be influenced by several factors such as type of microorganisms, culture media, age of culture, cellular concentration, rehydration methods, rate of dehydration, O_2 atmosphere and cryoprotectants.

The best results are obtained when the process is followed under low O_2 concentration and when the culture media are poor. This stress triggers a high sporulation which is a requisite for the lyophilization process.

CRYOPROTECTANT AGENTS (CPA)

The (CPA) are used to protect cells from damage during freezing. A suitable cryoprotectant agent for lyophilization should not form foam during the lyophilization process, and must easily rehydrate in cold water (Smith and Onions, 1983).

Recovery of lyophilized fungal strains

The recovery must be obtained by rehydration during 24 h in liquid or solid media (peptone 0.1 %, liquid malt or potato dextrose agar) or in distilled water, for 30 min.

Preservation in liquid nitrogen

This storage method assures a long-term viability of strains and prevents the genetic variability of fungal strains, because of its very low temperature (-196°C) which completely stops the metabolism of cell. It is used for fungi that do not sporulate in cultures or their spores do not survive the lyophilization process.

Cryoprotectants are also necessary to prevent damages of cells by freezing and thawing.

Various protocols depending on the capacity of sporulation of fungi, whether it grows better in liquid or in solid media, the use of cryoprotectants and others, have been reported (Stalpers *et al.*, 1987).

Periodical transfer

This is a very simple and low-cost technique that allows sub-cultures live for a longer time, reaching to a survival of 2-30 years depending on the fungal species. Although some fungi can adapt well the subsequent transfer, the drawback of this method is that morphological and physiological changes of fungi are quite common (for example, yeasts can vary flocculation conditions) (Castellani, 1967).

This traditional method of preservation depends on several factors:

a) Culture medium used

A minimum culture medium should be used in order to diminish the metabolism of the fungi, although some fungi requires more complex media.

DNA Sequence analysis

DNA sequence analysis offers unrivalled precision in molecular species identification, overcoming the potential shortcomes of RFLP. Among the variety of methods developed for sequencing, the dye terminator technique is currently the most efficient; limited only by the costly equipment and reagents. During the cycle sequencing PCR reaction, fluorescently labeled dideoxynucleotides are being incorporated into the newly synthesized DNA strands. Upon incorporation of a fluorescently labeled dideoxynucleotide the extension of the respective strand stops. Each of the four different dideoxynucleotides is labeled with another flourescent dye characterised by a different emission spectrum. The dye terminator labeled DNA fragments of different sizes are separated by gel electrophoresis in denaturing polyacrylamide gels. Denaturing conditions are essential in order to obtain an approximatively linear relationship of running time and fragment size, since the formation of secondary structures under non-denaturing conditions would strongly influence migration behaviour. The fragments are detected by incitation of the fluorochromes with laser light and by recording the emission spectra with a CCD-camera. The specialised software required for data collection is offered by the manufacturers of sequencing equipment (e.g., Applied Biosystems, Beckham Coulter). Reading lengths of about up to one kilobase (kb) are possible using standard capillary sequencers. The assemblage of overlapping or reverse sequences with specialised software packages (e.g., Vector NTI, Invitrogene Inc.) improves fidelity of the results.

Sequence information allows the unequivocal identification of fungal species, if sufficiently variable DNA regions are used (like the ITS regions or the 5´terminal domains of the LSU) and if reliable reference sequences are available. Furthermore, if identical reference sequences are lacking, the affinity of the sample can be deduced from phylogenetic reconstruction, if the query sequence can be aligned to a selection of candidate sequences.

Data analysis

Reference sequences can be retrieved from publicly accessible databases, e.g., at the National Center of Biotechnology Information (http://www.ncbi.nlm.nih.gov/), at the European Bioinformatics Institute (http://www.ebi.ac.uk/embl/) or at the DNA Database of Japan (http://www.ddbj.nig.ac.jp/). Newly submitted data are exchanged among these databases, any of them can offer the most recent body of information. The BLAST (Basic Local Alignment Search Tool) algorithm is a highly efficient bioinformatic tool which allows to search for most similar sequences in DNA databases. The quality of database entries is highly variable, since sequences are contributed by individual researchers. The submitted data may have various kinds of errors, from misidentifications, misspellings and outdated names to sequence errors (Hawksworth, 2004). It is highly recommended to consult the publications in which the reference sequences are cited and to exclude poorly documented reference data, whenever possible. It is further advisable to use specialised and more strictly edited databases, like UNITE (http://hermes.zbi.ee/) for complementary information (Kõljalg *et al.*, 2005).

It has to be emphasised that the best BLAST matches do not necessarily represent the most closely related organisms. If nearly identical reference sequences are lacking, the affinity of the query sequence needs to be inferred by phylogenetic reconstruction.

The inference of phylogenies from molecular data has inspired a plethora of articles, methods, algorithms, statistical tests and bioinformatic software implementations and can not be covered in this contribution (for an overview see http://evolution.genetics.washington.edu/phylip/software.html, a website maintained by Joe Felsenstein). For a given alignment, methods based on maximum likelihood, on maximum parsimony and on Monte-Carlo simulations are expected to result in similar topologies, if the data are not contradictory and if the appropriate models of evolution are chosen. Less computationally intensive methods like neighbour joining can be used as a first approximation and generally produce good results for closely related species.

CURRENT DEVELOPMENTS AND FUTURE PERSPECTIVES

The molecular analysis of mycorrhizal communities is still largely based on the analysis of single root tips. This analysis involves laborious preparation steps and the potential bias related to root tip preselection and morphotyping, which may select against rare and less conspicuous ectomycorrhizal types. It would be highly desirable to analyse the fungal community of whole soil samples, including ECM root tips and mycelia. This step was achieved by DNA extraction from soil samples, which is rather challenging due to purification difficulties. Filtration through sepharose or the repeated use of commercial kits are proposed to obtain sufficiently pure DNA. It was shown that there is good correspondence between data obtained from root tips and from soil extracts (Landeweert *et al.*, 2003).

Since PCR products obtained from soil DNA extracts are highly diverse, community profiling methods are necessary to analyse these products (Bidartondo and Gardes, 2005). DGGE and TGGE (Denaturing or Temperature Gradient Gel Electrophoresis), SSCP (Single-Strand Conformation Polymorphism), T-RFLP (Terminal Restriction Fragment Length Polymorphism) and cloning are possible options. DGGE and TGGE detect polymorphisms based on different melting behaviour and do not depend upon restriction enzymes. SSCP is based on differences in secondary structure which result in differential migration behaviour under non-denaturing electrophoresis conditions. This technique does not require restriction enzymes.

T-RFLP analysis uses PCR products generated with one or two fluorescently labeled primers that are cut with restriction endonucleases. The fluorescently labeled terminal fragments are analysed with sequencing equipment under denaturing conditions. T-RFLP was shown to detect 93% of ECM fungi in pooled ECM samples and may be regarded as effective, despite all limitations (Burke *et al.*, 2005).

DGGE, SSCP and T-RFLP allow the simultaneous analysis of mixed samples. In DGGE, DNA bands can be recovered from the gel and then sequenced, but resolution and reproducibility are limited. T-RFLP offers the high resolution and reproducibility of the sequencing technique, but single peaks can not be extracted from the gel for further analysis. For fragments with known sequences, the fragment lengths after restriction can be easily predicted. This approach allows to assign theoretical T-RF lengths to sequences deposed in public genebanks. However, different fungi can have the same T-RFLP length for a given restriction enzyme. It seems that theoretical approaches are not sufficient to interpret T-RFLP profiles correctly, even though the use of different enzymes can improve the taxonomic resolution of this teqhnique. Most substantial results can be obtained by comparing T-RFLP profiles to clone libraries obtained by transforming competent cells with a vector containing a population of fragments amplified by PCR. This step is all the more recommended since empirically determined T-RFLs may differ from the predicted length by a few bases.

Species specific quantification of fungal biomass in natural substrates is among the remaining challenges in mycology. Conventional fungal-specific biomarkers such as ergosterol or phospholipid fatty acids lack the required taxonomic resolution. Landeweert *et al.* (2003) compared three PCR based methods, PCR + DGGE, cloning + DGGE, and real time PCR with fungus specific primers for the quantification of ECM fungi in experimental microcosms. Using these methods of high taxonomic resolution, quantification can be relative at best, due to the complexity of factors influencing the results.

Future improvements in quantitative analysis will be achieved by the DNA/RNA microarray (DNA/RNA chip) technology. For the sake of taxonomic identification, this technology employs a set of hundreds or thousands hierarchically designed probes with different specificities on a single microarray (Loy and Bodrossy, 2006). DNA/RNA microarrays can be designed for phylogenetically informative genes as well as for functionally important genes, and will constitute a significant step towards the integration of the science of mycorrhizal symbioses and general soil microbial ecology.

REFERENCES

Agerer R., ed (1987-2002). *Colour Atlas of Ectomycorrhizae*, 1st-12th del., Einhorn-Verlag, Schwäbisch Gmünd, Germany.

Bidartondo M.I., Kretzer A.M., Pine E.M.,and Bruns T.D. (2000). High root concentration and uneven ectomycorrhizal diversity near Sarcodes sanguinea (Ericaceae): A cheater that stimulates its victims. *American Journal of Botany* 87: 1783-1788.

Bidartondo M.I. and Bruns T.D. (2005). On the origins of extreme mycorrhizal specificity in the Monotropoideae (Ericaceae): performance trade-offs during seed germination and seedling development. *Molecular Ecology* 14: 1549–1560.

Bidartondo M.I., Bruns T.D., Weiss M., Sergio C. and Read D.J. (2003). Specialized cheating of the ectomycorrhizal symbiosis by an epiparasitic liverwort. *Proceedings of the Royal Society of London. Biological Sciences.* 270: 835-842.

Bidartondo M.I. and Gardes, M. (2005). Fungal diversity in molecular terms: profiling, identification, and quantification in the environment. In: *The Fungal Community*, 3rd edition, Dighton J., White J. F., and Oudemans P., eds., pp. 215-239.

Burke D.J., Martin K.J., Rygiewicz P.T and Topa M.A. (2005). Ectomycorrhizal fungi identification in single and pooled root samples: terminal restriction fragment length polymorphism (TRFLP) and morphotyping compared. *Soil Biology and Biochemistry* 37: 1683-1694.

Chevalier G. And Frochot H. (1997). *La truffe de Bourgogne*. Ed. Pétrarque, Levallois-Perret, 258 p.

Danell E. and Camacho F. (1997). Successful cultivation of the golden chanterelle. *Nature* 385: 303.

Gardes M.,and Bruns T.D. (1993). ITS primers with enhanced specificity for basidiomycetes - application to the identification of mycorrhizae and rusts. *Molecular Ecology* 2: 113-118.

Gardes M. and Bruns T.D. (1996). Community structure of ectomycorrhizal fungi in a *Pinus muricata* forest: above- and below-ground views. *Canadian Journal of Botany* 74: 1572-1583.

Glen M.I.C., Tommerup I., Bougher N.L. and O'Brien P.A. (2002). Are Sebacinaceae common and widespread ectomycorrhizal associates of Eucalyptus species in Australian forests? *Mycorrhiza* 12: 243-247.

Gross E., Thomazini-Casagrande L.I. and Caetano F.H. (2004). Ultrastructural study of ectomycorrhizas on *Pinus caribaea* Morelet. var. *hondurensis* Barr. and Golf. Seedlings. *Acta Bot. Bras.* 18: 1-7.

Hall I.R., Wang Y. and Amicucci A. (2003). Cultivation of edible ectomycorrhizal mushrooms. *Trends in biotechnology* 21: 433-438.

Hawksworth D.L. (2004). ′Misidentifications′ in fungal DNA sequence databanks. *New Phytologist* 161: 13-14.

Horton T.R. and Bruns T.D. (2001). The molecular revolution in ectomycorrhizal ecology: peeking into the black-box. *Molecular Ecology* 10: 1855-1871.

Ingleby K, Mason P.A., Last F.T. and Fleming L.V. (1990). *Identification of Ectomycorrhizas.* ITE Research Publication No. 5. HMSO, London. 112 p.

Izzo A., Agbowo J. and Bruns T.D. (2005). Detection of plot-level changes in ectomycorrhizal communities across years in an old-growth mixed-conifer forest. *New Phytologist* 166: 619-630.

Joner E.J. and Leyval C. (2003). Phytoremediation of organic pollutants using mycorrhizal plants: a new aspect of rhizosphere interactions. *Agronomie* 23: 495-592.

Kennedy P., Izzo A.D. and Bruns T.D. (2003). There is high potential for the formation of common mycorrhizal networks between understorey and canopy trees in a mixed evergreen forest. *Journal of Ecology* 91: 1071-1080.

Kõljalg U., Dahlberg A., Taylor A.F.S., Larsson E., Hallenberg N., Stenlid J., Larsson K-H., Fransson P.M., Karen O. and Jonsson L. (2000). Diversity and abundance of resupinate thelephoroid fungi as ectomycorrhizal symbionts in Swedish boreal forests. *Molecular Ecology* 9: 1985-1996.

Kõljalg U., Larsson K-H., Abarenkov K., Nilsson R.H., Alexander I.J., Eberhardt U., Erland S., Hoiland K., Kjoller R., Larsson E., Pennanen T., Sen R., Taylor A.F.S., Tedersoo L., Vralstad T. and Ursing B.M. (2005). UNITE: a database providing web-based methods for the molecular identification of ectomycorrhizal fungi. *New Phytologist* 166: 1063-1068.

Kottke I., Beiter A., Weiss M., Haug I., Oberwinkler F. and Nebel M. (2003). Heterobasidiomycetes form symbiotic associations with hepatics: Jungermanniales have sebacinoid mycobionts while Aneura pinguis (Metzgeriales) is associated with a *Tulasnella* species. *Mycological Research* 107: 957–968.

Landeweert R., Leeflang P., Kuyper T.W., Hoffland E., Rosling A., Wernars K.,and Smit E. (2003). Molecular Identification of Ectomycorrhizal Mycelium in Soil Horizons. *Appl. Environ. Microbiol.* 69: 327-333.

Lee S.S. (1998). Root symbiosis and nutrition. In: *A Review of Dipterocarps -Taxonomy, ecology and silviculture*, Appanah S. and Turnbull J.M., Eds., CIFOR - Center for International Forestry Research, 99-114.

Lilleskov E.A., Bruns T.D., Horton T.R., Taylor D.L. and Grogan P. (2004). Detection of forest stand-level spatial structure in ectomycorrhizal fungal communities. *FEMS, Microbiology Ecology* 49: 319-332.

Loy A. and Bodrossy L. (2006). Highly parallel microbial diagnostics using oligonucleotide microarrays. *Clinica Chimica Acta* 363: 106-119.

Melin E. (1922). Untersuchungen über die *Larix*-Mykorrhiza. I. Synthese der. Mykorrhiza in Reinkultur. *Sv. Bot. Tidskr.* 16: 161-96.

Molina R., Massicotte H. and Trappe J.M. (1992). Specificity phenomena in mycorrhizal symbiosis: community-ecological consequences and practical implications. In: *Mycorrhizal functioning: an integrated plant-fungal process*, Allen M.F., Ed., Chapman and Hall, London, pp 357-423.

Mridha M.A.U. (2003). Application of Mycorrhizal Technology in Plantation Forestry in Bangladesh. *XII Word Forestry Congress 2003*, Québec City, Canada. http://www.fao.org/DOCREP/ARTICLE/WFC/XII/0961-B1.HTM

Rygiewicz P.T. and Armstrong, J.L. (1994). Ectomycorrhizal DNA: Isolation, RFLPs and Probe Hybridization. In: *Techniques in Mycorrhizal Research*, Norris J.R., Read D.J. and Varma A.K., Eds. Academic Press, London, pp. 253-293.

Saghai-Maroof M.A., Soliman K., Jorgensen R.A. and Allard, R.W. (1984). Ribosomal DNA spacer-length polymorphisms in barley: mendelian inheritance, chromosomal location, and population dynamics. *PNAS* 81: 8014-8018.

Selosse M.A., Weiß M., Jany, J.L. and Tillier A. (2002). Communities and populations of sebacinoid basidiomycetes associated with the achlorophyllous orchid *Neottia nidus-avis* (L.) L.C.M. Rich. and neighbouring tree ectomycorrhizae. *Molecular Ecology* 11, 1831-1844.

Simard S.W., Perry D.A., Jones M.D., Myrold D.D., Durall D.M., and Molina R. (1997). Net transfer of carbon between tree species with shared ectomycorrhizal fungi. *Nature* 388: 579-582.

Smith S.E. and Read D.J. (1997). *Mycorrhizal Symbiosis.* 2nd ed. Academic Press, London, 605 p.

Urban A., Weiß M. and Bauer R. (2003). Ectomycorrhizas involving sebacinoid mycobionts. *Mycological Research* 107: 3-14.

Urban A., Plattner-Neuner I., Krisai-Greilhuber I. and Haselwandter K. (2004). Molecular studies on terricolous microfungi reveal novel anamorphs of two *Tuber* species. *Mycological Research* 108: 749-758.

Vralstad T. (2004). Are ericoid and ectomycorrhizal fungi part of a comon guild? *New Phytologist* 164: 7-10.

Weiß M., Selosse M.A., Rexer K.H., Urban A. and Oberwinkler F. (2004). Sebacinales: a hitherto overlooked cosm of heterobasidiomycetes with a broad mycorrhizal potential. *Mycological Research* 108: 1003-1010.

10 SPPADBASE: The First On-Line Searchable Database of PCR Primers for Phytopathogenic Fungi

Stefano Ghignone[1] and Quirico Migheli[2]

[1] *Plant Protection Institute, Sect. Turin – CNR, V.le P.A. Mattioli, 25, I-10125 Turin, Italy.*

[2] *Department of Plant Protection - Center for Biotechnology Development and Biodiversity Research, University of Sassari, Via E. De Nicola 9, I-07100 Sassari, Italy.*

INTRODUCTION

The fast and unambiguous identification of microbial pathogens affecting plants or plant products is an essential prerequisite for obtaining high-quality and safe production. Ecologically friendly practice of the modern agriculture requires the adoption of diagnostic techniques able to detect minimum inoculum levels of pathogens in soil, seeds, transplants or crops, to limit the raise of epidemics and to address the adoption of rational and efficient control means. Moreover, there is an increasing public and official awareness of the potential threat of bio-terrorism directed against food and agriculture (Monke, 2004). Rapid detection techniques for bioweapon agents are a critical need for the first-responder community.

Among the nucleic acid-based diagnostic techniques, those involving the Polymerase Chain Reaction (PCR; Mullis and Faloona, 1987) are the most suited for early detection of phytopathogenic agents, due to their high sensitivity and the potential for automation. Many sequence source types could be selected and used as target for specific primer design. These may include, for instance, Random Amplified Polymorphic DNAs (Williams *et al.*, 1990; Welsh and McClelland, 1990), internal transcribed spacer (ITS) regions of the ribosomal RNA genes (White *et al.*, 1990) or other specific gene sequences. Primer sets can be designed to target specificity at the genus, species, or physiological race level, to distinguish a particular pathogen from closely related organisms.

A common and tedious task for researchers and technicians is to search for and retrieve bibliographic references of published and validated specific primer sets for a given pathogen querying the Internet, abstract collections and monthly journals' tables of contents. Very few examples of specific primer set collections for phytopathogenic agents have been released: a summary of primers for the diagnostic characterization of phytopathogenic bacteria seems to be the only one printed so far (Louws *et al.*, 1999). Moreover, among 719 molecular biology databases publicly available recorded by Galperin (2006) or among the 2470 BMC biomedical databases catalog available at http://databases.biomedcentral.com/, no online repository of primer sets of this kind is accessible. To overcome this lack of information, we released the first online searchable database of primer sets useful for the detection and identification of plant pathogenic fungi.

ACCESSIBILITY

"*The Database of PCR Primers for Phytopathogenic Fungi*" can be accessed at http://www.sppadbase.com (Fig. 10.1). SPPADBASE is an acronym that stands for "Specific Primers for Phytopathogenic Agents Data Base" and is a logical choice to host primer sets specific to other phytopathogens, such as viruses or bacteria, including microbial herbicides and biological agents relevant in agroterrorism.

Primer sets can be searched through a quick (Fig. 10.2(A)) or an advanced (Fig. 10.2(B)) search field. In the first case, users can search for primers typing the genus or species name of the target organism, the forward or the reverse primer name, the template DNA, the PCR technique, an author name or the year of publication; the use of word parts is allowed. In the second case, users can focus their searches by typing or selecting an appropriate search term; with this interface it is possible to retrieve primer sets using as search term the database ID (known from a previous search), the primer sequence, the GenBank accession numbers and the reference journal. The organism definition fields (with grey background) are combinable and allow retrieving primers for a specific pathogenic fungal agent, defining its subspecies rank, the subspecies epithet or other name modifiers.

The database records matching the query criteria are then tabled in a search results page (Fig. 10.3). For each set, the forward and the reverse primer couple, the target DNA, the PCR technique (if the data are available), and the bibliographic reference are displayed. Clicking on the reference, a pop-up window will open, showing the publication details in brief; if a PubMed ID is available, a link will open the corresponding page at NCBI. Clicking on the database ID of a selected set, will show the complete details (Fig. 10.4). The primer set detail pages are organized in the following sections:

1. **Internet resources**: This section allows queries with external search engines (Google, Altavista, Alltheweb) and other reference databases: the "NCBI Taxonomy Browser" shows the taxonomical lineage of the organism of interest, as it appears in the GenBank database, and provides direct access to the Entrez-deposited sequences obtained from each organism; the "CABI Bioscience Index Fungorum" provides the correct organism nomenclature based on a database of fungal names containing over 350,000 names of fungi at species level and below, derived from a number of published lists; the "Centraal Bureau voor Schimmelcultures" database provides the Anamorph/Teleomorph connections of the organism.
2. **Primer set**: This section shows the name and the sequences of the forward/reverse primers of the selected set, the amplicon size (if available), and a tool to perform BLAST similarity searches with the selected oligonucleotide sequence. If the primer set was designed for Real Time PCR experiments based on TaqMan™ chemistry, the probe name and sequence are also displayed.

Figure 10.1. "*The Database of PCR Primers for Phytopathogenic Fungi*" home page. Live database statistics are shown. Site pages are reachable trough the navigation bar.

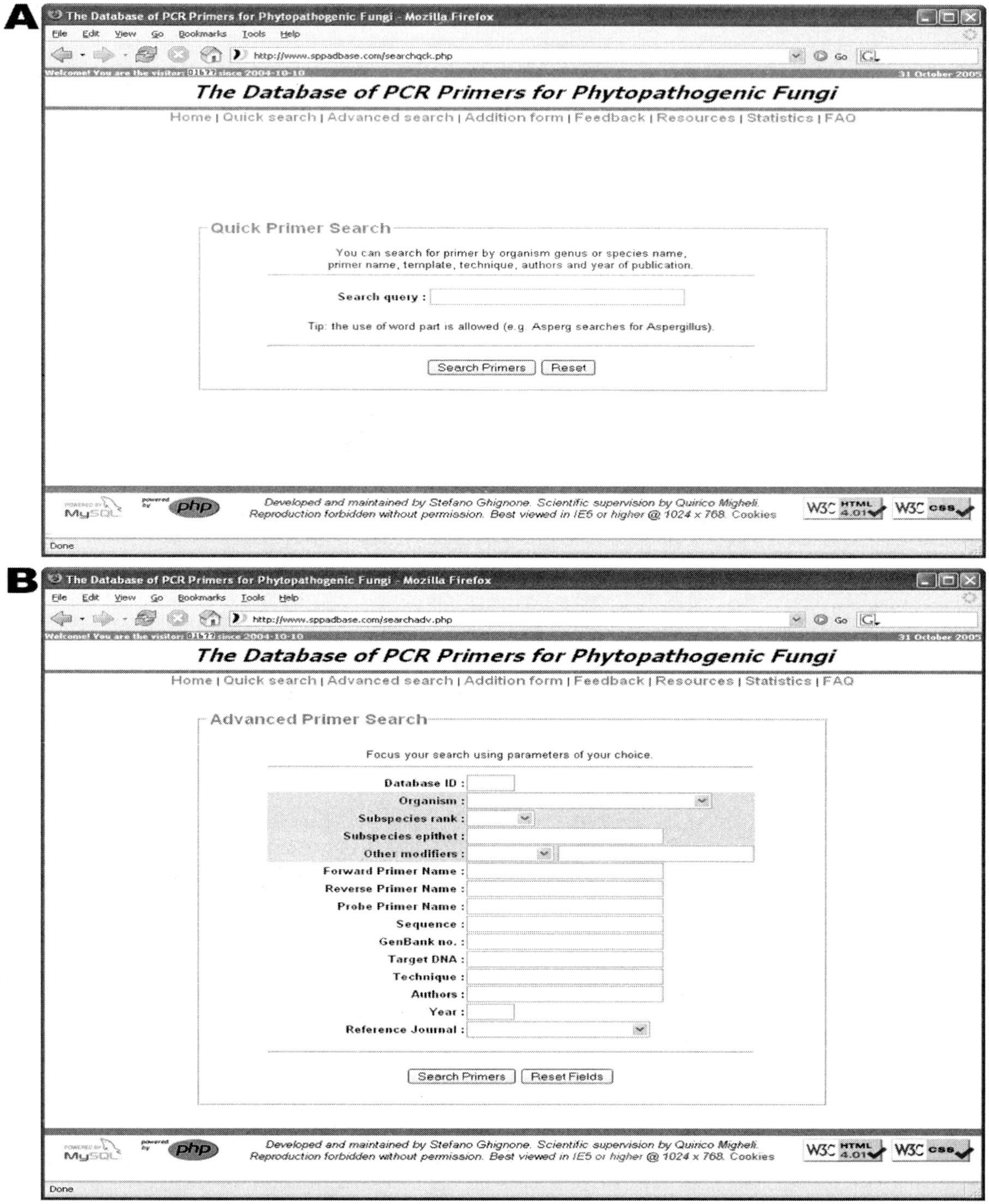

Figure 10.2. SPPADBASE **(A)** Quick search interface and **(B)** Advanced search interface.

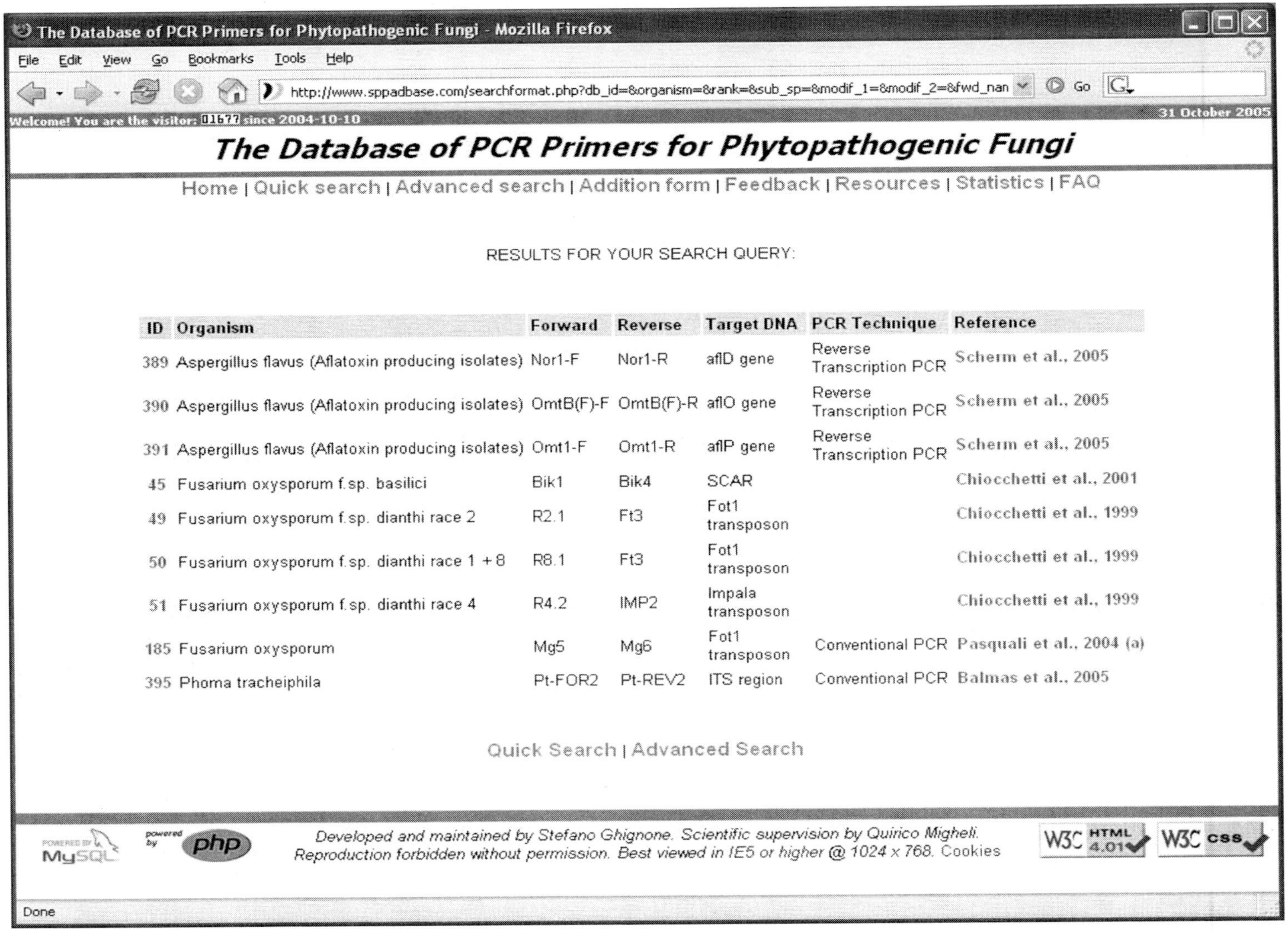

Figure 10.3. Example of a search result page. Target organisms are listed in alphabetical order. For each record, database ID, complete organism definition, forward and reverse primer name, target DNA, PCR technique and reference are given. Clicking on the reference, a pop up will open showing the publication details in brief. Clicking on the database ID of a selected set, the complete details will be shown.

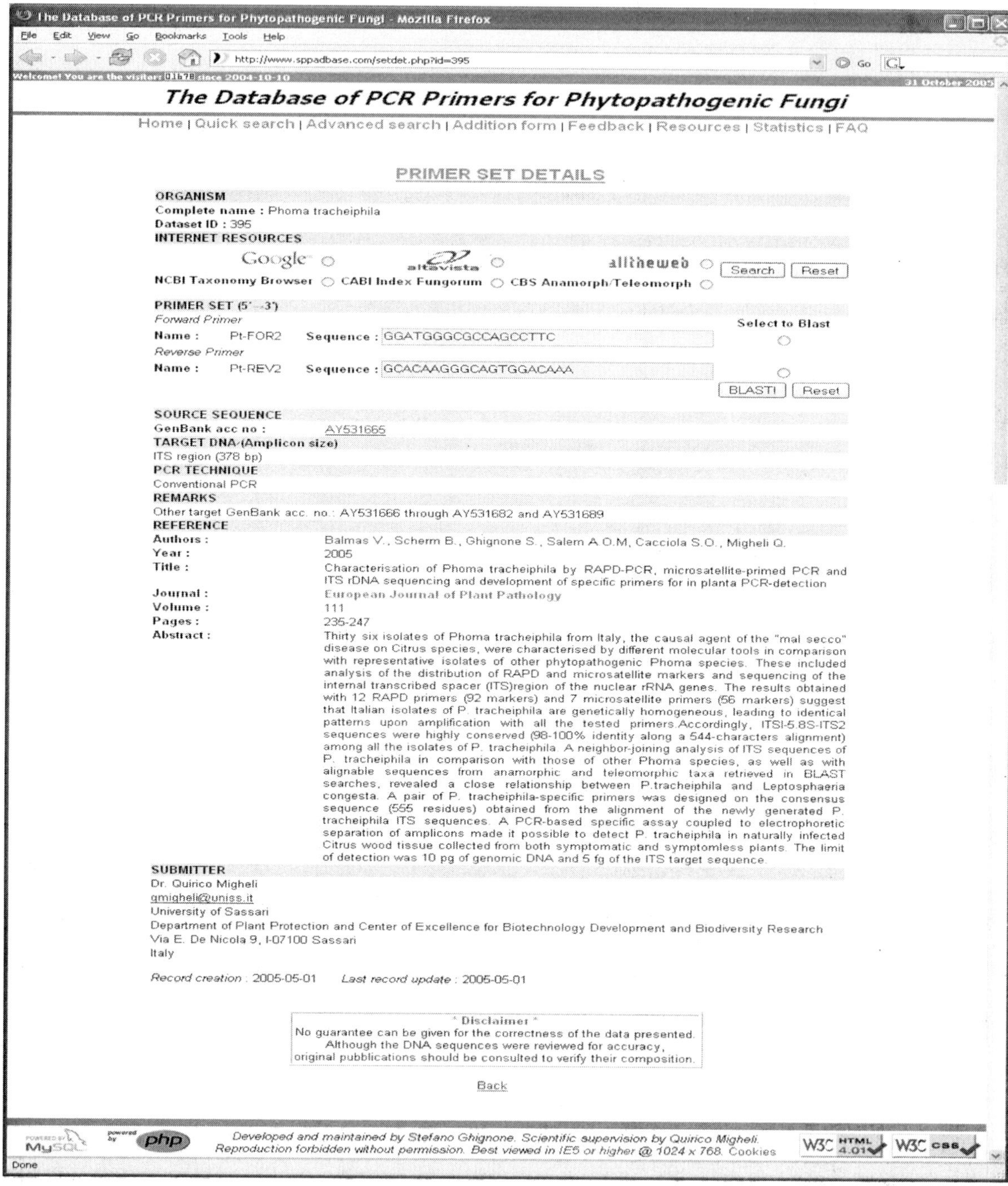

The Database of PCR Primers for Phytopathogenic Fungi - Mozilla Firefox

File Edit View Go Bookmarks Tools Help

http://www.sppadbase.com/setdet.php?id=395

Welcome! You are the visitor: 01678 since 2004-10-10 — 31 October 2005

The Database of PCR Primers for Phytopathogenic Fungi

Home | Quick search | Advanced search | Addition form | Feedback | Resources | Statistics | FAQ

PRIMER SET DETAILS

ORGANISM

Complete name : Phoma tracheiphila

Dataset ID : 395

INTERNET RESOURCES

Google ○ altavista ○ alltheweb ○ [Search] [Reset]

NCBI Taxonomy Browser ○ CABI Index Fungorum ○ CBS Anamorph/Teleomorph ○

PRIMER SET (5'→3') — Select to Blast

Forward Primer

Name : Pt-FOR2 **Sequence :** GGATGGGCGCCAGCCTTC ○

Reverse Primer

Name : Pt-REV2 **Sequence :** GCACAAGGGCAGTGGACAAA ○

[BLAST!] [Reset]

SOURCE SEQUENCE

GenBank acc no : AY531665

TARGET DNA (Amplicon size)

ITS region (378 bp)

PCR TECHNIQUE

Conventional PCR

REMARKS

Other target GenBank acc. no.: AY531666 through AY531682 and AY531689

REFERENCE

Authors : Balmas V., Scherm B., Ghignone S., Salem A.O.M, Cacciola S.O., Migheli Q.

Year : 2005

Title : Characterisation of Phoma tracheiphila by RAPD-PCR, microsatellite-primed PCR and ITS rDNA sequencing and development of specific primers for in planta PCR-detection

Journal : European Journal of Plant Pathology

Volume : 111

Pages : 235-247

Abstract : Thirty six isolates of Phoma tracheiphila from Italy, the causal agent of the "mal secco" disease on Citrus species, were characterised by different molecular tools in comparison with representative isolates of other phytopathogenic Phoma species. These included analysis of the distribution of RAPD and microsatellite markers and sequencing of the internal transcribed spacer (ITS)region of the nuclear rRNA genes. The results obtained with 12 RAPD primers (92 markers) and 7 microsatellite primers (56 markers) suggest that Italian isolates of P. tracheiphila are genetically homogeneous, leading to identical patterns upon amplification with all the tested primers.Accordingly, ITSI-5.8S-ITS2 sequences were highly conserved (98-100% identity along a 544-characters alignment) among all the isolates of P. tracheiphila. A neighbor-joining analysis of ITS sequences of P. tracheiphila in comparison with those of other Phoma species, as well as with alignable sequences from anamorphic and teleomorphic taxa retrieved in BLAST searches, revealed a close relationship between P.tracheiphila and Leptosphaeria congesta. A pair of P. tracheiphila-specific primers was designed on the consensus sequence (555 residues) obtained from the alignment of the newly generated P. tracheiphila ITS sequences. A PCR-based specific assay coupled to electrophoretic separation of amplicons made it possible to detect P. tracheiphila in naturally infected Citrus wood tissue collected from both symptomatic and symptomless plants. The limit of detection was 10 pg of genomic DNA and 5 fg of the ITS target sequence.

SUBMITTER

Dr. Quirico Migheli

qmigheli@uniss.it

University of Sassari

Department of Plant Protection and Center of Excellence for Biotechnology Development and Biodiversity Research

Via E. De Nicola 9, I-07100 Sassari

Italy

Record creation : 2005-05-01 *Last record update* : 2005-05-01

* Disclaimer *

No guarantee can be given for the correctness of the data presented. Although the DNA sequences were reviewed for accuracy, original pubblications should be consulted to verify their composition.

Back

Developed and maintained by Stefano Ghignone. Scientific supervision by Quirico Migheli. Reproduction forbidden without permission. Best viewed in IE5 or higher @ 1024 x 768. Cookies

Done

Figure 10.4. Example of a primer set details page. Primers (and probe, in case) sequences, link to source sequence, target DNA and amplicon size, PCR technique and publication details are shown, ordered in section. Query systems to search the Internet with the organism definition, to check the current organism nomenclature, to perform homology BLAST searches with primer/probe sequences are also available. In the lower side of the page, submitter details are displayed along with the record submission and modification dates.

3. **Source Sequence**: If the GenBank accession number of the primer's source sequence is available, it is displayed and linked directly to the NCBI GenBank flat file.
4. **Reference**: Complete reference details are shown here. If available, the article abstract and the link to the corresponding PubMed ID are displayed.
5. **Submitter**: Users are invited to submit basic data (authors, reference) of missing primer sets. Submitter details will appear in this section.

DATA SOURCES

Only published primer sets that do not require further amplification product analyses for the identification of the target organism have been considered for the construction of the live database. Primer set reference collection has been performed scanning the tables of contents of the most widely diffused phytopathology-related journals from the early 1990s to date. More than 30 reference sources are now monitored to maintain the reference database (Table 10.1).

Table 10.1. List of the phytopathology-related journals currently monitored to maintain the primers' database updated.

African Journal of Biotechnology	Journal of Plant Diseases and Protection
Applied and Environmental Microbiology	Journal of Plant Pathology
Applied Microbiology and Biotechnology	Letters in Applied Microbiology
Australasian Plant Pathology	Molecular Ecology
Biological Control	Molecular Genetics and Genomics
BMC Microbiology	Molecular Plant Pathology
Canadian Journal of Plant Pathology	Molecular Plant-Microbe Interactions
Canadian Journal of Botany	Mycologia
European Journal of Forest Pathology	Mycological Research
European Journal of Plant Pathology	Mycoscience
FEMS Microbiology Letters	Physiological and Molecular Plant Pathology
Forest Pathology	Phytopathology
Fungal Genetics and Biology	Plant Disease
International Journal of Food Microbiology	Plant Pathology
Journal of Applied Microbiology	Plant Pathology Bullettin
Journal of Food Protection	Systematic and Applied Microbiology
Journal of Microbiological Methods	Weed Science
Journal of Phytopathology	

The accuracy of the organism nomenclature, the taxonomical position and anamorph/teleomorph connection, as well as GenBank accession numbers reported can be checked querying external databases accessible from the primer set details page (Fig. 10.4). However, we encourage users to send us notification regarding missing or erroneous data. Furthermore, being this list far from exhaustive, users are invited to submit missing primer sets through an appropriate interface, providing detailed literature reference information and/or uploading an electronic copy of the published article. Submitters also are asked to provide personal details that will appear linked to the newly added primer set, following the system administration review, in the live database.

IMPLEMENTATION

"*The Database of PCR Primers for Phytopathogenic Fungi*" is a Web service implemented with PHP (http://www.php.net/), an open-source server-side scripting language, designed for generating HTML contents. MySQL (http://www.mysql.com/), an open-source database, is used as the data-management system.

The complete application is currently hosted on a Linux based machine, running an Apache 2.0 Web server.

A single script generates each HTML page code; each page code compliancy to the current HTML 4.01 Specification has been validated with the W3C MarkUp Validation Service (http://validator.w3.org/), as well as the Cascading Style Sheet (CSS) employed and the site links has been validated with the W3C CSS Validation Service (http://jigsaw.w3.org/css-validator/) and the Link Checker (http://validator.w3.org/checklink/), respectively.

The application has been optimized for the Microsoft Internet Explorer 5 browser or higher and it is best viewed at 1024x768 dpi screen resolution.

Application functionality has also been tested with other widely used Internet browsers such as the Gecko codebase class-based ones (e.g., Netscape, Mozilla), Firefox and Opera, running in Windows, Linux or Mac environment.

REFERENCES

Galperin M.Y. (2006). The Molecular Biology Database Collection: 2006 update. *Nucleic Acids Research* 34 (Database issue): D3-D5.

Louws F.J., Rademaker J.L.W. and de Bruijn F.J. (1999). The three Ds of PCR-based genomic analysis of phytobacteria: Diversity, detection, and disease diagnosis. *Annual Review of Phytopathology* 37: 81-125.

Monke J. (2004). Agroterrorism: threats and preparedness, Congressional Research Service Report RL32521 for US Congress. http://www.fas.org/irp/crs/RL32521.pdf

Mullis K. and Faloona F.A. (1987). Specific synthesis of DNA *in vitro* via a polymerase-catalyzed chain reaction, *Methods in Enzymology* 155: 35-40.

Welsh J. and McClelland M. (1990). Fingerprinting genomes using PCR with arbitrary primers. *Nucleic Acids Research* 18: 7213-7218.

White T.J., Bruns T., Lee S. and Taylor J. (1990). Amplification and direct sequencing of fungal ribosomal RNA Genes for phylogenetics. In: PCR - Protocols and Applications - A Laboratory Manual, Innis N., Gelfand D., Sninsky J. and White T. Eds., Academic Press, New York, 315-322.

Williams J.C.K., Kubelik A.R., Livak K.J., Rafalski J.A. and Tingey V. (1990). DNA polymorphisms amplified by arbitrary primers are useful as genetic markers. *Nucleic Acids Research* 18: 6531-6535.

11 *Claviceps*: A Biotechnologically Exploited Fungus

Archna Mehta and Pradeep Mehta

Microbiology Laboratory, Department of Botany, Dr. H. S. Gour University, Sagar 470003, Madhya Pradesh, India

INTRODUCTION

Claviceps is truly a higher organism in terms of its cellular structure and genetic organization, which appears to be identical to or closely resemble that of higher eukaryotes. Nevertheless, many attributes of *Claviceps* suggest that it may be a valuable model for the study of eukaryotic regulation. *Claviceps* can be easily manipulated because of its simple growth requirements. The large size and highly organized structure of *Claviceps* cells has consequences for regulation systems. Metabolic reserves are also commonly found in these eukaryotic cells thus providing for greater metabolic stability than found in prokaryotes. New techniques of analysis at the molecular level are revealing an astonishing elegance and economy in the living *Claviceps* cell and a gratifying unity in the principles by which cells function. The life of *Claviceps* can be considered as controlled repletion of biochemical events, which are time-, osmo-, buffer- and thermo-regulated stepwise around distinct borderlines in an anatomical and physiological architecture of biochemical events.

Ergot was formerly defined as the dried sclerotium of *Claviceps purpurea* (Fries) Tul. developed on plants of rye, cereals and in florets of grasses. Ergot alkaloids belong to the group of pharmacologically active compounds produced by a number of fungi mainly *Claviceps* sp. is often referred to as secondary metabolites. Ergot alkaloids have also been found in higher plants.

HISTORY

In early days, the medieval midwives used to collect the fungus from naturally infected plants for its use in the induction of childbirth and in the control of postpartum bleeding. Since then, many additional uses of ergot alkaloids have been reported. The role of sclerotium of this pyrenomycete (*Claviceps purpurea*) has undergone from important source of biologically effective substances. Apart from yeast, *Claviceps purpurea* is the first fungus (Fig. 11.1), which was biotechnologically exploited without its existence known. The ergot research dates back to the extraction of ergotinine, the first

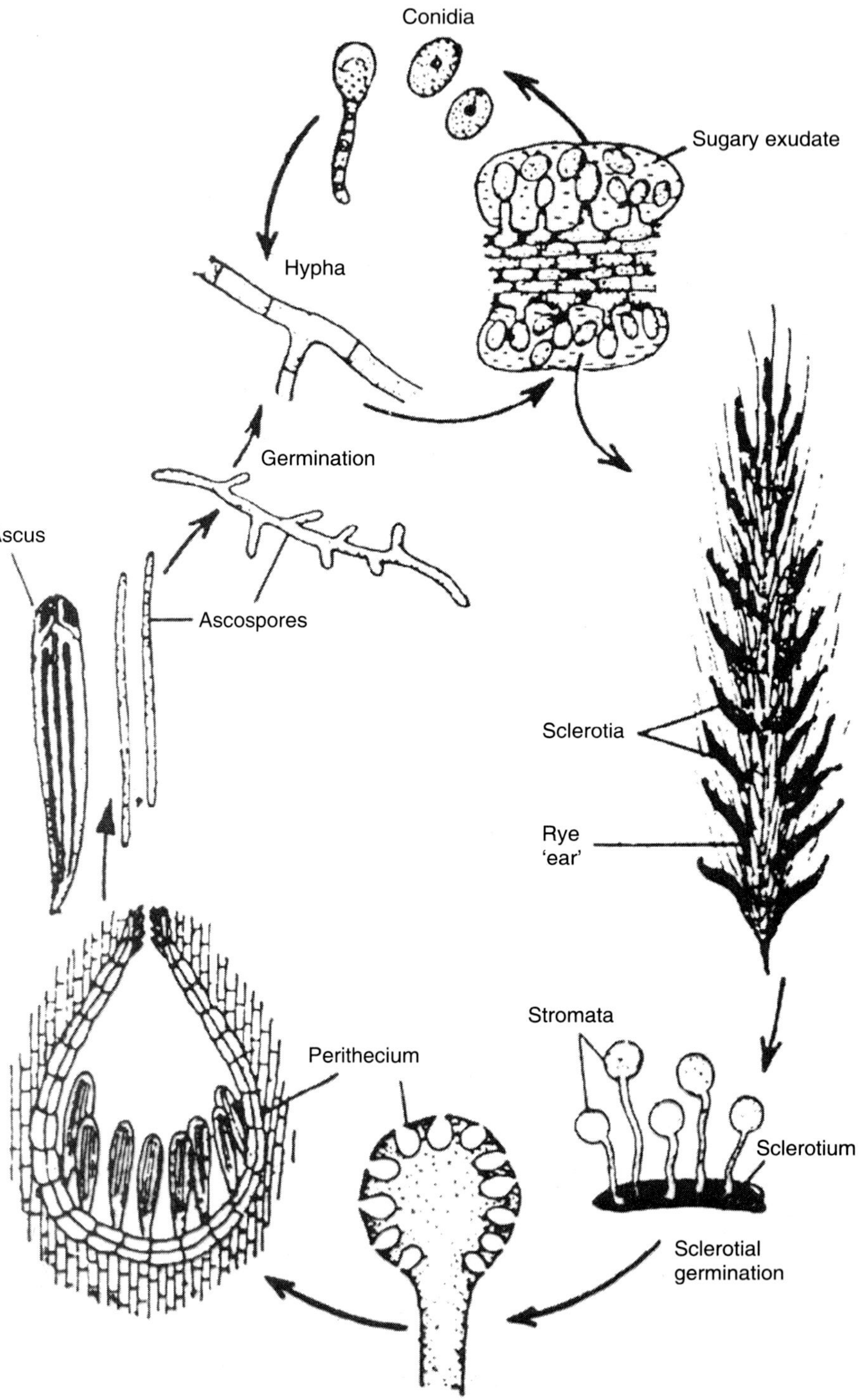

Fig. 11.1. Life cycle of *Claviceps purpurea*.

crystalline alkaloid from sclerotia in 1875, to the major pharmacological discoveries of ergotoxine in 1906 and the discovery of the ergotamine, the first clinically tested compound in 1918. Toxic effect of ergot compounds have been realized since the discovery that the consumption of infected cereals by *Claviceps* could cause severe illness (egotism), or even before it was evidently tested in obstetrics to enhance muscle contraction and was also used for abortions. During these early days, many deaths were brought about by ergot extract because neither its physiological action nor the appropriate dosage was known. Later with increasing understanding of pharmacology, it was found that chemical alterations of alkaloids improved their medical efficiency. Further, after a correlation between clinical indications and their therapeutically dosage have been developed, pure ergot alkaloids and their semi-synthetic derivatives are being used as drugs at low concentrations, extensively in medicine at present. Initially, chemical properties of the alkaloid were known while biologists studied the physiological aspects of the synthesis of alkaloids from the practical point of view, i.e. for both agronomists, who were trying to obtain grain with lowest possible alkaloid content and for pharmacologists, who, on the other hand, were interested in high alkaloid yields.

Since the supply of natural grown ergots is not sufficient, the biotechnological production of ergot alkaloids has two approaches: (i) the development of large-scale processes to exploit such knowledge, (ii) the acquisition of knowledge for producing microorganism. Several workers (Berde and Sturmer, 1978; Rehacek and Mehta, 1991, 1993; Mukherjee and Menge, 2000; Flieger *et al.*, 2003) have extensively reviewed the chemistry, biosynthesis, physiology, biochemistry, genetics, and therapeutical applications of ergot alkaloids. The present chapter describes the biotechnological exploitation of *Claviceps*.

SOURCES

In nature, the ergot alkaloids are mainly formed by *Claviceps*. More than 50 different alkaloids have been isolated from this pyrenomycete, which has about 50 species. *Claviceps* forms ergot sclerotia on various plants. More than 100 plants are attacked by ergot to form sphacelia but not the sclerotia of the fungus (Rehacek and Sajdl, 1990). Ergot alkaloids have also been found from organisms (Table 11.1) other than *Claviceps* (Flieger *et al.*, 1997; Kozlovsky, 1999).

ERGOLINE

Ergoline, a common part of ergot alkaloids, is formed from tryptophan (Taber and Vining, 1959), mevalonic acid (Groger *et al.*, 1961) and methionine (Baxter *et al.*, 1961). The ergot alkaloids constitute the largest known group of nitrogenous fungal metabolites. The common part of chemical structure in ergot alkaloids is the tetracyclic ergoline ring system (Fig. 11.2).

ERGOT GROUPS

The ergot alkaloids can be structurally separated into four main groups: (i) clavine alkaloids, (ii) lysergic acids, (iii) simple lysergic acid amides, and (iv) peptide alkaloids (Fig. 11.3; Table 11.2).

Table 11.1. Sources of ergot alkaloids

Fungi imperfectii	*Hypocreales*	*Zygomycetes*	*Basidiomycetes*	*Convolvulaceae*
Aspergillus clavatus	*Balansia claviceps*	*Rhizopus arrhizus*	*Corticum caeruleum*	*Argyreia nervosa*
A. flavus	*B. epichloe*	*R. nigricans*	*Lenzites trabea*	*Ipomoea argyrophylla*
A. japonicus	*B. obtecta*	*Cunninghamella blakesleana*	*Pellicularia filamentosa*	
A. tamarii	*B. strangulans*			*I. Hildebrandtii*
A. versicolor	*Epichloe typhina*	*Mucor hiemalis*		*I. tuberosa*
A. oryzae	*Neotyphodium coenophialum,*			*I. violacea*
A. fumigatus				*I. rubrocaerulea*
A. nidulans	*N. lolii,*			*I. piurensis*
Penicillium chermesinum	*Hypomyces*			*Rivea corymbosa*
	aurantius,			*Stictocardia tiliifolia*
P. concavorugulosum	*Sepedonium sp.*			
P. fumigatus				
P. viridicatum				
P. sizovae				
P. rugulosum				
P. rubrum				
P. kapuscinskii				
P. palitans				
P. aurantiovirens				
P. camembertii				
P. clavigerum				
P. crustosum				
P. griseofulvin				
P. roqueforti				

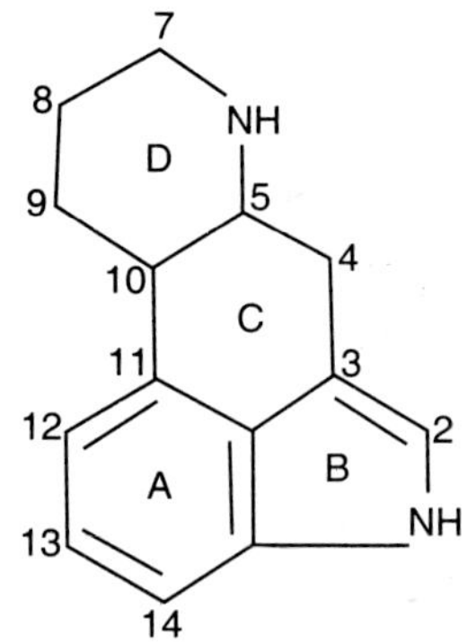

Fig. 11.2. Ergoline

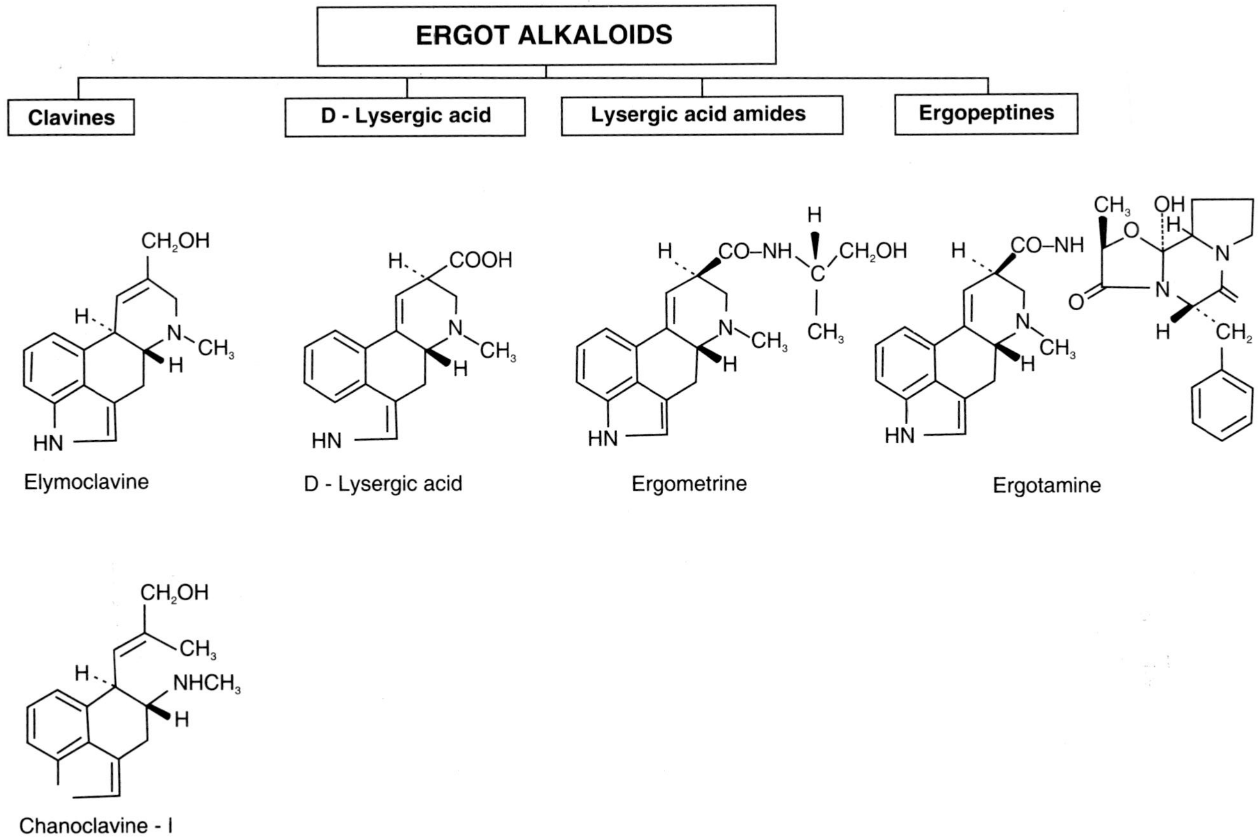

Fig. 11.3. Characteristic representative of the main groups of ergot alkaloids

Table 11.2. Types of Ergot Alkaloids

Clavines	*Lysergic acid and its simple derivatives*	*Ergopeptines*
Chanoclavine-I (secaclavine),	Ergometrine (ergonovine, ergobasine),	Ergotamine,
Chanoclavine-II,	Lysergic acid 2- hydroxyethylamide,	Ergosine,
Isochanoclavine-I,	Lysergic acid amide (ergine),	Ergovaline,
Norchanoclavine-I,	Lysergic acid	Ergostine,
Norchanoclavine-II,		Ergoptine,
Chanoclavine-I acid,		Ergonine,
Paliclavine,		Ergobutine,
Paspaclavine,		Ergocrystine,
Dihydrochanoclavine-I,		Ergocryptine,
Isodihydrochanoclavine-I,		Ergokryptine,
Rugolovasine A,		Ergocornine,
Rugolovasine B,		Ergobutyrine
Clavicipitic acid,		
Agroclavine,		
Elymoclavine,		
Molliclavine,		
Elymoclavine-O-β-D-fructofuranoside,		
Lysergine,		
Lysergol,		
Isolysergol,		
Lysergene,		
Setoclavine,		
Iisosetoclavine,		
Norsetoclavine,		
Penniclavine,		
Isopennniclavine,		
Festuclavine,		
Costaclavine,		
Pyroclavine,		
Dihydrolysergol-I, Dihydrosetoclavine,		
Fumigaclavine A,		
Fumigaclavine B,		
Roquefortine A(IsofumigaclavineA)		
Roquefortine B(IsofumigaclavineB)		
Cytoclavine,		
Aurantioclavine		

CLAVINES

The clavines (Fig. 11.4) are substituted 6,8-dimethylergolines and include a few members, namely the chanoclavines with a 6,7-seco D ring. The degree of oxidation is a criterion for further differentiation in the group of clavine alkaloids.

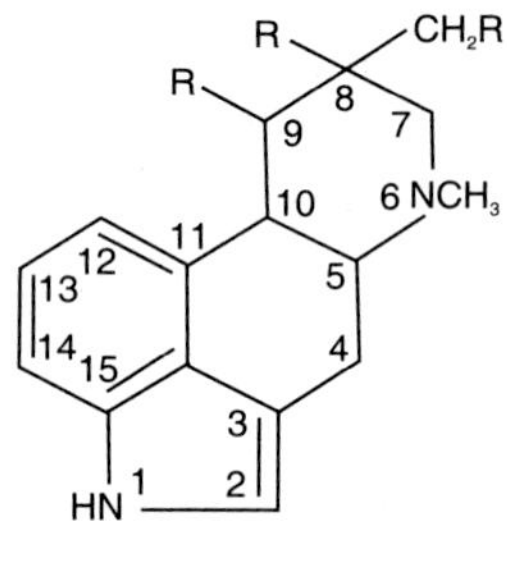

Fig. 11.4. Clavine

LYSERGIC ACID AND ITS SIMPLE DERIVATIVES

The derivatives of lysergic acid are mostly amides in which the amide part is a small peptide or a simple alkylamide. Non-peptide amides of lysergic acid found in ergot are ergometrine, lysergic acid 2- hydroxyethylamide, lysergic acid amide (ergine) and paspalic acid (Fig. 11.5).

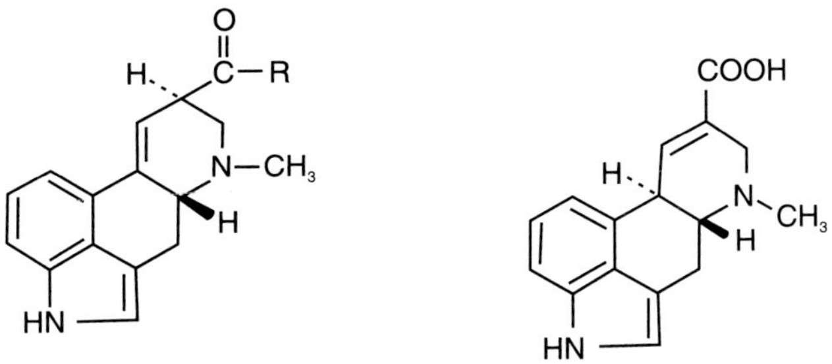

(A) **(B)**

Fig. 11.5. Simple derivatives of lysergic acid (A) and paspalic acid (B)

Name	*R*
Ergometrine (ergonovine, ergobasine)	CH_3 -HN – CH CH_2OH
Lysergic acid 2-hydroxyethylamide	CH_3 -HN – CH-OH
Lysergic acid amide (ergine)	$-NH_2$
Lysergic acid	-OH

Ergopeptines

Ergotamine was the first alkaloid to be isolated from ergot crop and became commonly used compound for therapeutics. Ergotamine and other classical peptide ergot alkaloids (ergopeptines) are composed of lysergic acid and an L- proline containing complex tripeptide moiety.

NEW ALKALOIDS

Few natural alkaloids have been described recently (Table 11.3) and many of them have only biogenetic importance (Rehacek and Sajdl, 1990).

Table 11.3. Some New Natural Ergot Alkaloids

Alkaloid	*Source*
Chanoclavine- II	*Claviceps* sp.
Isochanoclavine- I	*Claviceps* sp.
Aurantioclavine- I	*Penicillium aurantiovirens*
Cycloclavine	*Ipomoea hildebrandtii*
Cycloclavine	*Aspergillus japonicum*
Cividiclavine	*Penicillium citroviride*
Dihydrosetoclavine	*Claviceps paspali*
6- Norsetoclavine	*Claviceps* sp.
Elymoclavine- O- α-D fructofuranoside	*Claviceps* sp. SD 58
Epoxyagroclavine	*Penicillium corylophium*
Paliclavine	*C. paspali*
Paspaclavine	*C. paspali*
Clavicipitic acid	*C. fusiformis*
Ergoheptine	*Claviceps* sp.
Ergohexine	*Claviceps* sp.
Paspalic acid	*C. paspali*
Paspalicine	*C. paspali*
Paspaline	*C. paspali*
Ergocornom	*C. purpurea*
Ergobutine	*C. purpurea*
Ergobutyrine	*C. purpurea*
8- Hydroxyergotamine	*C. purpurea*
Ergosecaline	*C. purpurea*
Ergosine	*Ipomoea argyrophylla*
Dihydroergosine	*Sphacelia sorghi*
Ergostine	*C.purpurea*
Roquefortine A, B	*Penicillium roqueforti*
Rugulovasine A, B	*P. concavorugulosum*

PRODUCTION

There are 3 basic ways to get ergot alkaloids: (i) by isolation from the sclerotia of *Claviceps purpurea* collected from the field (95% of the world production), (ii) by extraction of saprophytic cultures of other ascomycetes especially of the *Claviceps* genus and (iii) by partial/total synthesis. Sclerotia of *Claviceps purpurea* collected from rye may contain 0.8% peptide alkaloids on dry weight basis which is an equivalent of about 500 mg/L alkaloid level in a fermentation broth of 6% mycelial dry weight (Udwardy, 1980). Since the naturally grown ergot is not sufficient, also as its production is labour intensive and weather dependent, the biotechnological production of ergot alkaloids has gained importance. In some cases, the saprophytic production of ergot alkaloids is more economical than parasitic cultivation on rye plants. In bioreactors, the process is more precisely controlled, not dependent on specific season. The production and isolation of alkaloids can take place on the same site and possible to apply some manipulations, which are not feasible in field conditions and above all clavines and simple lysergic acid derivatives are easier to manufacture than ergopeptines. In bioreactors, the alkaloid yield may exceed 20% of the mycelial dry mass while natural sclerotial yield is normally less than 1% or even less than 0.1% in some cases.

SAPROPHYTIC STRAINS OF *CLAVICEPS*

Saprophytic *Claviceps* strains are prerequisites for industrial production of ergot alkaloids, e.g.,

Clavines : *C. fusiformis* B 35, *C. fusiformis* w1, *C. fusiformis* F 27, *C. purpurea* Pepty 695/ch, *C. purpurea* 129, *C. purpurea* CP 7 and *C. paspali* SD 58

Simple lysergic acid derivatives: *C. paspali* NRRL 3027, *C. paspali* Li 342/SE 60

ERGOPEPTINES

(i) ergotamine: *C. purpurea* ATCC 15383, *C. purpurea* FI 275, *C. purpurea* OKI 56/1970
(ii) ergosine : *C. purpurea* MUT 168
(iii) ergocristine: *C. purpurea* ATCC 20 103 (IMI 131669, FIS 40), *C. purpurea* Ech 420
(iv) ergocornine and ergosine : *C. purpurea* FI 43/14
(v) ergotamine and ergotryptine : *C. purpurea* FI 32/17, *C. purpurea* IC/39/20
(vi) ergocornine and ergokryptine : *C. purpurea* Exy 20, *C. pu*rpurea OKI 00022, *C. purpurea* DM 838 (IMET PA 13), *C. purpurea* Pepty 695
(vii) dihydroergosine : *Sphacelia sorghi*

ERGOT ALKALOID BIOSYNTHESIS

Ergot alkaloid is derived from tryptophan, mevalonic acid and methionine (Birch *et al.*, 1960; Groger *et al.,* 1960; Taylor and Ramsted, 1960). Tryptophan has also been found to be a factor in the induction and depression of enzymes catalyzing alkaloid formation (Krupinski *et al.*, 1976). In *Claviceps purpurea* the formation of conidia is inversely dependent on the synthesis of alkaloids for those saprophytic strains which partially retain parasitic development, i.e., differentiation of the sphacelial phase to the conidial or the sclerotial phase or both (Mehta, 1984). In contrast, in those strains for which the conidial phase, the phase of vegetative growth and the alkaloid phase are not clearly separated, the conidia formation occurs simultaneously with alkaloid synthesis. Physiological conditions necessary for conidiation and germination of conidia are different from conditions controlling alkaloid synthesis (Rehacek *et al.*, 1974). The morphological development of the mycelium in submerged *Claviceps* cultures show significant differences between high-yielding and degenerated cultures (Flieger *et al.*, 1982). The character of fat hyphae and sclerotia like cells of submerged mycelium is reminiscent of the plectenchymatic structure of production sclerotia of parasitic cultures. Such sclerotia like cells influence the alkaloid production in *C. fusiformis* (Dickerson *et al.*, 1970). In *Claviceps* sp. high alkaloid production is accompanied with thick walled round cells having full cytoplasm with long, thin vacuolated cells (Mehta, 1984). Another important fact is that clavine alkaloids and simple derivatives of lysergic acid are extracellular products with high solubility in the culture medium and feedback their own biosynthesis (Flieger *et al.*, 1988) while peptide ergot alkaloids (ergopeptines) are mostly intracellular and are accumulated in lipid droplets (oleosomes) with no influence on the metabolism (Neumann *et al.*, 1979).

Clavine alkaloids lie at the beginning of the biogenetic order and classic ergot peptide alkaloids at the end. The oxidation order agroclavine – elymoclavine- lysergic acid is the main biogenetic pathway (Fig. 11.6). A variety of clavine alkaloids are formed due to action of various enzymes,

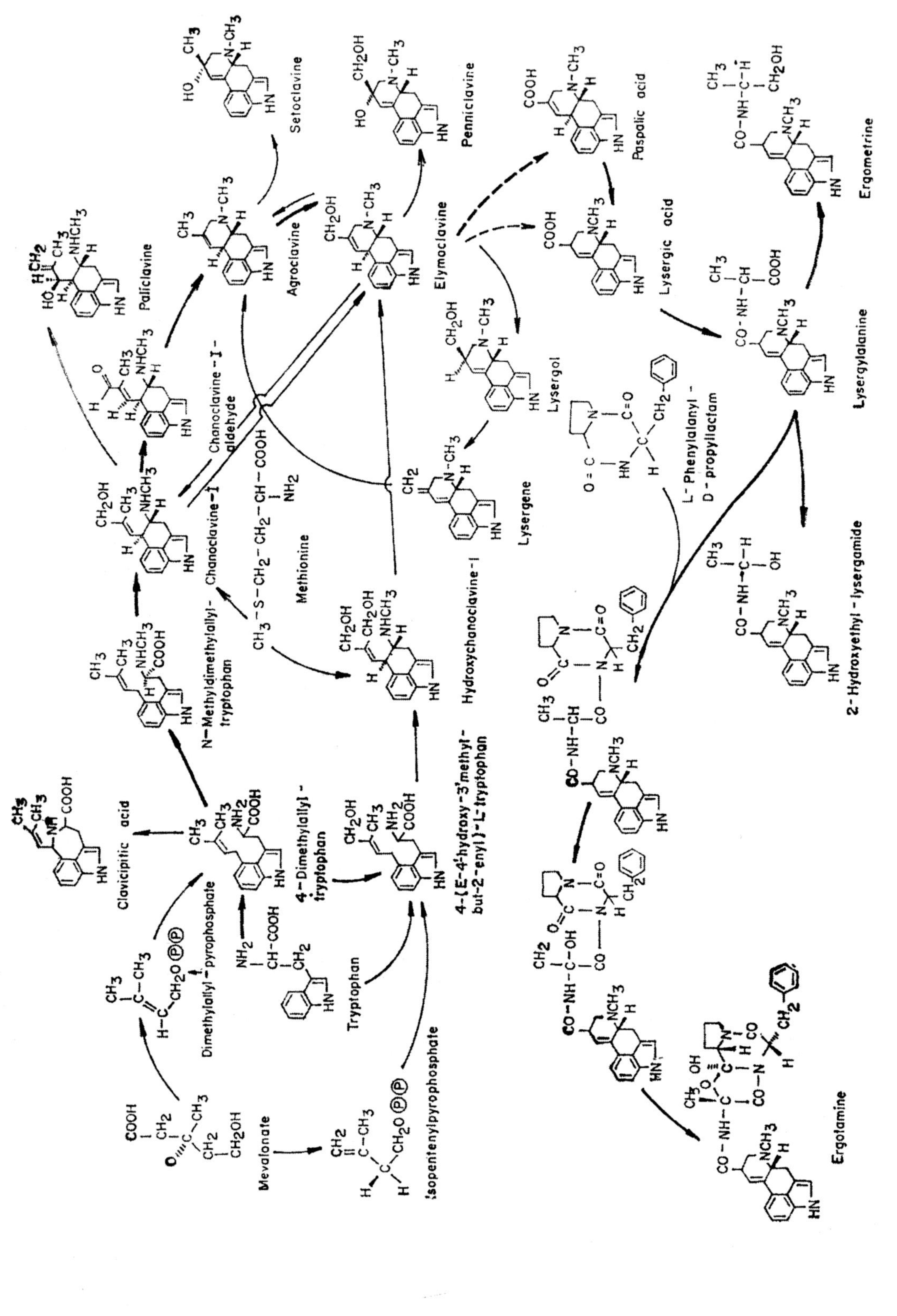

Fig. 11.6. Biogenesis of ergot alkaloids

which direct elymoclavine and agroclavine from the main biosynthetic route. Such type of shunt activity is particularly evident in strains that lack the complete pathway and are unable to synthesize substituted lysergic acid (Vining, 1980). Citrate is an intermediate, which has an important role in metabolic regulation. The region of formation of ergot alkaloids is not generally the same as where alkaloids are found in increased concentrations. In *Ipomoea*, the alkaloids are formed in leaves while they are found accumulated in seeds, which do not produce the alkaloids. The clavine alkaloids synthesis takes place in endoplasmic reticulum which later forms vacuoles containing lipoproteins (Vorisek and Rehacek, 1978). The formation of clavine alkaloids is a part of endogenous metabolism releasing the energy required for the maintenance of integrity and viability of cells under conditions of limited growth (Rehacek *et al.*, 1974). Clavine alkaloids probably may form a reservoir, which may lead to the formation of peptide alkaloids. Ergot alkaloids are not waste, physiologically inert products of cell metabolism (Rehacek *et al.*, 1974). In parasitic *Claviceps* cultures cytoplasmic oleosomes fuse while in young saprophytic cultures they do not coalesce possessing at least a superficial monomolecular phospholipid layer. In contrast, vacuolar oleosomes have no visible boundary. The enzyme activity inside the oleosomes is not surprising since they contain proteins (Yatsu *et al.*, 1971) and enzymes (Hislop *et al.*, 1974). Ergometrine added to a submerged culture of *Claviceps paspali* increases the alkaloid production by 120% (Rehacek *et al.*, 1974). Ergotamine and ergometrine reduce the intensity of conidia formation in a productive culture of *Claviceps purpurea* (Rehacek *et al.*, 1974). The effect of tryptophan on the enzyme system in the ergot alkaloid synthesis (Robbers *et al.*, 1972) suggests participation of the alkaloids in the feedback inhibition and repression of the enzyme-forming system.

Thorough research in the field of ergot alkaloid biosynthesis became possible only after the successful development of techniques for alkaloid production in saprophytic culture. Studies on metabolic pathways taking part in the synthesis have been made in detail while physiological formation of ergot alkaloids have been worked out negligibly. The biosynthesis of ergot alkaloids is correlated with distinct genotypes (Esser and Tudzynski, 1979). Most of the enzymes and co-enzymes of basic metabolism and alkaloid biosynthesis are common. In submerged culture, alkaloid synthesis takes place under conditions of decreased proliferation and clearly defined degradation and re-synthesis of nucleic acids and proteins (Rehacek *et al.*, 1972).

Ergot alkaloid synthesis requires a change in differentiation, which is considered to be a result of different expression of the genome under various conditions and at different times. Morphological differentiation is preceded by synthesis of membrane, i.e., protein and lipids. Normally, the metabolism of the early sequences of fungal differentiation is characterized by an increase in the relative importance of the pentose phosphate route and that of the glycoxylate cycle, NADP dependent functions predominate and there is a decrease in oxygen consumption. The vegetative fungal cells initially form a polarization gradient in the cytoplasm where respiration plays an important role in maintaining polarized growth. Degenerated strains are no longer able to use large amounts of nutrients while the alkaloid yield of higher producing mutants could be further improved by supplying them with higher carbon and nitrogen concentrations.

The aeration and agitation should meet the biochemical requirements. Cultural conditions suitable for metabolism of primary substances are not suitable for alkaloid metabolism (Taber and Vining, 1963). Submerged *Claviceps* cultures form alkaloids after using up the phosphate in the medium and accumulating polyols, lipids, polyphosphates and also inducers and precursors of alkaloids. Good alkaloid synthesis requires the presence of higher citrate level or some other intermediate of the Krebs cycle in the cell (Pazoutova *et al.*, 1981). Cell differentiation and the onset of the alkaloid synthesis are governed by both, the phosphate limitation (Vining and Nair, 1966) and citrate utilization. The synthesis of alkaloids is unfavorably influenced by an increased phosphate concentration in a fermentation medium (Robbers *et al.*, 1972). High production strains are usually tolerant to higher phosphate concentrations (Arcamone *et al.*, 1970). Phosphate plays a negative role in the induction of enzymes catalyzing the alkaloid synthesis (Krupinski *et al.*, 1976). Probably high phosphate levels seem to limit the synthesis of tryptophan. The rate of alkaloid synthesis and the activity of Krebs cycle were proved to be inversely proportional. The level of the Krebs cycle enzymes decreased in favour of glyoxylate enzymes. The absence of citrate or another Krebs cycle intermediate in the medium induced a specific complex of differentiation abberation, which resulted in 50% decrease of biomass, cessation of the alkaloid synthesis (agro- elymo-, chanoclavine) and medium acidification by lactate. Citrate is an intermediate, which has an important role in metabolic regulation. It inhibits phosphofructokinase and stimulates glucose-6-dehydrogenase. It seems that the inability of submerged *Claviceps* cultures to grow on hexoses in absence of Krebs cycle intermediates comes from the parasitic way of life of this pyrenomycete, where the level of citrate and malate in the phloem sap of the host plant is high. Therefore, in parasitic strains the impairment of mentioned regulatory relationships has not been excluded during the *Claviceps* evolution as a selection disadvantage. Cultural conditions affect the fermentation broth directly as well indirectly (Fig. 11.7). However, the effects of rheological characteristics have not really been considered appropriately in any of the published transport correlation (Rehacek, 1983).

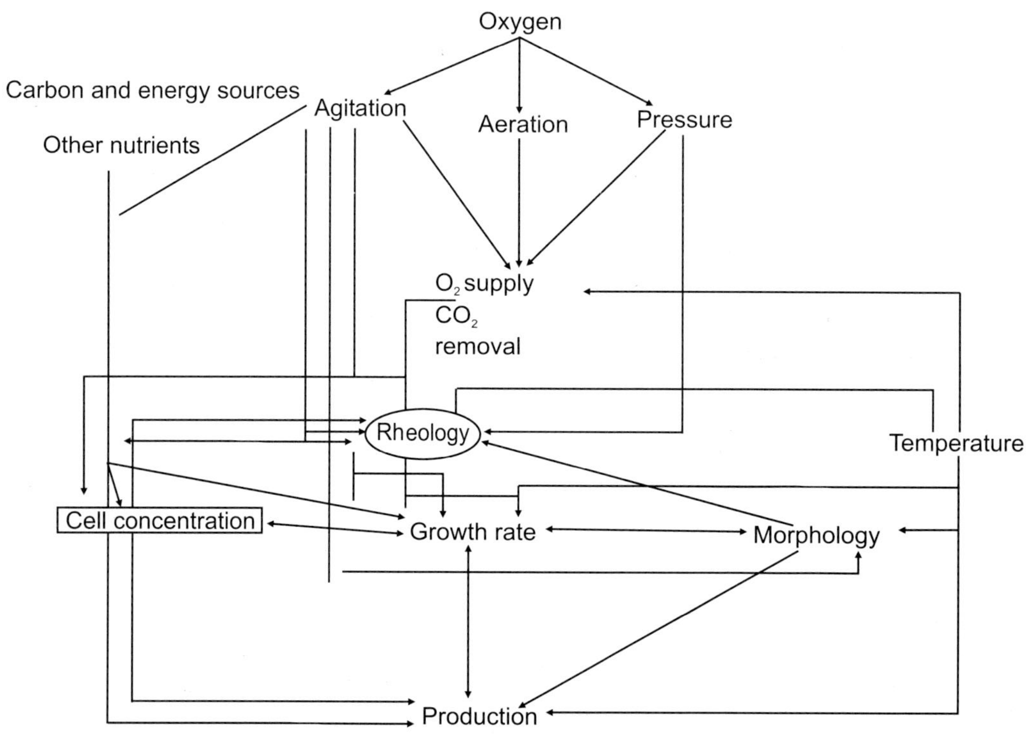

Fig. 11.7. Effect of cultivation on the culture growth rate and metabolite production (Rehacek, 1983).

CLAVICEPS STRAINS

The key to successful production of ergot alkaloids is obtaining the proper strain of the fungus. Slices of plectenchymatous tissues of the purified sclerotia are normally used for the preparation of saprophytic cultures of *Claviceps purpurea* producing ergot alkaloids. There are three main processes used for the preparation of *Claviceps* strains for saprophytic culture— (i) Plating of plectenchymatic tissue from the surface sterilized sclerotia on an agar growth medium (Mantle, 1969), (ii) Plating of honeydew drops containing conidia formed at early stage of *Claviceps* infection (Pazoutova *et al.*, 2002), (iii) Trapping of sexual ascospores ejected from fruiting bodies on germinated sclerotia. By this method monosporic culture can be obtained (Vasarhelyi *et al.*, 1980).

STRAIN IMPROVEMENT

The classical methods (selection pressure, mutagenesis and recombination) used for the strain improvement are more complicated with *Claviceps* due to incomplete information on cell nucleus. Strains used for saprophytic cultivation might be heterokaryotic and homokaryotic (Didek-Brumec *et al.*, 1991). Mutagenesis followed by subsequent selection of strain is an important technique in increasing the yield of alkaloids (Didek-Brumec *et al.*, 1987). An ergocristine producing *C. purpurea* strain showed 180-fold increase in alkaloid production after eight-step mutation-selection with different mutagens (Kobel and Sanglier, 1978). To increase the mutation frequency, the protoplasts prepared from spores of selected strains should be used (Zalai *et al.*, 1990). Hyphal fragments are rather unsuitable for mutagenesis due to the higher number of nuclei. Even protoplast formation from young mycelium and subsequent regeneration without any mutagenic treatment yielded strains with different properties (Schumann *et al.*, 1987). Strain improvement can have 3 phases—(i) the generation of desirable altered genotype, (ii) the recognition and isolation of such genotypes, and (iii) their acceptance in commercial processes. Removal of the feedback regulation of alkaloid synthesis may lead to strain improvement. However, the protoplast fusion technique is beginning to find useful applications either in producing improved mutant strains by intraspecific crosses or in formation of novel spectrum of products by interspecific hybrids (Socic and Gaberc-Porekar, 1992).

Many field isolates of *C. purpurea* produce significant amounts of ergot alkaloids only in the parasitic phase and not in axenic culture. Thus, the screening of different ergot strains from field surveys or various culture collections over the years has yielded interesting strains producing different specific alkaloids (or desired mixtures) at moderate rates in plants, but they were only suited (if at all) for parasitic production and not for submerge production of alkaloids. However, due to the lack of antibiotic or their activity that can be tested in quick bioassay, the screening is more time consuming than normal and requires a great deal of analytical work in order to select the desired clones. Pre-selection methods to narrow down the number of interesting isolates involve growing natural isolates or mutated clones under culture conditions that enforce sclerotia-like cell growth, such as by the addition of salt to the media (Udwardy, 1980) or by strictly using media imitating the natural physiology of parasitic culture (Strandova *et al.*, 1986). Therefore, ergot biotechnology strain improvement programs have been launched, aimed at an increased capacity of strains to produce high amounts of a specific alkaloid, and it is mainly achieved, as in other strain improvement programs of secondary-metabolite-producing fungi, by a stepwise mutagenesis selection process.

PRESERVATION AND MAINTENANCE

Cultures are the basic ingredients of microbiological work and at present there is much concern regarding the need to conserve genetic resources. Selection of a preservation method and the stability of desired properties of a microorganism are correlated with microorganism's identity, properties, use, restoration and growth conditions. Conservation and systematically performed selection of the isolates is the only way to eliminate the biological effects given by transfer of cultures, ageing, and other influences. Two main methods are applied for preserving sporulating cultures—(i) deep-freezing and (ii) maintaining cultures on rye grains or agar slants placed in refrigerator. Non-sporulating strains due to higher sensitivity to conservation procedure are frequently preserved as cultures on the agar plates. The most universal technique applied for preservation is keeping of lyophilized cultures and cultures frozen in liquid nitrogen (Pertot *et al.*, 1990). However, methods applied for long-term preservation have been recently reviewed (Hunter-Cervera and Belt, 1996).

SAPROPHYTIC CULTIVATION OF *CLAVICEPS* SP.

The first saprophytic cultivation of *Claviceps* on artificial nutritional media dates back to 1922 (Bonn, 1922). The first attempt to industrial production of ergot alkaloids was isolation of clavine alkaloids from submerged cultures of different *Claviceps* sp. (Abe and Yamatodani, 1955; Abe *et al.*, 1956). Later conditions for saprophytic production of simple derivatives of lysergic acid were developed using different strains of *C. paspali* (Arcamone *et al.*, 1961). Only 5 years later a new isolate of *C. purpurea* was found to produce ergotamine under submerged conditions (Amici *et al.*, 1966). Since that time, all types of ergot alkaloids for direct use as therapeutic agents or precursors for the preparation of semisynthetic drugs could be obtained by fermentation. Malinka (1999) described the successful submerged fermentations for the production of D-lysergic acid, the simple D-lysergic acid amides and D-lysergic acid peptides.

FERMENTATION

All technologies developed for industrial scale have the same aims, i.e., the maximum production of desired ergot alkaloids, the minimum amount of accompanying contaminants (other alkaloids, toxins, etc.), the shortest possible time of fermentation, the minimized cost of medium, energy, equipments, and labor. Different fermentation technologies have been employed for the production of ergot alkaloids, i.e., (i) stationary cultivation on liquid or solid medium (Trejo-Hernandez *et al.*, 1993) (ii) submerged cultivations (Kobel and Sanglier, 1986) also adapted to semicontinuous or continuous processes (Knopp and Rehm, 1984). In some cases, the immobilized microorganisms were used for the production of ergot alkaloids under condition of submerged fermentation (Komel *et al.*, 1985; Kren 1991).

SUBMERGED CULTIVATION

Several workers have got success in isolating strains of *Claviceps purpurea* and their cultivation under submerged conditions to produce ergot peptide alkaloids (ergopeptines), simple derivatives of lysergic acid and clavine alkaloids (Vorisek and Rehacek, 1976). Submerged fermentation in laboratory scale, i.e., shaker cultivation, is a primary step in getting knowledge of production, microorganism's physiology, biosynthesis, sporulation, stability, and influence of medium composition on production of alkaloids. In industrial scale, submerged fermentation in shaker culture is mostly used for preparation of sufficient amount of inoculum for further cultivation step. During the inoculum preparation an optimal state of the culture for biosynthesis of ergot alkaloids in the production step can be established (Socic *et al.*, 1986). Usually, the whole industrial process consists of four steps, i.e., (i) shaker culture, (ii) pre-inoculating fermentation, (iii) seed fermentation, and (iv) production fermentation followed by downstream processing (Malinka, 1999). Industrial production of alkaloid mixture (6 g/L) by submerged fermentation of different strains of *Claviceps* sp. (Pazoutova and Tudzynski, 1999) was reported in a number of patents (Kren *et al.,* 1988; Trinn *et al.*, 1990). In contrast to high production and well elaborated procedure of submerged cultivation, the industrial production of clavine alkaloids could not receive much attention due to limited amount of procedures leading to the preparation of desirable final products. Production of lysergic acid and its simple derivatives is of high technological and industrial interest due to relatively simple procedure of chemical modifications to semisynthetic ergot preparations. The basic studies on biosynthesis, physiology and production of simple derivatives of lysergic acid have been done on relatively limited number of *Claviceps* strains (Bumbova *et al.*, 1991). Flieger *et al.,* (1989 a, b), using *C. paspali* CCM 8061, developed a technology of aggressive bioconversion of clavine alkaloids to simple derivatives of lysergic acid with total production of nearly 6 g/L in batch cultivation and about 3 g/L in industrial fermentor. Different strains of *C. purpurea,* as the only producers of ergopeptines, have been described for their submerged production. Ergotamine producing strains and their cultivation are the best-studied processes due to direct therapeutical use of ergotamine (Table 11.4).

Table 11.4. Industrial submerged production (s) of ergot alkaloids (Rehacek, 1983)

Alkaloid	**Company**			
	A	B	C	D
Ergocornine	s	s	s	s
Ergocrystine	s	s	-	s
Ergokryptine	s	s	s	s
Ergotamine	-	s	s	-
Ergometrine	-	-	s	-
Lysergic acid	-	s	-	-
Paspalic acid	s	-	-	s

A Biochemie GmbH, Kundl/Tirol, Austria
B Fermitalia S.p.A., Milano, Italy
C Gedeon Richter, Budapest, Hungary
D Sandoz-Wander AG, Basal, Switzerland

PRODUCTION IN PARASITIC CULTURE

Bekesy developed the first large technology for the parasitic production of ergot on rye in 1934 (Minghetti and Perellino, 1999). For a long time, due to severe problems in early fermentation processes, the field cultivation was the major technology in Europe. Recently, the methods of cultivation, inoculation and harvesting have been improved (Nemeth, 1999). Presently, the field production technology has reached 1,000–2,000 kg sclerotia/ha (Nemeth, 1999). Since the alkaloid content in sclerotia of commercial strains averages 1% and more, an alkaloid yield of 8–10 kg per hectare can be achieved, even if losses during the purification process are taken into account (Udwardy, 1980). The main technical problem in this technology is the generation and storage of high quality inocula, which is mostly achieved by using deep-frozen conidia suspensions. However, several disadvantages of this method have led to a significant reduction of parasitic alkaloid production: (i) production is too dependent on environmental parameters, (ii) the areas in which the climate is adequate for such production are limited, (iii) protection of the valuable production strains is difficult, and (iv) the purification procedures are expensive. Therefore, the present proportion of alkaloids produced in parasitic culture worldwide is low, although exact figures are not available (Tudzynski *et al.*, 2001).

STATIONARY SURFACE CULTIVATION

The development of processes for the production of ergot alkaloids under conditions of stationary cultivation on liquid media started in early sixties (Adams, 1962). The stationary surface cultivation on agar slants was commonly used for the preparation of starting cultures. This technology when transformed to industrial scale showed some limitations mostly due to difficulties with control of aseptic conditions of large surfaces. Vlcek and Kybal (1974) developed technology for stationary cultivation of *C. purpurea* in plastic bags partially filled with inoculated liquid medium. This procedure was used for the production of ergotoxins and later adapted to the production of asexual spores of *C. purpurea* for field production.

SOLID SUBSTRATE FERMENTATION

Solid substrate fermentation describes the microbiological transformation of biological materials in their natural state, in contrast with liquid of submerged fermentation, which is carried out in dilute solutions or slurries. Recently, Robinson *et al.,* (2001) have proposed a solidstate (substrate) fermentation (SSF) for the production of enzymes and secondary metabolites to meet the global demand. Production of ergot alkaloids by SSF was found 3.9 times greater than that produced by *C. fusiformis* in SmF (Hernandez *et al.*, 1993). Products of secondary metabolites by submerged liquid fermentations (SmF) have been found to be less efficient than SSF. The use of antifoam chemicals and the sheer stresses caused by stirring prevented high yields in SmF. Better air circulation can also be achieved in SSF increasing ergot alkaloid yields further (Balakrishnan and Pandey, 1996). The solid substrate fermentation has been shown to produce a more stable product, requiring less energy in smaller fermenters, with easier downstream processing measures. Also, by removing the cost and trouble associated with anti-foaming chemicals and by maximizing yield production, SSF may be seen as a viable option for industrial scale production of ergot alkaloids.

BIOREACTOR DESIGN

Many ergot laboratories have used different bioreactors from time to time. In scaling-up, a 'conventional' stirred tank fermentor must be of the lowest possible cost and should satisfy all requirements (good heat, mass transfer, perfect sterilization, aseptic procedures, reliable foam control, good bulk mixing, simple, rapid and thorough cleaning methods, reliable and appropriate monitoring control systems, proven construction methods, appropriate materials of construction and multiple product application).

OPTIMIZATION AND SCALE UP PROCESS

The following facts should be taken into account for optimization and scale-up of fermentation processes for the production of ergot alkaloids (Kobel and Sanglier, 1978), (i) the production strains should be continually tested to maintain the optimal quality of selected production strain and thus maximally eliminate the biological effects given by transfer of cultures, ageing and other external influences, (ii) as bulk of the peptide alkaloids are mainly confined in mycelial cells, therefore large mycelial mass with high alkaloid content should be obtained (iii) long cultivation period in absence of antibiotics require very high standard of sterility in operation and equipment, (iv) balanced aeration and stirring are required due to the sensitivity of the *Claviceps* cultures to the stress and high oxygen tension, (v) the use of antifoam agents could cause considerable loss in alkaloid yield. Recently, the application of oxygen vectors to *Claviceps purpurea* cultivation has been published (Menge *et al.*, 2001). The classical problem of the large-scale fermentations is an optimal supply of oxygen to growing *Claviceps* sp. High oxygen demand in the exponential growth phase can be met by addition of different hydrocarbons (Gilmanov *et al.*, 1996) or perfluorocarbons (Menge *et al.*, 2001) to shake flask and/or stirred reactors. Perfluorocarbons were also applied successfully in cultivations of other microorganisms (Lowe *et al.*, 1998).

MODERN BIOTECHNOLOGY

Modern research into ergot alkaloids can draw on a very considerable number and variety of truly new means and methodologies, which are in continuous rapid evolution. A full range of features which involves (i) computerization of documentation (ii) automation of various experimental activities (iii) mathematical modeling (iv) calculation of atomic, molecular and crystal structures and their relevant physical, chemical, biological and therapeutical properties and (v) statistical design of experimentation and the interpretation thereof with translation into mathematical models.

GENETIC ENGINEERING

The recent identification of a cluster of genes (Tudzynski *et al.*, 1999) involved in ergot alkaloid biosynthesis in *C. purpurea* and the availability of molecular genetic techniques allow the development of strategies for rational drug design of ergoline-related drugs by enzyme engineering and by biocombinatorial approaches (Tudzynski *et al.*, 2001).

Protoplast fusion technique is beginning to find useful applications either in producing improved mutant strains by intraspecific crosses or in formation of novel spectrum of products by interspecific hybrids (Socic and Porekar, 1992). Relevant structural and regulatory genes of the alkaloid biosynthesis in *C. purpurea* form cluster of about 50 kbp in length (Tudzynski *et al.*, 1999), therefore, isolation and cloning of the entire pathway for more rapidly growing fungus would be difficult. Molecular genetic techniques have recently been successfully applied to *Claviceps* sp. (Pazoutova and Parbery, 1999; Tndzynski *et al.*, 1999) and have considerably widened the options for genetic research. They have been used for (i) detailed characterization of strains for evolutionary studies (ii) for a molecular analysis of the biosynthetic pathway and (iii) for detailed regulatory studies have been initiated. These molecular approaches will allow the design of recombinant strains with improved production properties (e.g., by enhancing expression of genes of interest or by creating defined detection mutants) or that produce new compounds, which are not accessible by chemical synthesis.

Ergot alkaloids are synthesized by the enzymatic transformation of small precursor molecules and hence they are not direct gene products. In biogenesis of *Claviceps* the relevant structural and regulatory genes have not been mapped, therefore, isolation and cloning of the entire pathway of structural genes for alkaloids would be difficult. Gene transfer via protoplast fusion has opened up new possibilities in both the cellular and molecular aspects of *Claviceps* genetic engineering. There are three potentially profitable avenues which may help to bridge the gap between academic and industrial genetics of *Claviceps*—(i) the construction of a plasmid that replicates autonomously in *Claviceps* and carries easily selectable markers, (ii) the development of transformation systems with drug resistance markers for use in species lacking available auxotrophic mutants and (iii) the study of genes controlling the regulation of entire pathway. Jungehulsing and Tudzynski (1997) have analyzed the genetic diversity in *C. purpurea* using RAPD markers. The ergot alkaloid production is a result of dynamic conversion, which takes place during the culture growth. The process of development inside the cell is determined by two dynamic systems. The first system is fast and is represented by enzymatically catalyzed reactions of small molecules while the second system is joined with cell structures determined by macromolecules (nucleic acids, proteins, etc). The second system is relatively slow and acts as a driving filter. Ergot alkaloids are not essential to the complex life cycle and genetic structure of *Claviceps*, which makes genetic manipulations difficult. Genetic recombination and gene cloning are not only important in strain development but it also offers future prospects for the systematic construction of strains capable of over-producing existing or structurally modified alkaloids.

MATHEMATICAL MODELING

The growth of the fungus and alkaloid formation in submerged batch fermentation can be described by mathematical model, which can be used in automatic process control and optimization of an industrial process. Grm *et al.*, (1980) while working with *C. purpurea* proposed a growth model based on morphological features of a cell population during the fermentation process. Pazoutova *et al.*, (1981), however, proposed another model accentuating the antagonistic effects of phosphate on growth and alkaloid production. Another possibility for process optimization lies in feeding the culture and thus prolonging the idiophase of the fermentation process (Milicic *et al.*, 1987). This process seems to be one of the perspectives in modern alkaloid production (Socic and Porekar, 1992). The activation-inhibition kinetics of clavine alkaloids production was evaluated for two *Claviceps fusiformis* strains (Flieger *et al.*, 1988) and it was found that feedback inhibition could be eliminated by combination of fermentation and separation units in a closed loop. Increased efficiency (more than 100%) of the fermentation process was also found when inducers of cytochrome P-450 were used (Rylko *et al.*, 1988).

USE IN MEDICINE

The ergot alkaloids have many potential therapeutic uses, some of which have been realized and others have not survived the critical test of time. Ergot alkaloids are of prime interest to pharmacologists because of their multiple actions on utrine and vascular muscle, inhibitors of prolactin secretion, as a tool for studying the mechanism of sympathetic nervous system, to produce stimulation of dopaminergic receptors and in the treatment of parkinsonism. The ergot and its semisynthetic derivatives have been widely used in many diseases (acromegaly, amenorrhea-galactorrhea, suppression of postpartum lactation, treatment of breast cancer and possibly cancer of prostrate, angina pectoris, enterics (Demain, 1999) and in infectious diseases like glaucoma and herpes zoster). A new group of alkaloids has been reported which interferes with DNA synthesis in mammalian cells. Ergot alkaloids can be chemically manipulated to form derivatives that possess altered pharmacological effects. Preparations containing ergot alkaloids or their semisynthetic derivatives are also applied for the treatment of migraine (ergotamine), high blood pressure (dihydrocornine, dihydrokryptine) and obstetrics (ergometrine). Ingestion of isoergine and lysergic acid diethylamide (LSD) are known to cause marked hallucinations (DeGroot *et al.*, 1998). Ergot alkaloids are not simple pharmacological or therapeutical entity as they also have wide range of biological effects (Berde, 1984). The clavine alkaloids cause apathy, drowsiness and sedation in many mammalian systems and have antimicrobial activity against a series of human pathogenic bacteria and *Candida albicans* (Eich *et al.*, 1984), hypotensive and oxytoxic properties and are able to inhibit prolactin secretion (Edwardson, 1968). Several other ergot derivatives influence the secretion of several hormones and act at the central nervous system. Ergot has 3 main types of actions—(i) central (ii) peripheral and (iii) neurohormonal. LSD is pharmacologically inactive but its various derivatives display pharmacological actions which include (i) sympatholytic effect-inhibition of the hormones adrenalin, noradrenalin and serotonin and thereby a dilation of veins and decrease in blood pressure and (ii) effect on uterus via contraction of smooth muscle. Nicergolin has been found to be of therapeutic value in senile dementia. Ergometrine and its semisynthetic derivatives like methylergometrine have been widely used in postpartum to prevent hemorrhage. Since early time, ergot alkaloids have shown their effects on the reproduction processes of animals, therefore ergot derivatives (ergotoxine and ergocornine) have been tried as antifertility agents in rat and can also terminate early pregnancy in both rat and mice (Sommer and Buchanan, 1955). Lisuride was also introduced into therapy as an antimigraine drug and has a high dopaminergic activity. Bromokryptine with dopaminergic effects on DA receptors are used in the therapy of Parkinson's disease and for curing acromegaly. Pergolide, a synthetic ergoline has been found to have a positive effect on parkinsonism (Liberman *et al.*, 1979). The ergot alkaloids can be chemically manipulated to form derivatives that possess altered pharmacologic effects. Moreover, after obtaining a correlation between clinical indications and therapeutical dosage, pure ergot alkaloids and their semisynthetic derivatives can be used in medicine extensively in future.

FUTURE PROSPECTS AND CONCLUSIONS

In nature, the ergot alkaloids are formed primarily by *Claviceps* sp. Ergot alkaloids are indole compounds having a tetracyclic ring called ergoline, which are formed from tryptophan, mevalonic acid and methionine. Saprophytic production of ergot alkaloids is more economical than parasitic cultivation on rye plant as in bioreactors the process is more precisely controlled, the cultivation is not limited to the specific season, the production and isolation of alkaloids can take place on the same site, also possible to apply some manipulations which are not practicable under field conditions. Clavines and simple lysergic acid derivatives are easier to manufacture than ergopeptines. Production of ergot alkaloids is due to action of multiple enzyme complexes. Biosynthesis of ergot alkaloids lies at the level of genetic information apparatus and continues at the level of physiological realization. Ergot alkaloids are not waste, physiologically inert, products of cell metabolism. Genes involved in alkaloid biosynthesis are under rigorous metabolic control. It is possible to understand the genetic control of alkaloid production and to find the genes responsible for specific alkaloids. Mathematical modeling extended by physiological studies can increase the production of alkaloids in bioreactors. The ergot alkaloids and its semisynthetic derivatives in addition to biological activity have been widely used as potential therapeutic agents in obstetrics and also to control various diseases (parkinsonism, acromegaly, cancer, blood pressure and migraine, etc). Ergot alkaloids can be chemically manipulated to form derivatives that possess altered pharmacologic effects. The search for new natural alkaloids involves the testing of non-traditional organisms. Mutation of a gene can also lead to the formation of new compounds and to structural modifications inside the cell. The series of ergot alkaloids can be widened by their semisynthetic anologues for use in medicine by exploiting *Claviceps* biotechnologically as the production of ergot alkaloids thrives on new knowledge generated by molecular biology, genetics, microbiology and biochemical engineering.

For modern production of ergot alkaloids and its semisynthetic derivatives the following few points need more attention: (i) shortening of the initial non-productive phase of alkaloid fermentation (ii) immobilization of cells (iii) construction of a plasmid which can carry selectable markers (iv) the development of transformation systems with drug resistance markers (v) study of membrane processes and vacuoles in the productive organism (vi) use of molecular genetic techniques (vii) genetic control of alkaloid production and to find the genes responsible for specific alkaloids (viii) concerted breeding of the strains concerned which requires a pre-requisite knowledge of the life cycle and ability to manipulate it in order to get controlled genetic recombination (ix) use of mathematical modeling for description of the phenomenon observed during culture growth and alkaloid production (x) possibility to use (a) solid state fermentation (b) perfluorocarbons as oxygen vectors (c) modern analytical methods (capillary electrophoresis and 2-dimentional fluorescence spectroscopy) in *Claviceps* fermentation.

REFERENCES

Abe M. and Yamatodani S. (1955). A new water-soluble ergot alkaloid, molliclavine. *Bull. Agric. Chem. Soc.* (Japan) 19: 161-162.

Abe M., Yamatodani S., Yamano T. and Kusumoto M. (1956). Isolation of two new water-soluble alkaloids pyroclavine and costaclavine. Bull Agric Chem Soc. (Japan). 20: 59-60.

Adams R.A. (1962). Process for the production of ergot alkaloids. US pat 3117917.

Amici A.M., Minghetti A., Scotti T., Spalla C. and Tognoli L. (1966). Production of ergotamine by a strain of *Claviceps purpurea* (Fr.) Tul. *Experientia*, 22: 415-418.

Arcamone F., Chain E.B., Ferreti A., Minghetti A., Pennela P., Tonolo A and Vero I. (1961). Production of a new lysergic acid derivative in submerged cultures by a strain of *Claviceps paspali* Stevens & Hall. *Proc. Royal. Soc.* B 155: 26-54.

Arcamone F, Caddinelli G., Feroni F., Penco S., Penella P and Pol C. (1970). Ergotamine production and metabolism of *C. purpurea* strain 275 FI in stirred fermentors. *Can. J. Microbiol.* 16: 923-931.

Balakrishnan K. and Pandey A. (1996). Production of biologically active secondary metabolites in solid-state fermentation. *J. Sci. Ind. Res.* 55: 365-372.

Baxter R.M., Kandel S. and Okany A. (1961). Studies on the mode of incorporation of mevalonic acid into ergot alkaloids. *Tetrahedron Lett.* 596-600

Berde B. (1984). The future of ergot research. *J. Roy. Soc. Med.* 77: 5.

Berde B. and Sturmer, E. (1978). Introduction to the pharmacology of ergot alkaloids and related compounds. In: Ergot Alkaloids and Related Compounds, Berde, B. and Schild, O. Eds., Springer Verlag Berlin 49: 1-28..

Bonn W. W. (1922). A preliminary study of *Claviceps purpurea* in culture. *Amer. Jour. Bot.* 9: 339-353.

Brich A. J., McLoughlin B. J. and Smith H. (1960). The biosynthesis of the ergot alkaloids. *Tetrahedron Lett.* 7: 1-3.

Bumbova-Linhartova R., Flieger M., Sedmera P. and Zima J. (1991). New aspects of submerged fermentation of *Claviceps paspali. Appl. Microbiol. Biotech.* 34: 703-706.

De Groot A., Dongen P.W.J.V., Vree T.B., Hekster Y.A. and Roomalen J.V. (1998). Ergot alkaloids: Current status and review of clinical pharmacology and therapeutic use compared with other oxytocics in obsterics and gynaecology. *Drugs.* 56: 523-535.

Demain A.L. (1999). Pharmaceutically active secondary metabolites of microorganisms. *Appl. Microbiol. Biotech.* 52: 455-463.

Dickerson A.G., Mantle P.G. and Szezyrbak C.K. (1970). Autolysis of extracellular glucans produced *in vitro* by a strain of *Claviceps fusiformis. J. Gen. Microbiol.* 60: 403-415.

Didek-Brumec M, Puc A, Socic H, and Alacevic M (1987). Isolation and characterization of a high yielding ergotoxins producing *Claviceps purpurea* strains. *Food Technol Biotechnol Rev* 25: 103-109.

Didek-Brumec M, Gaberc-Porekar V., Alacevic M., Druskovic B. and Socic H. (1991). Characterization of sectored colonies of a high-yielding *Claviceps purpurea* strains. *J. Basic. Microbiol.* 31: 27-35.

Edwardson J.A. (1968). The effect of agroclavine an alkaloid on pregnancy and lactation in the rat. *Brit. J. Pharma.* 33: 215.

Eich E., Eichberg D. and Muller W.E.G. (1984). Clavines: New antibiotics with cytostatic activity. *Biochem. Physiol.* 33: 523-526.

Esser K and Tudzynski P. (1979). Genetics of the ergot fungus *Claviceps paspali. Theor Appl Genet.* 53: 145-149.

Hernandez M.R.T., Lonsane B.K., Raimbault M., and Roussos S. (1993). Spectra of ergot alkaloids produced by *Claviceps purpurea* 1029 C in solid state fermentation system: Influence of the composition of liquid medium used for impregnating sugarcane pith bagasse. *Process Biochem* 28: 23-27.

Jungehulsing U. and Tudzynski P. (1997). Analysis of genetic diversity in *Claviceps purpurea* by RAPD markers. *Mycol Res* 101: 1-6.

Fleiger M., Sedmera P., Vokoun J., Ricicova A. and Rehacek Z. (1982). Separation of four isomers of lysergic acid hydroxyethylamide by liquid chromatography and their spectroscopic identification. *J. Chromatogr.* 236: 453-459.

Flieger M., Votruba J., Kren V., Pazoutova S., Rylko V., Sajdl P. and Rehacek Z. (1988). Physiological control and process kinetics of clavine production by *Claviceps purpurea* 129 and 129/35. *Appl. Microbiol. Biotechnol.* 29: 181-185.

Flieger M., Linhartova R., Rehacek Z., Saidl P., Stuchlik J., Malinka Z., Harazim P. , Cvak L. and Bremek J. (1989a). Industrial strain of microorganism *Claviceps paspali* Stevens et Hall. CCM 8061 CZ Pat 276430 (In Czech).

Flieger M., Linhartova, R., Rehacek, Z., Saidl, P., Stuchlik, J., Malinka, Z., Harazim, P., Cvak, L. and Bremek, J. (1989b). Process for production of lysergic acid simple derivatives. CZ pat 281 196 (In Czech).

Flieger M., Wurst M. and Shelby R. (1997). Ergot alkaloids – sources, structures and analytical methods. *Folia Microbiol.* 42: 3-30.

Flieger M., Mehta P. and Mehta. A. (2003). Biotechnological potential of ergot alkaloids. In: Handbook of Fungal Biotechnology, 2nd Edition. Arora, D.K., Bridge, P. D. and Bhatnagar, D. Eds., Marcel and Dekkar, New York, USA. 21: 91-99.

Gilmanov V., Flieger M. and Dymshits V. (1996). Influence of hydrocarbons on some *Claviceps* cultures. *J. Appl. Bacteriol* .81: 678-680.

Grm B., Mele M. and Kremser M. (1980). Model of growth and ergot alkaloid production by *Claviceps purpurea. Biotechnol. Bioengg.* 22: 255-270.

Groger D., Mothes K., Simon H., Floss H.G. and Weygand F.(1960). Uber den einbau von mevalonsaure in das ergolinsystem der clavine-alkaloide. *Z. Naturforsch.* 15: 141-143.

Groger D., Tylerm V.E. Jr. and Dusenberry J.E. (1961). Investigations of the alkaloid of paspalum ergot. *Lloydia.* 24: 97-102.

Hislop E.C., Barnaby V. M., Shellis C. and Laborad F. (1974). Localization of arabinofuranosidase and acid phosphate in mycelium of *Scerotinia fructigena. J. Gen. Microbiol.* 81: 79-99.

Hunter-Cervera J.C. and Belt A. (1996). *Maintaining Cultures for Biotechnology and Industry.* Academic Press, New York.

Kobel, H. and Sanglier, J.J. (1978). Formation of ergotoxine alkaloids by fermentation and attempts to control their biosynthesis. In: Antibiotics and other secondary metabolites Hutter, R., Leisinger, T., Nuesch, J. and Wehrli, W. Eds., Academic Press, New York 233-242.

Kobel, H. and Sanglier, J.J. (1986). Ergot alkaloids. In: Biotechnology. Rehm HJ and Reed G. Eds., Germany Weinheim 4: 569-609.

Komel R., Rozman D., Puc A. and Socic H. (1985). Effect of immobilization on the stability of *Claviceps purpurea* protoplasts. *Appl. Microbiol. Biotechnol.* 23: 106-109.

Knopp B. and Rehm H.J. (1984). Semicontinuous cultivation of immobilized *Claviceps purpurea. Appl. Microbial. Biotechnol.* 19: 141-145.

Kozlovsky A.G. (1999). Producers of ergot alkaloids out of *Claviceps* genus. In: Medicinal and aromatic plant-industrial profiles. Ergot, the genus *Claviceps*. Kren, V. and Cvak, L. Eds., Harwood Acedemic, Chur. 479-499.

Kren, V. (1991). Bioconversion of ergot alkaloids. In: *Advances in Biochemical Engineering/Biotechnology.* (Ed. Fiechter, A.). pp.123-144. Springer Verlag, Germany.

Kren, V., Sajdl, P., Rehacek, Z., Malinka, Z. and Bremek, J. (1988). Industrial strain of *Claviceps purpurea* (Fr.) Tul 88-EP/1988 for elymoclavine preparation. CZ Pat 272562

Kobel H. and Sanglier J.J. (1978). Formation of ergotoxine alkaloids by fermentation and attempts to control their biosynthesis. In: Antibiotics and other secondary metabolites. Hutter R., Leisinger T., Nuesch J. and Wehrli W. Eds., Academic Press, NewYork. 233-242.

Krupinski V.M., Robbers J.E. and Floss H.G. (1976). Physiological studies of ergot. Induction of alkaloid synthesis by tryptophan at the enzymatic level. *J. Bacteriol.* 125: 158-165.

Lowe K.C., Davey M.R and Power J.B. (1998). Perfluorochemicals: their applications and benefit to cell culture. *Trends Biotechnol.* 16: 272-277.

Liberman A.N., Neibowitz M., Neophytides A., Kuperschmit M., Mehl S., Kleinberg D., Serby M. and Goldstein M. (1979). Pergolide and lisuride for Parkinson's disease. *Lancet.* 2: 1129-1130.

Malinka Z. (1999). Saprophytic cultivation of *Claviceps*. In: Medicinal and aromatic plant-industrial profiles. Ergot, the genus *Claviceps*. Kren, V. and Cvak, L. Eds., Harwood Academic Publishers 321-371.

Mantle, P.G. (1969). Development of alkaloid production *in vitro* by a strain of *Claviceps purpurea* from *Spartia towsendii. Trans. Br. Mycol. Soc.* 52: 381-392.

Mehta P. (1984). Submerged fermentation of clavine alkaloids. *Ph.D. Thesis.* Czech Academy of Sciences, Praha, Czech Republic, pp. 202.

Menge M., Mukherjee J. and Scheper T. (2001). Application of oxygen vectors to *Claviceps purpurea* cultivation. *Appl. Microbiol. Biotechnol.* 55: 411-416.

Milicic S., Kremser M., Gaberc-Porekar V., Didek-Brumec M. and Socic H. (1987). Correlation between growth velocity and biosynthesis of ergot alkaloids in *Claviceps purpurea batch* fermentation. *Appl. Microbiol. Biotechnol.* 27: 117-120.

Minghetti A., Crespi-Perelino N. (1999). The history of ergot. In: Medicinal and aromatic plant-industrial profiles. Ergot, the genus *Claviceps. Kren, V. and Cvak, L.Eds.,* Harwood Acedamic Publishers. Chur.1-24.

Mukherjee J. and Menge M. (2000). Progress and prospects of ergot alkaloid research. *Adv. Biochem. Engg. Biotech.* 68: 1-20.

Nemeth E. (1999). Parasitic production of ergot alkaloids. In: *Medicinal and aromatic plant-industrial profiles. Ergot, the genus Claviceps.* Kren, V., Cvak, L. Eds., Harwood Academic Chur. 303-319.

Neumann D., Loseckem W., Maierm W. and Groger D. (1979). Localization of alkaloids in sclerotia and suspension cultures of *Claviceps purpurea* (Fr.) Tul. *Biochem. Physiol. Pflzen.* 174: 504-508.

Pazoutova S., Votrubam J. and Rehacek Z. (1981). A mathematical model of growth and alkaloid production in submerged culture of *C. purpurea. Biotech Bioengg.* 23: 2837-2849.

Pazoutova, S. and Parbery, D. (1999). The taxonomy and phylogeny of *Claviceps*. In: *Medicinal and aromatic plant-industrial profiles. Ergot, the genus Claviceps.* (Eds. Kren, V. and Cvak, L.). pp. 55-77. Harwood Acedemic, Chur.

Pazoutova S. and Tudzynskim P, (1999). *Claviceps* sp. PRL 1980 (ATCC 26245), 59 and Pepty 695/ch-I: Their True Story. *Mycol Res.* 103/8: 1044-1048.

Pazoutova S., Johnsonm N. and Rajasab A.H. (2002). *Heteropogon triticeus*, a new host of *Claviceps sorghi* in India. *J. Phytopathol.* 150: 1-4.

Pertot E., Gaberc-Porekar V. and Socic H. (1990). Isolation and characterization of an alkaloid-blocked mutant of *Claviceps paspali. J. Basic. Microbiol.* 30: 51-56.

Rehacek Z., Sajdl P., Kozova J., and Ricicova A. (1972). Physiological activities of ergoline alkaloids in submerged cultures of *Claviceps paspali* and *C. purpurea. Folia Microbiol* 17: 308-315.

Rehacek Z. (1972). Physiology of formation of some antibiotics and ergot alkaloids. D. Sc. Thesis. Institute of Microbiology, Czech Academy of sciences, Praha.

Rehacek Z. (1974). Ergot alkaloids and some problems of the physiology of their formation. Zbl. Bakt. II. 129: 20-49.

Rehacek Z., Kozova J., Sajdl P. and Vorisek J. (1974). Physiology of conidial formation in submerged cultures of *C. purpurea* (Fr.) Tul. producing alkaloids.*Can. J Microbiol* 20: 1323-1329.

Rehacek Z. (1983). New trends in ergot biosynthesis. Process Biochemistry. 22-33.

Rehacek Z. and Sajdl P. (1990). *Ergot alkaloids: Chemistry, Biological Effects, Biotechnology*. Czech Academy of Sciences, Academia, Praha, Czech Republic, pp. 376.

Rehacek Z. and Mehta P. (1991). *Claviceps* strains for saprophytic production of ergot alkaloids. *The Genet. Engg. Biotechnol.* London 11:6: 19-21.

Rehacek, Z. and Mehta, P. (1993). Biological effects of ergot alkaloids. In: *Fungal Ecology and Biotechnology*. Rai, B., Arora, D.K., Dubey, N.K. and Sharma P.D. Eds., Rastogi Publications, Meerut, India. 275-287.

Robbers J.E., Robertson L.W., Horneman K.M., Jindra A. and Floss H.G. (1972). Physiological studies on ergot. Further studies on the induction of alkaloid synthesis by tryptophan and its inhibition by phosphate. *J Bacteriol.* 112: 791-796.

Robinson T., Singh D. and Nigam P. (2001). Solid-state fermentation: A promising microbial technology for secondary metabolite production. *Appl Microbiol Biotech.* 55: 284-89.

Rylko V., Flieger M., Sajdi P., Rehacek Z., Malinka Z., Harazim P and Stuchlik J. (1988). Process for production of ergot alkaloids by *Claviceps* with induced increase of productive ability, CS pat appl 4725-88.

Schumann B., Maier W. and Groeger D. (1987). Characterization of some *Claviceps* strains derived from regenerated protoplasts. *Z. Naturforsch.* 42: 381-386.

Socic H., Gaberc-Porekar V., Pertot E., Puc A. and Milicic S. (1986). Developmental studies of *Claviceps paspali* seed cultures for the submerged production of lysergic acid derivatives. *J. Basic. Microbiol.* 26: 533-539.

Socic H. and Gaberc-Porekar V. (1992). Biosynthesis and physiology of ergot alkaloids. In: Handbook of Applied Mycology, Fungal Biotechnology. Arora, D.K., Elander, R.P. and Mukerje, K.G. Eds. Marcel Dekkar Inc NewYork. 4: 475-515

Sommer A.F. and Buchanan A.R. (1955). Effects of ergot alkaloids on pregnancy and lactation in the albino rat. *Amer. J. Physiol.* 18: 296-300.

Strandova K., Kybal J., Svoboda E., Spacil J., Krajicek A., Malinka Z and Harazim P (1986). Process of improvement of gens *Claviceps* microorganisms and new industrial strain of *Claviceps purpurea* (Fr.) Tul. CCM8043. CS Patent No. 267573.

Taber W.A. and Vining L.C. (1959). Tryptophan as a precursor of ergot alkaloids. *Chem. Ind.* London 1218-1219.

Taber W.A. and Vining L.C. (1963). Physiology of alkaloid production by *Claviceps purpurea* (Fr.) Tul. Correlation with changes in mycelial polyol, carbohydrate, lipid and phosphorous containing compounds. *Can. J. Microbiol.* 9: 1-14.

Taylor E.M. and Ramsted E. (1960). Biosynthesis of lysergic acid in ergot. *Nature.* 188: 494-495.

Trejo-Hernandez M.R., Lonsane B.K., Raimbault M. and Roussos S. (1993). Spectra of ergot alkaloids produced by *Claviceps purpurea* 1029 C in solid-state fermentation system. Influence of the composition of liquid medium used for impregnating sugarcane pith bagasse. *Process Biochem.* 28: 23-27.

Trinn M., Manczinger L., Polestyukne N.A., Kordik G., Pecsne R.A., Zalai K., Beszedics G., Ferenczy L., Nagy L., Robicsek K. and Szegedi M. (1990). Process for the elymoclavine production and preparation of new strain. HU Pat 209325 (In Hungarian).

Tudzynski P., Holter K., Correia T., Arntz C., Grammel N. and Keller U. (1999). Evidence for an ergot alkaloid gene cluster in *Claviceps purpurea. Mole. Genet* 261: 133-141.

Tudzynski P., Correia T. and Keller U. (2001). Biotechnology and genetics of ergot alkaloids. *Appl. Microbiol. Biotech.* 57: 593-605.

Udwardy N.E. (1980). Consideration of the development of an ergot alkaloid fermentation process. *Process Biochem.* 15: 5-8.

Vasarhelyi G., Tetenyi P., Zambo I., Kinicky M., Balassa-Barkanyi I., Kiss J., Recs-Razso A. and Zsoka-Somkuti E. (1980). Strain of *Claviceps purpurea*, virulent on rye for ergot alkaloids, especially ergocristine production and process for preparation of it. DE Pat. 3006989 (In German).

Vining L.C. and Nair P.M. (1966). Clavine alkaloid formation in submerged cultures of a *Claviceps* sp. *Can. J. Microbiol.*, 12: 915-931.

Vining L.C. (1980). Conversion of alkaloid and nitrogenous xenobiotics. In: Economic Microbiology. Rose, A.H. Ed., Academic Press, New York 523-573.

Vlcek V. and Kybal, J. (1974). Equipment for cultivation of microorganisms. CS Patent 172552 (In Czech).

Vorisek J. and Rehacek Z. (1976). Origin of sclerotia like cells in submerged *Claviceps purpurea* producing clavine alkaloids. *Arch. Microbial.* 107: 321-327.

Vorisek J. and Rehacek Z. (1978). Fine structure localization of alkaloid synthesis in endoplasmic reticulum of submerged *C. purpurea. Arch. Microbiol.* 117: 297-302.

Yatsu L.Y., Jacks T.J. and Heusarling T. P. (1971). Isolation of spherosomes/oleosomes from onion, cabbage and cotton seed tissues. *Plant Physiol.* 48: 675-682.

Zalai K., Kordik G., Manczinger I., Pecsne R.A., Beszedics G., Polestyukne N.A., Ferenczy I., Olasz K., Szegedi M. and Trinn M. (1990). Process for production of ergolen compounds, mainly ergometrine, and process of colour selection. HU Pat 209324 (In Hungarian).

12 Biotechnological Applications of Chitin and Chitosan of Microbial Origin: Progress and Future Perspectives

Jiří Kučera

Food Research Institute Prague (Výzkumný ústav potravinářský Praha), Radiová 7, CZ 10231 Praha 10, Czech Republic

INTRODUCTION

Chitin – the homopolymer of N acetyl glucosamine was discovered in the beginning of nineteenth century by Henri Braconnot in Botanical Garden in Nancy, France. Braconnot described chitin under name of fungine as a component of mushroom cells. The name chitin was introduced by Odier in the year 1823 (http://www.seaborne.com/index 11.asp), who replaced the older name. The name chitin is derived from Greek word, meaing 'tunic'. So, chitin was discovered a couple of years earlier than cellulose – the other natural polysaccharide used in many areas of human activities. Chitin could be deacetylated (by alkali or enzyme) giving chitosan, the polymer of glucosamine.

Although the first data concerning chitin was obtained from material of fungal provenience, the production and more detailed study was carried out on chitin of crustacean origin crab, lobster and shrimps, which contains as much as 20 – 40% of chitin (No and Meyers, 1995).

The first source of chitin used by Henri Braconnot was mushrooms (http://www.plasticstrends.net/articles/chitosan.htm), but later when it was recognized as major part of exoskeleton of crustacean and insects, crustacean chitin is the most favorable one. Chitin is primary constituent of crustacean shells, insect cuticles and fungal cell-walls, but chitosan is present in fungal cell walls only.

Mold chitin was reinvestigated as late as in 1979 (White *et al.*, 1979). Chitosan forms main part of fungal cell walls in some fungi like *Rhizopus, Absidia*, and *Mucor* genera. Traditional source of chitin are limited by fishing seasons and location of the krill processing factories. Fungal chitosan have, therefore, growing importance in last few years, as the other sources are limited.

Major source of chitin is still crustacean shells for industrial production. Chitin and chitosan are prepared in these cases using crab and shrimp shells obtained as wastes or by products in seafood industry. This shell contains 20–40 % (weight base) of chitin, most usually 14–27 %, in the mixture with proteins, calcium and small amount of pigments (Ashford *et al.*, 1977). Under current processing protocol calcium is removed from shrimp or crab shell by diluted hydrochloric acid, protein is removed with sodium hydroxide, and residue is usually bleached with potassium permanganate or hydrogen peroxide. Chitosan is usually produced by deacetylation with concentrated (40–50%) sodium hydroxide at elevated temperature. Crustacean waste supplies are limited by season and fishing industry location, while fungal mycelium is available in any company dealing with the production of fungal products like citric acid, enzymes or antibiotics. The waste mycelium can be utilized for production of chitin and chitosan, lowering by this way the expenses of fungal products preparation. For instance, annual world production of citric acid is estimated at 400,000 tons (Kristiansen and Mattey, 1999), which results in approx. 80,000 tons of waste mycelium of *Aspergillus niger* (Ali *et al.*, 2002). Furthermore, fungal mycelia have lower level of inorganic materials compared with crustacean shells, and as a consequence, no demineralization is needed during processing (Teng *et al.*, 2001a).

Among various applications of chitin and chitosan in agricultural, food, and pharmaceutical industries, the uses of these natural polymers as antimicrobials and bio-pesticides are most attractive. For example, chitosan shows strong antimicrobial activity against Gram negative, as well as Gram positive bacteria like *Escherichia coli, Salmonella typhimurium, Staphylococcus aureus,* and *Listeria monocytogenes* (Sudharshan *et al.*, 1992; Joen *et al.*, 2001; No *et al.*, 2002). Chitin and chitosan are also known to inhibit the growth of phytopathogenic fungi and bacteria in plant, activating its defense reactions (Shibuya and Mimami, 2001).

CHITIN BIOCHEMISTRY

Chitin is the most widespread polysaccharide in nature and it is estimated annually to be produced almost as much as cellulose. *Aspergillus niger* and *Mucor rouxii* are two of the most frequently used fungi for chitin and chitosan preparation (White *et al.*, 1979; Arcidiacono and Kaplan, 1992; Pochanavanich and Suntornsuk, 2002; Wu *et al.*, 2005). Some other fungi also serve as chitin and chitosan sources (Kendra *et al.*, 1989; Bernard and Latge, 2001). Edible mushrooms, namely *Agaricus bisporus*, are also convenient and commercially attractive source of chitin, as chitin could be produced from wastes of industrial mushroom processing (Cheung 1997; Tao *et al.*, 2004).

Despite mycelium of *Aspergillus niger* is available in large quantities as waste from citric acid production, mycelium of *Mucor rouxii* was studied more profoundly (Synowiecki and Ali Khateeb, 1997; Synowiecki and Ali Khateb, 2003; Chatterjee *et al.*, 2004; Tao *et al.*, 2005). Typically mycelium of *Mucor rouxii* is homogenized in 1 M sodium hydroxide and autoclaved at 120°C for 15 minutes. The alkali insoluble material is then separated and washed with water, followed by washing with ethanol (Synowiecki and Ali Khateeb, 1997; Chatterjee *et al.*, 2004). Chitosan is prepared by the way of acid or alkali deacetylation of chitin (Synowiecki and Ali Khateeb, 1997; Chatterjee *et al.*, 2004). The yield of chitin is usually near to 35%, which after hydrolysis, yield 6 to 8% of chitosan (calculated on the dry matter base) (Chatterjee *et al.*, 2004). Similarly, wastes from the processing of *Agaricus bisporus* is treated with 1 M NaOH at elevated temperature, washed, and deacetylated in acidic medium. The yield of chitin prepared from *Agaricus bisporus* waste is usually 11 to 27% on dry matter base.

Alternatively, mycelium as well as mushroom waste could be lyophilized and used independently on the cultivation process.

The mycelium, namely that from wastes after citric acid production, is formed in spherical pellets with the properties particularly beneficial for the use as filtration aids or as a sorbent for chromatography (Kučera, 2004). To maintain the spherical shape during conversion into chitin or chitosan, the structure should be stabilized by cross linking before deacetylation.

Chitin is composed largely of alternating *N* acetylglucosamine residues, which are linked by β (1-4) glycosidic bonds. X-ray diffraction analysis suggested that chitin is a polymorphic substance that occurs in three different crystalline modifications, termed α, β and γ chitin. They mainly differ in the degree of hydration, in the size of the unit cell and in the number of chitin chains per unit cell (Rudal and Kenchington, 1973). In the α form, all chains exhibit an anti-parallel orientation; in the ß form the chains are arranged in a parallel manner; in the γ form sets of two parallel strands alternate with single anti parallel strands. In addition, non-crystalline, transient states have also been reported in a fungal system (Vermeulen and Wessels, 1986).

The specific mechanism by which chitin is produced is still unknown. However, evidence suggests that chitin is synthesized through an asymmetric mechanism, accepting GlcNAc units from the cytosolic UDP GlcNAc pool and releasing the nascent chain into the extraplasmic phase (Ruiz-Herrera and Martinez Espinoza, 1999). Nevertheless, in fungal systems substantial data have accumulated indicating that chitin synthase activity of at least one chitin synthase isoform (CHS3p) is associated with specialized intracellular microvesicles, known as chitosomes, which exhibit a special lipid and protein composition (Bracker *et al.*, 1976; Hernandez *et al.*, 1981; Florez Martinez *et al.*, 1990). Electron microscopy has revealed that in the presence of UDP GlcNAc and activators, purified chitosomes synthesize microfibrils that crystallize in the lumen of the vesicles (Bracker *et al.*, 1976).

Chitin synthesis is influenced in different ways by other effectors as well, depending on the particular enzyme source. For instance, GlcNAc has been reported to stimulate chitin synthesis in fungi and also in some insects (Keller and Cabib, 1971; Quesada Allue *et al.*, 1976). By contrast, studies with microsomal fractions from *Stomoxys* showed almost complete inhibition of chitin synthesis with 1 mmol L^{-1} GlcNAc (Mazer *et al.*, 1980).

Chitin synthases and chitinolytic enzymes work hand in hand in remodeling chitinous structures. So far, we have discussed those enzymes that are relevant for chitin synthesis. The degrading enzymes include the chitinases {poly [1,4 (*N*-acetyl-ß-D glucosaminide)] glycanohydrolase; EC 3.2.1.14} and ß-*N*-acetylglucosaminidases (ß-*N*-acetyl-ß-D-hexosaminide *N*-acetyl hexose aminohydrolase; EC 3.2.1.52). These enzymes catalyze the hydrolysis of ß (1-4) glycosidic bonds of chitin polymers and oligomers. Some of them, including one insect enzyme, additionally catalyze transglycosylation reactions (Usui *et al.*, 1987; Fukamizo, 2000).

CHEMICAL AND PHYSICOCHEMICAL PROPERTIES OF FUNGAL CHITIN AND CHITOSAN

Chitin could be considered a derivative of cellulose in which hydroxyl in position 6 is substituted by acetyl amino group. By the same way chitosan is formally derivative of cellulose in which this hydroxyl is substituted by amino group. Real transformation of chitin or chitosan into cellulose will not be so simple. On the other hand, chitosan can be transferred by the well known reaction with nitrous acid into polysaccharide closely similar to cellulose. Chitosan is not present in shrimp and crab shell. Some fungi, for instance *Mucor rouxii* has relatively high level of chitosan in their cell wall, while others not, for instance *Aspergillus niger*; (Feofilova *et al.*, 1996; Bernard and Latge, 2001; Tao *et al.*, 2005). *Mucor rouxii* was found to produce enzyme chitin deacetylase which convert chitin to chitosan (Kafetzopoulos *et al.*, 1993). This fact may be the reason of the presence of chitosan in cell wall of this fungus.

Due to its structure, chitin and more chitosan are basic polysaccharides. The both polysaccharides, as a consequence, are soluble in strong mineral acids. In insoluble form chitin as well as chitosan could serve as anion exchanger (Nagasawa *et al.*, 1970; Nagasawa *et al.*, 1971). Acetyl amino group in chitin and amino group in chitosan are also strong metal chelators. This chelating ability is utilized in decontamination of water contaminated with heavy metal (Miyoshi *et al.*, 1992).

Physicochemical properties of chitin and chitosan extracted from mycelia of *Aspergillus niger* and *Mucor rouxii* were compared (Tao *et al.*, 2005). Glucosamine level in mycelium of *Aspergillus niger* was lower (11.10%) compared with that of *Mucor rouxii* (20.13%). From the data obtained during stepwise extraction of both polysaccharides from freeze dried mycelia mycelium, of *M. rouxii* appears to contain chitin as well as chitosan, whereas mycelium of *A. niger* contains only chitin. The degree of acetylation was determined to be 76.53% in *A. niger* and 50.07 in *M. rouxii*. The crystallinity of fungal chitin and chitosan was found to be less intense compared with that of shrimp ones (Tao *et al.*, 2005).

Wastes from *Agaricus bisporus* processing, which is also used as chitin and chitosan source contains glucosamine at level of 7.14 to 11.00 % on dry matter base depending on the time after harvest (Tao *et al.*, 2004). The yield of crude chitin from stalks stored at 25°C for 5 days was 27% consisted of 46% glucose amine and 21% neutral polysaccharides. Growth of chitin content in the mushroom during storage is probably done by the loss of water. Degree of acetylation of mushroom chitin is 75.8 to 87.6%, which means that the mushroom does not produce enzyme chitin deacetylase similarly like *A. niger.*

A variety of culture and processing protocols using *Mucor rouxii* were studied for their effects on biomass yield and chitosan molecular weight (Arcidiacono and Kaplan, 1992). Average molecular weight determined by gel chromatography ranged from 2.0×10^5 to 1.4×10^6 Unit. Chitosan yield ranged from 5 to 10% of total biomass dry-weight, comprising 30 to 40% of cell wall weight (Arcidiacono and Kaplan, 1992).

PREPARATION OF FUNGAL CHITIN AND CHITOSAN

Fungal chitin could be prepared from mycelium cultivated just for this purpose (White *et al.*, 1979; Arcidiacono and Kaplan, 1992; Synowiecki and Ali Khateeb, 1997; Jaworska and Konieczna, 2001; Teng *et al.*, 2001b; Pochanavanich and Suntornsuk, 2002; Synowiecki and Ali Khateeb, 2003; Chatterjee *et al.*, 2004) or from waste mycelium indigenous from other fungal cultivation processes (Kristiansen and Mattey, 1999; Ali *et al.*, 2002). Special fungal source of chitin and chitosan constitute mushroom waste (Tao *et al.*, 2004). Fungal mycelium, regardless of its origin, is processed principally by the same way as that from crustacean origin.

The first step in the chitin extraction is exhaustive washing of the mycelium to avoid all residue of cultivation medium. This step can help in the further purification. Fungal mycelium usually does not contain significant amount of calcium and, therefore, the washing with dilute hydrochloric acid is not necessary. Mycelium is then homogenized and proteins and lipids are extracted in alkaline medium (usually 1 M sodium hydroxide) at 95°C for 30 minutes. This operation results in alkali insoluble material, which was separated and extensively washed with water. This alkali insoluble material is then refluxed with 2% acetic acid for 6 hours to hydrolyze the acetyl group and produce soluble chitosan and insoluble chitin. After separation of residual chitin by centrifugation chitosan is precipitated after adjustment of pH to 10 with 1 M NaOH (Chatterjee *et al.*, 2004). In some cases different concentration of alkali or acid was used with similar effect (Synowiecki and Ali Khateeb, 1997; Synowiecki and Ali Khateeb, 2003). Modified process was also described. In this case, mycelium after washing is homogenized in 1 M NaOH and autoclaved at 120°C for 15 minutes. Afterward, the homogenate is washed with water followed by ethanol and refluxed in acetic acid as before (Chatterjee *et al.*, 2004). In the case of mushroom waste material the process is similar (Tao *et al.*, 2004).

Interesting new method for chitin extraction was described using lactic acid fermentation process. Lactic acid process is much gentler than conventional methods (using hydrochloric acid) and gives product of higher quality (Anon, 1993).

Fungal chitosan could also be used as chromatographic sorbent. Such sorbent is prepared by slightly modified way to conserve its microscopic structure (Kučera, 2004). The main advantage of fungal chitin and chitosan for use as sorbent in comparison with crustacean ones is its shape. Very nearly spherical pellets of fungal mycelium are formed during submerged cultivation (e.g., in the production of citric acid or fungal enzymes). The pellets grow from a single initial cell by elongation of fibers of diameter equal to cell thickness and the length ruled by the cultivation conditions. The fibers are coiled to form the pellets and its diameter is conditioned by the conditions of growth, such as aeration, stirring and, to a certain extent, by geometrical parameters of fermentor. This internal structure of pellets is very convenient for chromatographic utilization of this material. The structure is desirable to be kept. For this purpose, the mycelium was first treated with proteolytic enzyme in alkaline medium to open the cells, followed by the cross linking with glutaraldehyde and afterward hydrolyzed to chitosan in alkali (Kučera, 2004). The product is then suitable for the use as anion exchanger directly or could be modified for other chromatographic purposes.

THE USE OF CHITIN AND CHITOSAN

Chitin as well as chitosan, regardless of fungal or crustacean, is material of high value. Chitin and yet more chitosan contains reactive secondary (chitin) or primary (chitosan) amino groups in high level. This property makes them suitable for cell proteins immobilization. Crude chitin, for instance, was used to immobilize endo- and exo- proteases (Kise and Hayakawa, 1991; Ge *et al.*, 1996). Immobilized enzyme mixture prepared this way was used for proteolysis of casein in column packed bed reactor. This support material gives flow rate very convenient for large scale process. Similarly, fungal ß galactosidase (ß D galactoside galactohydrolase, E.C. 3.2.1.23) was immobilized on chitin and used in packed-bed reactor for modification of undesirable functional and nutritional properties derived from lactose content in milk and dairies (Illanes *et al.*, 1990).

Enzyme urease (urea amidohydrolase, E.C. 3.4.1.5) prepared from *Cajanus indicus*, has been also immobilized on glutaraldehyde treated chitin as the solid support. This immobilized enzyme could be used for lowering of blood urea level (Iyengar *et al.*, 1979; Krajewska *et al.*, 1990a; Krajewska *et al.*, 1990b).

The determination of biological oxygen demand (BOD) in wastewater plants may be difficult in many cases, because of its high time of consumption of five days. A determination within some minutes is possible by microbial BOD sensors. Such sensor was prepared using yeast *Trichosporon cutaneum* immobilized on chitin as biological component (Reiss *et al.*, 1998).

Many other immobilized enzymes of practical value were prepared using chitin or chitosan as support (see e.g. Muzzarelli *et al.*, 1976; Zihnioflu and Telefoncu, 1995; Magnin *et al.*, 2001).

Chitin as well as chitosan could be used for affinity chromatography. Some proteases are adsorbent on chitosan without derivatization (Shi *et al.*, 1996). Lectins specific for N-acetyl glucosamine and its polymers were also purified by chromatography on chitin. Similarly, chitinases of different origin were also purified (Alen, 1979; Abraham *et al.*, 1982; Balasubramanian and Manocha,1992; Ilyina *et al.*, 1994; Ilyina *et al.*, 1995). Some other proteins were chromatographed on chitin or chitosan after substitution with appropriate affinity ligands (Holme *et al.*, 1988).

Chitosan gels are being used in biotechnological research to immobilize enzymes or living cells, encapsulate cells or recover valuable protein by protein complex formation. Advantages of chitosan over other carriers for immobilization of microbial cells were reviewed (Anon, 1988).

In pharmacy, chitin and chitosan also received broad application. Chitin and chitosan are of great interest for their antihypercholesteremic and antihyperlipidemic activities, which have been demonstrated experimentally (Chobot *et al.*, 1995) as well as clinically (Chobot *et al.*, 1995). Derivatives of chitosan–chitosan sulfate has anti-coagulative and anti-aggregative effects. Both polysaccharides serve as good drug vehicles, namely promising effects were obtained with antineoplastics and nonsteroidal antiphlogistics (e.g. ibuprofen) (Chobot *et al.*, 1995). Very advantageous is the fact, that chitin and chitosan possess only negligible toxicity and alergenicity (Chobot *et al.*, 1995).

Chitin from mycelium of *Mucor mucedo* promotes the growth of fibroblast and provides a matrix for their anchorage, thus contributing to the granulating phase of the healing cascade (Chung *et al.*, 1994). Many other studies show chitin and moreover chitosan to be efficient remedy for wound healing (Hermann and Woodward, 1972; Shahabeddin *et al.*, 1990; Okamoto *et al.*, 1995; Drenda *et al.*, 1997; Kratz *et al.*, 1997; Madihally and Matthew, 1999).

Numerous studies have shown that chitosan, as a mucoadhesive polymer, is a potential enhancer for transmucosal drug delivery (Dodane *et al.*, 1999).

Chitin and chitosan were also used in cancer metastasis prevention (Murata *et al.*, 1990; Azuma and Saiki, 1994; Suzuki *et al.*, 1995). Chitin and chitosan lower the low density lipoprotein in blood serum (Yihua and Binglin, 1997), possess hypocholesteremic and hypolipidemic effects (Jennings *et al.*, 1988; Sugano *et al.*, 1988; Miura *et al.*, 1995; Lee *et al.*, 1999), and help to reduce body weight (Pitter *et al.*, 1999). Chitosan also improve liver and renal function (LeHoux and Grondin, 1993; Jing *et al.*, 1997).

In agriculture and food industry the antimicrobial properties of chitin and chitosan play very important role (El Ghaouth *et al.*, 1992; Guang Hua, 1992; Sudarshan *et al.*, 1992; Fang *et al.*, 1994; Irbe *et al.*, 1995). Chitosan is also used as coating material for fruits and vegetables (Savage *et al.*, 1994; Hanks *et al.*, 1995;). Chitosan was used for the separation of fat from cheese whey (Der Chyan Hwang and Srinivasan, 1995), for meat preservation (Purnama Darmadji and Izumimoto, 1994), for separation of coelia active peptides from brad wheat (Vincenzi *et al.*, 1993), and for flocculation of single cell proteins (Wang and Yam, 1995).

Chitosan is also efficient adsorbent for heavy metals from waste water (Iyengar and Venkobachar, 1990), as well as for adsorption of carcinogens (Hayatsu *et al.*, 1997) and dyes (Knorr, 1983).

There are some other applications of chitin and chitosan reviewed by various investigators (Hirano, 1996).

FUTURE PERSPECTIVES AND CONCLUSIONS

Chitin and chitosan are very useful natural polysaccharides with many applications in medicine, agriculture and food industry, as well as in control of water pollution. Waste from shrimp, crab and lobster processing industries, the traditional sources of chitin and chitosan, have several disadvantages. Crustacean waste supplies are limited by seasons and fishing industry location. Fungal mycelium, on the other hand, can be obtained by simple fermentation regardless of geographical location and seasons. Furthermore, fungal mycelia have lower level of inorganic materials compared with crustacean shells, and no demineralization is required during processing. Considering the significant amount of fungal based waste material accumulated in the mushroom production and fermentation industries, and the expenses involved in managing these waste, production of highly functional value added products may provide a profitable solution for industry. Fungal materials are believed, therefore, to be the most perspective future source of chitin and chitosan. Complexes of fungal mycelium with some beneficial compound (like III chromium, selenium, zinc, etc.) could be prepared by the cultivation of fungi in the medium containing these components.

New fermentation capacities are not necessarily be built for the production of fugal chitin, as these capacities exist as present fermentation factories, namely those producing citric acid and fungal enzymes.

The new developments in the field of fungal chitin and chitosan production and applications could be expected.

REFERENCES

Abraham A.K., Kolseth S. and Pihl A. (1982). Removal of wheat germ agglutinin increases protein synthesis in wheat germ extracts. *Eur.J.Biochem* 124: 383-388.

Ali S., Ikram ul Haq, Qadeer M.A. and Iqbal J. (2002). Production of citric acid by *Aspergillus niger* using cane molases in a stirred fermentor. *Elect. J. Biotechnol.* 5(3), Issue August 15.

Allen A.K. (1979). A lectin from the exudate of the fruit of the vegetable marrow (Cucurbita pepo) that has a specificity for beta-1,4-linked N-acetylglucosamine oligosaccharides. *Biochem. J.* 183: 133-137.

Anon (1988). Cell immobilization possible with chitosan gels. *Food Engineering* 60: 72.

Anon (1993). Chitin extraction from crays. *Food Technology in New Zealand* 28: 23.

Arcidiacono S. and Kaplan D.L. (1992). Molecular weight distribution of chitosan isolated from *Mucor rouxii* under different culture and processing conditions. *Biotechnol. Bioeng.* 39: 281-286.

Ashford N.A., Hattis D. and Murray A.E. (1977) Industrial prospects for chitin and protein from shellfish waste. MIT Sea Grant Report MISG 1977. Cambridge MA, Massachusetts Institute of Technology. p. 77

Azuma I. and Saiki I. (1994). Prevention of cancer metastasis in mice with fibronectin related substances. *Princess Takamatsu Symp* 24: 125-141.

Balasubramanian R. and Manocha M. S. (1992). Cytosolic and membrane bound chitinases of two mucoraceous fungi: a comparative study. *Can. J. Microbiol* 38: 331-338.

Bernard M. and Latge J. P. (2001). *Aspergillus fumigatus* cell wall: composition and biosynthesis. *Med. Mycol.* 39 Suppl 1: 9-7.

Bracker C.E., Ruiz Herrera J. and Bartnicki Garcia S. (1976). Structure and transformation of chitin synthase particles (chitosomes) during microfibril synthesis *in vitro*. *Proc. Natl. Acad. Sci. USA* 73: 4570-4574.

Chatterjee S., Adhya M., Guha A.K. and Chatterjee B.P. (2004). Chitosan from *Mucor rouxii*: production and physicochemical characterization. *Process Biochem.* 40: 395-400.

Cheung P.C.K. (1997). Dietary fibre content and composition of some edible fungi determined by two methods of analysis. *J.Sci.Food Agric.*73: 255-260.

Chobot V., Kremenák J. and Opletal. L. (1995). [Phytotherapeutic aspects of diseases of the circulatory system. 4. Chitin and chitosan]. *Ceska Slov. Farm.* 44: 190-195.

Chung L.Y., Schmidt R.J., Hamlyn P.F., Sagar B.F., Andrews A.M. and Turner. T.D. (1994). Biocompatibility of potential wound management products: fungal mycelia as a source of chitin/chitosan and their effect on the proliferation of human F1000 fibroblasts in culture. *J. Biomed. Mater. Res.* 28: 463-469.

Der Chyan Hwang and Srinivasan D. (1995). Selective precipitation and removal of lipids from cheese whey using chitosan. *J. Agr. Food Chem.* 43: 33-37.

Dodane V., Amin K. and Merwin J.R. (1999). Effect of chitosan on epithelial permeability and structure. *Int. J. Pharm.* 182: 21-32.

Drenda P., Lampe P., Górka Z., Bursig H. and Dylag. S. (1997). Evaluation of the usefulness of dressings made from chitosan and lyophilized human placenta on wound healing. *Wiad. Lek.* 50 Su 1 Pt 2: 252-256.

El Ghaouth A., Arul J., Grenier J., and Asselin A. (1992). Antifungal activity of chitosan on two postharvest pathogens of strawberry fruits. *Phytopathology* 82: 398-402.

Fang S.W., Li C.F. and Shih D.Y.C. (1994). Antifungal activity of chitosan and its preservative effect on low sugar candied kumquat. *J. Food Protect.* 57: 136-140.

Feofilova E.P., Nemtsev D.V., Tereshina V.M., and Kozlov V.P. (1996). Polyaminosaccharides of mycelial fungi: new biotechnological use and practical implications (review). *Appl. Biochem. Microbiol.* 32: 437-445.

Florez Martinez A., Lopez Romero E., Martinez J.P., Bracker C.E., Ruiz Herrera J. and Bartnicki Garcia S. (1990). Protein composition of purified chitosomes of *Mucor rouxii. Exp. Mycol.* 14: 160-168.

Fukamizo T. (2000). Chitinolytic enzymes: catalysis, substrate binding, and their application. *Curr. Protein Pept. Sci.* 1: 105-124.

Ge S.J., Bai H., Yuan H.S. and Zhang L.X. (1996). Continuous production of high degree casein hydrolysates by immobilized proteases in column reactor. *J. Biotechnol.* 50: 161-170.

Guang Hua (1992). Inhibition and inactivation of five species of food borne pathogens by chitosan. *J. Food Protect.* 55: 916-919.

Hanks A. L., Simpson B.K., Haggerty R.J. and Branen A.L. (1995). Biodegradable films from starch and chitosan. *IFT Annual Meeting* 1995, p. 291.

Hayatsu H., Tanaka Y. and Negishi K. (1997). Preparation of DNA chitosan columns and their applications: binding of carcinogens to the column. *Nucleic Acids Symp. Ser.* 139-140.

Hernandez J., Lopez Romero E., Cerbon J. and Ruiz Herrera J. (1981). Lipid analysis of chitosomes, chitin synthesizing microvesicles from *Mucor rouxii. Exp. mycol.* 5: 349-356.

Herrmann J.B. and Woodward S.C. (1972). An experimental study of wound healing accelerators. *Am. Surg.* 38: 26-34.

Hirano S. (1996). Chitin biotechnology applications. *Biotechnol. Annu Rev.* 2: 237-258.

Holme K.R., Hall L.D., Armstrong C.R. and Withers S.G. (1988). Synthesis and evaluation of chitosan based, affinity chromatography materials. *Carbohyd. Res.* 173: 285-291.

Illanes A., Zuniga M.E. and Ruiz A. (1990). [Immobilization of microbial lactase]. *Arch. Biol. Med. Exp. (Santiago)* 23:159-164.

Ilyina A.V., Tikhonov V.E., Varlamov V.P., Radigina L.A., Tatarinova N.Y. and Yamskov I.A. (1995). Preparation of affinity sorbents and isolation of individual chitinases from a crude supernatant produced by Streptomyces kurssanovii by a one step affinity chromatographic system. *Biotechnol. Appl. Biochem.* 21(Pt 2): 139-148.

Ilyina A.V., Varlamov V.P., Tikhonov V.E., Yamskov I.A. and Davankov V.A. (1994). One step isolation of a chitinase by affinity chromatography of the chitinolytic enzyme complex produced by *Streptomyces kurssanovii. Biotechnol. Appl. Biochem.* 19(Pt 2): 199-207.

Irbe R.M., Restaino L. and Frampton E.W. (1995). Characterization of the antimicrobial activity of chitosan. *IFT Annual Meeting* 1995, p. 14.

Iyengar L. and Venkobachar C. (1990). Waste product of food industry for metal pollution abatement. (In '*Pollution management in food industries*'. Conference. Mysore 570 013, India; Association of Food Scientists and Technologists). pp.135-149 [see FSTA (1991) 23 9E1]

Iyengar, L. and Rao. A.V. (1979). Urease bound to chitin with glutaraldehyde. *Biotechnol. Bioeng.* 21: 1333-1343.

Jaworska M.M. and Konieczna E. (2001). The influence of supplemental components in nutrient medium on chitosan formation by the fungus *Absidia orchidis. Appl. Microbiol. Biotechnol.* 56: 220-224.

Jennings C.D., Boleyn K., Bridges S.R., Wood P.J. and Anderson J.W. (1988). A comparison of the lipid lowering and intestinal morphological effects of cholestyramine, chitosan, and oat gum in rats. *Proc. Soc. Exp. Biol. Med.* 189: 13-20.

Jing S.B., Li L., Ji D., Takiguchi Y. and Yamaguchi T. (1997). Effect of chitosan on renal function in patients with chronic renal failure. *J. Pharm. Pharmacol.* 49: 721-723.

Joen Z.J., Park P.J. and Kim S.K. (2001). Antimicrobial effect of chitooligosaccharides produced by bioreactor. *Carbohyd.Polym.* 44: 71-76.

Kafetzopoulos D., Martinou A. and Bouriotis V. (1993). Bioconversion of chitin to chitosan: purification and characterization of chitin deacetylase from *Mucor rouxii. Proc. Natl. Acad. Sci. U S A* 90: 2564-2568.

Keller F.A. and Cabib E. (1971). Chitin and yeast budding. Properties of chitin synthetase from *Saccharomyces carlsbergensis. J. Biol. Chem.* 246: 160-166.

Kendra D.F., Christian D. and Hadwiger L.A. (1989). Chitosan oligomers from *Fusarium solani/pea* interactions, chitinase/ beta glucanase digestion of sporelings and from fungal wall chitin actively inhibit fungal growth and enhance disease resistance. *Physiological and Molecular Plant Pathology* 35: 215-230.

Khor E. (2001). Chitin: Fulfilling a Biomaterials Promise. Elsevier Science Federal Office of Consumer Protection and Food Safety, Oxford UK. ISBN: 0-080-44018-5

Kise H. and Hayakawa A. (1991). Immobilization of proteases to porous chitosan beads and their catalysis for ester and peptide synthesis in organic solvents. *Enzyme Microb. Technol.* 13: 584-588.

Knorr D. (1983). Dye binding properties of chitin and chitosan. *J. Food Sci.* 48: 36-37.

Krajewska B., Leszko M. and Zaborska W. (1990a). Development of urease chitosan membrane. *Polim. Med.* 20: 31-41.

Krajewska B., M. Leszko and Zaborska W. (1990b). Urease immobilized on chitosan membrane: preparation and properties. *J. Chem. Technol. Biotechnol.* 48: 337-350.

Kratz G., Arnander C., Swedenborg J., Back M., Falk C., Gouda I. and Larm O. (1997). Heparin chitosan complexes stimulate wound healing in human skin. *Scand. J. Plast. Reconstr. Surg. Hand. Surg.* 31: 119-123.

Kristiansen B. and Mattey M. Joan Linden (1998). Citric acid biotechnology. Taylor and Francis, London UK, pp. 189. ISBN: 0-203-48339-1

Kučera J. (2004). Fungal mycelium the source of chitosan for chromatography. *J. Chromatogr.* 808: 69-73.

Lee J.K., Kim S.U. and Kim J. H. (1999). Modification of chitosan to improve its hypocholesterolemic capacity. *Biosci. Biotechnol. Biochem.* 63: 833-839.

LeHoux J. G. and Grondin. F. (1993). Some effects of chitosan on liver function in the rat. *Endocrinology* 132: 1078-1084.

Madihally S.V. and Matthew H.W. (1999). Porous chitosan scaffolds for tissue engineering. *Biomaterials* 20: 1133-1142.

Magnin D., Dumitriu S., Magny P. and Chornet. E. (2001). Lipase immobilization into porous chitoxan beads: activities in aqueous and organic media and lipase localization. *Biotechnol. Prog.* 17: 734-737.

Mayer R.T., Chen A.C., and DeLoach J.R. (1980). Characterization of a chitin synthase from the stable fly, Stomoxys calcitrans (L.). *Insect Biochem.* 10: 549-556.

Miura T., Usami M., Tsuura Y., Ishida H. and Seino Y. (1995). Hypoglycemic and hypolipidemic effect of chitosan in normal and neonatal streptozotocin induced diabetic mice. *Biol. Pharm. Bull.*18: 1623-1625.

Miyoshi H., Shimura K., Watanabe K. and Onodera K. (1992). Characterization of some fungal chitosans. *Biosci., Biotechnol. Biochem.* 56: 1901-1905.

Murata J., Saiki I., Matsuno K., Tokura S. and Azuma I. (1990). Inhibition of tumor cell arrest in lungs by antimetastatic chitin heparinoid. *Jpn. J. Cancer Res.* 81: 506-513.

Muzzarelli R.A., Barontini G. and Rocchetti R. (1976). Immobilized enzymes on chitosan columns: alpha chymotrypsin and acid phosphatase. *Biotechnol. Bioeng.* 18: 1445-1454.

Nagasawa K., Watanabe H. and Ogama A. (1971). Ion exchange chromatography of dinucleoside 3' leads to 5' phosphates on chitosan impregnated cellulose thin layers. *J. Chromatogr.* 56: 378-381.

Nagasawa K., Watanabe H. and Ogamo A. (1970). Ion exchange chromatography of nucleic acid constituents on chitosan impregnated cellulose thin layers. *J. Chromatogr.* 47: 408-413.

No H.K. and Meyers S.P. (1995). Preparation and characterization of chitin and chitosan (A review). *J. Aquatic Food Products Technol.* 4: 27-52.

No H.K., Park Y.N., Lee H.S. and Meyers P.S. (2002). Antimicrobial activity of chitosans and chitosans oligomers with different molecular weights. *Int. J. Food Microbiol.* 74: 65-72.

Okamoto Y., Shibazaki K., Minami S., Matsuhashi A., Tanioka S. and Shigemasa Y. (1995). Evaluation of chitin and chitosan on open would healing in dogs. *J. Vet. Med. Sci.* 57: 851-854.

Pittler M.H., Abbot N.C., Harkness E.F. and Ernst E. (1999). Randomized, double blind trial of chitosan for body weight reduction. *Eur. J. Clin. Nutr.* 53: 379-381.

Pochanavanich P. and Suntornsuk W. (2002). Fungal chitosan production and its characterization. *Lett. Appl. Microbiol.* 35: 17-21.

Purnama Darmadji and Izumimoto. M. (1994). Effect of chitosan in meat preservation. *Meat Sci.* 38: 243-254.

Quesada Allue L.A., Marechal L.R. and Belocopitow E. (1976). Chitin synthesis in *Triatoma infestans* and other insects. *Acta Physiol.Lat.Am.* 26: 349-363.

Reiss M., Heibges A., Metzger J. and Hartmeier W. (1998). Determination of BOD values of starch containing waste water by a BOD biosensor. *Biosen.s Bioelectron.* 13: 1083-1090.

Rudal K.M. and Kenchington W. (1973). The chitin system. *Biol.Rev.* 48: 597-636.

Ruiz Herrera J. (1991). Fungal cell walls: Structure, Synthesis, and Assembly. CRC Press, Boca Raton, FL. pp. 256, ISBN: 0849366720.

Ruiz Herrera J. and Martinez Espinoza A.D. (1999). Chitin biosynthesis and structural organization *in vivo*. *EXS* 87: 39-53.

Savage P.J. and Savage G.P. (1994). The effect of coating apples on the quality of stored apples. *Proc. . Nutr. Soc. New Zealand*; 19, 129-133.

Shahabeddin L., Berthod F., Damour O. and Collombel C. (1990). Characterization of skin reconstructed on a chitosan cross linked collagen glycosaminoglycan matrix. *Skin Pharmacol.* 3: 107-114.

Shi Y.C., Jiang Y.M., Sui D.X., Li Y.L., Chen T., Ma L. and Ding Z.T. (1996). Affinity chromatography of trypsin using chitosan as ligand support. *J. Chromatogr.* A 742 :107-112.

Shibuya N. and Mimami E. (2001). Oligosaccharide signaling fpr defense responses in plant. *Physiological and Molecular Plant Pathol.* 59: 223-233.

Sudarshan N. R., Hoover D.G. and Knorr D. (1992). Antibacterial action of chitosan. *Food Biotechnol.* 6: 257-272.

Sugano M., Watanabe S., Kishi A., Izume M. and Ohtakara A. (1988). Hypocholesterolemic action of chitosans with different viscosity in rats. *Lipids* 23: 187-191.

Suzuki K., Nakamura T., Matsuura H., Kifune K. and Tsurutani R. (1995). A new drug delivery system for local cancer chemotherapy using cisplatin and chitin. *Anticancer Res.* 15: 423-426.

Synowiecki J. and Ali Khateeb N.A.A.Q. (1997). Mycelia of *Mucor rouxii* as a source of chitin and chitosan. *Food Chemistry* 60:605-610.

Synowiecki J. and Ali Khateeb N.A.A.Q. (2003). Production, properties, and some applications of chitin and its derivative. *Crit.Rev.Food Sci.* 43: 145-171.

Tao W., Zivanovic S., Draughon F.A. and Sams C.E. (2004). Chitin and chitosan value added products from mushroom waste. *J.Agric.Food Chem.*52: 7905-7910.

Tao W., Zivanovic S., Draughon F.A., Conway W.S., and Sams C.E. (2005). Physicochemical properties and bioactivity of fungal chitin and chitosan. *J. Agric Food Chem.* 53: 3888-3894.

Teng W.L., Khor E., Tan T.K., Lim L.Y. and Tan S.C. (2001). Concurrent production of chitin from shrimp shells and fungi. *Carbohydr. Res.* 332: 305-316.

Usui T., Hayashi Z., Nanjo F., Sakai K. and Ishido Y. (1987). Transglycosylation reaction of a chitinase purified from *Nocardia orientalis. Biochim.Biophys.Acta* 923: 302-309.

Vermeulen C.A. and Wessels J.G. (1986). Chitin biosynthesis by a fungal membrane preparation. Evidence for a transient non crystalline state of chitin. *Eur.J.Biochem.* 158: 411-415.

Vincenzi M., Dessi M.R., and Muzzarelli R.A.A. (1993). Separation of coeliac active peptides from bread wheat with the aid of methylpyrrolidinone chitosan. *Carbohyd. Polymers* 21: 295-298.

Wang Y. M. and Yam KL. (1995). Separation of single cell protein with chitosan by flocculation method. *IFT Annual Meeting* p. 282.

White S.A., Farina P.R. and Fulton I. (1979). Production and isolation of chitosan from *Mucor rouxii. Appl. Environm. Microbiol.* 38: 323-328.

Wu T., Zivanovic S., Draughon F.A., Conway W. S. and Sams C.E. (2005). Physicochemical properties and bioactivity of fungal chitin and chitosan. *J. Agric. Food Chem.* 53: 3888-3894.

Yihua Y.U. and Binglin H. E. (1997). A new low density lipoprotein (LDL) adsorbent. *Artif. Cells Blood Substit. Immobil. Biotechnol.* 25: 445-450.

Zihnioflu F. and Telefoncu A. (1995). Preparation and characterization of chitosan entrapped microsomal UDP glucuronyl transferase. *Artif. Cells Blood Substit. Immobil. Biotechnol.* 23: 545-552.

13 Biodiversity and Biotechnological Potentials of Fungi from the Marine Environment

Guangyi Wang, Quanzi Li, and Sanghwal Yoon

Hawaii Natural Energy Institute, Department of Oceanography, School of Ocean and Earth Science and Technology, University of Hawaii at Manoa, Honolulu 96822, HI

INTRODUCTION

Fungi have long been known to exist in the marine environment and are an understudied marine ecological group. There remains some debate as to what constitutes a "marine" species. However, it is generally agreed upon that obligate marine fungi are those that grow and sporulate exclusively in a marine or estuarine habitat while facultative marine fungi are those from freshwater or terrestrial milieus which are able to grow (and possibly sporulate) in the marine environment (Kohlmeyer and Kohlmeyer, 1979; Kohlmeyer *et al.*, 2003). Due to their unique living environments in the oceans, marine organisms must adjust metabolically in order to survive in their harsh environments. Accordingly, intensive evolutionary pressures have forced marine organisms to produce chemically diverse compounds and special structural materials that have a wide variety of possible ecological roles (Childress *et al.,* 1995 and Faulkner, 2000). Likewise, marine fungi are uniquely adapted for life in the sea and able to produce novel bioproducts (Biabani and Lattsch, 1998; Jensen and Fenical, 2002). Therefore, marine fungi recently have received a great deal of attention from marine mycologists, chemists, environmental scientists and engineers. In this chapter, we review recent progress on aspects of biodiversity, natural product bioprospecting and biotechnology of marine filamentous fungi.

BIODIVERSITY AND DISTRIBUTION OF MARINE FUNGI

Marine fungi play a significant role in marine ecosystems (Hyde *et al.*, 1998; Prasannrai and Sridhar, 2001), where they are the important intermediaries of energy flow from detritus to higher tropic levels (Hyde and Lee, 1995). They play an important role in nutrient regeneration cycles as decomposers of dead and decaying organic matter. Also, they are major decomposers of woody and herbaceous substrata entering marine ecosystems (Fell and Master, 1980). Some marine fungi also cause diseases of marine animals and plants while others form mutualistic symbiotic relationships with other organisms (Hyde *et al.,* 1998; Kubanek *et al.,* 2003). Consequently, their biodiversity has recently been an interesting area in marine mycology.

Early researchers primarily isolated marine fungi from drift wood, shoreline plants and sediments. Since the first report on marine fungi from mangrove plants (Crib and Crib, 1955), tremendous work has been carried out on the manglicolous fungi (Hyde and Lee, 1995; Jones and Alias, 1997). Currently, marine fungi have been isolated from all different marine habitats, including, but not limited to, sediments, sands, mangrove, algae, wood, marine corals, intertidal grasses, marine animals and plants. It was originally estimated that relatively few fungi are adapted to complete their entire life cycle in the sea water, with only 500-600 or so species considered truly marine and possibly with a similar number to be discovered (Edward *et al.,* 1998). However, some mycologists believe that the biodiversity of this ecological group is likely to be considerably underestimated. At least, the number of currently described marine fungi clearly supports that the above figures only represent a portion of the total number of marine fungal species (Edward *et al.,* 1998). Moreover, it is postulated that the description of marine fungi is still at the inventory stage and many new taxa wait to be identified (Hyde *et al.*, 1996). Marine fungi occur as saprophytes and parasites on seaweed, drift wood and higher marine organisms in littoral and neritic environments (Biabani and Laatsch, 1998). In general, filamentous ascomycetes and fungi imperfecti dominate the described marine fungi species.

Description of marine fungal diseases has received several excellent reviews (Kohlmeyer and Kohlmeyer, 1979; Sinderman, 1987; Noga, 1990; Rand, 1996). However, little is known about the biology of these pathogens. Recently, investigation of marine pathogenic fungi has yielded very interesting results. Several marine fungi are recently reported to cause diseases of rock oyster, corals, and sea fans (Raghukumar and Raghukumar, 1991; Raghukumar *et al.,* 1991; Smith *et al.,* 1996; Gieser *et al.,* 1998; Alker *et al.,* 2001; Raghukumar *et al.*, 2001). The most interesting case is the study of the sea fan pathogen *A. sydowii* (Gieser *et al.,* 1998). The pathogen is morphologically identical and forms a strongly supported clade with strains isolated from terrestrial materials, but only the marine isolates are pathogenic. Differences in metabolic profiles between the marine and the terrestrial isolates clearly support the theory that there are genetic differences among strains of different origins (Alker *et al.,* 2001). In addition, more than 80 marine fungi have been isolated from algae as parasites or symbionts (Paitra, 1997; Pivkin, 2000; Afiyatulove *et al.,* 2002; Kohlmeyer, 2002, 2003; Proksch, 2003). Although fungi associated with algae are known to be a prolific source of novel natural compounds (Jensen and Fenical, 2002; Proksch *et al.,* 2003) little is known about the ecological importance of these natural chemicals.

Geographically, marine fungi are widely distributed and occur in all climates and salinities. However, they are often uniquely adapted for life in the oceans (Biabani and Laatsch, 1998). Water temperature has been considered the single most important environmental factor influencing the composition of the marine mycobiota in a specific geographic location, but sea masses present no barriers to the dispersal of marine fungi (Jones, 1993; Hyde *et al.,* 1998). Recently, the investigation of the diversity of intertidal habitats of the south-west coast of India indeed reveals that the seawater temperature affects the number of fungi observed in a certain geographic location (Prasannrai and Sridhar, 2001). Nevertheless, temperature is also one of the major environmental factors that influence many marine diseases and dramatically affect the marine host-fungal pathogen interactions (Alker *et al.,* 2001). On the other hand, few fungi, such as *Corollospora maritima* and *Halosphaeria appendiculata*, are present in both temperate and tropical seas (Hyde and Lee, 1995; Hyde *et al.*, 1998). In addition, marine fungal species appear to respond to the salinity of the water, with different species being common at different salinities (Kohlmeyer, 1969). On the other hand, it has been observed that full-strength seawater is typically not optimal for growth of several fungal species (Abanat *et al.*, 1998; Edward *et al.*, 1998). Host preference has been observed in mangrove fungi. Therefore, the hosts present in a mangrove may shape the complexity of the fungal community in the habitat. However, no evidence supports the hypothesis that marine fungi are host-specific in mangroves (Hyde and Lee, 1995; Jones and Alias, 1997; Schmet and Shearer, 2003). To conclude, the information on the dispersal of marine fungi is fragmentary, and more thorough investigations are needed before we can draw an inclusive assessment of how environmental factors affect the distribution and diversity of fungal species in the oceans.

NATURAL PRODUCTS

Terrestrial fungi are known to produce a variety of chemically diverse natural compounds with interesting biological activities. Until recently, only modest attention has been devoted towards metabolites from marine fungi. To date, over a hundred new metabolites from marine fungi have been described. The majority described before 1998 is summarized by Biabani and Laatsch (1998). Some discovered compounds are reviewed by Faulkner (2001). More recently, the fungal metabolites were intensively reviewed by Jensen and Fenical (2002). These documented discoveries clearly show that marine fungi are a prolific source of structurally unique and biologically active natural substances. Many of the natural compounds from marine fungi are of pharmaceutical and agrochemical importance. Among these compounds, many have unique structures with diverse biological activities. Some of these metabolites have structural similarity to, and others are common to, those metabolites isolated from terrestrial fungi (Holler *et al.*, 2000; Jensen and Fenical, 2002; Proksch *et al.*, 2003).

Overall, both obligate and facultative marine fungi have been the targets of natural product chemists for natural product bioprospecting. To date, only a small fraction of the compounds described from marine fungi is from obligate marine fungi, while the majority of these compounds are from facultative marine fungi. Fungi associated with sponges are the single most prolific source of novel natural compounds. These compounds have been shown to possess antifungal, antialgal, and antibacterial activity, cytotoxicity and show inhibition of reverse-transcriptase and tyrosine kinase (Holler *et al.*, 2000). The sponge-associated fungus *Microsphaeropsis* sp. produces anthraquinones and betaenone derivatives as well as microsphaerones (Bracher and Schulti, 1997; Wang *et al.*, 2002). Anthraquinones and one betaenone derivatives inhibit protein kinases with the IC_{50} ranged from 18.5-54.0 μM (Bracher and Schulte, 1997). Furthermore, *Cladosporium herbarum* and *Hortaea werneckii* derived from a Mediterranean sponge produce new compounds of herbarins and hortein, respectively (Brauers *et al.*, 2001; Jadulco *et al.*, 2002). Nevertheless, one drawback is that a high percentage of marine fungi derived from sponges is of terrestrial origin (Holler *et al.*, 2000). Furthermore, marine algae are a well-known habitat of marine higher filamentous marine fungi (Kohlmeyer and Kohlmeyer, 1979). Recent discoveries suggest that marine algae are also a significant source of fungi with the ability to produce new natural compounds (Jensen and Fenical, 2002; Osterhage *et al.*, 2002). For instance, *Penicillium* sp. isolated from the green alga *Enteromorpha intestinalis* can produce a handful of new natural compounds with interesting biological activities in different fermentation conditions (Numata *et al.,* 1993; Takahasi *et al.,* 1996; Numata *et al.,* 1996; Iwamoto *et al.,* 1998). The marine fungus *Drechslera dematioidea* isolated from the inner tissues of the marine red alga *Liagora viscida* produces four classes of unusual irregular terpenoids (Jensen and Fenical, 2002; Osterhage *et al.,* 2003). The fungus *Epicoccum* sp. isolated from the marine brown alga *Fucus vesiculosus* produces new natural compounds with antioxidant activities (Abdel-latif *et al.,* 2003a and b).

Moreover, marine fungi isolated from other marine habitats contribute noticeably to the inventory of natural compounds. *Aspergillus* is one of the most interesting genus of marine fungi. Many strains of the fungus *Aspergillus* have been isolated from the marine environment (Jensen and Fenical, 2000; Jensen and Fenical, 2002). The marine *Aspergillus* isolates produce novel natural compounds that differ from those isolated from their terrestrial counterparts. *A. versicolor* isolated from sponge produces the new natural compounds, aspergillitine and aspergiones A-F (Lin *et al.,* 2003). The terrestrial *A. versicolor* are known to mainly produce xanthones, anthraquinones and other cyclic peptides (Bullock *et al.,* 1962; Aucamp and Holzapfel, 1969; Fukuyama *et al.,* 1975; Berger and Berger, 1982; Brauers *et al.,* 2001). A new polyketide, aspermytin A, is isolated from a cultured marine *Aspergillus* sp. induced neurite outgrouth in rat pheochromocytoma (PC-12) cells at concentration of 50 μM (Tsukamoto *et al.,* 2004).

Polyketides, nonribosomal, terpenoids and alkaloids are the majority of natural compounds discovered from marine environments. Most interestingly, rare novel sesterterpenoids have been isolated from marine fungi. The marine-derived *Aspergillus* strain CNM-713 is shown to produce the novel sesterterpene aspergilloxide (Cutue *et al.,* 2002). This marine isolate is identified to be *A. vericolor* using ITS primers for *Aspergillus* genera (Wang, unpublished data). The marine fungus *Halorosellinia oceanica* isolated from Samutsongkram, Thailand, produces several natural compounds with novel structures and interesting bioactivities (Chinworungsee, 2001; Chinworungsee *et al.,* 2002). Among these compounds, halorosellinic acid is an ophiobolane sesterterpene and exhibits antimalarial activity with IC_{50} 13 ug/ml and a weak antimycobacterial activity with the MIC of 200 μg/ml (Chinwounggsee *et al.,* 2001). Furthermore, the marine fungus *Fusarium heterosporum* isolated from the surface of driftwood collected in the Bahamas yields novel sesterterpenoids (neomangicols and mangicols) with significant anti-inflammatory, antimicrobial, and cytotoxic activities (Renner *et al.,* 1998; Renner *et al.,* 2000). In addition, many other reports prove that marine fungi produce natural compounds with novel structures and activities (Nagai *et al.,* 2002; Lin *et al.,* 2002; Abdel-latef *et al.,* 2003a, 2003b; Byun *et al.,* 2003; Li *et al.,* 2003; Liu *et al.,* 2003; Tan *et al.,* 2003).

It is undisputable that marine fungi are an important resource for novel natural products. However, a high percentage of described compounds from facultative marine fungi that belong to their terrestrial counterparts has been a real challenge for marine fungi oriented natural product bioprospecting. The current isolation methods intend to select for the fastest growing strains with no selection on dominant spores washed into oceans (Jensen and Fenical, 2002). Therefore, morphology based taxonomy for marine strains needs to be revaluated using molecular approaches. Furthermore, fermentation conditions appear to significantly affect the profiles of secondary metabolites from marine strains (Numata *et al.,* 1993;. Numata *et al.,* 1996; Takahashi *et al.,* 1996; Iwamoto *et al.,* 1998). Microbes that produce specific metabolites in response to a specific antagonist are fairly common (Burgess *et al.,* 1999). The marine fungus *Rosenvingea* sp. is reported to produce the new antibiotic pestalone in response to a marine bacterium (Cueto *et al.,* 2001). Metabolite profiling studies reveal that the marine-derived fungi *Phoma* sp. and *A. sydowii* differ significantly from their terrestrial counterparts with respect to their secondary metabolite content and quality (Osterhage *et al.,* 2000; Alker *et al.,* 2001). Therefore, information on biosynthetic pathways of marine fungi should shed light on some of these issues associated with marine fungi oriented natural products discovery. In particular, evolutionary studies of metabolic pathways of both marine fungus and its terrestrial counterpart may yield more clues to distinguish the two.

HYDROLYTIC ENZYMES

The enzymatic degradation of carbohydrates is the basis of many biotechnological processes. Filamentous fungi are widely used as sources of enzymes. Therefore, studies of enzymatic systems of filamentous fungi are of biotechnological interest. Terrestrial fungi degrade carbohydrates using a battery of well characterized hydrolytic enzymes (Mai *et al.,* 2004). There are three types of lignocellulose degradation by terrestrial fungi— White rot, brown rot, and soft rot. White rot uses extensive and non-enzymatic degradation of all wood components; brown rot can cause rapid cellulose and hemicellulose decay through non-enzymatic oxidation, but with limited lignin degradation; soft rot involves extensive enzymatic decay of cellulose and hemicellulose with limited lignin degradation (Eaton and Hale,1993). The detailed decay mechanism of each type has been the subject of several reviews (Eaton and Hale,1993; Mai *et al.,* 2004). The three types of wood decay fungi have been widely used in the wood industry (Mai *et al.,* 2004). Likewise, marine fungi also produce hydrolytic enzymes and can degrade lignocellulose as well as other environmental contaminants. Based on the features of morphology, soft rot and white rot decay have been noticed in wood samples colonized by marine fungi (Mouourous, 1989). Several genera have been discovered to show wood decay activity in marine and estuarine environments (Hyde *et al.,* 1998). Unlike many other marine borers, marine fungi have the capacity to tolerate low oxygen tensions and consequently are considered the major contributor to the lignocellulose turnover in the marine sediments (Holt *et al.,* 1983; Singh *et al.,* 1990; Hyde *et al.,* 1998). Furthermore, they are extremely important decomposers of wood in the upper intertidal regions where other marine borers do not exist (Hyde *et al.*, 1998). Marine fungi also play an important role to convert mangrove leaves, wood, sea grasses, and other lignocellulosic materials into detritus. Such detritus serve as a primary supply of food for detritivorous animals (Hodson and Moran, 1989). Therefore, marine fungi are the major player in the food web of marine ecosystem.

Cellulose, semicellulose and lignin are three major components of all wood plants. Marine fungi can use all three types of carbon compounds as substrate by secreting different hydrolytic enzymes. They have been reported to degrade all three major components of wood. Over 30 strains of marine fungi from diverse taxonomic groups are reported to have the ability to use the carboxymethylcellulose as carbon source and to produce enzymes involved in wood degradation (Roharmann *et al.,* 1992). The marine fungi *Julella aviceniae*, *Lignincola laevis*, *Nia vibrissa*, and *Stagonospora sp.* can use crystalline cellulose substrates and appear to produce cellobiohydrolase (Rohrman and Molitoris, 1992; Raghukumar *et al.,* 1994; Pointing *et al.,* 1998). Degradation of xylan has been used as a good assay for the hemicellulose degradation. More than 10 strains of marine fungi have the xylan degradation activity, and *Gongronella sp.*, *Aigialus mangrovei*, *Lophiostoma mangrovei*, and *Hypoxylon oceanicum* are very good xylan degraders (Raghukumar *et al.,* 1994). Moreover, decolourization of the polymeric dyes (PolyR-478 and Azure B) has been used in the growth studies of the lignin degradation by marine fungi (Raghukumar *et al.,* 1994; Pointing *et al.,* 1998). While using glucose or cellulose as the primary carbon source, four marine fungi *N. vibrissa, J. avicenniae, L. laevis* and *Stagonospora sp.* have been demonstrated to completely decolourize these dyes (Pointing *et al.,* 1998). Marine fungi have also been reported to produce manganese dependent peroxidase and laccase (Raghukumar *et al.,* 1994). Aggressive color dye colourizers such as *N. vibrissa, J. avicenniae, L. laevis* and *Stagonospora sp.* are highly active at all salinities, while several moderate or weak dye colourizing isolates show optimal activity at both low and higher salinities (Pointing *et al.,* 1998). The marine fungus *Phlebia* sp. MG-60 can efficiently degrade lignin and holocellulose when it is incubated with whole sugarcane bagasse (WSB) (Li *et al.,* 2002). Two marine fungal strains, *Monodictys pelagica* and *Nia vibrissa,* can mineralize lignin into CO_2 and 5-6% of the 14C-labeled lignin is mineralized by the end of a 30-day incubation period (Southerland *et al*., 1982). Moreover, the obligate marine fungus *Halosarpheia ratnagiriensis* and two facultative marine fungi *Sordaria fimicola* and *Flavodon flavus* also have the ability to mineralize ^{14}C-labeled lignin (Raghukumar *et al.,* 1996). *F. flavus*, a white-rot fungus, can mineralize about 24% of lignin within 24 days in sea water medium. Several marine fungi produce laccase, one of the major lignin-modifying enzyme present in many genera of wood-rotting fungi, and other lignocellulose modifying enzymes (Raghukumar *et al.,* 1994, 1996, 1999, 2001)

Marine fungi have been reported to produce other hydrolytic enzymes. The marine fungus *Beauveria bassiana* has the ability to produce chitinase. The production yield can reach 248 units/g in the medium containing prawn waste and sea water in solid state fermentation (Suresh and Chandrasekheran, 1998,1999). In addition, L-glutaminase is a potential anticancer agent and a flavor enhancing agent (Roberts *et al.,* 1970). Salt tolerant and heat stable L-glutaminase is in great demand in the food industry (Moriguchi *et al.,* 1994). Marine bacteria, *Pseudomonas fluorescens, Vibrio cholerae,* and *V. costicola,* produce this enzyme in both submerged and solid state fermentation (Chandrasekaran, 1997). Interestingly, the marine fungus *Beauveria* sp. isolated from marine sediments can produce salt tolerant L-glutaminase and represent a potential for continuous commercial production of this interesting enzyme (Sabu *et al.,* 2000; Sabu, 2002).

With the ability to produce so many interesting hydrolytic enzymes, marine fungi have great potentials of application in the biotechnological industry. The potential of applying marine fungi in bioremediation has been reviewed by Raghukumar (2001). In particular, research on bioremediation of colored pollutants using marine fungi has made significant progress. For example, the marine isolate of *A. niger* has a high capacity for rapid decolourization of textile wastewater and absorb 90-95% dyes (Azo-reactive-sulfuric, pigmentous) (Payman and Mahana, 1998). Several marine fungi have been successfully used to treat highly colored waste-waters from paper and pulp mills, textiles and dyestuff industries, distilleries and tanneries efficiently (Raghukumar *et al.,* 1996; Raghukumar, 2000; Raghukumar and Rivonkar, 2001). Furthermore, the prawn waste is chitinous solid waste of the shellfish producing industry and has raised some environmental concerns. As a potent producer of chitinase, *B. bassiana* has been proven to have great potential in treating shellfish waste chitin (Suresh and Chandrasekaran, 1998,1999). The marine fungus *Phlebia* sp. MG-60 can efficiently degrade lignin and holocellulose when it is incubated with whole sugarcane bagasse (WSB) (Li *et al.,* 2002). After being treated with *Phlebia* sp. MG-60, bagasses fiber and WSB can be used as raw material in the pulp and paper industry and animal feeding, respectively. Finally, several marine fungi isolated from the deep sea can degrade commercially produced thermoplastic polymer (poly-beta-hydroxybutyric acid (PHB)) in solid medium (Gonda *et al.,* 2000).

CONCLUSIONS

Inventory data for the marine fungi in several countries are available, but marine fungi from many regions and types of marine habitats remain to be investigated (Hyde *et al.,* 1998). There are many gaps in our understanding of the biodiversity of fungi in the marine environment. Coordinated efforts between marine mycologists, marine chemists and engineers are needed to study the biodiversity and biotechnological potentials. Marine fungi isolated for natural bioproducts bioprospecting require further effort for species identification. Currently, the plate method has been commonly used to isolate marine fungi. This method has a tendency to select non-marine species (Jensen and Fenical, 2002). One of the biggest challenges for research related marine fungi is to verify whether or not a fungus is of indigenous marine origin. The solution for the challenge is to inspect the isolates on marine substrate from the marine habitat for fruiting structures (Kohlmeyer and Kohlmeyer, 1979; Hyde *et al.,* 2000). A marine fungus identified based on its morphology needs to be carefully examined using molecular approaches. Sometimes, the marine and terrestrial isolates are morphologically identical, but genetically different. That is the case for *A. sydowii.* The marine strains are pathogenic to sea fans, but the terrestrial strains are not (Geiser *et al.,* 1998; Alker *et al.,* 2001).

At this point, collection and culture of symbiotic fungi have been very time-consuming, tedious, and even impossible. This may ascribe to the nature of their association with hosts. Some species only grow in their hosts while others can be easily outgrown by the faster growing terrestrial anamorphs (Kohlmeyer and Kohlmeyer, 2003). However, the environmental genomic approach could provide a solution for the identification of symbiotic and other unculturable marine fungi. In some cases, some marine fungi may need specific growth conditions. Adapted engineering equipment has been used to cultivate fungi from the deep sea (Lorenz and Molitoris, 1997). Therefore, reliable cultivation methods and instruments may facilitate the discovery of new species in the marine environment.

Natural products bioprospecting from marine fungi have made noticeable progress. Marine fungi have harvested significant attention from drug discovery. However, isolation of authentic obligate marine fungi is a major challenge for novel natural compound oriented drug discovery. The high proportion of ubiquitous genera and the relatively low proportion of new compounds suggest that fungi isolated from marine habitats are of terrestrial origin (Holler *et al*., 2000). Consequently, about 30% of new compounds from the "marine fungi" are truly from terrestrial fungi, such as *Aspergillus* and *Penicillium* (Jensen and Fenical, 2002). The fundamental differences at the molecular level between marine fungi and their terrestrial counterparts still need to be addressed in future research. Studying the biosynthetic pathway of these natural compounds should shed light on the evolutionary history of marine fungi. Completely sequencing the genomes of marine fungi may give more information about the genetic base for their life style in the oceans. The ecological importance of some natural compounds isolated from fungi originating from marine animal, algae, and plants is still unknown.

So far, studies on hydrolytic enzymes from marine fungi have been focused on qualitative analyses. Although a handful of marine fungi are shown to produce interesting hydrolytic enzymes, more quantitative data is needed to better understand both their role in the environment and the development of their potential applications in biotechnology. Several marine fungi have been successfully used in environmental bioremediation (Raghukumar *et al.,* 1996; Raghukumar, 2002). The scale-up step can still present a significant challenge in future research. The collaboration of mycologists, environmental scientists and engineers would facilitate the commercial applications for some of these strains. In particular, metabolic engineering approaches can significantly improve biotechnological traits in some of these interesting strains. Their tolerance and growth in the presence or absence of sea salts and at alkaline pH certainly give marine fungi an edge over their terrestrial counterparts in bioremediation (Raghukumar, 2002). In general, research on this area is still in the infant stage. Continuous screening may yield the "super" biotechnological strains. With new techniques and engineering tools applied in bioproducts bioprospecting from the marine environment, application of marine strains and bioproducts biotechnological industries should be seen in the near future.

REFERENCES

Abbanat D., Leighton M. , Maiese W. , Jones E.B.G. , Pearce C. and Greestein, M. (1998). Cell wall active antifungal compounds produced by the marine fungus *Hypoxylon oceanicum* II-15g 256. *Journal of Antibiotics* 51: 296-302.

Abdel-Lateff A., Fisch K.M., Wright A.D. and Konig G.M. (2003a). A new antioxidant isobenzofuranone derivative from the algicolous marine fungus *Epicoccum* sp. *Planta Medica* 69: 831-834.

Abdel-Lateff A., Klemke C., Konig G.M. and Wright A.D. (2003b). Two new xanthone derivatives from the algicolous marine fungus *Wardomyces anomalus. Journal of Natural Products* 66: 706-708.

Afiyatullov S.S., Kalinovsky A.I., Kuznetsova T.A., Isakov V.V., Pivkin M.V., Dmitrenok P.S. and Elyakov G.B. (2002). New diterpene glycosides of the fungus *Acremonium striatisporum* isolated from a sea cucumber. *Journal of Natural Products* 65: 641-644.

Alker A.P., Smith G.W. and Kim K. (2001). Characterization of *Aspergillus sydowii* (Thom et Church), a fungal pathogen of Caribbean sea fan corals. *Hydrobiologia* 460: 105-111.

Aucamp P.J., and Holzapfel C.W. (1969). Toxic substances from *Aspergillus versicolor.* African Chemical Institute 22: 35S-40S.

Berger-Deguee, M., and Berger Y. (1982). Structure of Versicolorone Isolated from *Aspergillus versicolor. Phytochemistry* (Oxford) 21: 1449-1451.

Biabani, M.A.F., and Laatsch H. (1998). Advances in chemical studies on low-molecular weight metabolites of marine fungi. *Journal fur Praktische Chemie-Chemiker-Zeitung* 340: 589-607.

Bracher, F., and Schulte B. (1997). An enantiodivergent synthesis of both enantiomers of the macrocyclic lactone curvularin. *Liebigs Annalen*: 1979-1982.

Brauers G., Ebel R., Edrada R., Wray V., Berg A., Grafe U. and Proksch P. (2001). Hortein, a new natural product from the fungus *Hortaea werneckii* associated with the sponge *Aplysina aerophoba. Journal of Natural Products* 64: 651-652.

Bullock E., Roberts J.C. and Underwood J.G. (1962). Studies in mycological chemistry. Part XI. The structure of isosterigmatocystin and an amended structure for sterigmatocyctin. *Journal of the Chemical Society*: 4179-4183.

Burgess J.G., Jordan E.M., Bregu M., Mearns-Spragg, A. and Boyd K.G. (1999). Microbial antagonism: a neglected avenue of natural products research. *Journal of Biotechnology* 70: 27-32.

Byun H.G., Zhang H.P., Mochizuki M., Adachi K., Shizuri Y., Lee W.J. and Kim S.K. (2003). Novel antifungal diketopiperazine from marine fungus. *Journal of Antibiotics* 56: 102-106.

Chandrasekaran, M. (1997). Industrial enzymes from marine microorganisms - the indian scenario. *Journal of Marine Biotechnology* 5: 86-89.

Childress, J.J. (1995). Are there physiological and biochemical adaptations of metabolism in deep-sea animals. *Trends in Ecology & Evolution* 10: 30-36.

Chinworrungsee M., Kittakoop P., Isaka M., Chanphen R. , Tanticharoen M. and Thebtaranonth Y. (2002). Halorosellins A and B, unique isocoumarin glucosides from the marine fungus *Halorosellinia oceanica. Journal of the Chemical Society* Perkin Transactions 1.: 2473-2476.

Chinworrungsee M., Kittakoop P., Isaka M., Rungrod A., Tanticharoen M. and Thebtaranonth Y. (2001). Antimalarial halorosellinic acid from the marine fungus *Halorosellinia oceanica. Bioorganic & Medicinal Chemistry Letters 11*: 1965-1969.

Cribb A.B., and Cribb J.W. (1955). Marine fungi from Queensland I, University of Quensland Paper, Department of Botany: 77-81.

Cueto M., Jensen P.R. and Fenical W. (2002). Aspergilloxide, a novel sesterterpene epoxide from a marine-derived fungus of the genus *Aspergillus*. Organic Letters 4: 1583-1585.

Cueto M., Jensen P.R., Kauffman C., Fenical W., Lobkovsky E. and Clardy J. (2001). Pestalone, a new antibiotic produced by a marine fungus in response to bacterial challenge. *Journal of Natural Products* 64: 1444-1446.

Eaton R.A. and Hale M.D. (1993). *Wood decay pests and prevention*, London: Chapman and Hall.

Edwards J., Chamberlain D., Brosnan G., West D., Stanley M.S., Clipson N.J.W. and Hooley P. (1998). A comparative physiological and morphological study of *Dendryphiella salina* and *D. arenaria* in relation to adaptation to life in the sea. *Mycological Research* 102: 1198-1202.

Faulkner, D.J. (2000). Marine pharmacology, Antonie van Leeuwenhoek International Journal of General & Molecular Microbiology 77: 135-145.

Faulkner, D.J. (2001). Marine natural products. *Natural Product Reports* 18: 1-49.

Fell J.W. and Master I.M. 1980. in Contributions to the 3rd International Marine Mycology Symposium, Morehead City, N.C., USA, Sept. 9-15, 1979.

Fukuyama K., Tomitake T., Katsube Y., Hamasaki T., Hatsuda Y., Tanaka N., Ashida T. and Kakudo M. (1975). The crystal and molecular structure of sterigmatin, a metabolite of *Aspergillus versicolor. Bulletin of the Chemical Society of Japan (Tokyo)* 48: 1639-1640.

Geiser D.M., Taylor J.W., Ritchie K.B. and Smith G.W. (1998). Cause of sea fan death in the west indies. *Nature* 394: 137-138.

Gonda K.E., Jendrossek D. and Molitoris H.P. (2000). Fungal degradation of the thermoplastic polymer poly-beta-hydroxybutyric acid (PHB) under simulated deep sea pressure. *Hydrobiologia* 426: 173-183.

Hodson R.E. and Moran M.A. (1989). Microbial transformation of mangrove detritus: fate of soluble, structural lignocullulosic components. *In Recent Advances in Microbial Ecology,* (Hattori, T., Maruyama, Y., Morita, G.Y., and Uchida, a., Eds.) Japan Scientific Press.

Holler U., Wright A.D., Matthee G.F., Konig G.M., Draeger S., Aust H.J. and Schulz B. (2000). Fungi from marine sponges: diversity, biological activity and secondary metabolites. *Mycological Research* 104: 1354-1365.

Holt D.M., and Jones E.B.G. (1983). Bacterial degradation of lignified wood cell walls in anaerobic aquatic habitats. *Applied & Environmental Microbiology* 46: 722-727.

Hyde K.D., Jones E.B.G., Leano E., Pointing S.B., Poonyth A.D. and Vrijmoed L.L.P. (1998). Role of fungi in marine ecosystems. *Biodiversity & Conservation* 7: 1147-1161.

Hyde K.D., and Lee S.Y. (1995). Ecology of mangrove fungi and their role in nutrient cycling - what gaps occur in our knowledge. *Hydrobiologia* 295: 107-118.

Hyde Kevin D. and Pointing Stephen B. (2000). *Fungal diversity research series. Marine mycology: A practical approach*, Fungal Diversity Press, Pokfulam Road, Hong Kong, Hong Kong.

Iwamoto C., Minoura K., Hagishita S., Nomoto K. and Numata A. (1998). Penostatins f-i, novel cytotoxic metabolites from a *Penicillium* species separated from an enteromorpha marine alga. *Journal of the Chemical Society Perkin Transactions 1.*: 449-456.

Jadulco R., Brauers G., Edrada R.A., Ebel R., Wray V. Sudarsono and Proksch P. (2002). New metabolites from sponge-derived fungi *Curvularia lunata* and *Cladosporium herbarum. Journal of Natural Products* 65: 730-733.

Jensen P.R., and Fenical. (2000). Marine microorganisms and drug discovery; current status and future potential. In Drugs from the Sea, (N, F., Ed.) S. Karger AG, Basel pp. 6-29.

Jensen, P.R., and Fenical W. (2002). Secondary metabolites from marine fungi. In Fungi In Marine Environemnts, (Hyde Kevin, D., Ed.) Fungal Diversity Press, Hong Kong pp. 293-315.

Jones, E.B.G. (1993). Tropical marine fungi. In Aspects of Tropical Mycology, (S., I., Frankland, J.C., Watling, R., and Whalley, A.J.S., Eds.) Cambridge University Press, Cambridge pp. 73-89.

Jones E.B.G., and Alias S.A. (1997). Biodiversity of mangrove fungi. In Biodiversity of Tropical Microfungi, (Hyde, K.D., Ed.) Hong Kong University Press, Hong Kong pp. 71-92.

Kohlmeyer, J. (1969a). Ecological notes on fungi in mangrove-D forests. *Transactions of the British Mycological Society* 53: 237-250.

Kohlmeyer, J. (1969b). Marine fungi of Hawaii including the new genus *Helicascus*. *Canadian Journal of Botany* 47: 1469-1487.

Kohlmeyer J. and Kohlmeyer E. (1979). Marine mycology the higher fungi. In, Kohlmeyer, J. And E. Kohlmeyer. Marine Mycology: The Higher Fungi. Xiv+690p. Academic Press: New York, N.Y., USA: London, England. Illus. . 1979. Xiv+690p.

Kohlmeyer J. and Volkmann-Kohlmeyer B. (2002). Fungi on Juncus and Spartina: New marine species of *Anthostomella*, with a list of marine fungi known on Spartina. *Mycological Research* 106: 365-374.

Kohlmeyer J. and Volkmann-Kohlmeyer B. (2003). Marine ascomycetes from algae and animal hosts. *Botanica Marina* 46: 285-306.

Kohlmeyer J. and Volkmann-Kohlmeyer B. (2003). Mycological research news. *Mycological Research* 107: 385-387.

Kubanek J., Jensen P.R., Keifer P.A., Sullards M.C., Collins D.O. and Fenical W. (2003). Seaweed resistance to microbial attack: A targeted chemical defense against marine fungi. *Proceedings of the National Academy of Sciences of the United States of America* 100: 6916-6921.

Li X., Kondo R. and Sakai K. (2002). Biodegradation of sugarcane bagasse with marine fungus *Phlebia* sp MG-60. *Journal of Wood Science* 48: 159-162.

Li X. F., Jeong J.H., Lee K. T., Rho J.R., Choi H.D., Kang J.S. and Son B.W. (2003). Gamma-Pyrone derivatives, kojic acid methyl ethers from a marine-derived fungus *Alternaria* sp. *Archives of Pharmacal Research* 26: 532-534.

Lin W. H., Brauers G., Ebel R., Wray V., Berg A., Sudarsono, and Proksch P. (2003). Novel chromone derivatives from the fungus *Aspergillus versicolor* isolated from the marine sponge *Xestospongia exigua. Journal of Natural Products* 66: 57-61.

Lin Y.C., Li H.J., Jiang G.C., Zhou S.N., Vrijmoed L.L.P. and Jones E.B.G. (2002). A novel gamma-lactone, eutypoid-A and other metabolites from marine fungus *Eutypa* sp.(#424) from the South China Sea. *Indian Journal of Chemistry Section B-Organic Chemistry Including Medicinal Chemistry* 41: 1542-1544.

Liu Z.M., Jensen P.R. and Fenical W. (2003). A cyclic carbonate and related polyketides from a marine-derived fungus of the genus *Phoma*, *Phytochemistry* 64: 571-574.

Lorenz R. and Molitoris H.P.(1997). Cultivation of fungi under simulated deep sea conditions. *Mycological Research* 101: 1355-1365.

Mai C., Kues U. and Militz H.(2004). Biotechnology in the wood industry. *Applied Microbiology & Biotechnology* 63: 477-494.

Moriguchi M., Sakai K., Tateyama R., Furuta Y. and Wakayama M. (1994). Isolation and characterization of salt-tolerant glutaminases from marine *micrococcus luteus* k-3. *Journal of Fermentation & Bioengineering* 77: 621-625.

Mouzouras R. (1989). Soft rot decay of wood by marine fungi. *Journal of the Institute of Wood Science* 11: 193-201.

Nagai K., Kamigiri K., Matsumoto H., Kawano Y., Yamaoka M., Shimoi H., Watanabe M. and Suzuki K. (2002). YM-202204, a new antifungal antibiotic produced by marine fungus *Phoma* sp. *Journal of Antibiotics* 55: 1036-1041.

Noga E.J. (1990). A synopsis of mycotic diseases of marine fishes and invertebrates. In Pathology in marine science, (Perkins, F.O., and Cheng, T.C., Eds.) Academic press, London, UK pp. 143-160.

Numata A., Takahashi C., Ito Y., Minoura K., Yamada T., Matsuda C. and Nomoto K. (1996). Penochalasins, a novel class of cytotoxic cytochalasans from a penicillium species separated from a marine alga - structure determination and solution conformation. *Journal of the Chemical Society Perkin Transactions* 1: 239-245.

Numata A., Takahashi C., Ito Y., Takada T., Kawai K., Usami Y., Matsumura E., Imachi M., Ito T. and Hasegawa T. (1993). Communesins, cytotoxic metabolites of a fungus isolated from a marine alga. *Tetrahedron Letters* 34: 2355-2358.

Osterhage C., Konig G.M., Holler U. and Wright A.D. (2002). Rare sesquiterpenes from the algicolous fungus *Drechslera dematioidea. Journal of Natural Products* 65: 306-313.

Osterhage C., Schwibbe M., Konig G.M. and Wright A.D. (2000). Differences between marine and terrestrial *Phoma* species as determined by HPLC-DAD and HPLC-MS. *Phytochemical Analysis* 11: 288-294.

Payman M.A., and Mahnaz M.A. (1998). Decolourization of textile effluent by *Aspergillus niger* (marine and terrestrial). *Fresenius Environmental Bulletin* 7: 1-7.

Pietra F. (1997). Secondary metabolites from marine microorganisms - bacteria, protozoa, algae and fungi - achievements and prospects. *Natural Product Reports* 14: 453-464.

Pivkin M.V. (2000). Filamentous fungi associated with holothurians from the Sea of Japan, off the Primorye coast of Russia. *Biological Bulletin* 198: 101-109.

Pointing S.B., Vrijmoed L.L.P. and Jones E.B.G. (1998). A qualitative assessment of lignocellulose degrading enzyme activity in marine fungi. *Botanica Marina* 41: 293-298.

Prasannarai K., and Sridhar K.R. (2001). Diversity and abundance of higher marine fungi on woody substrates along the west coast of India. *Current Science* 81: 304-311.

Prasannarai K., and Sridhar K.R. (2003). Fungal assemblage and diversity on periodically sampled intertidal woody litter. *Indian Journal of Marine Sciences* 32: 329-333.

Poksch P., Ebel R., Edrada R.A., Wray V. and Steube K. (2003). Bioactive natural products from marine invertebrates and associated fungi. *Sponges*: 117-142.

Raghukumar C. (2000). Fungi from marine habitats: an application in bioremediation, *Mycological Research* 104: 1222-1226.

Raghukumar C. (2002). Bioremediation of coloured pollutants by terrestrial versus facultative marine fungi. In Fungi In Marine Environemnts, (Hyde K.D., Ed.) Fungal Diversity Press, Hong Kong pp. 317-344.

Raghukumar C., Chandramohan D., Michel F. C. and Reddy C.A. (1996). Degradation of lignin and decolorization of paper mill bleach plant effluent (bpe) by marine fungi. *Biotechnology Letters* 18: 105-106.

Raghukumar C., D'Souza T. M., Thorn R.G. and Reddy C.A. (1999). Lignin-modifying enzymes of *Flavodon flavus*, a basidiomycete isolated from a coastal marine environment. *Applied & Environmental Microbiology* 65: 2103-2111.

Raghukumar C. and Lande V. (1988). Shell disease of rock oyster *Crassostrea cucullata*. Diseases of Aquatic Organisms 4: 77-81.

Raghukumar C., and Raghukumar S. (1991). Fungal invasion of massive corals. *Marine Ecology* 12: 251-260.

Raghukumar C., Raghukumar S., Chinnaraj A., Chandramohan D., Dsouza T. M. and Reddy C.A. (1994). Laccase and other lignocellulose modifying enzymes of marine fungi isolated from the coast of india. *Botanica Marina* 37: 515-523.

Raghukumar C. and Rivonkar G. (2001). Decolorization of molasses spent wash by the white-rot fungus *Flavodon flavus*, isolated from a marine habitat. *Applied Microbiology & Biotechnology* 55: 510-514.

Raghukumar C., Sharma S. and Lande V. (1991). Distribution and biomass estimation of shell-boring algae in the intertidal at Goa India. *Phycologia* 30: 303-308.

Rand T.G. (1996). Fungal diseases of fish and shellfish. In The Mycota: A treatise on the biology of fungi with emphasis on system for fundamental and applied research, (Esser, K., and Lemke, P.A., Eds.) Springer-Verlag, Berlin, Germany pp 298-313.

Rberts J., Holcenberg J.S. and Dolowy W.C. (1970). Antineoplastic acitivity of highly purified bacterial glutaminase, *Nature* 227: 1136-1137.

Renner M.K., Jensen P.R. and Fenical W. (1998). Neomangicols - structures and absolute stereochemistries of unprecedented halogenated sesterterpenes from a marine fungus of the genus *Fusarium*. *Journal of Organic Chemistry* 63: 8346-8354.

Renner M.K., Jensen P.R. and Fenical W. (2000). Mangicols: Structures and biosynthesis of a new class of sesterterpene polyols from a marine fungus of the genus *Fusarium*. *Journal of Organic Chemistry* 65: 4843-4852.

Rohrmann S. and. Molitoris H. P. (1992). Screening for wood-degrading enzymes in marine fungi, *Canadian Journal of Botany* 70: 2116-2123.

Sabu A., Keerthi T.R., Kumar S.R. and Chandrasekaran M. (2000). L-Glutaminase production by marine *Beauveria* sp under solid state fermentation. *Process Biochemistry* 35: 705-710.

Sabu A., Kumar S.R. and Chandrasekaran M. (2002). Continuous production of extracellular L-glutaminase by Ca-alginate-immobilized marine *Beauveria bassiana* BTMF S-10 in packed-bed reactor. *Applied Biochemistry & Biotechnology* 102: 71-79.

Schmit J.P. and Shearer C.A. (2003). A checklist of mangrove-associated fungi, their geographical distribution and known host plants. *Mycotaxon* 85: 423-478.

Sindermann C.J. (1970). *Principal diseases of marine fish and shellfish*. Academic Press, New York.

Singh A.P., Nilsson T. and Daniel G. (1990). Bacterial attack of *Pinus sylvestris* wood under near anaerobic conditions. *Journal of Institute of Wood Science* 11: 237-49.

Smith G.W., Ives L.D., Nagelkerken I.A. and Ritchie K.B. (1996). Caribbean sea-fan mortalities. *Nature* 383: 487.

Suresh P.V. and Chandrasekaran M. (1998). Utilization of prawn waste for chitinase production by the marine fungus *Beauveria bassiana* by solid state fermentation. *World Journal of Microbiology & Biotechnology* 14: 655-660.

Suresh P.V. and Chandrasekaran M. (1999). Impact of process parameters on chitinase production by an alkalophilic marine *Beauveria bassiana* in solid state fermentation. *Process Biochemistry* 34: 257-267.

Sutherland J.B., Crawford D.L. and Speedie M.K. (1982). Decomposition of carbon-14 labeled maple and spruce Lignin by marine fungi. *Mycologia* 74: 511-513.

Takahashi C., Numata A., Yamada T., Minoura K., Enomoto S., Konishi K., Nakai M., Matsuda C. and Nomoto K. (1996). Penostatins, novel cytotoxic metabolites from a *Penicillium* species separated from a green alga. *Tetrahedron Letters* 37: 655-658.

Tan L.T., Cheng X.C., Jensen P.R. and Fenical W. (2003). Scytalidamides A and B, new cytotoxic cyclic heptapeptides from a marine fungus of the genus *Scytalidium*. *Journal of Organic Chemistry* 68: 8767-8773.

Tsukamoto S., Miura S., Yamashita Y. and Ohta T. (2004). Aspermytin A: a new neurotrophic polyketide isolated from a marine-derived fungus of the genus *Aspergillus*. *Bioorganic & Medicinal Chemistry Letters* 14: 417-420.

Wang C.Y., Wang B.G., Brauers G., Guan H.S., Proksch P. and Ebel R. (2002). Microsphaerones A and B, two novel gamma-pyrone derivatives from the sponge-derived fungus *Microsphaeropsis* sp. *Journal of Natural Products* 65: 772-775.

14 Aeromycological Aspects of Mycotechnology

Donát Magyar

Plant Protection Institute of the Hungarian Academy of Sciences, Department of Plant Pathology, H-1022 Budapest, Herman Ottó u. 15, Hungary

INTRODUCTION

The study of airborne microbes led Pasteur to disprove the theory of spontaneous generation. He had demonstrated that microbes are normally abundant in the air, and only a minute quantity of air needs to gain access for a microbial population to develop and putrefaction to take place. Pasteur's simple statement opened the way for the modern microbiology. Fermentation and food preservation were the first topics where aeromicrobiology was applied. After recognizing the necessity of such studies, the surveys of organic and inorganic dust, including moulds was housed in Paris, in the palace of Observatoire Montsouris. Pierre Miquel developed here some methods for capturing and culturing airborne spores. Recognition of epidemiological aspects of some devastating human diseases, cholera and other fevers in 1870s yielded extensive hunt after infectious agents. Interest in the fungal airspora waned when it became clear that the particles (mainly fungal spores) found in the air samplers not correlated with the pathogens of epidemic diseases. After this first golden age, the appreciations for aerobiology were kindled two more times. The well-developed methods of Pasteur and Miquel sunk into oblivion, in the unproductive spaces, when only a few mycologists worked using simple methods like settling plate.

The observation of epidemics caused by plant pathogenic fungi in the 1930s inspired the second renaissance of aeromycology. Attention of many scientists turned to study of the dispersal of airborne spores. The volumetric trap constructed by Hirst was in general use, results were quantitative and reproducible. The word 'aerobiology' was first coined by Fred C. Meier, to describe a research project on microbial life in the upper air. The discipline was established by a symposium of the American Association for the Advancement of Science (Moulton, 1942). Biometeorology of fungal spores was extensively studied by Gregory, Meredith, Sreeramulu, Kramer and Pady. These works described in detail the changes in the counts of spores in connection with the weather, but they were hardly comparable lacking statistical validation. Mycologists, like Buller, Ingold and Webster underlined the role of air as the main agent of dispersal in fungi, and the necessity of its research.

Then the third golden age of aeromycology arrived in the end of the last century, owing to the increased incidence of the allergy. Aerobiological networks developed worldwide, using standardized methods to determine the concentration of allergenic particles, mainly pollen grains and a few fungal spore types – frequently *Alternaria*, *Cladosporium* and *Epicoccum,* which were also published in the 'daily pollen bulletins'. Huge databases accumulated in the decade-long operation of the networks, and statistical data processing was needed to create predictive models.

As species of the Kingdom Fungi lack mechanisms of active motion, the transport via air-flow is their main dispersal route and fungi represent dominant element of aerobiota. Airborne fungal propagules are found in great variety of types, sizes and number, which vary with time of day, season, weather and geographical location. Many fungal propagules lose their viability by exposure to different environmental factors (such as UV radiation, dehydration and high temperature), thus among the particles of fungal origin a distinction is generally made between viable and non-viable. Of those that survive, some may be harmful causing diseases of plants, humans and animals, or indiscriminately deteriorate artistic products, foodstuffs or waste matters. Aeromycology as a discipline has been established for the reason to understand the distribution, ecology, biometeorology and deposition patterns of airborne fungal propagules. It relates to interactions of airborne fungi with plants and with elements of phytosphere as well as to their role in human life. Most of our knowledge comes from plant pathology or from epidemiological studies of infectious diseases and allergy in man. Volumetric air sampling allows not only to determine the concentration of biological particles (spores) in a given period of time but also to establish the hours of the day in which they are more liable to cause infection. This information may be used in estimating the incidence of disease symptoms, the duration and seriousness of the disease.

Development in research methods offers new possibilities for the aeromycologist. Molecular biology and computerized morphometry are new challenges for the next decades. Quality control of fungal bioaerosols was applied successfully in public health, plant protection and conservation of products, but it stays behind in many other fields where its use can be expedient. Surveying technological processes and its efficiency in harmony with protection of natural systems should be the forthcoming assignment we have to accomplish.

The aim of the present chapter is to discuss the role of aeromycology in mycotechnology.

METHODS

No one method of aerobiology is ideal for all needs in mycotechnology. The aim of the survey should be clarified before a method is selected. Different techniques are relevant to viable and total spore count, number of single spores or spore clumps as dispersal units, sizing of bioaerosol components, concentration dynamics regarding time. Before deciding sampling method, we should take the time to answer some questions. Do the spores resist desiccation, long-range dispersal or special environments or treatments? Are spores microscopically distinct or culture is necessary for identification? Are expected concentrations to be small or large? Different methods are needed:

- to investigate the whole airborne mycobiota or the presence of a particular fungus in the air
- to detect the presence of spores in clean areas
- to monitoring the spread of a manufactured bioaerosol
- to relate spore concentration to the complaints
- to determine human or plant health hazards
- to elaborate selective isolation techniques.

AIR SAMPLING

Several hundred instruments have been designed for sampling particles in the atmosphere. Air samples are generally collected by sedimentation, impaction onto adhesive surfaces, and filtration through a membrane filter or impingement into a liquid medium. Modern airborne sampling employs a measured volume of air with the aid of an electric or battery-powered pump. The air is then directed through a chamber, or series of chambers, guiding the spores on a specific trajectory to a solid agar disc or adhesive medium (impactor samplers), a liquid buffer (impinger samplers), or a filter (filtration samplers). Isokinetic sampling is ideal: in this case wind speed and suction speed are equal and correct numbers of large and small particles are sampled by the volumetric trap (Fig 14.1). If air speed is too high in the intake, and it exceeds the wind velocity, large particles should escape due to their inertia. When aspiration rate is too low, large particles are blown into the intake tube by wind and there are too many large particles in the sample (Booth, 1971). Apparatus for sampling airborne particles should be:

- equipped with a highly retentive sampling surface upon which particles are impacted.
- free from sharp bends, because large particles tend to be impacted on their walls. Spores over 10 μm in diameter must be sampled directly and cannot be ducted around corners on the way to the sampler without heavy wall-losses. Intake edge nozzle should be feathered.
- airflow dimensions and distance from sampling surface are also particularly important variables.

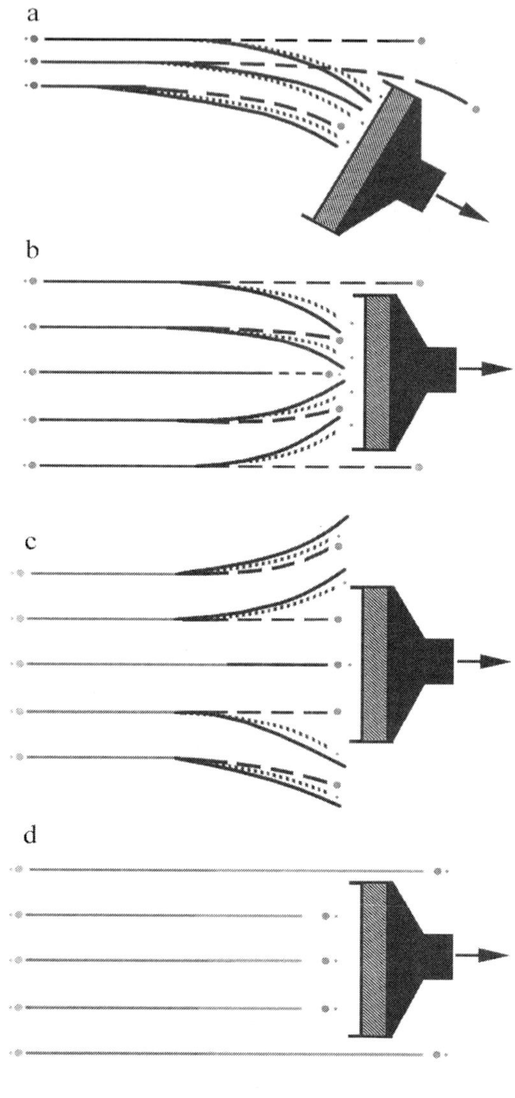

Fig 14.1. Behaviour of large and small spores at the intake of a suction trap. a: Intake is misaligned with respect to wind direction. b: Intake is wind-oriented, but intake speed is well above that of the ambient flow rate. c: Intake is wind oriented, but intake speed is well below ambient flow rate. d: Intake is wind oriented and speeds at intake and in ambient are equivalent (isokinetic).

Many variants of spore trap were developed for sampling viable or non-viable spores (Fig. 14.2). The culture plate methods are based on isolation, cultivation and subsequent observation of morphological traits and cultural

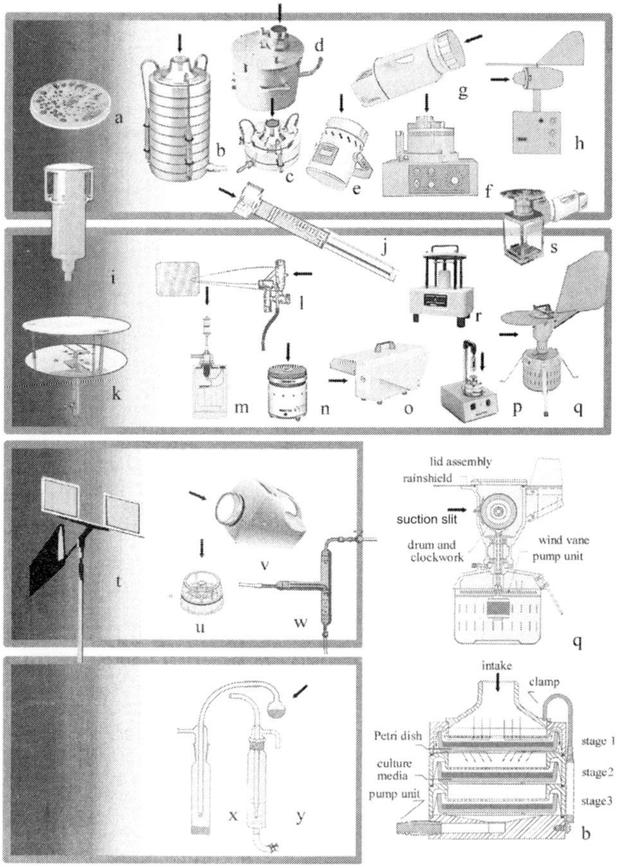

Fig 14.2. Spore traps. a: Gravity Petri dish, b: 9-stage Andersen sampler, c: 1-stage Andersen sampler, d: BIAP, e: MAS, f: Brunswick STA sampler, g: SAS, h: FOA, i: Tauber trap j: RCS, k: Durham sampler, l: Cascade impactor, m: PARTRAP, n: Burkard Personal Volumetric Air Sampler, o: portabe Hirst-type trap, p: Cyclone sampler, q: 7-day recording Hirst-type trap, r: Rotorod, s: suction sampler equipped with settling chamber and leaf portions, t: Cour trap, u: filter cassettes, v: Sartorius air sampler, w: Pasteur's gun-cotton filter, x: Porton-impinger with pre-impinger, y: AGI. Non volumetric devices (a, i, k, t) are in left side. a-h: Samplers with culture media. i, j, s-Samplers with alternative culture media. k, l, n, o, p: Instruments collecting particles on microscope slides. m ,r, q: Instruments collecting particles on microscope tapes. t-w: filters, x, y: liquid impingers. Arrows show the air intake.

characteristics of colonies, such as their colour and texture, conidiophore structure and conidial size and shape. Malt extract agar (MEA) and rose bengal agar (RBA) have been recommended for the collection and enumeration of fungi. Application of selective media is recommended to detect the presence of a particular fungus. To inhibit bacterial contamination, antibiotics, such as streptomycin, may be added. Incubation conditions for fungi are 25°C or room temperature with natural light. Impactors with culture plate may have range from a single slit to more than 400 holes. Cultural techniques are time-consuming and laborious, days or even weeks may be required to identify the fungi. Additionally, not all fungal species can be cultivated *in vitro*; cultivable organisms comprise a small fraction of the total of potentially pathogenic units in air. Spores, that are slow-growing should have limited detection threshold on a non-selective media. Aggregated spores may be inaccurately counted as a single colony. One microorganism might inhibit the growth of others at the same location by overrun or by metabolites. Morphological characteristics vary due to differences in growth and culture conditions; however lack of differential colony morphology in non-sporulating colonies often occurs. Culturing analyses may not reflect the true fungal community in a sample (Table 14.1). To trap obligate plant parasitic fungi suction samplers may equipped with leaf portions (Schwarzbach, 1979) maintained in dilute solutions of benzimidiazole in water (Wolfe, 1967). Many species of non-cultivable fungi should be detected by Hirst-type spore traps.

Table 14.1. Important airborne fungal taxa in mycotechnology and techniques proposed for sampling. C: culture plate techniques, H: Hirst-type air samples on microscope slides/tapes for microscopic examination, M: Molecular methods

Genus	*Method*	*Genus*	*Method*
Agaricus	H	*Mycosphaerella*	H
Albugo	H	*Nectria*	H
Alternaria	H	*Nigrospora*	H
Armillaria	M	*Oidiopsis*	H
Ascochyta	H	*Oidium*	H
Aspergillus	M,C	*Ophiobolus*	H
Botryosphaeria	M	*Paecilomyces*	M,C
Botrytis	H,C	*Penicillium*	M,C
Bovista	H	*Peronospora*	H,M
Bremia	H	*Pestalotia*	H
Candida	M,C	*Pestalotiopsis*	H
Cercospora	H	*Phialophora*	C
Cercosporella	H	*Phoma*	M
Cladosporium	H	*Phomopsis*	M
Claviceps	H	*Phragmidium*	H
Colletotrichum	H	*Phyllosticta*	M
Coprinus	H	*Phytophthora*	H,M
Coryceps	H,M	*Plasmopara*	H,M
Cucurbitaria	H	*Pleospora*	H
Cunninghamella	M,C	*Pleurotus*	H
Curvularia	H	*Polyporus*	H
Cylindrocarpon	H	*Polystigmina*	M
Cylindrosporium	M	*Polythrincium*	H
Cytospora	M	*Pseudopeziza*	M
Diaporthe	H	*Puccinia*	H
Diplocarpon	M	*Pyrenophora*	H
Diplodia	H	*Pyricularia*	H,C
Drechslera	H	*Pythium*	H,C
Endothia	H	*Ramularia*	H
Exophiala	M	*Rhabdocline*	M
Epicoccum	H	*Rhizopus*	H
Fusarium	M, C	*Rhytisma*	M
Fusicladium	H	*Rosellinia*	H
Gibberella	H	*Saccharomyces*	M,C
Gloeosporium	H	*Sclerotinia*	H,M
Gnomonia	H	*Septoria*	M
Guignardia	H	*Sphaeropsis*	H
Helminthosporium	H	*Stemphylium*	H
Heterosporium	H	*Stereum*	M
Humicola	H, C	*Stigmina*	H
Kabatiella	H	*Taphrina*	M
Kabatina	M	*Thielaviopsis*	H,C
Kuehneola	H	*Trichoderma*	C
Leptosphaeria	H	Uredinales	H
Lophodermium	H	*Uromyces*	H
Macrophomina	M	Ustilaginales	H
Metschnikowia	M	*Valsa*	H
Monilia	H, C	*Venturia*	H
Mucor	H, M	*Verticillium*	C, M

Several works have been achieved to compare advantages and disadvantages of different samplers. For further details see the studies of Whyte, 1981, Aylor, 1993, Mehta *et al.*, 1996, McCartney *et al.*, 1997, Pahl *et al.*, 1997, Tomas *et al.*, 1997, Cundill, 1998, Frenz, 1999, 2000, Aizenberg, 2000, Levetin *et al.*, 2000, Portnoy *et al.*, 2000, Riediker, 2000, Bellin and Schillinger, 2001, Schultz, 2001, Takahashi *et al.*, 2001, Tavora *et al.*, 2003, Wüst *et al.*, 2003.

SEDIMENTATION

Sedimentation methods are simple, cheap, but do not measure spore concentration accurately, because the volume of air sampled remain unknown. Sedimentation is semi-quantitative due to Stoke's law; the result is influenced by turbulence. Although, for air sampling sedimentation methods were used almost exclusively for many years, and still used extensively, they can no longer recommended now when quantitative volumetric instruments are available. Several simple traps were designed for qualitative work (conical funnels, settling box, aeroconiscopes, vertical microscope slides, vertical cylinder). The most frequently used devices are listed below:

Gravity Petri dish is an early method to get a picture of the airspora composition exposing an open Petri dish with 9 cm in diameter for 20 min with a mould medium. Concentration as number of colony forming units only per agar surface (CFU/m^2) can be expressed.

Gravity slide is an ordinary microscope slide exposed for a period of 24 hour. The adhesive coating of the slide is usually glycerine jelly or petroleum jelly. It has serious defects as quantitative method (selects large particles).

Durham sampler (Durham, 1946) contains a gravity slide with a roof. The trap is open to the wind, but protected from rain.

Cour trap is used for outdoor studies mainly in Mediterranean areas. The collector holds 400 cm^2 sterile gauze filters impregnated with silicone oil and fitted to a vane. The apparatus includes a totalizing anemometer, indicating the wind flow through the filter during exposure period.

Tauber trap (Tauber, 1974) consists of a cylindrical container covered with an aerodynamically shaped collar through which particles enter. Particles are caught by a layer of glycerol (or Petri dishes containing a nutrient medium) in the bottom of the container. The trap can be left unattended for long periods.

Bi- and multidirectional spore traps were designed by Rajasab and Ramalingam (1989) to study the splash dispersal of conidia. About 2 ml of 50% formaldehyde solution may be added to the collection bottle to inhibit conidial germination.

IMPACTION SAMPLERS

Principle of impaction is used to separate a spore from an air stream based on the inertia of the spore. Particles are collected onto adhesive tapes/microscope slides, or surface of a solid medium (agar). Aerodynamic diameter of a particle is an important feature, defined as the diameter of a hypothetical sphere of unit density (1g/cm^{-3}) that has the same settling velocity as the particle. Spores larger than a particular aerodynamic size will be impacted onto the trapping surface while smaller particles proceed through the sampler (Hinds, 1982).

Hirst-type trap (Hirst, 1952, Burkard trap, Burkard Manufacturing Co. Ltd., Rickmansworth,

England; VPPS-2000, Lanzoni Co. Ltd., Bologna, Italy) is optimized for accurate quantitative work and morphological identification of pollen grains, fungal spores or other airborne microparticles. A plastic strip (MELINEX®) or microscope slide is thinly smeared with glycerine jelly as an adhesive to catch the particles from the instreaming air suctioned into the device by a vacuum pump (10 L of air are aspirated in every minute). If the plastic strip is used, it is applied round a drum, and rotated by a clockwork, past the slit (14 mm long and 1.7 mm wide) where air stream enter. A 7-day-long continuous operation is possible, thus an hour corresponds to a distance of 2 mm on the strip. A wind vane enables the device always to face the prevailing wind direction. It is also possible to change suction slit, allowing the passage of smaller particles only. A roof protects the suction slit from the entrance of raindrops. Operating in perpetual mode this equipment provides a complete record of diurnal and seasonal changes in the airspora. The spores of obligate plant pathogens such as rust, smuts and powdery mildews are recorded often by this apparatus. Nonviable spores are also detected, which, although no longer infectious, still exhibit the allergenic and toxigenic properties. Identification is visual under the microscope. Two counting methods are in common use: one single longitudinal traverse near the middle of the strip, or twelve bihourly (4mm) transverse traverses. The advantages and disadvantages of these methods are described by Warner and Emberlin (1998) and Rogers and Muilenberg (2001). Modifications of this trap are: Partrap (Fiorina, 1998) and Burkard Personal Volumetric Air Sampler.

Rotorod and Rotobar (Perkins, 1957, Sampling Technologies Inc., Los Altos, USA) samplers consist of a motor equipped with a U-shaped holder at the end of a shaft. The collecting part of the device has different forms (acrylic rods, microscope slides, plastic flags). Collecting parts are smeared with silicone grease or glycerine jelly as adhesive. 120 *l* of air can sweep out per minute at 2400 rpm. The sampler collects particles in a size range of 1 to 100 μm. A suitable sampling time is between 5 to 30 minutes, because overloading may occur. Batteries with decreasing voltage can also modify results.

Cascade impactor (May, 1945) consist a series of collector stages and each succeeding stage collects smaller particles within a discrete range of size. The nozzle diameter gets smaller and smaller from one stage to another and hence the aerosol velocities get higher and higher. All particles not impacted on the upper stages are collected by the final one. Sampling surfaces are of various types such as filters, microscope slides, or Petri dishes with media. The sampling rate of 17.5 L/min May's design of the cascade impactor represents a historical milestone in air sampling techniques since many air samplers have been designed to size aerosols in a similar manner to the human respiratory tract (e.g., Marple-Miller Five Stage Cascade Impactor).

Andersen sampler (Andersen, 1958, Andersen Three Stage Viable Cascade Impactor, Lanzoni Co. Ltd., Bologna, Italy) is an extensively used device for viable sampling. Airflow provided by a vacuum pump is drawn through a succession of three (or more) stages, each perforated with 400 holes. The diameter of holes in each stages progressively decrease (1.20 mm, 0.70 mm, 0.35 mm), and the air speed reaches successively higher intake velocities from stage to stage. Airborne particles are impacted inertially according to their size (>7 μm, 7-3.3 μm, <3.3 μm) onto culture media in Petri dishes, placed directly under the stages. Successively smaller particles are impacted in lower stages of the sampler. Maximum collection efficiency takes place at constant air stream of 28.3 litres per minute. Sampling time will vary with the expected spore load of the air between 5 and 20 minutes. The particle distribution on the stages of the sampler reflects the particle deposition in the different regions of the respiratory tract. The probability that the next spore will impact an unoccupied hole decreases with sampling time and intensity. Six (or more) stage samplers are also used in bacteriology.

Sieving plate samplers (Burkard Portable air sampler for agar plates; SMA Microportable P100, Veltek Associates Inc. Phoenixville, USA; Merck Air Sampler MAS-100, Merck KGaA, Darmstadt, Germany; Surface Air System SAS-Super 100, International Pbi SpA Milan, Italy) are used, when sizing of airspora is not requested. The spores are collected onto culture medium in Petri dish through a stage with holes. The same process takes place in simplified Andersen sampler with a Petri dish and media in the lower stage only.

Slit-to-Agar (STA) samplers (Bourdillon *et al.,* 1941, C.F. Casella and Co. Ltd., London, UK; BIAP Sampler, Humlebæk, DK; Brunswick STA Biological Air Sampler LabTrader, Vista, USA). In the STA samplers air is sucked at high velocity (33.3L/min) through a narrow slit (0.5 mm) and the entrained particles are caught on the surface of agar medium in a Petri dish. During sampling Petri dish is rotated. Agar surface and slit distance is constant (1mm). The FOA slit-sampler equipped with wind vane, therefore is relevant for outdoor sampling as well.

Centrifugal samplers (Reuter Centrifugal Sampler– RCS, RCS-Plus, Biotest Diagnostics Co. Denville, USA). An air stream (with a flow rate of 40-100L/min) is generated by the rotor placed on the front of the instrument. Airborne particles are separated directly onto a special agar strip. This apparatus is used very extensively for routine microbiological sampling of indoor air.

Cyclones: In the Burkard designed cyclone sampler (Burkard Manufacturing Co. Ltd., Rickmansworth, England) spores are collected directly into 1.5 ml Eppendorf vials. Subsequently, spores can be transferred onto agar or analysed by immunological techniques. Continuous sampling can be carried out over long periods without overloading. The technique is not reliable for slime spores (Larson *et al.*, 2001). Cyclone dust collectors are used in industry as well. These devices retain large particles only, because of their low pressure drop.

Filtration samplers: Filtration technique is ideal for minute samplers measuring personal exposition (Aerocell filter cassette, Button Sampler, Nasal Sampler). The spore traps of the 19th century were mostly air filters such as Sieving filters, Pasteur's gun-cotton filter, Frankland's glass powder filter, Petri's sand tube filter, and some solvable filters, e.g., Buller and Lowe's sugar powder filter.

Filters are available in both fibrous and membranous forms with pore sizes of 0.01 to 10 μm (Costar Nuclepore, Cambridge, USA; Gelman Sciences Inc. Ann Arbor, USA; Millipore Corporation, Bedford, USA). Particles are intercepted by the strainers, when their diameter is larger than the pore size; the overall efficiency of membrane filters is approximately 100%. The filter may be weighted to determine collected mass. For gravimetric analysis, non-hygroscopic materials (glass fibres, silver, and polyvinyl chloride membranes) are the usual choices. For microscopy, cellulose ester or polycarbonate membranes are suitable. Filters may induce more particle bounce than adhesive surfaces. On membrane filters, cells would desiccate and become inappropriate for germination and cultivation.

Sartorius air sampler (Satrorius MD8 AirPort, Sartorius Ltd. Epsom Surrey, UK) uses gelatine membrane filter method. Because of its high flow rates, this device is ideal for validation and monitoring extremely low counts of airborne microbes for the food, beverage industries and hospitals. The inherent high moisture content of the gelatine membrane filters help to prevent any drying out of collected microorganisms during a sampling period. After sampling, the gelatine membrane filter is placed directly on an agar nutrient plate. The gelatine dissolves on the moist surface so that the microorganisms are in direct contact with the nutrients.

Liquid impingers: Impingement collection fluids from liquid impingers are plated directly or after serial dilution on agar, or the entire volume of fluid is pressed through a membrane filter. With this method extreme ranges of airborne concentration may be detected. During sampling, due to the violent agitation of the fluid, dispersion units or clusters of spores are broken up.

The inlet tube of the **All Glass Impinger-30** (AGI-30, Ace Glass Inc., Vineland, USA) is curved to simulate particle collection in the nasal passage (Cox, 1987). Respirable and non-respirable particles can be separately collected in the collection fluid, and in the inlet tube. Larger particles (over 15 µm) collected by the tube wall are washed with a known amount of fluid after sampling. Impingers such as AGI-30 or the Greenberg-Smith impinger use a simple salt solution (e.g., 0.3 mM phosphate-buffered dilution water). The glass device must be sterilized by autoclaving. Other impingers available in market are: Multistage liquid impinger, Rosebury-Henderson capillary impinger, Portion impinger, Multi-Slit Large volume air sampler, SKC BioSampler.

Electrostatic precipitation: Dust–laden air is conducted through the instrument, where particles are charged near the entrance. Petri dishes of culture medium are placed on metal electrodes which are oppositely charged to 7000 volts from a half-wave rectifier, thus particles are attracted to the sampling surface. The principle is widely used in industry to remove dust from air.

In the **Litton** sampler (LVS/10K sampler, Litton Systems Inc., Minneapolis, USA) aspired particles are negatively charged by the intake corona, and directed on to a positively charged collector plate. A continuous flow of liquid removes the particles from the plate into a container. This device is unique in its large volume sampling rate (11 m^{-3}/min). The sampler is suitable for sampling air for spores present in extremely low number, but also may be used to remove fungi from air.

Thermal precipitation: A heated (100°C) wire placed in a slow air stream produces a dust-free space around itself. It should be little used for aeromycological studies.

Natural spore traps: Some of the artificial spore traps mentioned above are developed to measure the exposition of the human respiratory tract or plant canopy. However, none of these models reach the complexity of the nature, therefore natural traps itself should be studied in particular cases. Phyllosphere, water, animal lung and hair, all may act as natural spore traps and play an important role by filtering air or binding airborne spores, pollen grains, and other particles. They are generally characterized by large and self-clearing surface which captures airborne microbes.

Exposed animals: A group of animals are exposed to detect the presence of certain pathogenic fungi in the air (Booth, 1971). Symptoms are observed on living animals, or the lungs of the animals are subsequently removed, homogenized in saline, and portions cultured out.

Exposed plants: Pathogen free host plants should be used to catch and "culture" viable spores of obligate plant pathogenic fungi. Dense vegetation filters spores efficiently both by impaction and slowing airstreams to allow sedimentation. Dust load of air in urban areas without trees are higher (65-90%) than in parks (14.4-22%). A single tree can filter 66-80% of the dust in air stream passing through its canopy, this amount of dust may reach a weight of some hundreds kilograms (Kovács, 1985). Tree leaves after washing down by rain can recover their original filtering capacity. Air has a concentration C of particles per m^{-3}. The air flows from the leaf samples with a certain efficiency (E). The number of spores floating through an imaginary unit area frame oriented at right angles to the leaf surface is expressed by area dose (AD). This AD is given by the multiplication of C with the mean wind speed. The number of spores deposited per unit area of leaf surface is the "trap dose" (TD). The leaf traps with a certain efficiency: E=TD/AD. Hirst *et al.,* (1971) studied the deposition of fluorescent powder dispersed as an aerosol on different plant leaves (*Ilex, Poa,*

Raphanus, Solanum) in a small wind tunnel. Deposition was greatest on the small obstacles such as stems, petioles and longer spines of the holly (*Ilex*) leaves. There was much less deposition on the larger laminae of the leaves, particularly where these inclined more nearly parallel to wind direction. Experiments with fungal spores showed similar results. Impaction is most efficient for large spores blown fast toward small objects. Small-spored fungi (4-5 μm in diameter) fail impact on objects only 1 mm diameter by wind speed of 2 m/sec (the usual maximum in vegetation). In other studies, filtering efficiency was the higher in *Rosa rugosa* (other examined plants are listed in descending order of their efficiency: *Acer campestre*, *Carpinus betulus, Lonicera* sp., *Ligustrum vulgare*) (Kovács, 1985).

Carter (1965) found that spore deposition of *Eutypa armeniaceae* is high on thin stems of apricot. When sedimentation under gravity occurred, spores were found on the adaxial surface of leaves, when airflow was generated, spores impacted on both abaxial and adaxial surfaces of the leaf surfaces. Wet and sticky surfaces (covered by honeydew, rainwater, plant sap, etc.), are good natural spore traps, and are retentive for spores (Chamberlain and Chadwick, 1972). Wax structures and plant hairs on leaf surface projecting from leaf surface may add to the trapping efficiency and assist retention of some spore types (Forster, 1977) but the hydrophobic feature of this leaves eliminate the efficiency of wet deposition of spores ("Lotus-effect"). Microscopic examination of leaves shows that spores tend to occur in preferred patterns in relation to leaf anatomy. The most active zone of the trees, concerning the impaction and resurfacing of spores is the top of their canopy.

SPORE IDENTIFICATION

The identification of spores is important both in future and conventional aeromycology because hundreds of spore types are comparable within a single sample representing 1 m^{-3} of air. Accurate morphological methods will open new doors to aerobiological studies and practices. Over the identification of the spores, their morphological modifications (intact or broken arthrospore chains, desiccation of spores, clumping in ascospores, injuries on spore wall, etc.) can be studied as well. The development of computerized morphology may become a "competitor" technique against methods based on molecular biology.

MICROSCOPY

Bright-field or light microscopy is used for simple observation or sizing. For nearly invisible, hyaline spores phase-contrast microscopy is applied. Fluorescence microscopy uses ultraviolet or near-ultraviolet source of illumination that causes fluorescent compounds in a specimen to emit light. The importance of this method is to combine viable counts with total number of specially stained bioaerosols (Palmgren, 1986). A magnification of 1000× is normally used for identifying spores (100× oil immersion objective with 10× ocular lenses). When identifying the three-dimensional surface structure of spores using a light microscope, it is important to focus slowly through the individual layers. Different surface ornamentations are studied with the so-called LO-analysis: at high focus the raised elements appear lighter (L= lux; light) while the holes and depressions appear darker (O= Obscuritas; darkness). At low focus the holes and depressions appear lighter than the raised elements which now appear darker (Moore *et al.*, 1991). An eyepiece micrometer (reticule), calibrated with the stage micrometer, is needed for measuring spore sizes. A stage Vernier scale or England finder should be also useful to find a certain particle in the sample. Stains such as lactophenol cotton blue, periodic acid-Schiff stain, or potassium hydroxide (10% KOH) may be used, however, in recognizing fungal spores their original colour is particularly important. Air samples on microscope slides or strips are mounted with a medium that is prepared with polyvinyl alcohol and glycerine or lactic acid, and also a very small amount of phenol.

MORPHOLOGY

Fungal spores occur in a fantastic scale of shapes from the simple globose to the exceptionally complex. There has been a tendency in the past to try to describe spores using single-word terms. The habit of using these short terms to describe unique, complex, three-dimensional shapes must be abandoned. Instead, spores should be illustrated by standardized descriptive biological terms. A complete description may include some photographs with bar and sporograms (Baral, 1992; Dominguez de Toledo, 1994), spore size, colour, spore wall, apiculum, symmetry, polarity, number of septa, ornamentation, chemical reactions and shape, according to Saccardo's spore groups ("Saccardo's code" Fig. 14.3.). Identification of spores should proceed to the most specific

Shape		Septa		Colour				Description	Example
Name	Code	Name	Code	Name	Code	Name	Code		
				phaeo	P	hyalo	H		
		amerospores	0	Phaeosporae	e0P	hyalosporae	e0H	unicellular, globular or slightly elongated	
		didymospores	1	phaeodidymae	e1P	hyalodidymae	e1H	bicellular	
Ellipsoidal	e	phragmospores	2,3,...	phaeophragmiae	e7P	hyalophragmiae	e7H	multicellular, transverse septa number defined	
		phragmospores	≡	phaeophragmiae	e≡P	hyalophragmiae	e≡H	multicellular, transverse septa number undefined	
		dictyospores	#	hyalodictyae	e # P	hyalodictyae	e # H	muriform, transverse and longitudinal septa	
Filiform	f	scolecospores			fP		fH	unicellular or septate, elongated	
Branched	b	staurospores			sP		sH	unicellular or septate, branched	
Helical	h	helicospores			hP		hH	unicellular or septate, spiral or helicoid	

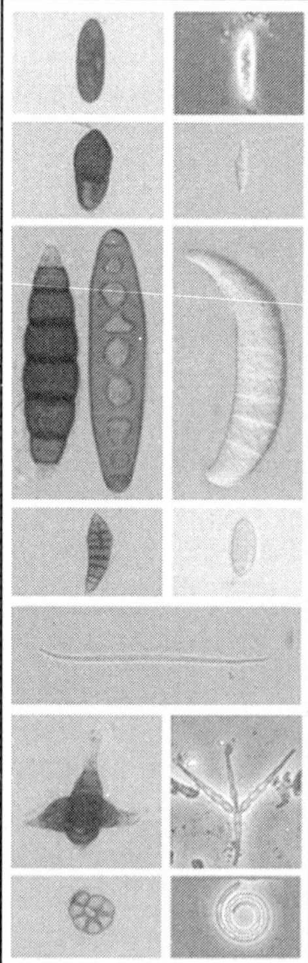

Fig. 14.3. Saccardo's spore groups.

classification. Free spores found in air samples can be identified morphologically on different levels, depending on their characteristics (Fig. 14.4) Rarely spores can be identified to species level. Some

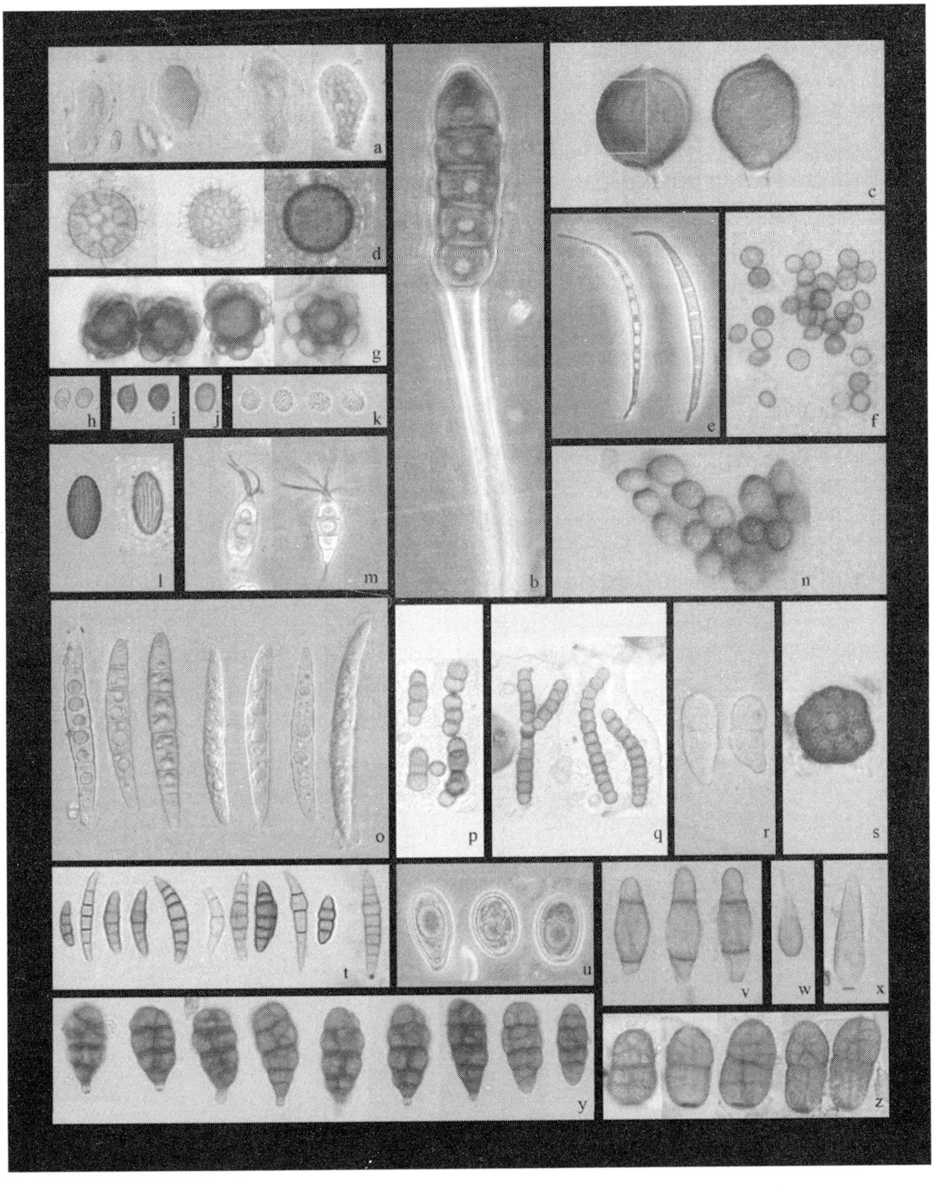

Fig. 14.4. Airborne fungal spores important in mycotechnology. a) *Cronartium/Melampsora* spp., b) *Phragmidium* sp., c) *Uromyces lilii* (Link) Kunze, d) *Tilletia* sp., e) *Fusarium equiseti* (Corda) Sacc., f) *Humicola grisea* Traaen g) *Urocystis* spp., h) *Agaricus bisporus* (J.E. Lange) Pilát, i) *Cortinarius* sp., j) *Ganoderma* sp., k) *Lactarius trivialis* (Fr.) Fr., l) *Ascobolus* spp., m) *Pestalotiopsis* sp., n) *Rhizopus* sp., o) *Leptosphaeria maculans* (Sowerby) P. Karst., p) *Dendryphion comosum* Wallr., q) *Torula* sp., r) *Trichothecium roseum* (Pers.) Link, s) *Epicoccum* sp., t) *Leptosphaeria* spp. u) *Melanconium juglandinum* Kunze, v) *Pollaccia elegans* Servazzi, w) *Spilocaea* sp., x) *Fusicladium* sp., y) *Alternaria* sp. z) *Stemphylium* sp. (bar=10 μm).

spores may have to be placed in groups of multiple genera or families. The author proposes standardization in the nomenclature of spore identification levels in the Table 2.

Table 14.2. Standardization of spore identification levels proposed by the author.

Spores can be identified as:

Genus (e.g., *Diplodia* spp.): when accurate identification is possible on the level of genus.

Genus? species (e.g., *Pleospora* ? *herbarum* (Pers.) Rabenh): when the identification is unsure on species level, but we have enough information to exclude the majority of other species.

Genus?? species (e.g., *Dictyodesmium* ?? *ulmicola* (Ellis & Kellerm.) S. Hughes): when the identification is unsure on species level. We don't have enough information to exclude the majority of other species, but the spore strongly resembles a given species.

***Genus species*-type** (e.g., *Alternaria tenuissima*-type): when the identification of genus is certain. The identification of species is not possible by spore morphology, but the morphological description is simplified by referring to a well-illustrated species.

Genus/Genus (e.g., *Aspergillus/Penicillium*): when two genera could not be differentiated.

***Genus*-type** (e.g., *Colletotrichum*-type): when the identification of genus is uncertain, but the morphological description is simplified by referring to a well-illustrated fungus.

Genus? (e.g., *Stereocaulon*?): when the identification is unsure on genus level. We don't have enough information to exclude the majority of other genus, but the spore strongly resembles a given genus.

Family (e.g., Diatrypaceae): when the identification of family is certain.

Unknown sp. (e.g., unknown 1-septate, hyaline, elliptical spores): when morphological identification is not possible on any taxonomical level. Unknown spores should be illustrated for future identification. (It is advisable that use of terms with undefined meanings (e.g., aff., cf.) should be avoided in spore description).

Morphological variants. (e.g., clumped conidia of *Botrytis*) Description may be expanded to the frequent forms of desiccated, deformed, clumped or immature spores of a given fungus*.

*Dehydratation during airborne transport may cause spores to collapse inwards, often assuming characteristic shapes and crease patterns (Fig. 14.5). The spores produced by fungi in culture may differ in appearance from those sampled from the air.

Spore size, roughness affect the dispersal of spore by the rate at which spores fall through the air, and their ability to impact on stems and other obstacles. Morphology of spores may be indicative of its ecology, dispersal or sporogenesis as well.

IMAGE ANALYSIS

Image analysis techniques are varying between simple manual morphometry and complex automated particle recognition algorithms (Bernier and Landry, 2000; Boucher *et al.*, 2002; Ronneberger *et al.*, 2002). Image analysis methods are often microorganisms– and sample-specific, the results of which are difficult to transpose to other studies. The image analysis steps (Fig. 6) usually are:

Pre-processing. The separation of the object (spore) from other particles and background should be started by conventional microscopic techniques. Two separate images from the same object are often necessary (e.g., staining, fluorescent image, phase contrast image; or images captured with different camera exposure time) (Takahashi *et al.*, 2000; Walker *et al.*, 2002).

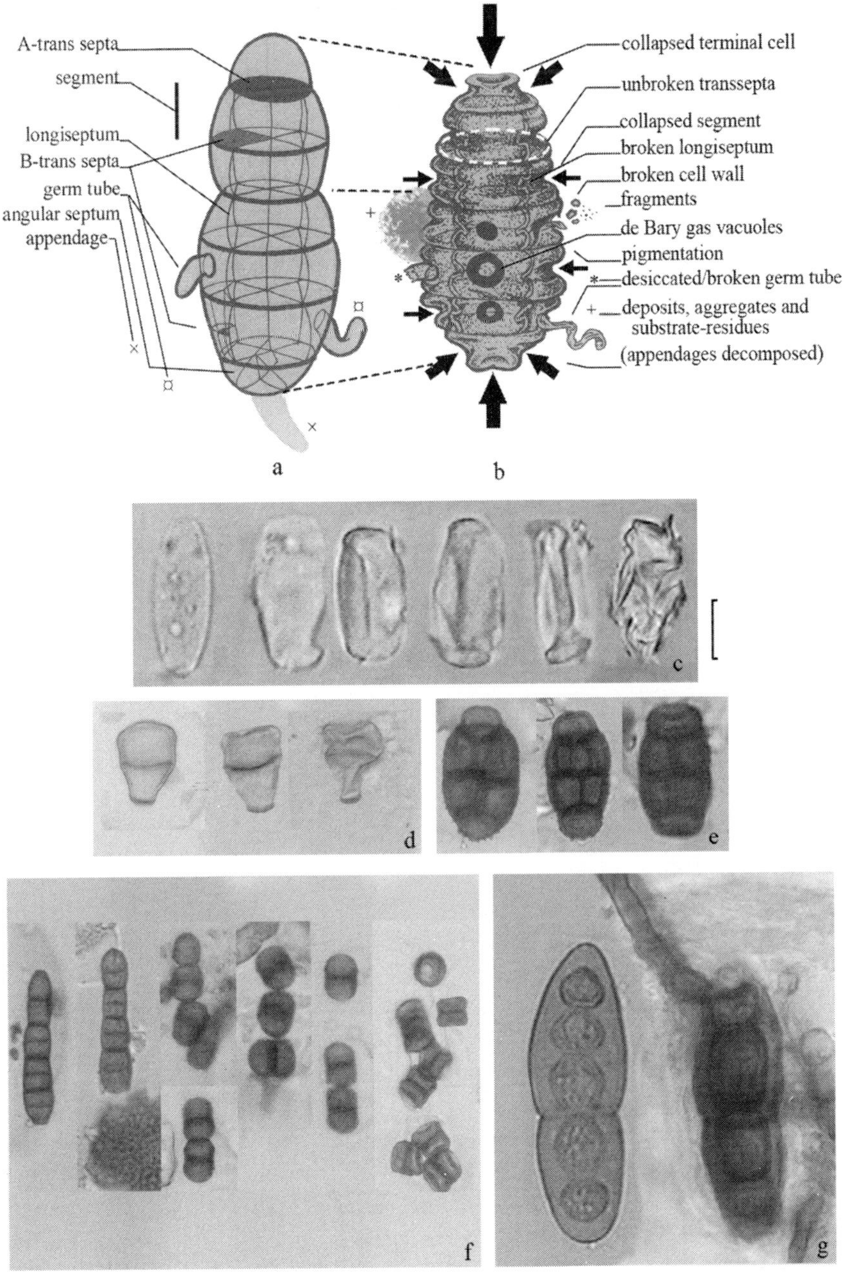

Fig 14.5. Alteration of fungal spore morphology during airborne transport. a) turgid spore, b) spore after longer aerial transport or resurfacing, showing crease patterns and other characteristic alternation in morphology. Creasing in c) amerosporae (*Oidium* sp.), d) didymosporae (*Polythrincium trifolii* Kunze), e) dictyosporae (*Pithomyces chartarum* (Berk. & M.A. Curtis) M.B. Ellis). f) Disruption of conidial chains (*Ampulliferina persimplex* B. Sutton), Characteristic changes after resurfaceing (*Splanchnonema quercicola* M.E. Barr).

Image acquisition is used to create digital image suitable for digital processing and storage. The picture must be calibrated by a microscope graticule slide.

Image segmentation is used to select the objects of interest from the background. Objects must be fully distinguished from both background and other objects.

The most common step of segmentation is tresholding, when the image is transformed into a binary image. This process analyses all tones and determines all those below a selected level are treated as black, and those above as white, and hence termed background. On binary image some correction should be needed (e.g., make a high contrast between the particles and background, remove any obvious materials manually or automatically on the basis of size or shape).

Morphometry by automatic measurement of the detected objects. Morphological feature measurements, including length-width ratio, area, perimeter, feret diameter, roundness, curvature, Fourier descriptions, eccentricity, sphericity, etc. (Benyon *et al.,* 1999; Liu *et al.*, 2001; Boucher *et al.,* 2002) should be computed for thousands of objects. The techniques provide digital data sets from each particle, including surface characteristics, texture and roughness (Dörge *et al.*, 2000; Jones, 2000; Gottwald *et al.*, 2001; Tian *et al.*, 2004).

Classification. Multivariate statistical methods are needed to analyze the data of numerous discrete measurements. When observations are classified, the results can be subsequently interpreted easier by graphic presentation.

Image archivation. One of the most valuable sources for identification is a microphotograph collection. Spores from air samples and reference materials from fungal collections will become comparable easily. Spore galleries are useful for retrospective identification as well. It is advisable that both line-drawings and more than one photographs from a given spore type should be illustrated. Accurate measurement data are also needed.

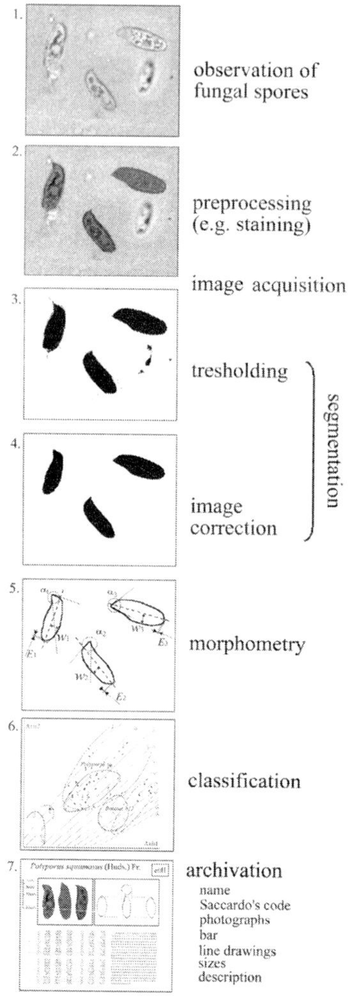

Fig. 14.6. General steps of image analysis of fungal spores.

Automated identification keys for objects are also in use. However, major difficulty in developing these techniques is to determine which combination of morphological feature measurements is sufficiently characteristic and to minimize the variance to gain adequate differentiation between each species from all the others. It is requested to measure a large number of individuals in each taxa, especially for distinguishing between species with similar spore shape. Search for specific characters should be achieved on the complete 3D microphotographs of the spore.

IMMUNOIDENTIFICATION OF AIRBORNE SPORES

Microscopy is unreliable for the detection of small, nondescriptive spores produced by many fungi.

Particles originated from certain fungal taxa can be difficult to identify and quantify on purely morphological ground. Measuring airborne quantity of ergosterol may be the modest but still adequate technique to determine the fungal proportion of airborne biomass. It is followed by the antibody-based blotting or fixation methods, which rely on the production of antibodies against specific antigen on the target particle. The specific antisera could be used either via fluorescent antibody technique or enzyme-linked immunosorbent assays (ELISA). ELISA methods enabled the counting of amorphous airborne particles carrying allergen and allowed the ready association of the released allergen with the morphologically identifiable particle (Razmovski *et al.,* 2000). Adhesive tape from a Hirst-type trap may be laminated with a protein-binding membrane. Both the particle and associated immunostained allergen released from the particles can be visualized through the transparent tape for final microscopical identification. Patient's serum may be used to detect specific spores on sampler slides. When IgE from allergenic patients is used, the density of immunostaining in the halo near the allergen carrying particle was found to be proportional to the level of specific IgE applied. Spore trapping formats have been developed to use with ELISA to detect and quantify spores of plant pathogenic fungi (Kennedy *et al.*, 2000). Immunological techniques have the potential to be applied in routine monitoring, if appropriate antibodies were found (O'Meara, 2000). The application of immunoassay is restricted by the difficulties of developing antibodies showing the required specifity.

APPLICATIONS OF METHODS OF MOLECULAR BIOLOGY IN AEROMYCOLOGY

Several DNA-based methods have also been developed for studying fungal aerosols. By using polymerase chain reaction (PCR) the presence of the target organism can be detected from a few cells, with primers designed to amplify DNA sequences of a specific species. Between universal fungal primers and species specific primers a variety of detection level are developed to rapidly screen samples and thus promote early diagnosis. Culturing of the targeted microorganism prior to DNA extraction is often not necessary: PCR based probes can be very sensitive as they amplify small amount of target DNA that can then be detected. It is necessary to validate each detection system on a wide range of species to avoid false positive results due to cross-amplifications. The "measureless" diversity of fungal species and their often low concentration in air samples are new challenges in aeromycology. Non-cultivable, non-viable fungal particles, which may still cause health problems are also detectable (Olsson *et al.,* 1996; Wakefield, 1996). Using the nested PCR assays after spore wall disruption of *Penicillium roqueforti*, the DNA from just one spore was detected by Williams *et al.,* (2001). The presence of nontarget bioaerosols and organic dust may inhibit the PCR processes, thereby reducing the potential sensitivity of the assay. However, it was also possible to detect the DNA from ten *P. roqueforti* conidia against a background of 4,500 other airborne spores. Calderon *et al.,* (2002) expanded the methods for the conventional impaction-type (Rotorod and Hirst-type) spore traps. An advantage of the new technique that the one-half of the air sample is retained for conventional microscopical analysis and to determine spore concentration, meanwhile the other half of the air sample is processed using spore wall disruption, DNA purification and PCR assay protocol. Quantitative studies should be achieved also by "real-time" PCR equipment, which allows a direct estimate of the initial amount of target DNA (Haugland *et al.,* 1999; Schweigkofler *et al.*, 2004). This latest technique offers an additional advantage that the DNA extracts can also be used to detect the new strains, genotypes or mating types.

AIR QUALITY

Indoor air quality

The biological components of enclosed spaces may be originated from the outdoor environment, however many intramural sources may exist. The correlation between the number and strength of sources and the average concentration of airborne particles is evident. There is some evidence that spores and VOCs (volatile organic compounds) from fungi can contribute to the "Sick Building Syndrome", allergy and mycoses (Tuomi *et al.*, 2000).

Ventilation systems may reduce the microbial concentration by filtering and reducing viability of spores. Spore cloud may be diluted, but not completely removed by artificial ventilation. On the other hand, ventilation systems rapidly circulate spores or transport them from any point of source throughout entire buildings (Philips, 1965).

The spore cloud may be distributed throughout a large facility in a short time even without any mechanical ventilation. Within a building thermal rise or chimney effect are effective in spore distribution. Spore transport may be facilitated by inside pressure as well. Outside air movement affect movement indoors forcing air thought cracks on the windward side and sucking it out on the leeward. Wall openings and gaps at structural joints (electrical outlets, cable boxes, etc.) have an average width of 1000 μm; therefore dispersal of fungal spores may occur via through-wall gaps.

Air circulates in indoor environment because of thermal convection as well. Thermal gradients between walls, windows and heaters generate mixing (Daws, 1967). Convection may be very effective, carrying spores from the first to fourth floor halls within 5 min and into rooms within 20 min (Christensen, 1950). Aerosols are moved by heat sources even by heat outputs from a lamp. The heat from the human body may be sufficient to alter the local convection system drawing air up from floor level (Lewis *et al.,* 1969). Occupants contribute to turbulence through their activities. Spores may become dispersed by sweeping, bed-making, and building repair work, giving spore concentrations 10-17 times greater than before disturbance (Maunsell, 1954; Reponen *et al.,* 1992).

Outdoor air quality

The concentration and composition of the community of the airborne fungal spores (airspora or aeromycobiota) outdoors vary according to time of day, season, geographical location, weather conditions, human activities and vegetation (Lacey, 1981). The dispersal of airborne spores – often referred as the concept of 'aerobiology pathway' involves three stages (Edmonds, 1979): (1) liberation, (2) transport, (3) deposition (Fig. 14.7). The journey of an airborne particle does not always finish here: a subsequent phase, (4) resurfacing, may also occur. Concentration dynamics and airspora (aeromycobiota or spore composition of the air) are strongly influenced by each stages of the pathway (Ingold, 1971; Lacey, 1981, 1996).

Liberation

In order to become airborne, spores must not only be discharged but also have to cross the laminar boundary layer (that surrounds all surfaces) then should enter the turbulent boundary layer above.

This may be accomplished through a variety of launching mechanisms. Some are passive, utilizing the energy from the environment, and some are active, utilizing the energy generated by the fungus. Frequent passive liberation mechanisms are: shedding under gravity or under convection currents, deflation, mechanical disturbance and rain tap and puff mechanism. Rain splash occurs when a raindrop falls onto a surface covered by a thin film of liquid, and forms a crater, with many (100-5000) secondary droplets produced at its periphery. Hygroscopic movements in conidial fungi, squirting mechanisms of Ascomycetes, and ballistospore discharge in Basidiomycetes are active mechanisms. The varied details of the dispersal processes have a great deal of information (Lacey, 1981, 1984, 1996). The classification of liberation types according to Lacey (1981) is inserted in Fig 14.7.

Transport

Airborne spores are mainly concentrated in the troposphere. Dispersal is closely related depending on the size, shape, roughness, density, and electrostatic charging of individual spores and air viscosity, air movement, turbulence, layering, convection, wind gradients close to the ground, and patterns of atmospheric circulation. Turbulence dilutes spore clouds according to the theories of eddy diffusion. Eddies dilute spore cloud as it travels downwind, scattering it both vertically and horizontally, or move it bodily. Air movements range from small turbulent eddies to large frontal systems (thousands of kilometres long and hundreds of kilometres wide) and jet streams in the stratosphere and in the upper troposphere where spores are rapidly transported. Gregory (1961) classified distinct zones of troposphere regarding the aerial dispersal of fungi.

Horizontal Transport

1. **The laminar boundary layer** is characterized by a thin layer of still air surrounding the earth surface and all objects. Above this motionless film there is an other layer, which flows in streamlines without turbulence.
2. **The local eddy layer** represents eddies caused by small roughness or cup-shaped depressions on surface.
3. In **the turbulent boundary layer** projecting objects cause eddies which break away from the surface and travel downwind. In this layer diurnal changes are less pronounced.
4. **The transitional or frictional turbulence layer** reach up to the height of 500-1000 m, in which turbulence diminishes with altitude. The top of this layer marks the limit to which dust and spores may be carried upwards by turbulence.
5. **In the convective layer** airborne particles can be transported only by convection. During summer, warm air bubbles carrying a considerable number of spores may form in areas of 1.25 km^{-2} every 6 to 15 minutes when the sun heats the ground.

Vertical Transport

1. The **microscale** horizontal transport should be limited to very small time and space, i.e., time scales on the order of an hour or less in duration and space scales of a few hundred meters.

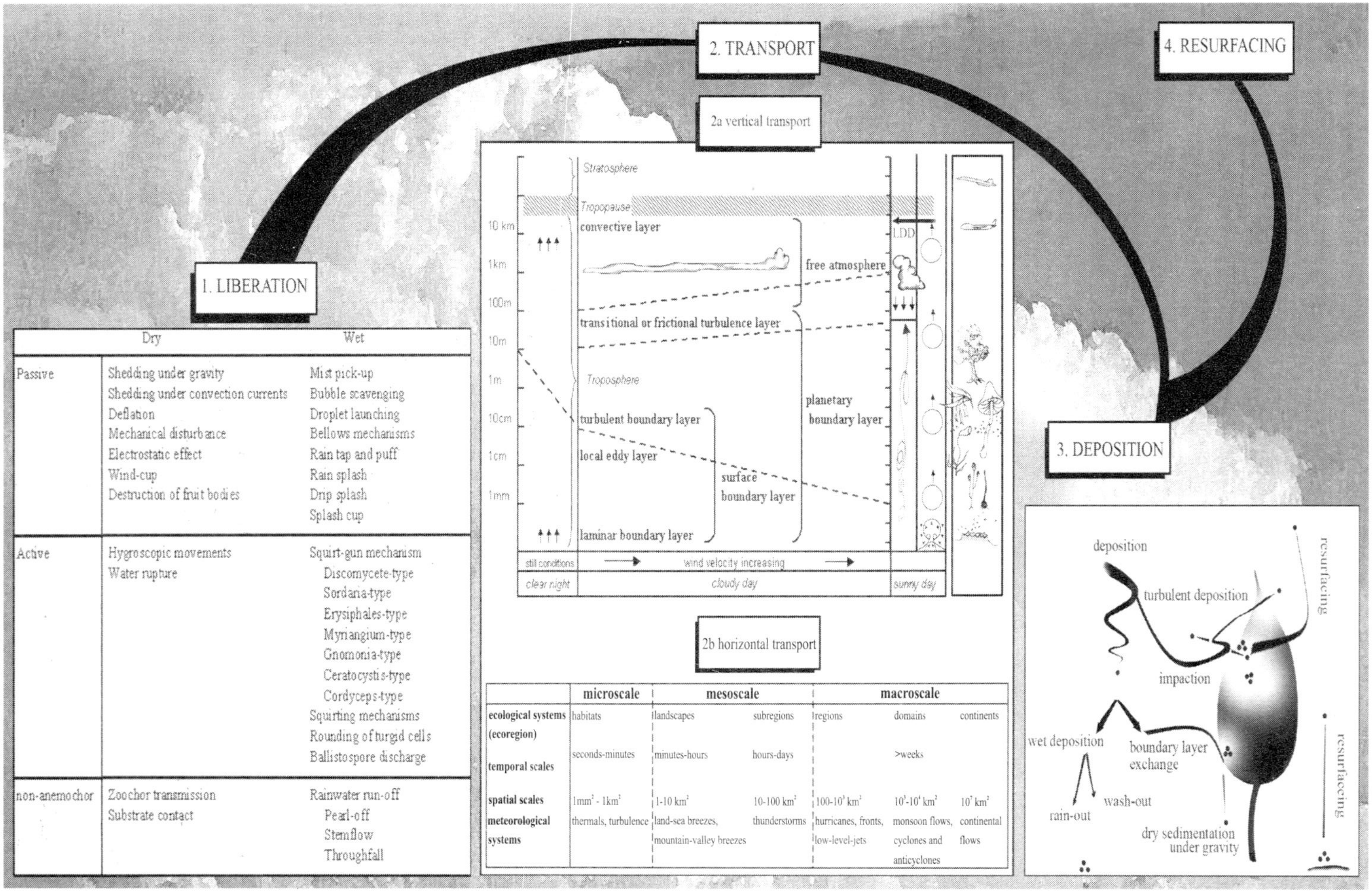

Fig. 14.7. The aerobiology pathway (liberation, transport, deposition, and resurfacing) modified after Edmond's concept (1979), including Lacey's (1981) classification of liberation mechanisms, vertical transport after Gregory (1961), horizontal transport after Gage (1999).

The time scale is set as the longest interval during which meteorological circumstances are unchanging while the space scale is limited to the distance over which homogenous turbulence persist.

2. For **mesoscale transport**, the appropriate time and space scales are days and a few hundred kilometres. The wind speed and direction vary, and surface roughness is not uniform where transport occurs.
3. **Macroscale transport** includes time and space scales great to include global circulation patterns. The dispersion is unpredictable. One of the approaches toward understanding the long distance dispersal is trajectory analysis, based on circulation patterns as derived from worldwide synoptic observations. Many examples of long distance dispersal (LDD) are in evidence (Brown and Hovmøller, 2002), e.g., in India (Nagarajan and Singh, 1976) between Paraguay and Argentina (Waller, 1979), from Angola to Brazil (Bowden *et al.,* 1971), across the Tasman Sea from Australia to New Zealand (Viljanen-Rollinson and Cromey, 2002), between North Africa and South-East Europe (Tuboly and Vörös, 1962).

Deposition

Spores are heavier than air and so tend to sediment, but upward movements by convection and by turbulence impede this process. The Stoke's law describes the relationships between terminal velocity (*Vt*) and size of smooth spheres 1-50 µm in diameter in viscous fluids. The simplified equation is appropriate to determine the *Vt* of many spores in the air. Value of *Vt* varies, because of spore shape (deviation from the ideal sphere), drag (by spore surface roughening), degree of hydration, aggregation of spores (with trapped air between cells), de-Bary gas vacuoles within spores, vesiculate spores and pollen grains (e.g., *Pinus, Urocystis*). Several different processes of deposition are known (sedimentation under gravity in still air, impaction, washing off by rain, boundary layer exchange, and turbulent deposition). Rain removes rapidly spores from air by impaction on raindrops, by capturing on cloud droplets, or even by forming cloud nuclei.

Resurfacing

Not all spores that land become securely attached but some are resuspended again by the agitation of strong wind or by rain splash. The demarcation of the resurfacing phase is important to introduce the concept of multiple-stage dispersal of spores. Observations of Carter (1965) in *Eutypa armeniaceae* become the classic description of multiple-stage dispersal. Ascospores of *Pleospora herbarum* discharged from the perithecium in wet weather can be rediscovered in the air samples several months later, during dust-storms, when deposited spores are resuspended into the air-stream. The possibility of resurfacing of spores is scarcely considered, however, such events may be hazardous and unexpected.

ECOLOGY OF THE AEROMYCOBIOTA

Meteorological circumstances strongly affect the airspora by sporulation of fungi, the diurnal periodicities, and dispersal. The large-spored leaf and stem pathogenic fungi appear to be specialized impactors, while the spheroid, minute spored moulds appear to be specialized penetrators (Gregory, 1961). Spores have been grouped into hydrophobic dry spores (xerospores) and readily wettable slime spores (gloiospores) (Fig. 14.8). Wettable spores are carried within rain droplets, but non-wettable spores are found on their surface. Early studies showed that aeromycobiota should also be divided into dry airspora and damp airspora (Hirst, 1953) according to the meteorological conditions (Fig. 14.9), but biometeorology of many fungal species is still poorly understood.

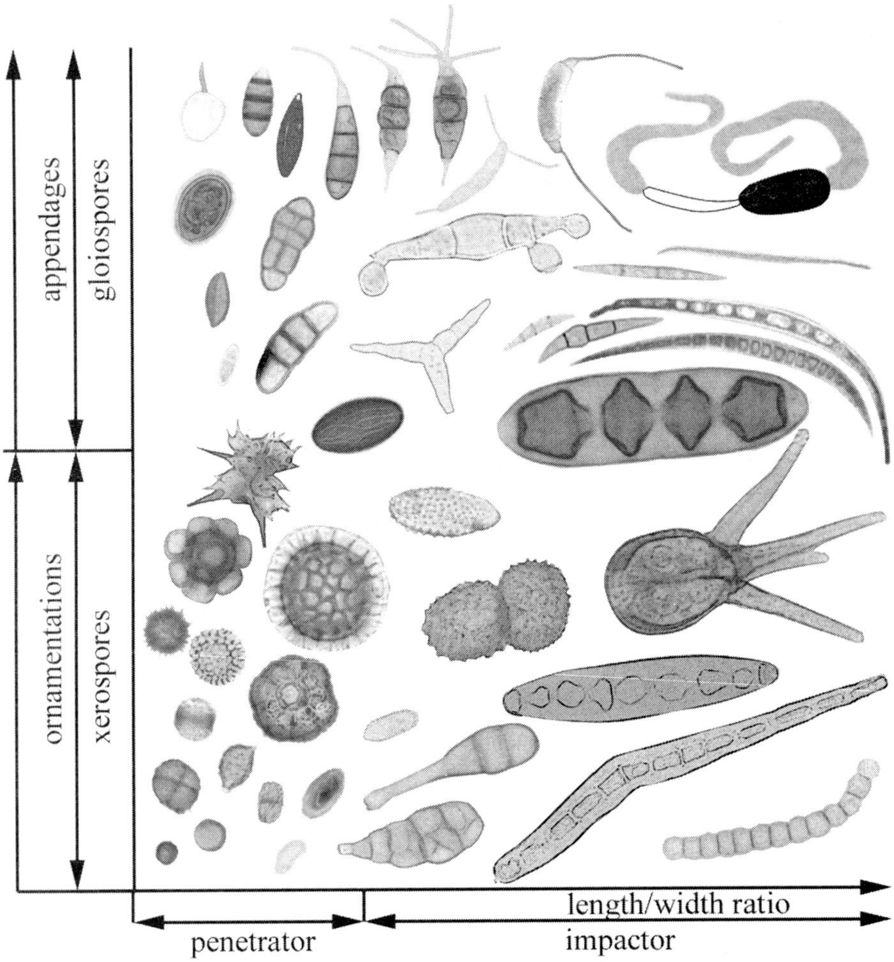

Fig. 14.8. Morphological adaptation of fungal spores to dispersal. Terms are used according to Mason (1937) and Gregory (1961).

DRY WEATHER

Effect of Temperature

Positive effect of temperature on dematiaceous hyphomycete spores (*Alternaria*, *Cladosporium*, *Epicoccum*, *Torula*, etc.) is reported in several studies (Beaumont *et al.*, 1985; Palmas and Consentino,

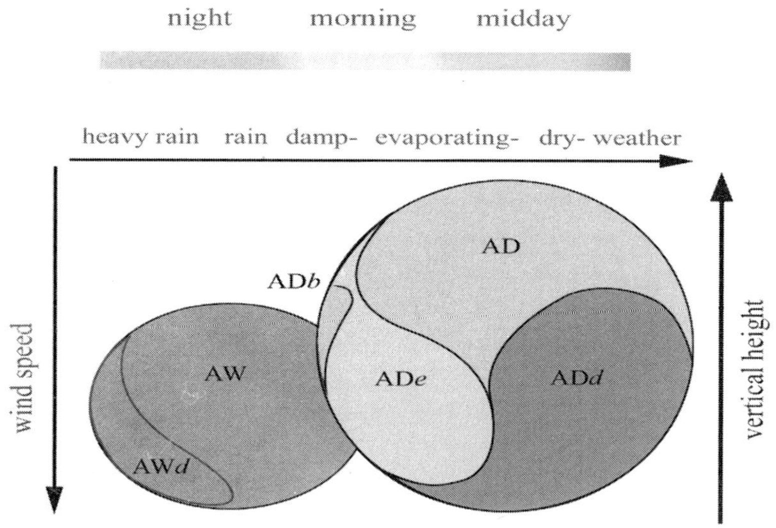

Fig. 14.9. Comprehensive **model for the airborne fungal biota**. In the figure first axis correspond to the temperature and humidity, second axis displays the wind speed. Circadian rhythm is also indicated. Dry (AD) and wet (AW) aeromycobiota described by Hirst (1953) were distinguished, but in addition four subdivisions are determined (1. spores liberated by evaporation (AD*e*), 2. spores teared off by windstorms (AD*d*), 3. basidiospores are present after rainy days (AD*b*), 4. splashborne spores (AW*d*). The arrow in the left side of the model indicates approximately the effect of wind velocity on airspora (by the Beaufort scale). In the right side of the figure, arrow shows the vertical distribution. Near ground relative abundance of the airborne spores are high, this gradient may concern especially for the large spores of the AD*d* and the spores dispersed by splash droplets

1990; Ballero *et al.,* 1992; Fernández-González *et al.,* 1993; Giner and Garcia, 1995; Ho *et al.,* 1995; Calderón *et al.,* 1997; Paredes *et al.,* 1997; Angulo-Romero *et al.,* 1999; Burch and Levetin, 2002; Stenett and Beggs, 2004). These echinate and dark-brown pigmented spores are known as "dry airsopra" (Hirst, 1953) or "dry weather spora" (AD). Many of them are saprotrophic or capable to trigger allergic symptoms. Relative abundance of *Cladosporium* species is high (80-95%). Peak period of their spore concentration is found around midday hours (Hasnain, 1993; Katial *et al.,* 1997; Lim *et al.,* 1998). Large mixing depth of air masses and convection should generate this maximum.

The effect of high temperature on the ascospore number should be both negative (Burge, 1986; Burch and Levetin, 2002) and positive (Hasnain, 1993; Levetin, 1998). Lin and Li (2000) showed that ascospore counts were decreased in summer, but increased in winter by temperature. Analysis of the whole year (Li and Kendrick, 1994; Levetin, 1998) evidenced the positive effect of the temperature. The critical point of this question should be the season.

Effect of evaporation

Results of several laboratory experiments on spore discharge of fungal species under controlled meteorological conditions showed active spore discharged in case of decreasing R.H. (de Bary,

1887; Meredith, 1961, 1962, 1965, 1966; Ingold, 1971). The spores of these *in vitro*-studied species occur in outdoor air samples frequently, when weather is considered to be similar to the circumstances generated artificially. Active spore liberation mechanism (e.g., hygroscopic movements) of these fungi requires evaporation. Spores of several plant pathogenic fungi (*Botrytis,* Peronosporaceae, etc.) should be cited in this group. Concentration of spores reaches its maximum at the morning hours (Hirst, 1953; Panzer *et al.,* 1957; Waggoner and Taylor, 1958; Hamilton, 1959; Sreeramulu, 1959; Dutzmann, 1985). The reduction of R.H. should be triggered by various factors (temperature/solar radiation, wind).

Ascospores (*Chaetomium*, Xylariaceae, etc.) may also correlate positively with drying. In this case ascospores are not ejected into the air after saturation, but a mass of spores (cirrus) is formed covering the surface of the fungal fruit body. After the cirrus is dried spores are liberated. Incipient desiccation of the perithecium of some ascomycetes (e.g., *Claviceps purpurea*) may also exert pressure on its content so the rate of discharge increases.

Effect of wind

The role of wind in transportation of spores is unambiguous. Moderate wind speed (0.4-2.0 m/s, Gregory, 1961) is enough to detach spores from the colony of many fungal species, but maximum number of the removed spores varies among species (e.g., between 0.4 and 1.0 m/s: *Alternaria tenuis*, Rotem, 1994; between 0.5 and 1.0 m/s: *Erysiphe graminis*, Aylor *et al.,* 1981; at 0.5 m/s: *Aspergillus fumigatus*, Pasanen *et al.,* 1991; between 1.8 and 2.3 m/s: *Puccinia recondita* f. sp. *tritici*; between 1.8 and 2.8 m/s: *Puccinia striiformis*, Geagea *et al.,* 1997; between 3.0 and 5.0 m/s: *Spilocea pomi*, Wiseman, 1932).

Fungal spores may be deflated by higher wind speed from the ground (3.0-5.4 m/s) than from the phyllosphere (0.5-2.0 m/s) (Jones and Harrison, 2004). The efficiency of deflation should be reduced when spores are slimy or the substrate is wetted (Ward and Manners, 1974; Jones and Harrison, 2004). In this case, spores are cemented to each other and to the surface (e.g., Zoberi, 1961; Pady *et al.,* 1969; Cohen and Rotem, 1970; Rotem, 1994; Geagea *et al.,* 1997). Concentration should grow in still air when mixing of air masses is low. Spore free air currents may reduce the number of spores floating above the source. High wind speed reduced spore concentration of *Alternaria, Cladosporium, Botrytis, Ganoderma, Cladosporium, Entomophtora,* and uredospores (Hamilton, 1959; Lopez and Salvaggio, 1983; Giner *et al.,* 2001). Wadia *et al.* (1998) showed that spore counts of *Passalora personata* were decreased by steady wind, but intermittent wind gusts caused high concentration. Continuous wind in 3-6 days exhausted the colonies. Stronger wind gusts tear off the strongly attached mycoparticles, sclerotia and immature conidia. Spores should not be produced until new conidiophores grow replacing the broken ones (Langenberg *et al.,* 1977). The erosion of fungal colonies is stronger on the upper part of the plant canopy than those near the ground (Aylor, 1990; Rotem, 1994).

Qualitative changes occur often in the composition of aeromycobiota in gusty weather. These air samples contain some broken, desiccated, immature or germinating spores and soil particles, and pollen grains from an earlier blooming season. Aggregates of spores, pollen grains and plant residuals may be collected from strong air blasts. Aggregates carry much more antigenic material than single particles, and the type of symptoms of allergy may altered as well (Lacey, 1991).

Some spores became airborne hardly, because their dispersal is not primary anemophilous, like coprophylic fungi. Staurosporae, "giant spores" (*Bipolaris*, *Podosphaera*, *Phragmidium*, and *Tetraploa*), fruit bodies, sclerotia and fungal particles aggregated with pollens and plant debris may also present. Rotem (1964) observed immature conidia of *Alternaria solani* in the windstorms of Negev desert, some of them were attached on plant debris. Meredith (1966) mentioned that conidiophores with immature conidia, hyphal fragments, and plant debris with spores were collected by a Hirst-type air sampler. Hyphal fragments of phylloplane fungi should be dispersed due to scrubbing action of host plants which takes place during wind currents (Tilak and Pande, 2005). Airborne plant debris carried the oospores of *Peronosclerospora sorghi* (Bock *et al.,* 1997). Li and Kendrick (1994) reported that big sized spores of *Drechslera, Pithomyces, Epicoccum* and *Nigrospora* returned into the air in winter in correlation with wind. Daily maximum of spore number should evolve in certain geographical regions, where diurnal high winds occur regularly (Gregory, 1961; Ogden *et al.*, 1969). Concentration of some of these spore types (e.g., *Helminthosporium*, *Tetraploa*), often shows this double-peak circadian pattern (Lacey, 1981). The double-peak circadian concentration curve of *Alternaria helianthi* is evolved by the convection in the midday hours and by the wind gusts at the afternoons (Allen *et al.*, 1983). If strong gusts are presented in night time, the circadian rhythm of spores should be altered, and a midnight concentration peak may be present, as Meredith (1966) showed in his study on the big sized *Alternaria porri* conidia.

Larger fungal and plant particles were collected frequently near the ground than in a vertical height of 150 cm (Magyar *et al.*, 2003), the strong turbulent currents may bounce them into the upper air layers. In cold nights the mixing depth of the air is low, the spores liberated in this period may reach the aspiration level of the trap only around midday. This group deserves attention from point of view of public health: unexpected outbreak of allergy should provoke by off-season resuspendation of allergens in dust and sand storms (Farkas *et al.*, 1999). Increased antigen content of the spore (and spore-pollen) aggregates should be emphasized.

WET WEATHER

Effect of precipitation and relative humidity on airspora

The community of the spores of fungal species present in humid circumstances in the air was entitled damp or wet weather spore (AW) by Hirst (1953). Precipitation correlates positively with the majority of Ascomycetes and Coelomycetes (Meredith, 1962; Royes, 1987; Hasnain, 1993; Turott and Levetin, 2001; Burch and Levetin, 2002). The quantity of moisture required for spore discharge should be various in different fungal taxa, consequently the peak of concentration may be present some minutes after the start of rainfall, or 5-10 hours later. Sometimes fog is enough to generate spore discharge (Gregory and Stedman, 1958; Carter and Moller, 1961; Hirst and Stedman, 1962; Arseniuk *et al.*, 1997). Spore counts of some taxa of Ascomycetes were reduced when washing off by intensive rainstorms occured (Meredith, 1962; Burge, 1986; Horner *et al.*, 1992; Turott and Levetin, 2001). Studies showed (Gregory and Steadman, 1958; Turott and Levetin, 2001) that asci may exhausted in rainfalls lasting many days. Number of subsequent dry days reduced the ascospore concentration (McCartney and Lacey, 1990). Burge (1986) showed that airborne spore concentration of *Leptosphaeria* and *Ophiobolus* were not decreased during rain. Rain splash spores are present in the air samples when intensive rainfall occurs.

Some components of the AW group should be included into night-time airspora. The laminar boundary layer is thickened at night-time, which should be broken through by spores launched by active liberation mechanisms.

The spore concentration of basidiospores of mushrooms is often determined by rains of the previous days, followed by rising temperature. This effect is probably owing to the growth of mushroom fruit bodies. Diurnal maximum of basidiospore number was measured in night and early dawn, because of their active ballistospore discharge mechanism (Lacey, 1981).

The correlation of dry airspora concentration with relative humidity (R.H.) is often negative. Impaction of spores on the wet surfaces increases by wind speed, but spores should resuspend into the air when adhesion diminished by drying air flow. In saturated air gain in weight of spores accelerates sedimentation (Weinhold, 1955). Madelin and Johnson (1992) showed that spores had larger aerodynamic diameters when aerosolized at 95-98% R.H. than at 40%. The role of rainfall is dual: the first large drops of the intensive rainstorms increase the concentration of dry fungal spores (Ainsworth, 1952; Hirst, 1953; Harvey, 1967; Rantio-Lehtimäki, 1977; Hjelmroos, 1993; Kurkela, 1997; Venables *et al.,* 1997; Allitt, 2000). A study by Hirst and Stedman (1963) showed the possibility of the dry dispersal of fungal spores by incident raindrops in case of some rust, smut and conidial fungi. In prolonged rain, droplets wash-off the dry airspora (Hamilton, 1959; Ho *et al.,* 1995; Katial *et al.,* 1997; Fernández *et al.,* 1998; Lim *et al.,* 1998).

Dry fungal spores are infrequent at night hours (Palmerley and Benedict, 1977; Rotem, 1994), because of low wind velocity and high saturation, which sticks the spores on the surfaces (Dutzmann, 1985; Savary, 1986; Ballero *et al.,* 1992).

SPORE DISPERSAL CONTROL

Aerobiological knowledge is necessary in control and elimination of mycotecnological products. Escape of spores of mycoherbicides, mutants, etc. from industrial processes, laboratories or treated crops may occur. It is necessary to know which factors affect the dispersal of the given fungus. Most important features are:

- aerobiology and other dispersal characteristics
- spore morphology, aerodynamical size and shape
- wind speed and direction
- viability of spores
- the possibility of role of insect vectors
- multiple stage dispersal and resurfacing of living spores, and
- long distance dispersal

Reservoir establishment on hidden alternative host plants or hybridization with wild-strain fungi in buffer area should also take into account.

Indoor air control

There are various separation techniques between outdoor air and indoor air of mycotechnological workplaces. In providing aerosol control, a ventilation system which includes ventilating zones can be ideal, where each zone may be isolated by air locks or pressure gradients. The simplest and most frequently used technique is to maintain the space or the volume at a positive pressure to the outdoors to reduce infiltration of uncontrolled air. Rooms with spore handling should be maintained at negative pressure so that bioaerosols generated there should not be transport elsewhere. Workplaces of low spore concentration should be maintained at positive pressure to keep out alien fungal aerosols. It is customary to filter incoming air. Leakage of the costly treated air is minimized in a good construction. When intramural sources are present, the principal method of removal of such contamination is dilution by ventilation. UV irradiation also can be applied to eliminate microorganisms in a ventilating duct. Correct use of unidirectional airflow technique (Whitfield *et al.*, 1962), wherein a stream of clean air emanating directly from a high-efficiency particulate air (HEPA) filter (Zeterberg, 1973) can be a successful control measure. It is used to provide particulate-free air in industry and surgical spaces. Generally a building heating, ventilation and air conditioning (HVAC) system mixes outdoor air with recirculated air, then filters, heats or cools the air mixture, and distributes it via ductwork throughout the building. In most cases, filters are located before the heat exchanger section. Recirculation of the water of HVAC system is not recommended. Some filters of HVAC are not efficient enough to remove 1-2 μm fungal particles. Organic dust in filters, heating/cooling coils and humidifiers may become moist allowing fungal growth on the device itself. If the relative humidity indoors is over 70%, materials containing carbon may absorb enough moisture to support fungal growth. Water spray humidifier containing stagnant water, cooling system condensate pans, etc. may all amplify fungal spore concentration. Consequently, indoor air may be contaminated by bioaerosols produced in areas beyond the heat exchanger section, such as heating/cooling coils, humidifiers, and water spray systems.

Outdoor air control

Separation between open-air areas by timing of activities (e.g., planting) or by distance (separation zones) is a tool to reduce the threat of impact of spores on non-target organisms. The spread of fungal spores from local sources results a hollow curve (with counts decreasing rapidly at first, and then gradually flattening out and becoming almost parallel with the distance axis). This explains Kerling's (1949) statement that the strength of infection decreases with the distance from the source of inoculum. Gregory (1952) concluded that 99.9% of spores should fall to the ground within the first 100 meters. However, a higher limiting value could be established on the basis of phytopathological considerations, below which the probability of infection is negligible. Good points of reference for dispersal of mycotecnological products are the studies referring on pollen drift in genetically modified organisms (GMOs). Pollen drift of GMOs has received more attention in the recent years, and some of the conclusions should be considered in mycotechnology. The pollen concentration of GMOs reduces 99% within various distances (12-500 m) from the source, which depends on many factors (wind speed, entomophil or anemophil pollen transport, pollen shape and size, etc.). The separation distance for dry-spored wind dispersed fungi may be similar than the separation distance for anemophylous plants (6000 m).

The interest for the use of physical barriers (wind shelters) in preventing the spore transport is renewing from time to time. The subject of natural spore traps, and its favourable effect has been discussed yet. Row of trees or hedges have been employed as protective screens in phytopathology.

A barrier crop can also act as an 'absorber' of the airspora. Jones and Brooks (1952) conducted experiments with tree and crop barriers adjacent to a crop of maize. The results indicated that a single row of trees was effective in reducing the amount of outcrossing by 50% in the plants situated immediately behind the barrier, but was much less effective at greater distances from the barrier. It was concluded that the tree barrier was less effective in reducing outcrossing than an area of barrier crop occupying an area of equal size to the trees. The study concluded that the percentage of outcrosses occurring in successive rows at different distances of isolation indicates that the first five rows adjacent to the source of contamination function as catchments of airborne contaminating pollen (Jones and Brooks, 1950).

The uncertainty of the outcome of such attempts was pointed out by Schrödter (1952), on the basis of studies described by Rugieri (1948) in Italy, and projects reported by Grebenšèiskov (1951) in the former Soviet Union. Some studies and theoretical modelling showed that the usual shelter strips are scarcely suited to hinder the dispersal of spores. Although the concentration decreases within the rows of trees, a recover of concentration directly behind the wind shelter was measured (Fig. 14.10). It should be concluded that wind shelters has some merit, especially in reducing spore concentration within dense vegetation, however their role in separating infected and non-infected areas is much less than suspected.

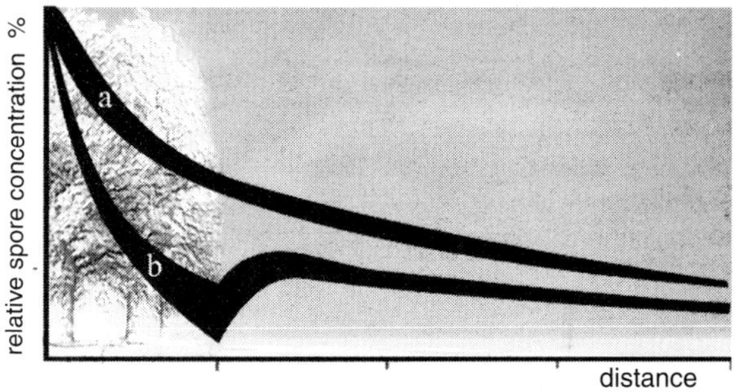

Fig. 14.10. Effect of tree barriers on spore dispersal (redrawn after Schrödter 1960). Change of relative spore concentration in and behind tree barriers a) outside and b) inside of the wind shelter.

TOWARDS THE STANDARDIZATION

There is an increasing demand in critical procedure standardization with analytical precision in all branches of sciences, especially in those areas that provide data for important economic decisions on industrial scale. The methods should be reproducible, traceable, and records should be kept available for subsequent inspection. In this chapter there are tesselated some suggested "building stones" for design protocols concerning indoor air quality measurement of fungal components. Air sampling makes possible to compare indoor quality with outdoor one, to test hypothesis about source or amplification sites, to check as HVAC had the desired effect and also to compare measured concentrations with standards. The proposed points are based on laboratory guides and standards (Lacey, 1995; Rogers and Muilenberg, 2001; ISO 8573-7:2003) as well as on experience of the author.

Indoor air sampling

- Before sampling, an assessment of the bioaerosol status of the building should be established. A fact-finding initial walkthrough is needed to gain a visual appreciation for possible contaminant sources before air sampling. The investigator should study the structure and maintenance of the building, look for possible sources of the indoor bioaerosol. Building management and maintenance personnel should be involved into the investigation.
- The sampling is done with an impaction-type sampler preferably by a viable spore trap (e.g. Andersen sampler, MAS 100, RCS).
- Accurate airflow rates are very important in calculating the concentration of airborne spores. All types of samplers should be calibrated before and after working to ensure the flow rate is within the standard. Calibration is performed by air-flow meters provided by the manufacturer of the trap.
- The samplers should be cleaned inside with a detergent at least after changing sampling condition. Drying of the device after cleaning with liquids should be also necessary, when minute holes are blocked by the detergent solution. For drying a clean holder is necessary.
- Close quickly the aspirating part to the sampler immediately after removing the Petri dish lid. After removal of the Petri dishes their covers are replaced and numbered at their bottom. The plate numbers, type of media, starting time, sampling height and time (length of time when plates are exposed in the sampler), laboratory incubation temperature and time must be recorded in the log. Note the observation that could influence the measurement.
- Unopened, unexposed, fresh media blanks, that are handled in every way the same as field samples, including labelling, except that no air is drawn through the sampler are incubated in order to check for sterility of the media. The number of media blanks is to provide two media blanks for every 10 samples (Jensen *et al.*, 1994).
- It is strongly advisable that samples at each site be taken at multiple times during the day. It is recommended to sample:
  - before, during, and after the sampling area is occupied,
  - when air condition system is activated and inactivated,
  - at morning, midday and afternoon.
- The indoor sites should include areas associated with complaints and, if possible, at least one area not associated with complaints. It may be possible to determine if amplification of microorganisms is occurring at the indoor air by comparing microbial profiles at the sites associated with complaints to those at sites associated with no complaints and to those at the outdoor air sites. The levels measured in problem areas can be compared with concentrations determined at other times and other locations.
- For the appropriate interpretation and for the comparison of microbial findings indoor, air samples must be taken at an outdoor site (preferably at the outdoor air intake of the building). The outdoor air sampling should be done at the same time and the same way as the indoor air sampling. If the types of bioaerosols indoors are not similar to those outdoors and the amounts are higher, this indicates a potential problem. The presence of fungal species in indoor air samples and not similarly presented in concurrent outdoor samples is evidence of an indoor source. The spore level of enclosed spaces is normally lower than outdoor concentration.

- Inquination of air ventillation systems should be proved with sequential sampling: first samples are taken in the system (inside the tube or close to the inlet); other samples are taken at further subsequent points in the room. While concentration of fungi gradually increases near the source, emitting sources should be identified (The carbon dioxide level should also be an indicator of ventilation adequacy).
- It is recommended that air sampling be done at multiple times. Each sampling must be repeated at least once at intervals 2 hours apart.
- At the end of the incubation time, count the CFU on Petri dish as early as 48 hours after the beginning of incubation, and then recounted every 24 hours for 10-14 days.
- Quantitative fungal amplification (high spore concentration) of the indoor air should be considered, when the number of colony forming units is above 200 per cubic meter of air and significantly exceeds that detected in the outdoor air. The number of fungi per cubic meter of air is determined.
- Qualitative fungal amplification occurs when specific fungal taxa that absent in the outdoor samples are found in the indoor air, the air quality may be considered inappropriate (even in case of low concentration of certain undesirable fungal species, which are contaminants in mycotechnological processes, potential human pathogenic fungi, e.g. *Aspergillus* species, etc.). The most predominant fungi should be identified at least genus level.
- Negative results do not indicate that air is free of fungi. For proper interpretation of negative results must claim attention for the possibility that the number of fungal spores was below the detection limits of the method.
- Visible marks of fungal contamination must be weight on the evaluation of air sampling data.
- Results documentation and sample archivation on photographs (e.g. UNIVID, agar plate archivation system, Lab-Intern Kft. Budapest, Hungary) should be necessary for further verification.

Outdoor air sampling

- The sampling is done with an impaction-type sampler (e.g. Hirst-type spore trap, Andersen sampler).
- Apparatus for use outdoors should be so pivoted that the intake aperture is directed into the main air stream by a vane (If the intake faces vertically upward it cannot be operated isokinetically and sampling errors become increasingly significant as particle diameters become greater than 5 μm).
- Care must be taken to evaluate proper location for the sampler.
- For general aerobiological monitoring the trap must be placed at a height above nearly tree tops. These samples will be representative to area with a radius of several kilometres. The particle composition of air samples collected near the ground is strongly influenced by local sources.
- The sampler must be placed as far as possible away from any shadowing obstacles on chimneys, abutments, frontage decoration and other rooftop structure.
- The flow rate of the sampler must be checked and adjusted if it is necessary.
- Instructions for sample preparation will be provided by the manufacturer of the samplers or by the manuals (Lacey, 1995; Rogers and Muilenberg, 2001).

SURVEY OF APPLICATION OF AEROBIOLOGY IN MYCOTECHNOLOGY

Changes of interrelationships between material sciences and fungal ecology resulted some adaptations of aerobiological knowledge in mycotechnology. The need of this special issue is emerged from the technological development which is hardly followed by the environmental researches. Some subjects of mycotechnology were studied by aerobiologists in the recent years, e.g. food production – including mushroom cultivation, and post-harvest management, fermentation industry – and application of mycoherbicides. In this chapter some aspects related to aerobiology are enumerated.

Food production

Mould spores deposited on crops before harvesting may develop in storage and their growth affects appearance and nutritional value of the product, and mycotoxins may be present in foods storage containers. Increasing concentration of airborne fungi is indicative of the fungal amplification even in hidden sources. Deposition on foods that are to be stored is principally an economic problem in the commercial canning industry. In bakeries, airborne conidia may be deposited on bread and pastries. In the frozen food industry unavoidable and usually undetected instances of excessive contamination by airborne microorganisms can cause the product to be susceptible to deterioration after melting.

In a study performed by Barui and Chanda (2000) eight rooms in a milk dairy were monitored. The highest spore quantity was recorded in the packaging location, and the minimum in the butter cold store. Results indicate that the optimum temperature was responsible for high growth of fungi. High spore concentration in the packaging section opens the chance of contamination in the processed milk. Similar results were obtained by Picco and Rodolfi (2004) when airborne fungal populations of some areas (dedicated to the fermentation, storage, and bottling of wine) were analysed. The highest fungal counts were observed in the bottling areas of industrial producers and the fermentation areas of family-run wineries. Some potential wine contaminant species (*Aspergillus niger A. ochraceus, A. terreus, Penicillium chrysogenum, P. citreonigrum, P. crustosum,* and *P. viridicatum*) and some potential allergenic species for winery workers were detected.

Concentration of airborne spores should be high with disturbance of mouldy substrates, for instance in cork factories (Lacey, 1973), in moist barley silos (Lacey, 1971) in cattle sheds (Baruah, 1961), rice mill (Savino and Caretta, 1992), biofilter of composting facilities (Reinthaler *et al.*, 1987).

Glasshouses may act as important spore sources. Glass windows by filtering short wavelength radiations prevent the reduction of spore numbers. Heating of glasshouses in sunlight leads to strong convention currents.

Grain dust is a complex material composed of plant fragments, actinomycetes, fungi, insect debris, mites, etc. and causes occupational lung diseases. Airborne conidia contribute both to the spoilage of stored food, agricultural produce and manufactured goods and to the decomposition of waste materials, composts.

Biological control

Aerobiology plays a crucial role in the monitoring of biological control agents, mostly to optimize and trace their dispersal, to assess the risk of non-target infections and to determine the width of safety zones. Some mycoherbicides are applied as aerial sprays. The environment is not always feasible to disease development where the pathogen may be limited in its dispersal. Factors, such as low inoculum levels, poor spore dispersal mechanisms and unfavourable environmental conditions often limit disease.

Escape and aerial dispersal of spores may create an additional disease risk in down-wind areas. A research was performed on the use of *Sclerotinia sclerotiorum*, as a biocontrol agent for *Cirsium arvense* in pasture (de Jong. *et al.*, 2002a, b). To determine the width of a safety zone for a pasture subjected by this fungus, plumes of dispersing ascospores were simulated in computer models. For quantifying escape and aerial dispersal of ascospores from a treated pasture, micro- and meso-scale transport models were also proposed. A hypothetical safety zone of 300 m was determined as a minimum isolation distances between a biocontrol site and a susceptible crop.

A successful biofungicide in commercial use, AQ10TM, is based on the hyperparasitic fungus *Ampelomyces quisqualis*. Passive dispersal of this hyperparasite by powdery mildew conidia occurs over large distances (Philipp *et al.*, 1984).

Basidiospores of *Chondrostereum purpureum,* another biocontrol agent are released during periods of prolonged rainfall and high relative humidity. The risk of infection to non-target trees was assessed by series of field experiments and simulation by Gaussian plume model. Based on the results, it was concluded that the risk of use of *C. purpureum* is acceptable when applied at least 500 m away from fruit orchards (de Jong *et al.*, 1991). Very little research has been concerned with dispersal of biological control agents, and perhaps there is an under-appreciation of the potential role of dispersal, especially the multiple-stage dispersal of the fungal spores.

Industry

In industry, fungal fermentations may become contaminated by undesirable fungi, and the manufactured goods may deteriorate. Some fungi, such as *Amorphotheca (Cladosporium) resinae* may grow in oil, causing corrosion and blockages in pipes. Many kind of fungi are commonly found where organic material is handled and where relative humidity is elevated. Species of *Cladosporium, Penicillium, Aspergillus,* and *Alternaria* are frequent as readily able to grow and sporulate in humidifier reservoirs. Spores released by these fungi can accumulate on surfaces over time, eventually becoming resuspended by human activity or other disturbances (Levy *et al.*, 1999). To support the development of pharmaceuticals, a range of tests using aerobiological methods (definition of doses, biological contamination, filter efficiency, biosafety) may be needed.

FUTURE TRENDS

The necessity to discuss "industrial aerobiology" is evolved by the rapid development of biotechnology and worldwide incidence of the use of methods and products of biotechnology. An interdisciplinary approach is important to resolve new problems in the industrialization of mycotechnology. No one method of aerobiology is ideal for all needs in mycotechnology, but future subjects may create new devices for monitoring air. The author's intention is to emphasize some possible future trends with a little speculation, and also use some of the current ones as a suggestion for future developments.

- Networks of air samplers established for allergenic pollen monitoring, would be suitable for assessing the risk and development of plant epidemics.
- With the development of molecular techniques markers it will be possible to measure the dispersal of the released isolate into the local population. The detectable number of identified species present in the air samples will increase.
- Image analysis techniques should be helpful to study morphological variants of airborne spores providing more detailed studies on the environmental effects.
- Automatic air monitoring systems with alarm via telecommunication network will be widely used in urban areas, crops and storages, to provide early and highly representative detection of fungal contamination.
- A methodology for studying natural spore traps should be elaborated, adopting the tools of surface sampling. Free spores found on phylloplane should be related to the spore composition of the air of a given environment.
- Before applying new fungal species in industry, or releasing them as biocontrol agents, standardized methods including aerobiological techniques will be necessary to reveal the dispersal characteristics.

The contributions of aeromycological applications include different fields of not only industry but also the global ecosystem. Human population vastly increase its demands on the ecosystem for food and energy. No doubt there will arise many challenges for microbiologists with respect to air quality. It is hoped that a central objective linked to internationally standardized methods will result in a fuller appreciation of aeromycology.

REFERENCES

Ainsworth G.C. (1952). The incidence of air-borne *Cladosporium* spores in the London region. *Journal of General Microbiology* 7: 358-361.

Aizenberg V., Reponen T., Grinshpun S.A. and Willeke K. (2000). Performance of Air-O-Cell, Burkard, and Button samplers for total enumeration of airborne spores, *AIHAJ* 61(6): 855-864.

Allen S.J., Brown J.F. and Kochman J.K. (1983). The infection process, sporulation and survival of *Alternaria helianthi* on sunflower. *Annual Applied Biology* 102: 413-419.

Allitt U. (2000). Airborne fungal spores and the thunderstorm of 24 June 1994. *Aerobiologia,* 16: 397-406.

Andersen A.A. (1958). A new sampler for the collection, sizing and enumeration of viable airborne bacteria. *Journal of Bacteriology* 87: 471-484.

Angulo-Romero J., Mediavilla-Molina A. and Dominguez-Vilches E. (1999). Conidia of *Alternaria* in the atmosphere of the city of Cordoba, Spain, in relation to meteorological parameters. *International Journal of Biometeorology* 43: 45-49.

Arseniuk E., Goral T. and Scharen A.L. (1997). Seasonal patterns of spore dispersal of *Phaeosphaeria* spp. and *Stagonospora* spp. *Plant Disease* 82: 187-194.

Aylor D.E. (1990). The role of intermittent wind in the dispersal of fungal pathogens. *Annual Review of Phytopathology* 28: 73-92.

Aylor D.E. (1993). Relative collection efficiency of Rotorod and Burkard spore samplers for airborne *Venturia inequalis* ascospores. *Phytopathology* 83:1116-1119.

Aylor D.E., McCartney H.A. and Brainbridge A. (1981). Deposition of particles liberated in gusts of wind. *Journal of Applied Meteorology* 20: 1212-1221.

Ballero M., Gioannis N., de Goretti G., Lombardini S. and Frenguelli G. (1992). Comparative study about airborne spores in Cagliari and Perugia. *Aerobiologia* 8: 141-147.

Baral H.O. (1992). Vital versus herbarium taxonomy: morphological differences between living and dead cells of Ascomycetes and their taxonomic implications. *Mycotaxon* 44(2): 333-390.

de Bary A. (1887). Comparative morphology and biology of the Fungi, Mycetozoa and Bacteria, Oxford, Clarendon Press. pp. 525.

Baruah H.K. (1961). The air spora of a cowshed. *Journal of General Microbiology* 25: 483-491.

Barui N.C. and Chanda S. (2000). Aeromycoflora in the Central Milk Dairy of Calcutta, India. *Aerobiologia* 16: 367-372.

Beaumont F., Kauffman H.F., van der Mark T.H., Sluiter H.J. and de Vries K. (1985). Volumetric aerobiological survey on conidial fungi in the North-East Netherlands: I. Seasonal patterns and the influence of meteorological variables. *Allergy* 40: 173-180.

Bellin P. and Schillinger J. (2001). Comparison of field performance of the Andersen N6 single stage and the SAS sampler for airborne fungal propagules. *Indoor Air* 11(1): 65-68.

Benyon F.H.L., Jones A.S., Tovey E.R. and Stone G. (1999). Differentiation of allergenic fungal spores by image analysis, with application to aerobiological counts. *Aerobiologia* 15: 211-223.

Bernier T. and Landry J.A. (2000). Algorithmic recognition of biological objects. *Canadian Agricultural Engineering* 42(2): 101-109.

Bock C.H., Jeger M.J., Fitt B.D.L. and Sherrington J. (1997). Effect of wind on the dispersal of oospores of *Peronosclerospora sorghi* from sorghum. *Plant Pathology* 46: 439-449.

Booth C. (1971). Methods in microbiology. Academic press, London, pp. 401.

Boucher A., Hidalgo P., Thonnat M., Belmonte J., Galán C., Bonton P. and Tomczak R. (2002). Development of a semi-automatic system for pollen recognition. *Aerobiologia* 18: 195-201.

Bourdillon R.B., Lidwell O.M. and Thomas J.C. (1941). A slit sampler collecting and counting air-borne bacteria. *Journal of Hygiene* 41: 197-224.

Bowden J., Gregory P.H. and Johnson C.G. (1971). Possible wind transport of coffee leaf rust across the Atlantic Ocean. *Nature* 229: 500-501.

Brown J.K.M. and HovmØller M.S. (2002). Aerial dispersal of pathogens on the global and continental scales and its impact on plant disease. *Science* 297: 537-541.

Burch M. and Levetin E. (2002). Effects of meteorological conditions on spore plumes. *International Journal of Biometeorology* 46: 107-117.

Burge H.A. (1986). Some comments on the aerobiology of fungus spores. *Grana* 25:143-146.

Calderón C., Lacey J., McCartney A. and Rosas I. (1997). Influence of urban climate upon distribution of airborne Deuteromycete spore concentrations in Mexico City. *International Journal of Biometeorology* 40(2): 71-80.

Calderon C., Ward E., Freeman J. and McCartney A. (2002). Detection of airborne fungal spores sampled by rotating-arm and Hirst-type spore traps using polymerase chain reaction assays. *Journal of Aerosol Science* 33(2): 283-296.

Carter M.V. (1965). Ascospore deposition in *Eutypa armeniaceae. Australian Journal of agricultural Research* 16: 825-836.

Carter M.V. and Moller W.J. (1961). Factors affecting the survival and disseminations of *Mycosphaerella pinodes* (Berk. & Blox.) Vestergr. in South Australian irrigated pea fields. *Australian Journal of agricultural Research,* 12: 879-888.

Chamberlain A.C. and Chadwick R.C. (1972). Deposition of spores and other particles on vegetation and soil. *Annals of Applied Biology* 71:141-158.

Christensen C.M. (1950). Intramural dissemination of spores of *Hormodendrum resinae. Journal of Allergy* 21:409-413.

Cohen Y. and Rotem J. (1970). The relationship of sporulartion to photosynthesis in some obligatory and facultative parasites. *Phytopathology* 60: 1600-1604.

Cox C.S. (1987). The aerobiological pathway of microorganisms, New York, John Wiley & Sons, 172-205.

Cundill P. (1998). A comparison of results from two designs of sedimentation pollen trap. *Grana* 37(1):49-52.

Daws L.F. (1967). Movement of air streams indoors. 31-59. In: Airborne microbes, Gregory P.H. and Monteith J.L. Eds., Cambridge Univ. Press, Cambridge, England.

Dominguez de Toledo L.S. (1994). Suggestions for describing and illustrating fungal spores. *Mycotaxon* 52:259-270.

Dörge T., Carstensen J.M. and Frisvald J.C. (2000). Direct identification of pure *Penicillium* species using image analysis. *Journal of Microbiological Methods* 41: 121-133.

Durham O.C. (1946). The volumetric incidence of atmospheric allergens. IV: A proposed standard method of gravity sampling, counting and volumetric interpolation of results. *Journal of Allergy* 18: 231-238.

Dutzmann S. (1985). Zur Analyse der Beziehung zwinschen Klimadaten und Sporenproduktion sowie Sporenverbreitung von *Erysiphe graminis* f. sp. *hordei. Zeitschrift für Pflanzenkrankheiten und Pflanzenschutz* 92: 629-642.

Edmonds R. L. (1979). Aerobiology– The ecological systems approach, Stroudsburg, Dowden, Hutchinson & Ross Inc, pp. 386.

Farkas I., Erdei E., Fehér Z., Magyar D., Páldy A. and Bánházi A. (1999). The effect of a dust storm on April 4, 1998 in Hungary on the aerobiological and morbidity data. *Epidemology* 10(4): 248.

Fernández D., Valencia M.V., Molnár T., Vega A. and Sagues E. (1998). Daily and seasonal variations of *Alternaria* and *Cladosporium* airborne spores in León (North-West, Spain). *Aerobiologia* 14: 215-220.

Fernández-González D., Suárez-Cevera M., Díaz-González T. and Valencia-Barrera R.M. (1993). Airborne pollen and spores of León (Spain). *International Journal of Biometeorology* 37: 89-95.

Fiorina A. (1998). A personal sampler to monitor airborne particles of biological origin. *Aerobiologia* 14: 299-301.

Forster G.F. (1977). Effect of leaf-surface wax on the deposition of airborne propagules. *Transactions of the British mycological Society* 68: 245-250.

Frenz D.A. (1999). Comparing pollen and spore counts collected with the Rotorod Sampler and Burkard spore trap, *Annals of Allergy. Asthma and Immunology* 83(5): 341-349.

Frenz D.A. (2000). The effect of wind speed on pollen and spore counts collected with the Rotorod Sampler and Burkard spore trap, *Annals of Allergy. Asthma and Immunology* 85(5): 392-394.

Gage S.H., Scott A.I. and Manuel Colunga-G. (1999). Ecological scaling of aerobiological processes. *Agricultural and Forest Meteorology* 249-261.

Geagea L.W., Huber L. and Sache I. (1997). Removal of urediniospores of brown (*Puccinia recondita* f. sp. *tritici*) and yellow (*Puccinia striiformis*) rusts of wheat from infected leaves submitted to a mechanical stress. *European Journal of Plant Pathology* 103: 785-793.

Giner M.M. and Garcia J.S.C. (1995). Daily variations of *Alternaria* spores in the city of Murcia (semi-arid southeastern Spain). *International Journal of Biometeorology* 38: 176-179.

Giner M.M., Garcia J.S.C. and Camacho C.N. (2001). Airborne *Alternaria* spores in SE Spain (1993-98)-Occurrence patterns, relationship with weather variables and prediction models, *Grana* 40(3): 111-118.

Gottwald S., Germeier C.U. and Ruhmann W. (2001). Computerized image analysis in *Fusarium* taxonomy. *Mycological Research* 105: 206-214.

Grebenšèsikov I. (1951). Über biologische Schädlingsbekämpfung (an Beispielen aus der neueren sowjetischen Literatur), *Urania* 14: 329-335.

Gregory P.H. (1952). Fungus spores. *Transactions of the British mycological Society* 35: 1-18.

Gregory P. H. (1961). The microbiology of the atmosphere, London, Intersience Publishers Inc. 252.

Gregory P.H. and Stedman O.J. (1958). Spore dispersal in *Ophiobolus graminis* and other fungi of cereal foot rots. *Transactions of the British mycological Society* 41(4): 449-456.

Hamilton E.D. (1959). Studies on the air spora, *Acta Allergologica* 13: 143-175.

Harvey R. (1967). Air-spora studies at Cardiff I. *Cladosporium. Transactions of the British mycological Society* 50(3): 479-495.

Hasnain S.M. (1993). Influence of meteorological factors on the air spora, *Grana* 32: 184-188.

Haugland R.A., Vesper S.J. and Wymer L.J. (1999). Quantitative measurement of *Stachybotrys chartarum* conidia using real time detection of PCR products with the TaqMan (TM) fluorogenic probe system. *Molecular and Cellular Probes* 13(5): 329-340.

Hinds W.C. (1982). Aerosol technology, New York, John Wiley & Sons, pp. 165.

Hirst J.M. (1952). An automatic volumetric spore trap. *Annals of Applied Biology,* 39: 257-265.

Hirst J.M. (1953). Changes in atmospheric spore content: diurnal periodicity and the effects of weather. *Transactions of the British mycological Society* 36: 375-393.

Hirst J.M. and Stedman O.J. (1962). The epidemiology of apple scab (*Venturia inaequalis* (Cke.) Wint.). III. The supply of ascospores. *Annals of Applied Biology* 50: 551-567.

Hirst J.M. and Stedman O.J. (1963). Dry liberation of fungus spores by raindrops. *Journal of General Microbiology* 33: 335-344.

Hirst J.M.and Stedman O.J. (1971). Patterns of spore dispersal in crops. In: Ecology of leaf surface micro-organisms, Preece T.F. and Dickinson C.H. Eds., London, Academic Press, pp. 640.

Hjelmroos M. (1993). Relationship between airborne fungal spore presence and weather variables, *Grana* 32: 40-47.

Ho T.M., Tan B.H., Ismail S. and Bujang M.K. (1995). Seasonal prevalence of air-borne pollen and spores in Kuala Lumpur, Malaysia. *Asian Pacific Journal of Allergy and Clinical Immunology* 13: 17-22.

Horner W.E., O'Neil C., Lehrer S.B. (1992). Basidiospore aeroallergens. *Clinical Reviews of Allergy* 10: 191-211.

Ingold C.T. (1971). Fungal spores, their liberation and dispersal, Oxford, Clarendon Press, pp. 302.

Jensen P.A., Lighthart B., Mohr A.J. and Shaffer B.T. (1994). Instrumentation used with microbial bioaerosol, In: Atmospheric microbial aerosols: theory and applications, Linghthart B., Mohr A.J. Eds., New York, Chapman & Hall 284.

Jones A.S. (2000). Image analysis applied for aerobiology, 2nd European Symposium on Aerobiology, Vienna, Austria, 5-9 September, M101.

Jones M.D. and Brooks J.S. (1950). Effectiveness of distance and border rows in preventing outcrossing in corn, Oklahoma Agricultural Experimental Station, Technical Bulletin No. T-38.

Jones M.D. and Brooks J.S. (1952). Effect of tree barriers on outcrossing in corn, Oklahoma Agricultural Experimental Station, Technical Bulletin No. T-45.

Jones A.M. and Harrison R.M. (2004). The effects of meteorological factors on atmospheric bioaerosol concentrations- a review. *Science of the Total Environment* 326: 151-180.

de Jong, M.D., Bourdot G.W., Powell J. and Goudriaan J. (2002a). A model of the escape of *Sclerotinia sclerotiorum* ascospores from pasture. *Ecological Modelling* 150: 83-105.

de Jong M.D., Bourdot G.W., Hurrell G.A., Saville D.J., Erbrink H. and Zadoks J.C. (2002b). Risk analysis for biological control–simulating dispersal of *Sclerotinia sclerotiorum* (Lib.) de Bary ascospores from a pasture after biological control of *Cirsium arvense* (L.) Scop. *Aerobiologia* 18: 211-222.

de Jong M.D., Wagenmakers P.S. and Goudriaan J. (1991). Modelling the escape of *Chondrostereum purpureum* spores from a larch forest with biological control of *Prunus serotina. Netherlands Journal of Plant Pathology* 97: 55-61.

Katial R.K., Zhang Y., Jones R.H. and Dyer P.D. (1997). Atmospheric mould spore counts in relation to meteorological parameters. *International Journal of Biometeorology* 41: 17-22.

Kennedy R., Wakeham A.J., Byrne K.G., Meyer U.M. and Dewey F.M. (2000). A new method to monitor airborne inoculum of the fungal plant pathogens *Mycosphaerella brassicicola* and *Botrylis cinerea. Applied and Environmental Microbiology* 66(7): 2996-3003.

Kerling I.C.P. (1949). Attack of peas by *Mycosphaerella pinodes* (Berk. et Blox.) Stone. *Tijdschr Plantenzickten* 55: 41-68.

Kovács M. (1985). A nagyvárosok környezete, Budapest, Gondolat Press, pp. 108.

Kurkela T. (1997). The number of *Cladosporium* conidia in the air in different weather conditions. *Grana* 36: 54-61.

Lacey J. (1971). The microbiology of moist barley storage in unsealed silos. *Annals of Applied Biology* 69: 187-212.

Lacey J. (1973). Air spora of a Portuguese cork factory. *Annals of Occupational Hygiene* 16: 223-230.

Lacey J. (1981). The aerobiology of conidial fungi. 373-415. In: The biology of conidial fungi, Vol. 1., Cole G.T. and Kendrick W.B. Eds., New York, Academic Press, pp. 660.

Lacey J. (1984). Aerial dispersal and the development of microbial communities. 140-170. In: Microbial ecology, Lynch J.M. and Poole N.J. Eds., Oxford, Blackwell Scientific Publications, pp. 266.

Lacey J. (1991). Aggregation of spores and its effect on aerodynamic behaviour, *Grana* 30: 347-445.

Lacey J. (1995). Airborne pollens and spores– A guide to trapping and counting, Harpenden, British Aerobiology Federation pp. 60.

Lacey J. (1996). Spore dispersal– its role in ecology and disease: The British contribution to fungal aerobiology. *Mycological Research* 100(6): 641-660.

Langenberg W.J., Sutton J.C. and Gillespie T.J. (1977). Relation of weather variables and periodicities of airborne spores of *Alternaria dauci. Phytopathology* 67: 879-883.

Larson C., Francl L.J. and Friesen T. (2001). Evaluation of the Burkard cyclonic spore sampler for collection efficiency of ascospores. *Plant Disease* 85(12): 1249-1252.

Levetin E. (1998). Aerobiology and biometeorology of basidiospores and ascospores, The 5th International Congress on Aerobiology, Perugia S19, 53.

Levetin E., Rogers C.A. and Hall S.A. (2000). Comparison of pollen sampling with a Burkard Spore Trap and a Tauber Trap in a warm temperate climate, *Grana* 39(6): 294-302.

Levy J.I., Nishioka Y., Gilbert K., Cheng C-H. and Burge H.A. (1999). Variables in aerosolizing activities and airborne fungal concentrations in a bakery. *American Industrial Hygiene Association Journal* 60: 317-325.

Lewis H.E., Foster A.R., Mullan B.J., Cox R.N. and Clark R.P. (1969). Aerodynamics of the human microenvironment. *Lancet* 1: 1273-1277.

Li D.W. and Kendrick B. (1994). Functional relationships between airborne fungal spores and environmental factors in Kitchener-Waterloo, Ontario, as detected by Canonical correspondence analysis. *Grana* 33: 166-176.

Lim S.H., Chew F.T., Dali S.D.B.M., Tan H.T.W., Lee B.W. and Tan T.K. (1998). Outdoor airborne fungal spores in Singapore. *Grana* 37(4): 246-252.

Lin W-H. and Li C-S. (2000). Associations of fungal aerosols, air pollutants, and meteorological factors. *Aerosol Science Technology* 32: 359-368.

Liu J., Dazzo F.B., Glagoleva O., Yu B. and Jain A.K. (2001). CMEIAS: A computer-aided system for image analysis of bacterial morphotypes in microbial communities. *Microbial Ecology* 41: 173-194.

Lopez M. and Salvaggio E. J. (1983). Climate-weather-air pollution 1203-1214. In: Allergy: principle and practice, Middleton E., Reed C.E., and Ellis E.F. Eds., Toronto, C.V. Mosby St. Louis, pp. 1440.

Mason E.W. (1937). Annotated account of fungi received at the Imperial Mycological Institute, List II, Fasc. 3., Kew Surrey, IMI, pp.

Madelin T.M. and Johnson H.E. (1992). Fungal and actinomycete spore aerosols measured at different humidities with an aerodynamic particle sizer. *Journal of Applied Bacteriology* 72: 400-409.

Magyar D., Szécsi Á.and Barasits T. (2003). Andersen és Hirst típusú levegőmintavevők összehasonlítása növény- és közegészségügyi vizsgálatokban. *Környezeti Ártalmak és Légzőrendszer* 13: 239-250.

Maunsell K. (1954). Concentration of airborne spores in dwelling under normal conditions and under repair. *International Archives of Allergy and Applied Immunology* 5: 373-376.

May K.R. (1945). The cascade impactor: an instrument for sampling coarse aerosols. *Journal of Scientific Instruments* 22: 187-95.

McCartney H.A. and Lacey M.E. (1990). The production and release of ascospores of *Pyrenopeziza brassicae* Sutton et Rawlinson on oilseed rape. *Plant Pathology* 39: 17-32.

McCartney H.A, Fitt B.D.L. and Schmechel D. (1997). Sampling bioaerosols in plant pathology. *Journal of Aerosol Science* 28(3): 349-364.

Mehta S.K., Mishra S.K. and Pierson D.L. (1996). Evaluation of three portable samplers for monitoring airborne fungi. *Applied and Environmental Microbiology* 62(5): 1835-1838.

Meredith D.S. (1961). Spore discharge in *Deightoniella torulosa* (Syd.) Ellis. *Annals of Botany* 25: 271-278.

Meredith D.S. (1962). Some components of the air-spora in Jamaican banana plantations. *Annals of Applied Biology* 50: 577-592.

Meredith D.S. (1965). Violent spore release in *Helminthosporium turcicum*. *Phytopathology* 55: 1099-1102.

Meredith D.S. (1966). Spore dispersal in *Alternaria porri* (Ellis) Neerg. on onion in Nebraska. *Annals of Applied Biology* 57: 67-73.

Moore P.D., Webb J.A. and Collinson M.E. (1991). Pollen analysis (second edition), Oxford, Blackwell Scientific Publications, pp. 216.

Moulton F.R. (1942). Aerobiology, Washington, American Association for the Advancement of Science, pp. 289.

Nagarajan S. and Singh H. (1976). Preliminary studies on forecesting wheat stem rust appearance. *Agricultural Meteorology* 17: 281-289.

Ogden E.C., Hayes J.V. and Raynor G.S. (1969). Diurnal patterns of pollen emission in *Ambrosia, Phleum, Zea* and *Ricinus*. *American Journal of Botany* 56: 16-21.

Olsson M., Sukura A., Lindberg L.A. and Linder E. (1996). Detection of *Pneumocystis carinii* DNA by filtration of air. *Scandinavian Journal of Infectious Diseases* 28(3): 279-282.

O'Meara T., Shawpan V., Mohamed H., Poulos L. and Tovey E. (2000). Use of nasal samplers and halogen assay to measure levels of personal domestic exposure to mite and pet allergens and IgE-based responses of subjects with atopic diseases. *Journal of Allergy and Clinical Immunology* 105(1): 1096.

Pady S.M., Kramer C.L. and Clary R. (1969). Periodicity in spore release in *Cladosporium*. *Mycologia* 60(1): 87-98.

Pahl O., Phillips V.R., Lacey J., Hartung J. and Wathes C.M. (1997). Comparison of commonly used samplers with a novel bioaerosol sampler with automatic plate exchange. *Journal of Aerosol Science* 28(3): 427-435.

Palmas F. and Consentino S. (1990). Comparison between fungal airspore concentration at two different sites in the south of Sardinia. *Grana* 29: 87-95.

Palmerley R.A. and Benedict W.G. (1977). Patterns of conidial release of *Helminthosporium turcicum* on sweet corn under controlled environmental conditions. *Canadian Journal of Botany* 55: 1991-1995.

Palmgren U., Ström G., Malmberg P. and Blomquist G. (1986). The nuclepore filter method: a technique for enumeration of viable and nonviable airborne microorganisms. *American Journal of Industrial Medicine* 10(3): 325-327.

Panzer J.D., Tullis E.C. and van Arsdel E.P. (1957). A simple 24-hour slide spore collector. *Phytopathology* 47: 512-514.

Paredes M.M., Martínez J.F., Silva I., Tormo R. and Muñoz A.F. (1997). Influencia de los parámetros meteorológicos en la dispersión de esporas de las especies de *Alternaria* Nees ex Fr. *Boletín de Sanidad Vegetal Plagas* 23: 541-549.

Pasanen A.L., Pasanen P., Jantunen M.J. and Kalliokoski P. (1991). Significance of air humidity and air velocity for fungal spore release into the air. *Atmospheric Environment* 25: 459-462.

Perkins W.A. (1957). The Rotorod sampler. Second semiannual report of the Aerosol Laboratory, Stanford University, Palo Alto, pp. 66.

Philipp W.D., Grauer U. and Grossmann F. (1984). Erganzende Untersuchungen zur biologischen und integrierten Bekampfung von Gurkenmehltau unter Glas durch *Ampelomyces quisqualis*. *Zeitschrift für Pflanzenkrankheiten und Pflanzenschutz* 91: 438-443.

Philips G.B. (1965). Safety in the chemical laboratory. XIII. Microbiological hazards in the laboratory, I. *Journal of Chemical Education* 42: A43-44, A46-A48.

Picco A.M. and Rodolfi M. (2004). Assessments of indoor fungi in selected wineries of Oltrepo Pavese (Northern italy) and Sottoceneri (Switzerland). *American Journal of Enology and Viticulture* 55(4): 355-362.

Portnoy J., Landuyt J., Pacheco F., Flappan S., Simon S. and Barnes C. (2000). Comparison of the Burkard and Allergenco MK-3 volumetric collectors. *Annals of Allergy Asthma and Immunology* 84(1): 19-24.

Rajasab A.H. and Ramalingam A. (1989). Splash dispersal in *Ramulispora sorghi* Olive and Lefebreve, the causal organism of sooty stripe of sorghum. *Proceedings of the Indian Academy of Sciences*, 99(4): 335-341.

Rantio-Lehtimäki A. (1977). Research on airborne fungus spores in Finland. *Grana* 16: 163-165.

Razmovski V., O'Meara T.J., Taylor D.J.M. and Tovey E.R. (2000). A new method for simultaneous immunodetection and morphologic identification of individual sources of pollen allergens. *Journal of Allergy and Clinical Immunology* 105(4): 725-31.

Reinthaler F.F., Marth E., Eibel U., Enayat U., Feenstra O., Friendl H., Köck M., Pichler-Semmelrock F.P., Priding G. and Schlacher R. (1987). The assessment of airborne microorganisms in large-scale composting facilities and their immediate surroundings. *Aerobiologia* 13: 167-175.

Reponen T., Lehtonen M., Raunemaa T. and Nevalainen A. (1992). Effect of indoor sources on fungal spore concentrations and size distributions. *Journal of Aerosol Science* 23: 663-666.

Riediker M., Keller S., Wuthrich B., Koller T. and Monn C. (2000). Personal pollen exposure compared to stationary measurements. *Journal of Investigational Allergology and Clinical Immunology* 10(4): 200-203.

Rogers C. and Muilenberg M. (2001). Comprehensive guidelines for the operation of Hirst-type suction bioaerosol samplers, The Pan-American Aerobiology Association, www.geocities.com/paaaorg.

Ronneberger O., Schultz E. and Burkhardt H. (2002). Automated pollen recognition using 3D volume image images from fluorescence microscopy. *Aerobiologia* 18: 107-115.

Rotem J. (1964). The effect of weather on dispersal of *Alternaria* spores in a semi-arid region of Israel. *Phytopathology* 54: 628-632.

Rotem J. (1994). The genus *Alternaria*, St Paul, APS Press, pp. 326.

Royes I.J. (1987). Some components of the air spora in Jamaica and their possible medical applications. *Grana* 26: 151-157.

Rugieri (1948). Fattori che conditionano o contribuiscono allo sviluppo del "mal secco" degli agrumi e metodi di lotto contra il medesimo. *Annali della Sperimentazione Agraria* 2: 1-51.

Savary S. (1986). Relative humidity and wind velocity associated with diurnal rhythmicity of aerial dispersal of *Puccinia arachnidis* urediniospores. *Netherlands Journal of Plant Pathology* 92: 115-125.

Savino E. and Caretta G. (1992). Airborne fungi in an Italian rice mill. *Aerobiologia* 8: 267-275.

Schrödter H. (1952). Zur phytopatologischen Problematic von Windschutzanlagen. *Nachrichtenblatt Deutscher Pflanzenschutzdienst* 6(32): 91-92.

Schrödter H. (1960). Dispersal by air and water–The flight and landing 169-227. In: Plant pathology: an advanced treatise, vol. 3: The diseased population epidemics and control, Horsfall J.G. and Dimond A.E. Eds., Oxford, Academic Press, pp. 227.

Schultz E. (2001). Sampling efficiency of different particle samplers and the microscopic analysis of air borne pollen. *Gefahrstoffe Reinhaltung der Luft* 61(6): 261-265.

Schwarzbach E. (1979). A high throughput jet trap for collecting mildew spores on living leaves. *Phytopathologische Zeitschrift* 94: 165-171.

Schweigkofler W., O'Donnell K. and Garbelotto M. (2004). Detection and quantification of airborne conidia of *Fusarium circinatum*, the causal agent of pine pitch canker, from two California sites by using a real-time PCR approach combined with simple spore trapping method. *Applied and Environmental Microbiology* 70(6): 3512-3520.

Sreeramulu T. (1959). The diurnal and seasonal periodicity of spores of certain plant pathogens in the air. *Transactions of the British mycological Society* 42(1): 177-184.

Stennett P.J. and Beggs P.J. (2004). *Alternaria* spores in the atmosphere of Sydney, Australia, and relationships with meteorological factors. *International Journal of Biometeorology* 39(2): 106-112.

Takahashi Y., Kawashima S., Fujita T., Ito C., Togashi R. and Takeda H. (2001). Comparison between real-time pollen monitor KH-3000 and Burkard sampler. *Arerugi* 50(12): 1136-42.

Takahashi Y., Nagoya T. and Nitta H. (2000). Autofluorescent image analysis in pollen counting, 2nd European Symposium on Aerobiology, Vienna, Austria 5-9 September, M102.

Tauber H. (1974). A static non-overload pollen collector. *New Phytologist* 73: 359-369.

Tavora L.G.F., Gambale W., Heins-Vaccari E.M., Arriagada G.L.H., Lacaz C.S., Santos C.R. and Levin A.S. (2003). Comparative performance of two air samplers for monitoring airborne fungal propagules. *Brazilian Journal of Medical and Biological Research* 36(5): 613-616.

Tian C-M., Shang Y-Z., Zhuang J-Y., Wang Q. and Kakishima M. (2004). Morphological and molecular phylogenetic analysis of *Melampsora* species on poplar in China. *Mycoscience* 45: 56-66.

Tilak S.T. and Pande B.N. (2005). Current trends in aeromycological research. 281-510. In: Fungi: Diversity and Biotechnology, Rai M.K. and Deshmukh S.K. Eds., Jodhpur, Scientific Publishers.

Tomas C., Candau P. and Minero F.J.G. (1997). A comparative study of atmospheric pollen concentrations collected with Burkard and Cour samplers, Seville (Spain), 1992-1994. *Grana* 36(2): 122-128.

Tuboly L. and Vörös J. (1962). A dohányperonoszpóra és leküzdésének lehetõségei. *Mezõgazdasági világirodalom* 4: 46-53.

Tuomi T., Johnsson T. and Reijula K. (2000). Mycotoxins and associated fungal species in building materials from water-damaged buildings. *Proceedings of Healthy Buildings* 3: 371-376.

Turott C. and Levetin E. (2001). Correlation of spring spore concentrations and meteorological conditions in Tulsa, Oklahoma. *International Journal of Biometeorology* 45: 64-74.

Venables K.M., Allitt U., Collier C.G., Emberlin J., Greig J.B., Hardaker P.J., Highham J.H., Laing-Morton T., Maynard R.L., Murray V., Strachan D. and Tee R.D. (1997). Thunderstorm-related asthma-the epidemic of 24/25 June 1994. *Clinical and Experimental Allergy* 27: 725-736.

Viljanen-Rollinson S.L.H. and Cromey M.G. (2002). Pathways of entry and spread of rust pathogens: implications for New Zealand's biosecurity. *New Zealand Plant Protection* 55: 42-48.

Wadia K.D.R., McCartney H.A. and Butler D.R. (1998). Dispersal of *Passalora personata* conidia from groundnut by wind and rain. *Mycological Research* 102: 355-360.

Waggoner P.E. and Taylor G.S. (1958). Dissemination by atmospheric turbulence: spores of *Peronospora tabacina. Phytopathology* 48: 46-51.

Wakefield A.E. (1996). DNA sequences identical to *Pneumocystis carinii* f sp *carinii* and *Pneumocystis carinii* f. sp. *hominis* in samples of air spora. *Journal of Clinical Microbiology* 34(7): 1754-1759.

Walker R.F., Ishikawa K. and Kumagai M. (2002). Fluorescence-assisted image analysis for freshwater microalgae. *Journal of Microbiological Methods* 51: 149-162.

Waller J.M. (1979). The recent spread of coffee rust (*Hemileia vastarix*) and attempts to control it 275-283. In: Plant Health, Ebbels D.L. and King J.E. Eds., Oxford, UK, Blackwell Scientific Publications.

Ward S.V. and Manners J.G. (1974). Environmental effects on the quantity and viability of conidia produced by *Erysiphe graminis. Transactions of the British mycological Society* 62: 119-128.

Warner F.E. and Emberlin J. (1998). An evaluation of fungal spore counting techniques for the Hirst-type spore trap tape. MICRO 98, International Microscopy Conference and Exhibition of the Royal Microscopical Society, London, pp.

Weinhold A.R. (1955). Rate of fall of uredospores of *Puccinia graminis tritici* Erikss & Henn. as affected by humidity and temperature, Technical Report of Botany and Plant Pathology, Department Colorado, pp. 104.

Whyte W. (1981). The Casella Slit Sampler or the Biotest Centrifugal Sampler– Which is the more efficient. *Journal of Hospital Infection* 297-299.

Whitfield W.J. (1962). A new approach to clean room design, Sandia National Laboratories, technical report no. SC-4673 RR, Sandia Co., Albuquerque, New Mexico, pp.

Williams R.H., Ward E. and McCartney H.A. (2001). Methods for integrated air sampling and DNA analysis for detection of airborne fungal spores. *Applied and Environmental Microbiology* 67(6): 2453-2459.

Wiseman R. (1932). Untersuchungen über die überwinterung des Apfelschorfpilzes *Fusicladium dendriticum* (Wallr.) Fckl. im tötem Blatt sowie die Ausbreitung der Sommersporen (konidien) des Apfelschorfes. *Landwirtschaftliches Jahrbuch der Schweiz* 46: 619-679.

Wolfe M.S. (1967). Physiologic specialization of *Erysiphe graminis* f. sp. *tritici* in the United Kingdom, 1964-5. *Transactions of the British mycological Society* 50: 631-640.

Wüst G., Friedl H., Haas D., Köck M., Pichler-Semmelrock F., Reinthaler F.F., Schlacher R. and Marth E. (2003). A comparison between Andersen (ACFM) and Reuter Centrifugal Sampler (RCS-plus) for indoor sampling of airborne moulds. *Aerobiologia* 19(2): 125-128.

Zeterberg J.M. (1973). A review of respiratory virology and the spread of virulent and possibly antigenic viruses via air conditioning systems II. *Annals of Allergy* 31: 291-299.

Zoberi M.H. (1961). Take-off of mould spores in relation to wind speed and humidity. *Annals of Botany* 25: 53-64.

15 Bioactive Sesquiterpenes Produced From Fungi: Possibilities and Limitations

Wolf-Rainer Abraham

GBF – National Research Center for Biotechnology, Chemical Microbiology, Mascheroder Weg 1, D-38124 Braunschweig, Germany.

INTRODUCTION

Fungi and bacteria are known to produce a wealth of secondary metabolites which can be used in a wide range of applications. Since the beginning of the era of penicillins in the middle of the 20th century a still growing number of antibiotics, immunosuppressive or phytotoxins are produced on industrial scale owing to the fact that many microorganisms can be grown in fermentors for the production of secondary metabolites. Here, fungi are still underexplored although a number of metabolites were isolated from fungi which found their way into applications as natural products, starting material for pharmaceuticals or as lead structures for the development of pharmaceutical products.

Fungi usually occur in microbial communities where they are under constant threat of other organisms feeding on them. As a consequence, fungi developed a number of strategies for protection, one of them is the production of toxins. The fungal subdivision Basidiomycotina produce toxic sesquiterpenes, many of them derived from the protoilludane skeleton. This skeleton is transformed and rearranged to a multitude of compounds.

There appears to be continuing interest in the overall chemistry of fungi because this group is among the world's greatest unexplored resources for biodiversity as well as chemodiversity. A still increasing number of fungal metabolites have been described and hundreds of terpenes have been isolated from fungi, the majority belonging to the group of sesquiterpenes. Within the fungal sesquiterpenes, some large groups of biosynthetically related sesquiterpene can be discerned, one of them is formed via the humulene-protoilludane biosynthetic pathway. Furthermore, the sesquiterpenes belonging to the groups of trichothecane and the eremophilane deserve some discussions. Additionally, there are sesquiterpenes like dermatolactone (Mayer *et al.*, 1996) or nidulal (Erkel *et al.*, 1996) and volatile sesquiterpenes which do not fit into this scheme and which will be discussed separately. This chapter focuses on bioactive sesquiterpenes remarkable for their chemotaxonomical, ecological and pharmaceutical implications but it is outside of the scope of this chapter to give a complete overview of all sesquiterpenes produced by fungi. A review on sesquiterpenes useful in medicinal chemistry appeared recently (Abraham, 2001).

SESQUITERPENES DERIVED FROM THE HUMULENE SKELETON

Fungal sesquiterpenes formed via the Humulene-protoilludane biosynthetic pathway are characteristic for the subdivision Basidiomycotina. Only one species of the Ascomycotina is known to date to form sesquiterpenes possessing the protoilludane skeleton (Hanssen *et al*., 1986). A number of reviews appeared covering this field of natural products (Anke, 1989), including their chemical syntheses (Ayer and Brown, 1981), or focusing on the genus *Lactarius* (Vidari and Vita-Finzi, 1995). An annual update of the literature published in the field of natural sesquiterpenes is currently given by Fraga, 2005.

Secondary metabolites with the Humulene skeleton are rare in fungi instead this skeleton is further transformed to bicyclic and tricyclic sesquiterpenes. In the sub-division Basidiomycotina both cis- and trans-configurated cyclobutyl containing caryophyllanes are known. Furthermore, caryophyllane can be cyclised by *Poronia punctata* to the rare punctaporonane skeleton. *Collybia confluens* can form collybial 32, another tricyclic skeleton only known from this compound.

An important pathway starting from Humulene produces the tricyclic sesquiterpene protoilludane. At least five different pathways are progressing via protoilludane as an intermediate. All of them are based on the rearrangement of the strained cyclobutane moiety. A pathway involving protoilludane is the biosynthesis of cerapicane. This skeleton is only known from an ascomycete but it is the intermediate for the group of metabolites possessing the sterpurane carbon framework. The cerapicane skeleton can be opened up yielding the bulleranes only known from *Cyathus bulleri*. Another pathway leads to sesquiterpenes of the marasmane type which itself is the intermediate for further rearrangements (Froborg and Magnusson, 1978). Migration of the geminal dimethyl group leads to the hydrogrammanes with the metabolite hydrogrammic acid, opening of the strained cyclopropyl ring gives the lactaranes and the norsesquiterpenes with the glutinopallane skeleton can be derived by contraction of the cyclopentane part and cleavage of a methyl group. The cyclobutane ring in protoilludane can be enlarged to a cyclopentane realized in ceratopicane from where the hirsutanes are formed by rearrangements of two methyl groups. Protoilludane can also be rearranged to the illudanes with its spirocyclopropane ring (Shirahama *et al*., 1978) or to the illudalanes. Cleavage of a C-C-bond in protoilludane yields the interesting fomannosane skeleton. As in the marasmane-hydrogrammane or the illudane-isoilludane pathway pleurotellane is formed from hirsutane via methyl migration. An important group of sesquiterpenes formed by a complex series of skeletal rearrangements is based on isolactarane. The cyclopropyl moiety of isolactarane can be opened resulting in the formation of the merulane metabolites currently only known from *Merulius tremellosus*. In the following sections for the most important sesquiterpene skeletons an overview over some of the known fungal metabolites, their biological activities and the producing species will be given.

HUMULENES

Only very few secondary metabolites with Humulene skeleton are known from fungi (Fig. 15.1). One of them is 4,15-dihydroxy-5,11-dioxo-humula-diene (**1**) isolated from an endophytic fungus from the tree *Taxus brevifolia* (Pulici *et al.*, 1996). The merosesquiterpene xenovulene A (**2**), a novel GABA-benzodiazepine receptor binding compound produced by *Acremonium strictum* possesses a very unique structure of a condensed Humulene (Ainsworth *et al.*, 1995). Humulene and its derivatives are mainly formed by plants. Some of these compounds are enjoyed by humans, e.g., humuleneepoxides, probably formed from the Humulene of hop during the brewing process by yeasts, are important fragrances of beer (Tressl *et al.*, 1988). It has been speculated that some of the fragrances of beer are produced by fungi during the fermentation process. The fermentation of humulene with *Chaetomium cochlioides* led to a rich variety of oxygenated Humulenes (Abraham and Stumpf, 1987), including the di-epoxy-humulenes (**3-7**) and the triepoxide (**8**). Interestingly, the first epoxidation of humulene by this fungus is not enantioselective while the following epoxidations are highly enantioselective leading to some epimeric di-epoxides.

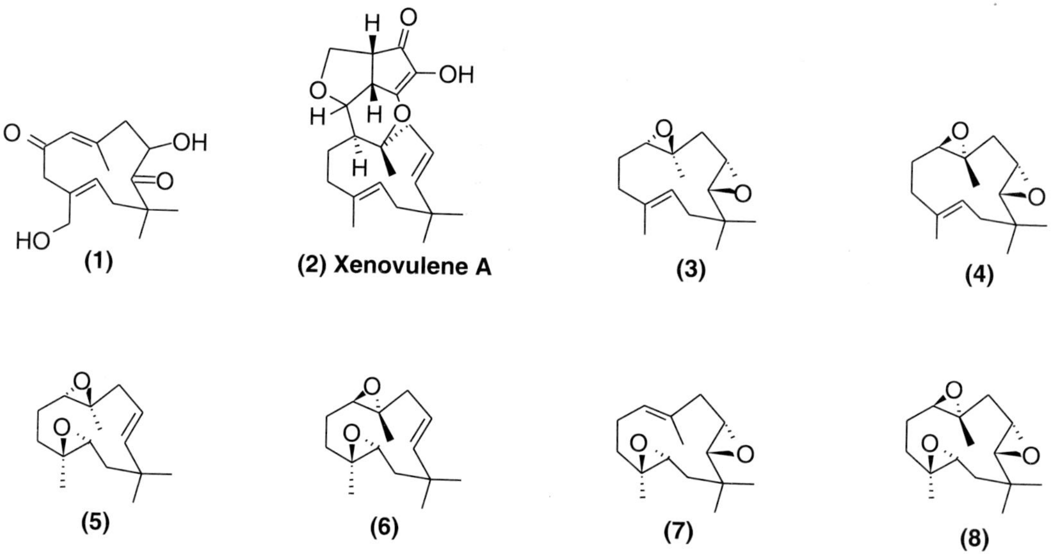

Fig. 15.1. Humulene sesquiterpenes and biotransformation products from fungi.

BISABOLANES AND OTHER MONOCYCLIC SESQUITERPENES

A number of monocyclic sesquiterpenes have been described from fungi (Fig. 15.2), several of them with very interesting biological activities: A novel immunosuppressant, FR65814 (**9**), was isolated and characterized from *Penicillium jensenii* together with fumagilol (**10**), a carcinolytic agent (Hatanaka *et al.*, 1988). From a *Sporothix* sp. chlovalicin (**11**) and ovalicin (**12**) have been characterized (Hayashi *et al.*, 1996). Bisabolanes contribute to the odor of fungi. From a 28 days old liquid culture of *Lepista irina* the bisabolane lepistirone (**13**) was obtained (Abraham *et al.*, 1991). A closely related compound (**14**) but with higher unsaturation was isolated from *Cystostereum* *murraii* (Abraham and Hanssen, 1987). Clovalicin is an interleukin-6 inhibitor which may be of interest for autoimmune diseases. The basidiomycete *Cheimonophyllum candidissimum* is the source of the bisabolanes cheimonophyllons A - E (**15**) and cheimonophyllal (**16**), which are cytoxins and mutagenes (Stadler *et al.*, 1994).

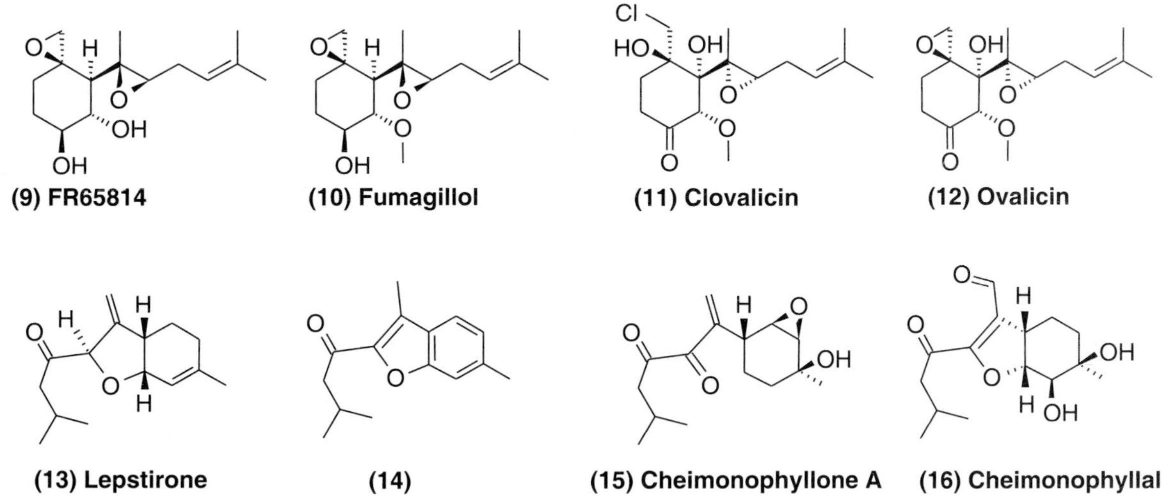

Fig. 15.2. Some bioactive fumigallanes and bisabolanes.

CARYOPHYLLANES AND RELATED SESQUITERPENES

From *Hypholoma fasciculare* and other *Hypholoma* sp. a cytotoxic, antiviral and vasodilative compound was isolated and named naematolin (Ito *et al.*, 1967) (Fig. 15.3). Naematolin (Backens *et al.*, 1984) (**23**) is a caryophyllenediol derivative while naematolone is its dehydro derivative. Naematolin showed no weak antibacterial activity and mutagenicity but naematolone showed only weak antibacterial activity and no mutagenicity in the Ames test. From *Lactarius camphoratus* a sesquiterpenepoxide was isolated and identified as 12-hydroxycaryophyllene-4,5-epoxide (Daniewski *et al.*, 1981) (**20**) also formed by fermentation of caryophyllene (**17**) with *Diplodia gossypina* (Abraham *et al.*, 1990). No biological activities were reported for this metabolite. *Lactarius subumbonatus* Lindgr. common in Italy is characterized by a strong, liquorice-like smell. From the fruit bodies of *L. subumbonatus* 6-hydroxycaryophyllene (S)-6-hydroxystearate (Clericuzio *et al.*, 1999) was isolated. The toxicity of this metabolite was tested against *Artemia salina*. While the ester showed almost no activity the corresponding caryophyllene alcohol (**18**) had an LD_{50} of 11 ppm. The activity of the alcohol and the lack of activity of the ester points to the ester as a prodrug which is activated by stress as has been reported for many protoilludane esters.

Two caryophyllane sesquiterpenes possessing the unusual cis ring junction of caryophyllene and named walleminol and walleminone (**19**) have been isolated from *Wallemia sebi* (Frank *et al.*, 1999). From *Poronia punctata*, a fungus belonging to the Ascomycotina, the caryphyllene triol punctaporonin B (Edwards *et al.*, 1989) and its ether punctaporonin G (Edwards *et al.*, 1989a) (**21**) with conjugated double bonds were isolated. Both compounds are only weak antibiotics. *Pestalotiopsis* sp. growing on the tree *Taxus brevifolia* produced the caryophyllenes pestalotopsin A (**22**) and pestalotiopsin B (Pulici *et al.*, 1996a).

From the fungus *Poronia punctata*, an ascomycete, a group of four closely related metabolites were isolated, all showing low antibiotic activities. These sesquiterpenes comprise the punctaporonins A (**29**), D (**30**), E (**31**), and F (Poyser *et al.*, 1986) shown in Fig. 15.3. The skeleton of punctaporonin C 25 (Anderson *et al.*, 1984) does not belong to the punctaporonane class of sesquiterpenes and represents another skeleton probably derived from a caryophyllene precursor. Biotransformation of caryophyllene (**17**) with the fungus *Diplodia gossypina* led to a multitude of derivatives, among them the diepoxide (**26**), sesquiterpenes (**24**) and (**27**) with a new skeleton with a contracted ring and the punctaporonane derivative (**25**) with an isomeric fusion between the 5- and 6-membered rings compared to the natural punctaporonanes (Abraham *et al.*, 1990a; Fig. 3). By using *Chaetomium cochlioides* instead of *Diplodia gossypina*, however, the isomer (**28**) possessing the relative stereochemistry of natural punctaporonanes (Abraham *et al.*, 1990) was obtained. These biotransformations produced a number of novel sesquiterpenes and proved that fungi can form the punctaporonanes enzymatically from caryophyllene.

The skeleton of collybial (**32**) isolated from the culture broth of *Collybia confluens* is unique and so far only found in this species (Simon *et al.*, 1995). Its biosynthesis can be explained via a

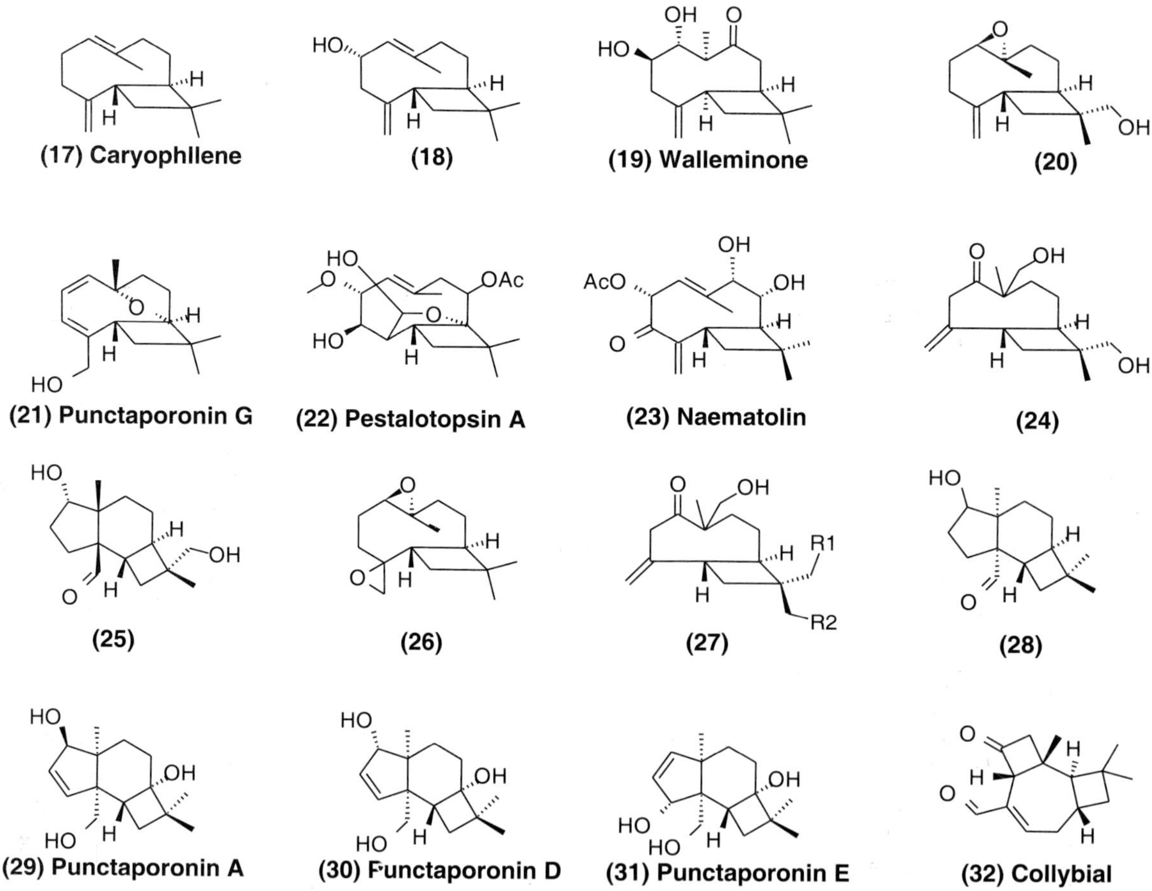

Fig. 15.3. Caryophyllane sesquiterpenes as natural and biotransformation products.

caryophyllane intermediate. Collybial exhibits a weak antifungal activity and rather weak activities against most other bacteria with the exception of *Bacillus subtilis*. Collybial is remarkable because of its antiviral activity inhibiting the propagation of the vesicular stomatitis virus in baby hamster cells. The sesquiterpene also shows some cytotoxicity. It is assumed that collybial achieves its antiviral, cytotoxic and antibiotic activities in a rather unselective manner due to the α,β-unsaturated aldehyde.

PROTOILLUDANES AND RELATED SESQUITERPENES

A huge group of fungal sesquiterpene metabolites possess the protoilludane skeleton. There exists a number of excellent reviews on this subject (Vidari and Vita-Finzi, 1995; Nasini and Vajna de Pava, 1997), therefore only a selection of results will be discussed here. From *Laurilia tsugicola* tsugicoline A – D (**33**) (Arnone *et al.*, 1995) and E (Arnone *et al.*, 1999) (**34**), protoilludanes oxidized in the cyclobutane moiety, were obtained (Fig. 15.4). Long chain fatty acid esters are found in violascenol (**35**) and its 6-ketostearoyl ester (**36**) from *Lactarius violascens* (Vidari *et al.*, 1998). Radulones A (**37**), Radulones B, and radudiol, the illudalane radulactone and the illudane radulol have been isolated from *Radulomyces confluens*. Radulone A (**37**) is cytotoxic and antimicrobial and a potent inhibitor of human platelet aggregatetion. The other four compounds show weak antimicrobial and

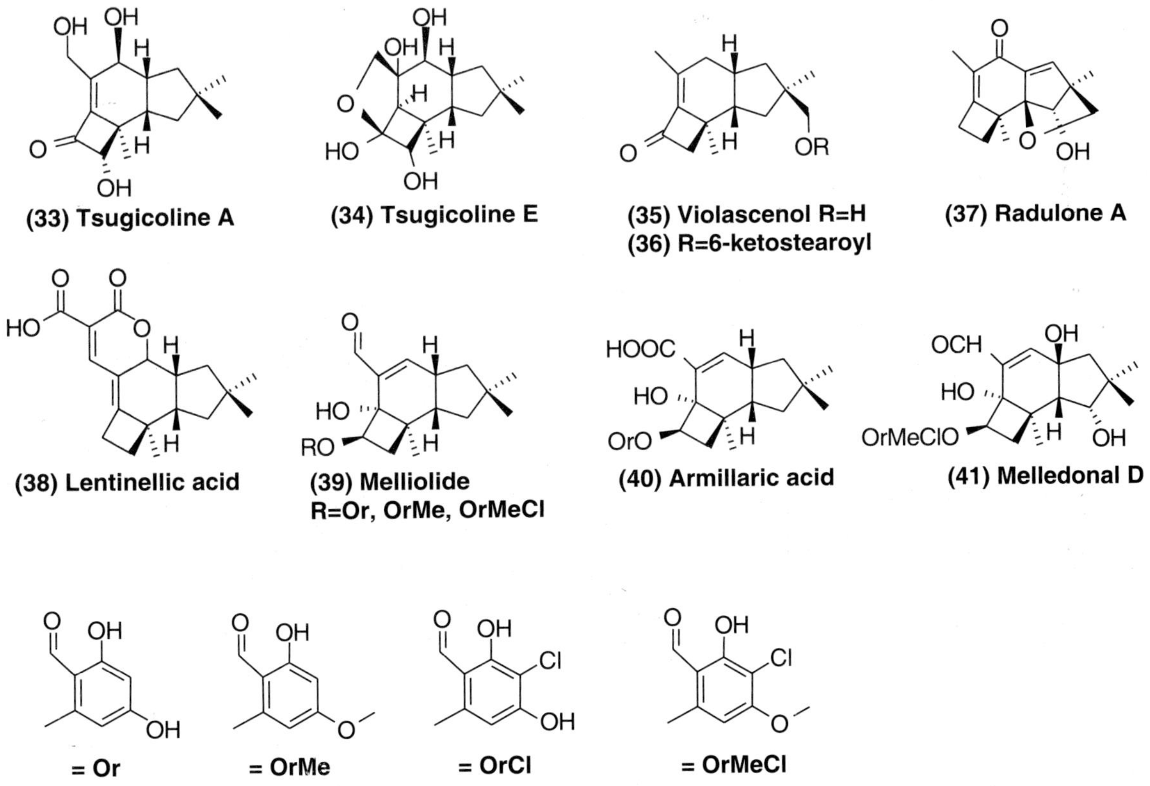

Fig. 15.4. Secondary metabolites with the protoilludane skeleton.

cytotoxic activity (Fabian *et al.*, 1998). Sulcatin, a norprotoilludane, sulcatin B, armillol and 5-epi-armillol (Arnone *et al.*, 1992) was found in *Laurilia sulcata* (Arnone *et al.*, 1987). A precursor of sulcatin may be lentinellic acid (**38**) found in *Lentinellus* species (Staerk *et al.*, 1988). While the antifungal activity of lentinellic acid is quite low it exhibits antibacterial and cytotoxic activities.

The tree pathogen, *Armillaria mellea*, is known to produce chemical defence substances to eliminate its bacterial predators and is the source of a multitude of antibacterial esters of protoilludanols and aromatic acids. These acids are mainly orselinate (Or), everninate (OrMe) and their chlorinated derivatives (Fig. 15.4). One of these esters is armillyl orselinate (Donnelly *et al.*, 1982) and the chlorinated derivative arnamiol (Donnelly *et al.*, 1986). Melleolide (**39**) (Midland *et al.*, 1982) is also known as the everninate armillarin and chloreverninate armillaridin (Yang *et al.*, 1984). From the same species armillaripin (Yang *et al.*, 1990), armillarilin and the chloreverninate ester armillarinin were reported (Yang *et al.*, 1990a) as well as armillaric acid (**40**) (Obuchi *et al.*, 1990). Melleolides E - H together with melledonal D (**41**) and melledonal E were extracted from *Clitocybe elegans* and *Armillaria mellea* (Arnone *et al.*, 1988).

Fomannosin (**42**) a phytotoxic and bactericidal sesquiterpene possessing a seco-protoilludane skeleton named fomannosane was characterized from the wood-rotting fungus *Fomes annosus* (Kepler *et al.*, 1967 (Fig. 15.5). Its absolute configuration was determined by x-ray studies (Cane *et al.*, 1977) and its biosynthesis was demonstrated to go via humulene and protoilludane intermediates (Cane *et al.*, 1978). From *Omphalotus olearius* and *O. nidiformis* (Burgess and Barrow, 1999) a second metabolite of this class of sesquiterpenes was reported, illudosin (**43**) (Arnone *et al.*, 1991). It has no antifungal activity and was inactive against *E. coli* but was active against *Bacillus* spp. and *Sarcina lutea*.

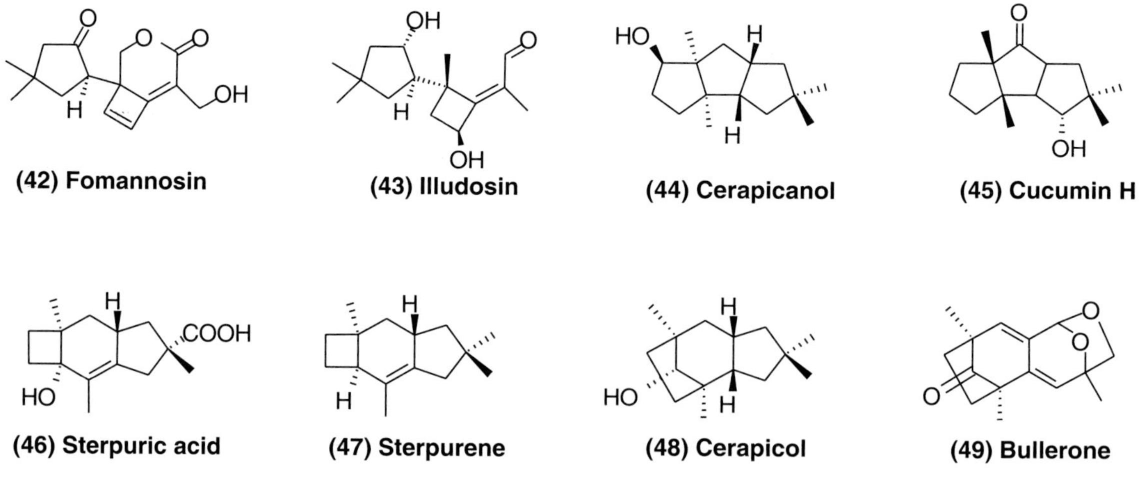

Fig. 15.5. Ceratopicanes and related sesquiterpenes.

CERATOPICANES

A triquinane intermediate has been postulated in the course of the biosynthesis of the hirsutane sesquiterpenes from the precursor sesquiterpene protoilludane but not found in any metabolite. In *Ceratocystis piceae*, a species which does not belong to the Basidiomycotina but to the Ascomycotina it was finally found. This is still the only fungus outside the Basidiomycotina possessing protoilludane derived sesquiterpenes (Hanssen *et al.*, 1986). The novel sesquiterpene skeleton was named ceratopicane and the sesquiterpene alcohol ceratopicanol (**44**). Co-occurring in the fungus is cerapicol (**48**) with another novel skeleton together with ceratopicane derived directly from protoilludane (Fig. 15.5). The absolute configuration of ceratopicanol was confirmed by total synthesis starting from, (R)-(+)-limonene (Mehta and Karra, 1991). No biological activity could yet be found for ceratopicanol. The basidiomycete *Macrocystidia cucumis* harboured the α,β-unsaturated ketone cucumin H (**45**), a second metabolite of the ceratopicane series with no antimicrobial or cytotoxic activities (Hellwig *et al.*, 1998).

CERAPICANES AND DERIVED SESQUITERPENES

For the biosynthesis of sterpuranes and bulleranes a tricyclic intermediate between protoilludane and those sesquiterpenes had been postulated. No metabolite with this postulated carbon skeleton was detected in Basidiomycotina but an alcohol with this carbon framework was isolated from the ascomycete *Ceratocystis piceae* together with ceratopicanol (**44**) and Δ^6-protoilludene (Hanssen and Abraham, 1988). It came as a complete surprise to isolate two long postulated carbon skeletons, cerapicane and ceratopicane, from one strain which, furthermore, was the first Ascomycotina producing sesquiterpenes derived from protoilludane. The alcohol was named cerapicol (**48**) and the hydrocarbon cerapicane, however, no antibiotic, phytotoxic or other biological activities were found for cerapicol.

Stereum purpureum causes a leaf disease common on fruit trees. When grown in liquid culture this fungus produced sterpuric acid (**46**), hydroxysterpuric acid, and its ethylidene acetal (Ayer *et al.*, 1981). The skeleton is novel and has been named sterpurane. The elucidation of its biosynthesis via protoilludane and cerapicane, the isolation of 7,12-dihydroxysterpurene (Abell and Leech, 1987), and the determination of its absolute configuration using Mosher's esters was accomplished by Abell and Leech (1988). The hydrocarbon 1-sterpurene (**47**) and the triol were isolated from the same fungus as well as the 4,12- and 5,12-dihydroxysterpurenes (Xie *et al.*, 1992). In a biotransformation reaction with the fungus *Diplodia gossypina* the protoilludane derivative tsugicoline A **33** was converted to highly oxygenated sterpurenes and proving that this type of sesquiterpene is a precursor of sterpuranes (Arnone *et al.*, 1997a).

A sesquiterpene possessing the structure of a seco-cerapicane was detected in *Cyathus bulleri* and named bullerone (**49**). No biological activity was reported for this acetal (Ayer and McCaskill, 1987).

MARASMANES

Sesquiterpenes of the marasmane (Fig. 15.6) type were one of the first sesquiterpenes isolated from fungi (Kavanagh *et al.*, 1949). Marasmic acid (**50**) was described as an antibacterial compound from *Marasmius conigenus* (Dugan *et al.*, 1966), its 9-hydroxy derivative, detected in another basidiomycete, displayed less antifungal, cytotoxic and phytotoxic activity than marasmic acid (Anke *et al.*, 1989). Velutinal (**51**) (Favre-Bonvin *et al.*, 1982) is protected by fatty acid esters (Clericuzio and Sterner, 1997) representing prodrugs which are cleaved to velutinal in case of an injury of the fruit body (Hansson and Sterner, 1991). The released velutinal is enzymatically or chemically converted to

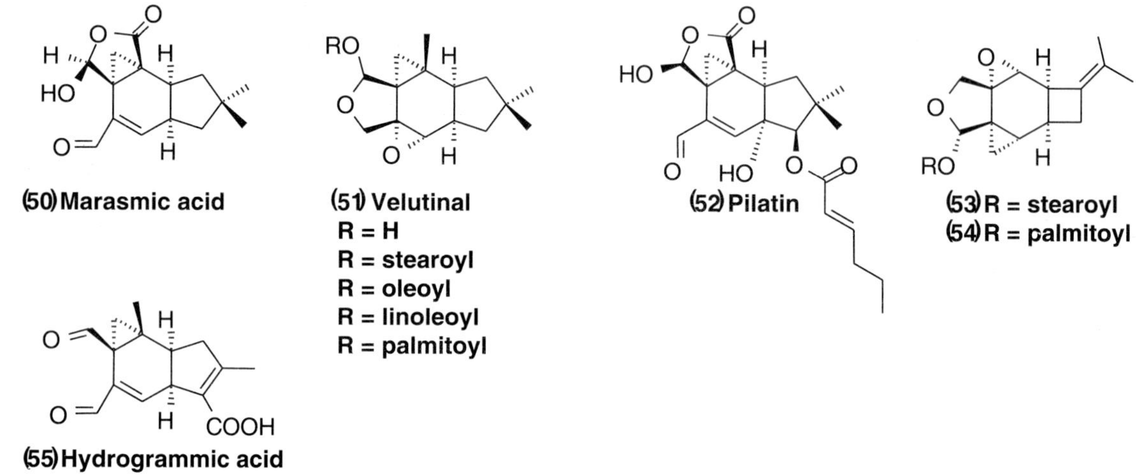

Fig. 15.6. Bioactive marasmanes.

isovellerol, isovellerdiol, and cupreal. *Lactarius vellereus* and *L. controversus* contain 13-Hydroxy-marasman-7,8-en-5-oic-γ-lactone and *L. pallidus* produced lactarotropene and lactaropallidine. From cultures of *Flagelloscypha pilatii* the antibiotically active pilatin (**52**) was characterized which is a higher oxidized derivative of marasmic acid, is highly cytotoxic by causing frameshift mutations in *Salmonella typhimurium* and inhibits the growth of bacteria and fungi (Heim *et al.*, 1988). From the basidiomycete *Clitocybe hydrogramma* (Arnone *et al.*, 1997) the marasmane sesquiterpenes 10-hydroxy-isovelleral and hydrogrammic acid (**55**) were isolated. Hydrogrammic acid, a rearranged marasmane sesquiterpene where one of the geminal methyl groups migrated, may arise directly from 10-hydroxy-isovelleral via a Wagner-Meerwein rearrangement. Both 10-hydroxy-isovelleral and hydrogrammic acid (**55**) show antibacterial activity against *Bacillus* species but were inactive against *Escherichia coli* and the fungi tested. Two norsesquiterpene fatty acid esters (**53**) and (**54**) were isolated from *Lactarius glutinopallens* possessing a novel skeleton with a 3-6-4-tricyclus. The parent hydrocarbon was named glutinopallane (Favre-Bonvin and Gluchoff-Fiasson, 1988) and is probably derived from marasmane by contraction of its cyclopentane part and the removal of a methyl group.

HIRSUTANES

Stereum hirsutum was reported to contain antibiotic compounds (Heatley *et al.*, 1947), later the main compound was identified as hirsutic acid C (Comer *et al.*, 1967) (**56**) which possesses a hirsutane skeleton (Comer *et al.*, 1965) (Fig. 15.7). *Stereum complicatum* yielded hirsutic acid C and complicatic acid (**57**) where the hydroxy group of hirsutic acid C (**56**) is oxidized to the ketone (Mellows *et al.*, 1973). The biosynthesis of complicatic acid and hirsutic acid runs via Humulene, protoilludane and ceratopicane skeletons (Feline *et al.*, 1974). Complicatic acid (**57**) showed antimicrobial activity against *Staphylococcus aureus*. From *Stereum consors* the parent hydrocarbon hirsutene was isolated and the proposed structure confirmed by chemical synthesis (Nozoe *et al.*, 1976).

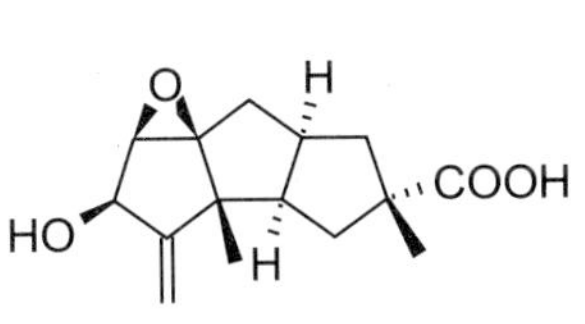

(56) Hirsutic acid

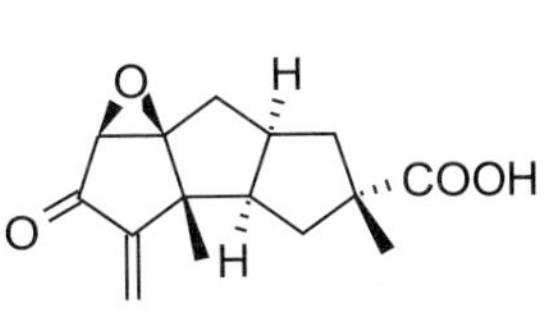

(57) Complicatic acid

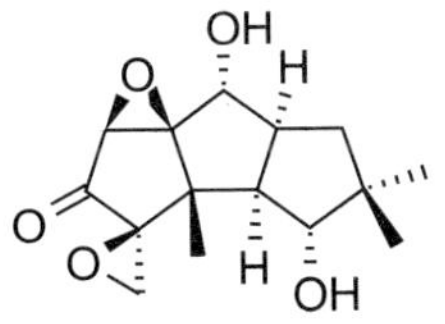

(58) Coriolin A

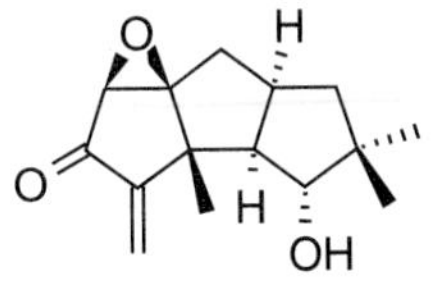

(59) Hypnophillin

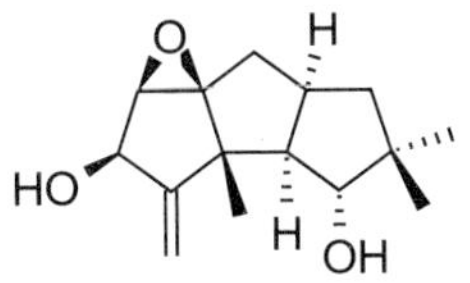

(60) Dihydro-hypnophillin

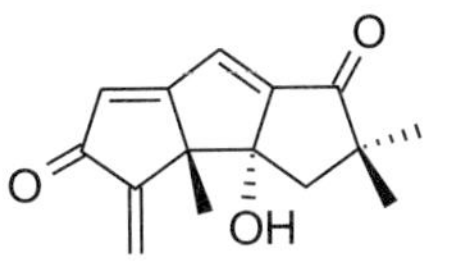

(61) Incarnal

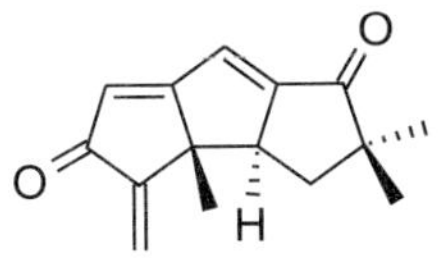

(62) Cucumin A

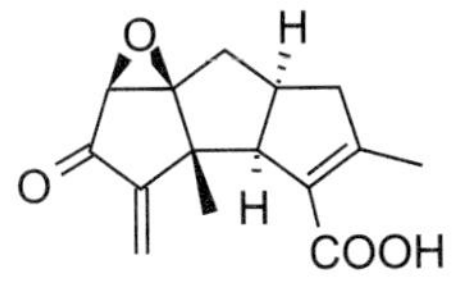

(63) Pleurotellic acid

Fig. 15.7. Biologically active hirsutanes.

Coriolus consors produces the remarkable hirsutane sesquiterpene antibiotics coriolins which possess antibacterial and antitumor activities. Coriolin A (**58**) is a highly oxidized sesquiterpene for which the absolute configuration has been determined by x-ray structure analysis (Nakamura *et al.*, 1974). Biosynthetic studies revealed the close relationship to hirsutic acid C (**56**) and a biosynthesis via humulene, protoilludane and cerapicane precursors (Tanabe *et al.*, 1974).

Lentinus crinitus, a mushroom widespread in Eastern Africa, contained the hirsutanes hypnophillin (**59**), 1-desoxy-hypnophilin together with their corresponding alcohols dihydro-hypnophillin (**60**) and dihydro-1-desoxy-hpnophillin (Abate and Abraham, 1994). The ketones are active against fungi. Furthermore, hypnophilin showed antibacterial activities against *Bacillus cereus*, *Staphylococcus aureus*, and *Mycobacterium smegmatis* and dihydrohypnophillin was active against malaria (Rukachaisirikul *et al.*, 2005).

Gloeosteretriol and incarnal (**61**) are antibacterial active sesquiterpenes from *Gloeostereum incarnatum* (Takazawa and Kashino, 1991). Incarnal is active against the Gram-positive bacteria *Staphylococcus aureus* and *Bacillus subtilis* but not against the Gram-negative *Escherichia coli* and *Pseudomonas aeruginosa.* From the antimicrobial and cytotoxic extract of *Macrocystidia cucumis* arthrosporone and eight triquinane sesquiterpenes named cucumin A-H were isolated and their structure elucidated. Cucumin A (**62**) is the 2-desoxy derivative of incarnal (**61**) while cucumin B is a dihydro-cucumin A and cucumin C an epoxidized cucumin B and cucumin H (**45**) a ceratopicane. Cucumin A, B and C display activities against *Bacillus subtilis* and *B. brevis*, antifungal activity against *Nematospora coryli*, *Penicillium notatum* and *Mucor mihei* and were also highly cytotoxic (Hellwig *et al.*, 1998).

From cultures of *Pleurotellus hypnophilus* the hirsutane hypnophilin (**59**) and the sesquiterpenes pleurotellic acid (**63**) and pleurotellol were isolated (Giannetti *et al.*, 1986). Pleurotellic acid and hypnophillin contain a rearranged hirsutane skeleton called pleurotellane which may originate from hypnophillin by a Wagner-Meerwein rearrangement. Pleurotellic acid and pleurotellol as well as hypnophillin (**59**) are antimicrobial active metabolites.

LACTARANES

Basidiomycotina are a rich source for lactarane sesquiterpenes. From *Lactarius subvellereus* subvellerolactone A (Zhang and Feng, 1997) **64**) and from *Russula brevipes* russulactarorufin (Suri *et al.*, 1997) were isolated. *L. blennius* contained blennin A (**65**), B, C (Vidari *et al.*, 1976), and D (De Bernardi *et al.*, 1980). From *L. rufus* lactarorufin B and from *L. necator* anhydrolactarorufin A, lactarorufin N, and 8-epi-pipertriol (Daniewski *et al.*, 1988) were obtained. Different *Lactarius* sp. contained vellerolactone and pyrovellerolactone (**66**) (Magnusson and Thoren, 1973) (Fig. 15.8). Lactaranes are not confined to the genus *Lactarius* and from *Russula sardonia* furosardonin A (**67**) and sardonialactone A (**68**) were characterized (Andina *et al.*, 1980)). Several *Lactarius* species produce piperdial (**69**), piperalol (Sterner *et al.*, 1985), lactarorufin D and E, lactarolide A, B, 3-O-ethyllactarolide A and B (De Bernardi *et al.*, 1979). *L. scrobiculatus* proved to be a rich source for lactaranes and the furoscrobiculins A-D, an ether epimeric to furanether A, lactaroscrobiculide B (Battaglia *et al.*, 1980) (**70**), 2,9-epoxylactarotropone, lactarotropone its hydration product, 3-epi-furandiol (Bosetti *et al.*, 1989), and scrobicalol (Pang *et al.*, 1992) (**71**) were described from this species. *Lactarius mitissimus* yielded furantriol 169 (Daniewski *et al.*, 1990), *L. necator* and *L.*

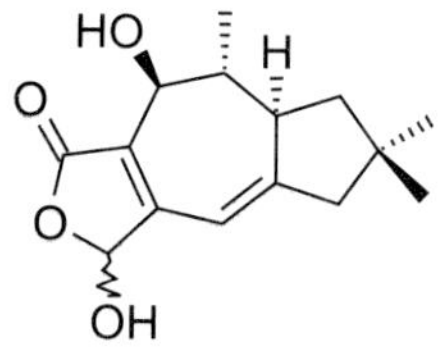

(64)Subvellero-lactone A

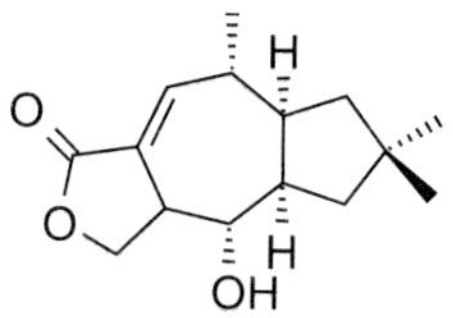

(65)Blennin A

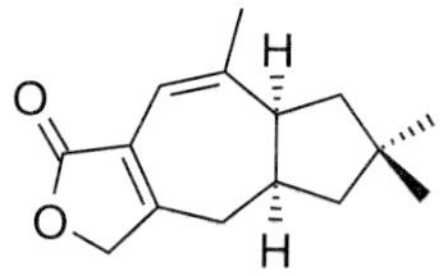

(66)Pyrovellerolactone

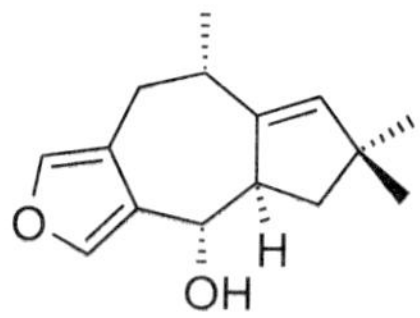

(67)Furosardonin A

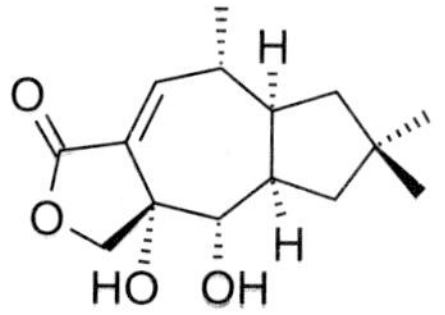

(68)Sardonia-lactone A

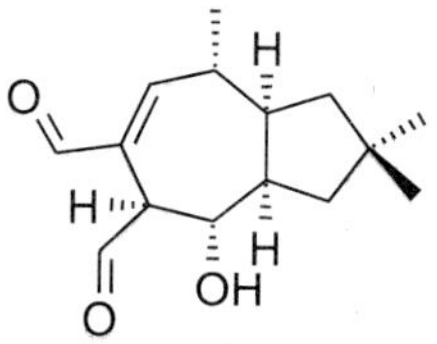

(69)Piperdial

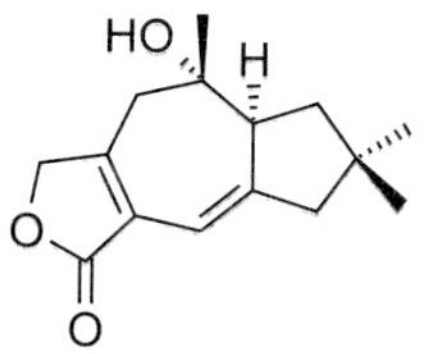

(70)Lactaro-scrobiculide B

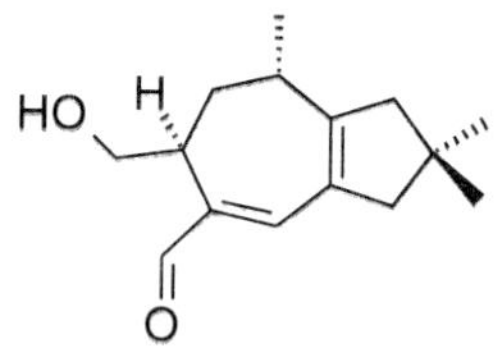

(71)Scrobicalol

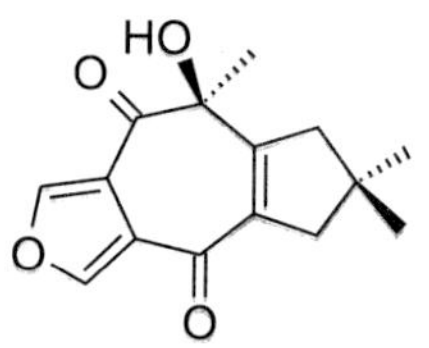

(72)Bisanhydro-lactorufin A

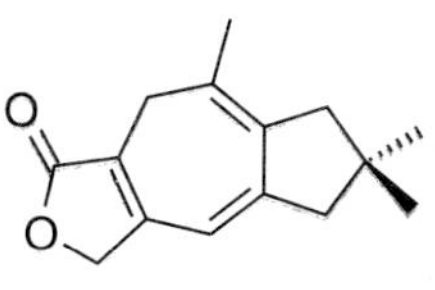

(73)Desoxy-lactarotropone

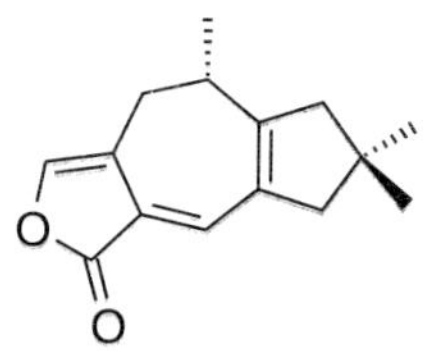

(74)Trienlactone

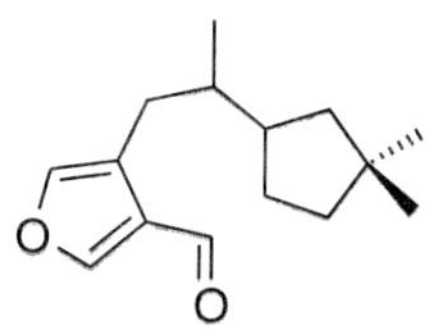

(75)Lactaral

Fig. 15.8. Lactaranes from fungi.

circellatus epi-piperalol (Sterner, 1989), and *L. vellereus* bisanhydrolactarorufin A (**72**), the desoxylactarotropone (Daniewski *et al.*, 1991) (**73**), and deoxy-3-epi-lactaroscrobiculide B (Daniewski *et al.*, 1992). The chemical basis of hot-tasting and yellowing of *L. chrysorrheus* and *L. scrobiculatus* are the trienelactone (**74**) and chrysorrhedial (De Bernardi *et al.*, 1993).

From *Lactarius vellerus* and *L. pergamenus* the secolactarane aldehyde lactaral (**75**) was isolated (Magnusson and Thoren, 1974). From *Lentinellus cochleatus* desoxylactarorufin A, blennin A, blennin C (De Bernardi *et al.*, 1986), and lentinellone have been obtained (Wunder *et al.*, 1996). Deoxylactarorufin A, blennin A (**65**), and blennin C are potent inhibitors of leukotriene biosynthesis in human peripheral blood leukocytes.

ISOLACTARANES AND RELATED COMPOUNDS

The milky juice of *Lactarius rufus* contains the isolactarane lactone isolactarorufin (Daniewski *et al.*, 1976; Konitz *et al.*, 1977) (**76**) and *Stereum purpureum* the pentacyclic isolactarene sterepolide (**77**) and dihydrosterepolide (Ayer and Saeedi-Ghomi, 1981)). *Merulius tremellosus* produces among other isolactaranes merulidial (**78**) (Giannetti *et al.*, 1986a) whose absolute configuration was established by total asymmetric synthesis. Merulidial is strongly mutagenic in the Ames test (Quack *et al.*, 1978)) and exhibits strong antibiotic effects against several bacteria and fungi as well as cytotoxic activity (Fig. 15.9).

Meruliolactone (**79**) possessing the novel sesquiterpene skeleton of merulane was found in the culture filtrate of *Merulius tremellosus* beside merulidial, the condensation product between merulidial

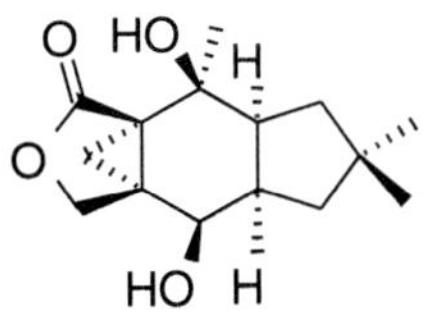
(76)Isolactarorufin

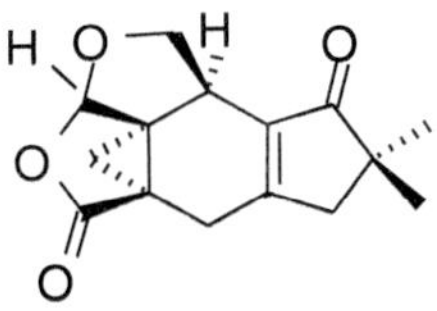
(77)Sterepolide

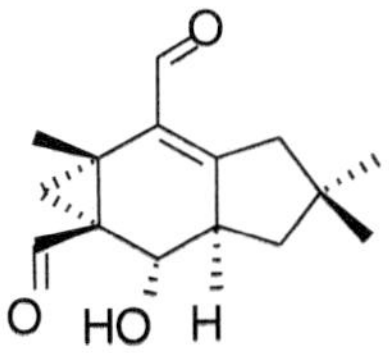
(78)Merulidial

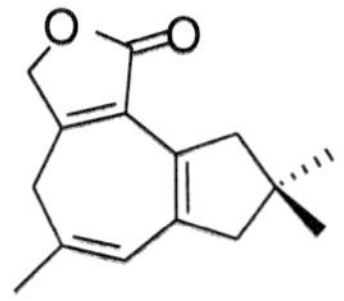
(79)Meruliolactone

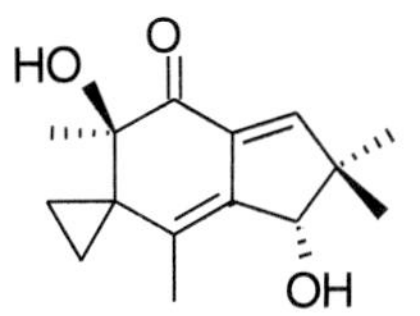
(80)Illudin M

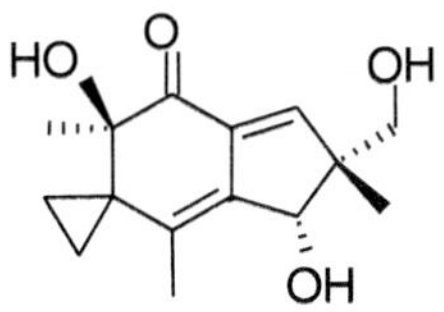
(81)Illudin S

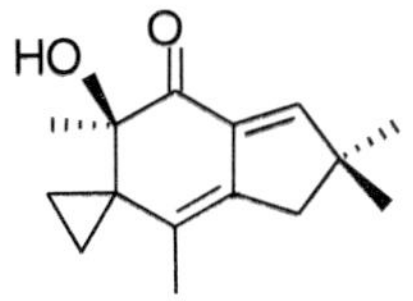
(82)6-Desoxyilludin M

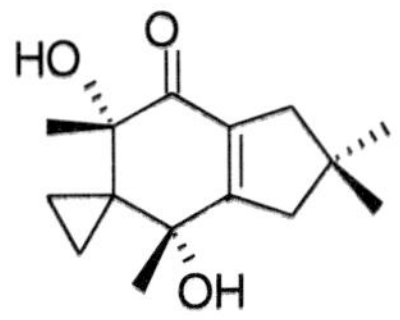
(83)Illudin A

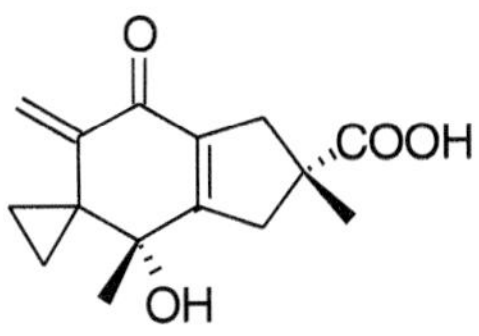
(84)Illudinic acid

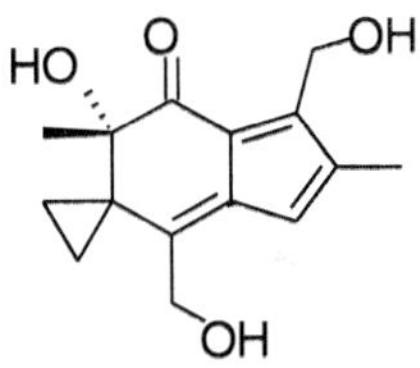
(85)Leaianafulvene

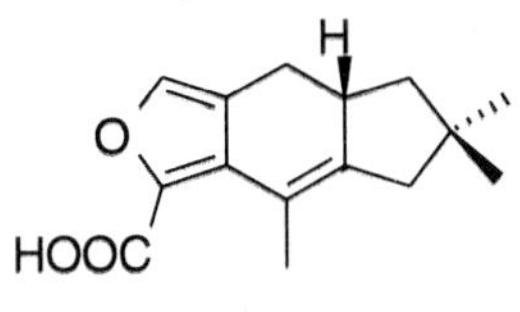
(86)Tsugicoline F

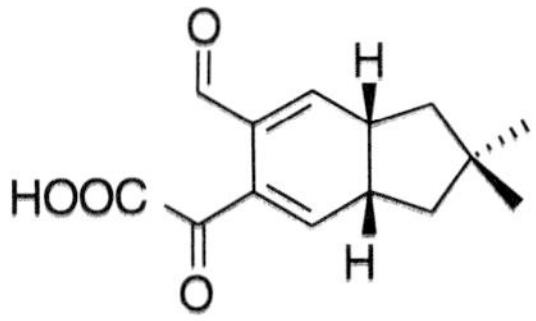
(87)Clavicoronic acid

Fig. 15.9. Some illudanes with reported bioactivity.

and a triketide. It is weakly antibiotic against *Bacillus subtilis*, *Mucor miehei* and *Paecilomyces varioti*, weakly phytotoxic but not cytotoxic (Jonassohn *et al.*, 1994).

ILLUDANES AND RELATED SESQUITERPENES

Illudin M (**80**) and illudin S (**81**) are antibiotics from *Clitocybe illudens* possessing the novel carbon skeleton of illudane (McMorris and Anchel, 1965) which were already reported in 1950 (Anchel *et al.*, 1950). Illudin S was also isolated from the luminescent mushroom *Lampteromyces japonicus* and from *Omphalotus olivascens* and *Clitocybe subilludens* (Nair *et al.*, 1983). The antitumor antibiotic 6-deoxyilludin M (**82**) was isolated from *Pleurotus japonicus* (Hara *et al.*, 1987). *Clitocybe illudens* produces illudin A (**83**) and illudin B and illudalenol (Arnone *et al.*, 1991a). Illudins C - E were obtained from *Clitocybe illudens*, illudins C_2 and C_3 from *Coprinus atramentarius* (Lee *et al.*, 1996), and illudins F 202, G 203, and H 204 from *Omphalotus nidiformis* (Burgess *et al.*, 1999). Illudinic acid (**84**) is an antibiotic against *Staphylococcus aureus* and has a pronounced cytotoxicity (Bernard *et al.*, 1997). Almost all illudanes reported so far exhibit antimicrobial and cytotoxic activities. Anticancer drugs based on 6-deoxyilludin M (**82**) are currently in clinical trials.

Leaianafulvene (**85**) was isolated from cultures of *Mycena leaiansa* possessing a rearranged isoilludane skeleton. This orange-yellow sesquiterpenoid fulvene derivative exhibits weak antibacterial activities especially against *Acinetobacter calcoaceticus* and strong cytotoxic and mutagenic activities in the Ames test (Harttig *et al.*, 1990). In cultures of *Panaeolus retirugis* two new biologically active illudane sesquiterpenes were detected (Ma *et al.*, 2004).

Tsugicolines F-I were produced by the fungus *Laurilia tsugicola* (Arnone *et al.*, 1998). Tsugicolines F (**86**) and G are both phytotoxic and antibacterial against *Bacillus* spp. and *Sarcinea lutea*. Clavicoronic acid (**87**) from *Clavicorona pyxidata* is an inhibitor of RNA-directed DNA polymerases. It is a noncompetitive inhibitor of avian myeloblastosis virus and Moloney murine leukemia virus reverse transcriptases. Furthermore, clavicoronic acid is antiviral but not cytotoxic and very weakly antimicrobial active (Erkel *et al.*, 1992).

TRICHOTHECENES

A number of fungal genera among them *Fusarium* (Moss and Thrane, 2004), *Verrucaria*, and *Myrothecium* produce metabolites containing the trichothecene sesquiterpene skeleton. The common structural feature of trichothecenes is the 12,13-epoxytrichothec-9-ene ring system (Sweeney and Dobson, 1999), and the metabolites can be divided into two main categories, namely, nonmacrocyclic trichothecenes (Grove, 1988) which are highly diverse in *Fusarium* (Brown *et al.*, 2002) and macrocyclic trichothecenes. The nonmacrocyclic trichothecenes can be divided further into two subgroups based on the presence or absence of a carbonyl function at C-8 (Desjardins *et al.*, 1993). The macrocyclic trichothecenes are classified as roridins (mainly C_{29} compounds) or verrucarins (mainly C_{27} compounds) (Tamm, 1974). The trichothecenes have generated a great deal of interest due to their wide range of bioactivities, and in particular as antibacterial, antifungal, antimalarial (Isaka *et al.*, 1999), antitumor, and antiviral agents. T-2 toxin (**88**), diacetoxyscirpenol (**89**), nivalenol (**90**), and vomitoxin (**91**) belong to the more important members of the nonmacrocyclic trichothecene mycotoxins (Fig. 15.10). All four have been detected as naturally occurring contaminants of agricultural commodities which are known to poison livestock and poultry.

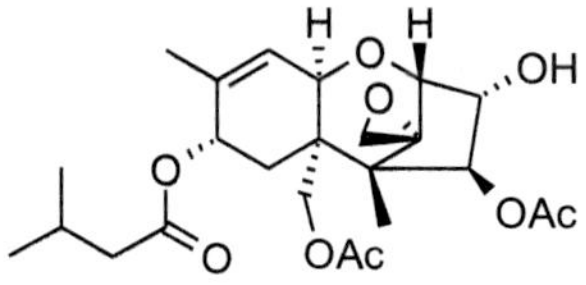

(88)T2 toxin

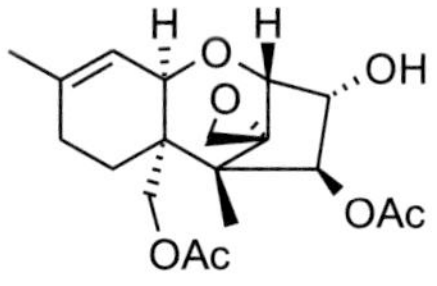

(89)Diacetoxyscirpenol

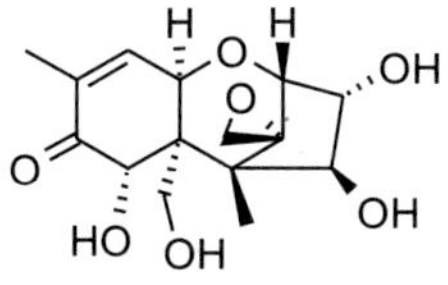

(90)Nivalenol

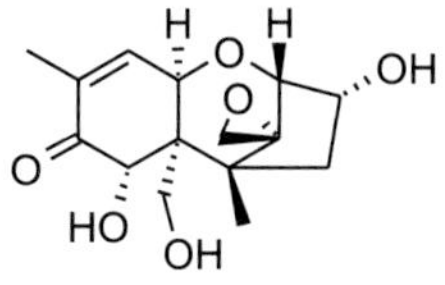

(91)Vomitoxin

Fig. 15.10. Non-macrocyclic and toxic trichothecenes.

Several of the macrocyclic trichothecenes of the roridin and verrucarin series as verrucarin A (**92**) verrucarin B (**93**), or verrucarin J (**94**) and the roridins A (**95**), D (**96**), E (**97**) or H (**98**) are active against some cancer lines or against malaria. There exist a number of derivatives where the sesquiterpene moiety has been hydroxylated, e.g., the 8α-hydroxy-verrucarin J called verrucarin L (Jarvis *et al.*, 1981). Some metabolites possess cyclic diesters, e. g., mytoxin B (**99**) which has been isolated from *Myrothecium roridum* together with 14′-hydroxymytoxin B and 16-hydroxyroridin E (Fig. 15.11). The latter two compounds showed potent cytotoxic activity against primary soft-tissue sarcoma cells (Alvi *et al.*, 2002). From *Myrothecium verrucaria* roridin L, M and verrucarin M have been characterized which are antitumor antibiotics (Murakami *et al.*, 2001).

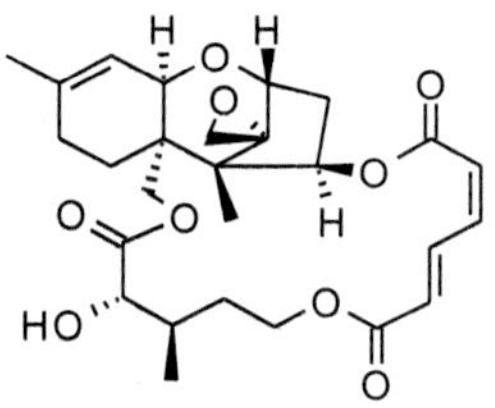

(92)Verrucarin A

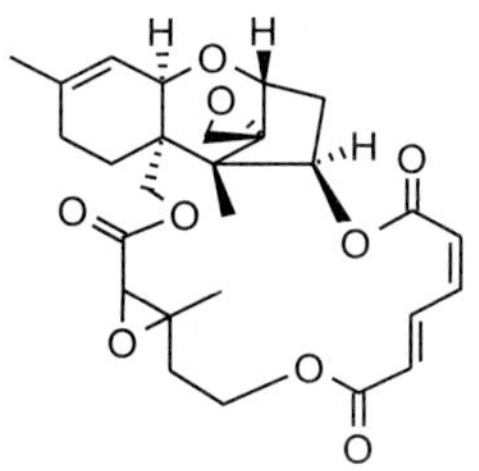

(93)Verrucarin B

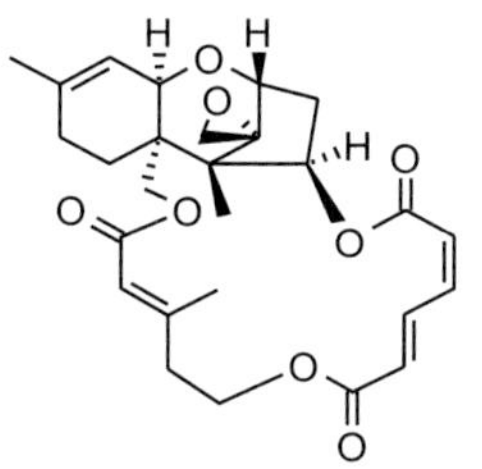

(94)Verrucarin J

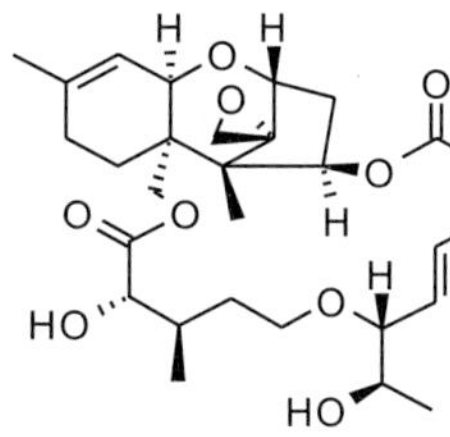

(95)Roridin A

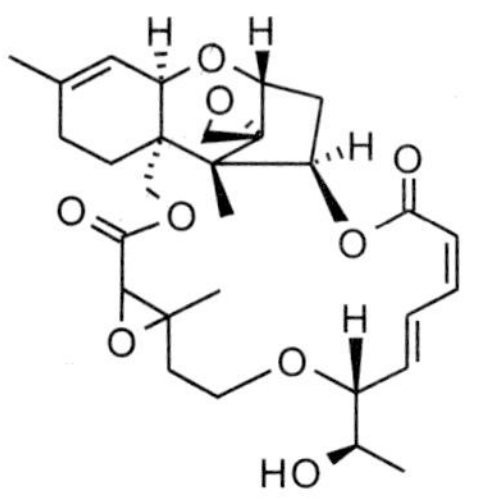

(96)Roridin D

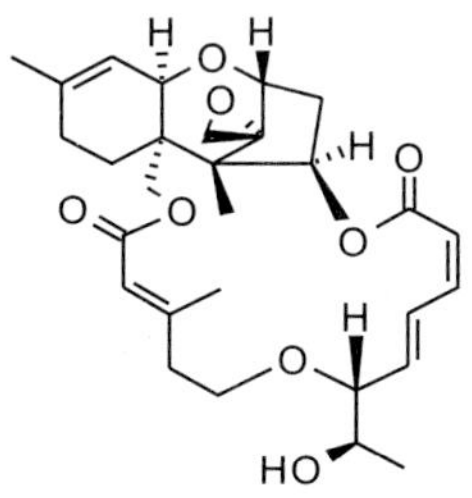

(97)Roridin E

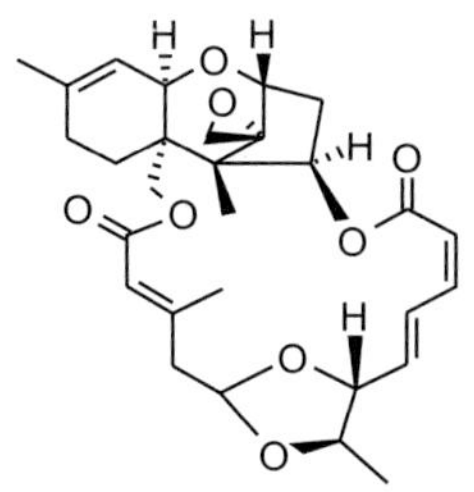

(98)Roridin H

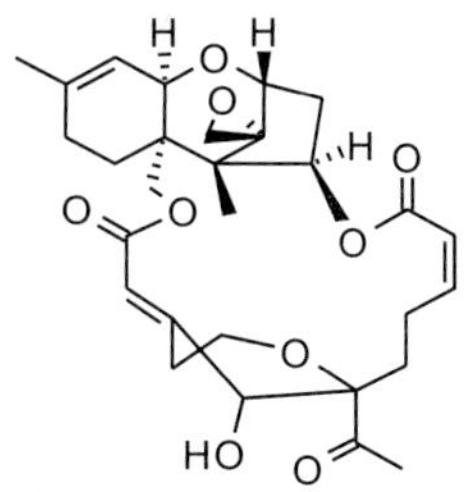

(99)Mytoxin B

Fig. 15.11. Anticancer macrocyclic trichothecenes.

EREMOPHILANES

Sesquiterpenes with the eremophilane skeleton have been described from a number of fungi (Fig. 15.12). The eremofortin A — D (Moreau *et al.*, 1980) (**100-103**) and PR toxin (Wei *et al.*, 1975)) (**107**) have been described from *Penicillium roquefortii*, sporogen AO-1 (**108**) from *Aspergillus oryzae* (Tanaka *et al.*, 1984), phomenone (**111**) from *Phoma exigua* var. *non-oxydabilis* (Riche *et al.*, 1974) and *Hansfordia pulvinata* (Tirilly *et al.*, 1983) and phaseolinone **112**) from *Macrophomina phaseolina* (Dhar *et al.*, 1982). Dacrymenone (**109**) from *Dacrymyces* sp. is closely related to eremofortin B (**101**) and exhibits cytotoxic and antifungl activities (Mierau *et al.*, 2003). The phytotoxin bipolaroxin (**113**) has been found in *Bipolaris cynodotis* (Sugawara *et al.*, 1985), dendryphiellins E and G together with noreremophilanes in *Dendryphiella salina* (Guerriero *et al.*,

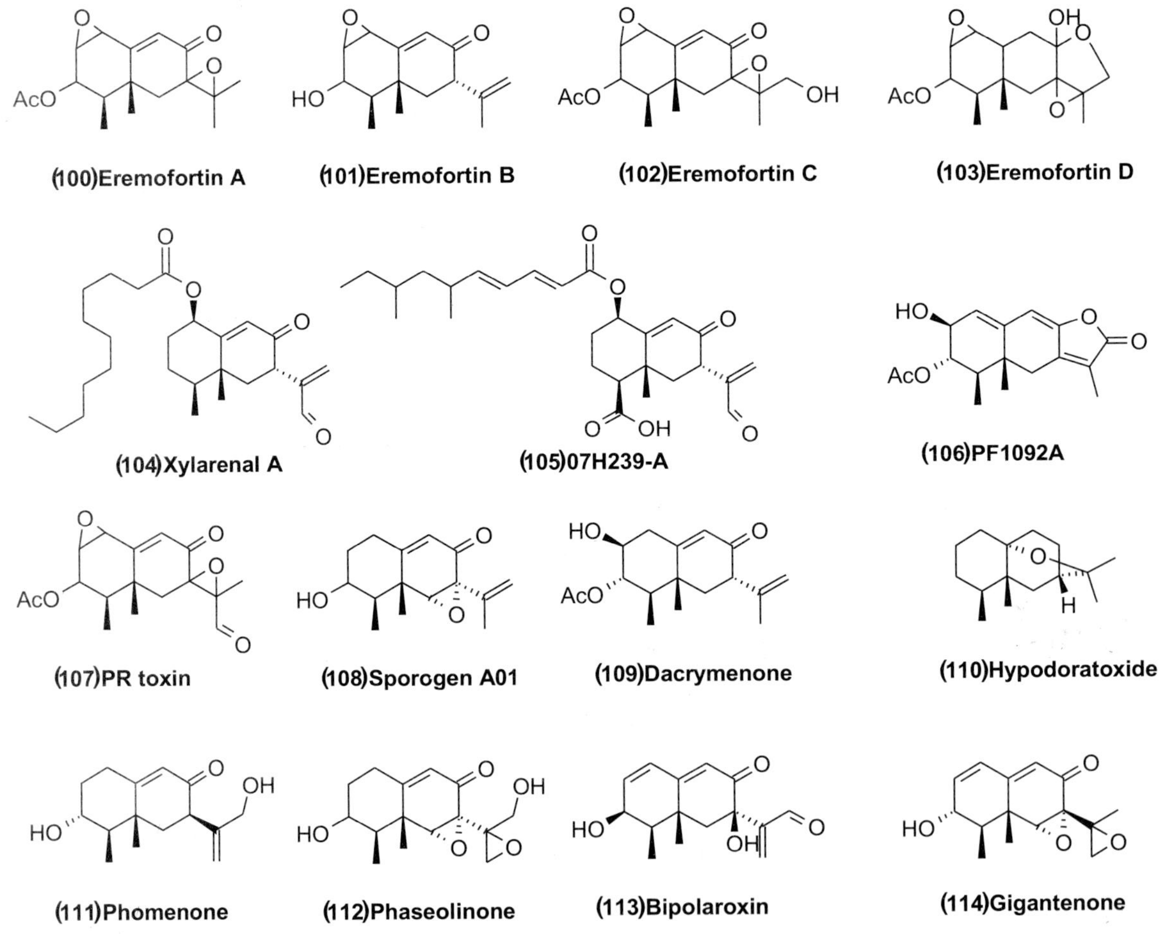

Fig. 15.12. Bioactive eremophilane derivatives from fungi.

1989) and gigantenone (**114**) in *Drechslera giganteagigantea* (Kenfield *et al.*, 1989). *Hypomyces odoratus* has a camphorous odor and produces the volatile eremophilanether hypodoratoxide (Kühne *et al.*, 1991) (**110**).

Hypodoratoxide reduces the germination rate of seeds, is a phytotoxine and displays antifungal activities (Urbasch *et al.*, 1991). In search for antiobesity drugs two eremophilane esters, xylarenals A (**104**) and xylarenals B, have been isolated from *Xylaria persicaria* which specifically block the NPY Y5 receptor but failed to block the other NPY receptors (Smith *et al.*, 2002). Closely related to xyleranal A is the cytotoxin 07H239-A (**105**) from a marine fungus (McDonald *et al.*, 2004). From the screening for HIV-1 integrase inhibitors integric acid, another eremophilane sesquiterpene closely related to 07H239-A was discovered in a *Xylaria* sp. It inhibited 3'-end processing, strand transfer and disintegration reactions catalyzed by HIV-1 integrase with IC_{50} values of 3-10 ìM (Singh *et al.*, 1999). Rather different from these group of sesquiterpenes are the eremophilanes PF1092A (**106**) – PF1092C produced by *Penicillium oblatum*. All three compounds inhibit binding of progesterone to the progesterone receptor (Tabata *et al.*, 1997).

VOLATILE SESQUITERPENES FROM FUNGI

Several fungi are known for their smell and Ewald Sprecher and his group focused on the volatile compounds responsible for the olfactory sensations of certain fungi. *Leptographium lundbergii* produces the sesquiterpene alcohols leptographiol (**117**), isoleptographiol (**118**) and isoafricanol (**119**) with the very rare africane (Abraham *et al.*, 1986). From *Lentinus lepideus* three different driman ethers, (**120 - 122**) have been identified possessing a characteristic anise-like odor (Abraham *et al.*, 1988) (Fig. 15.13).

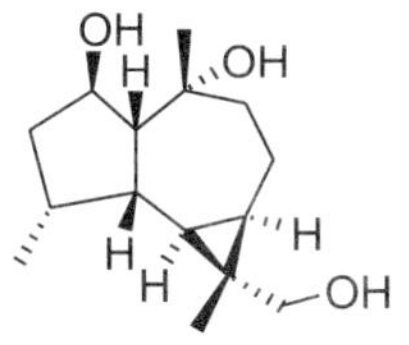

(115)2,13-Dihydroxyledol

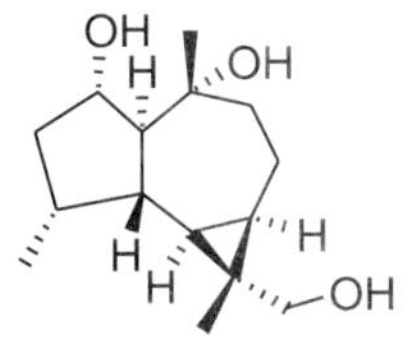

(116)2,13-Dihydroxy-globulol

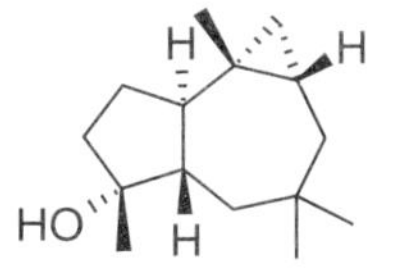

(117)Leptographiol

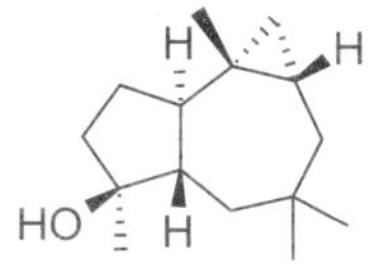

(118)Isoleptographiol

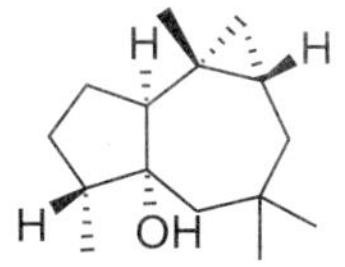

(119)Isoafricanol

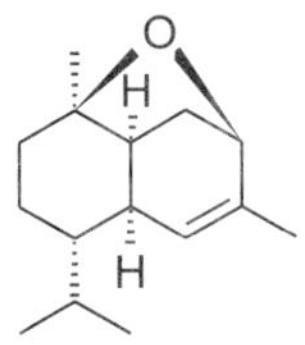

(120)Lepideusether

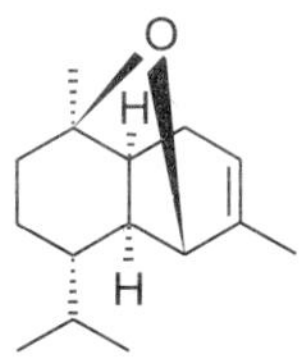

(121)Isolepideusether

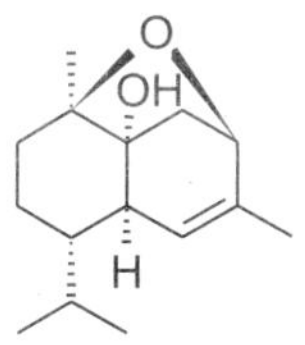

(122)10-Hydroxy-lepideusether

Fig. 15.13. Some unusual volatile sesquiterpenes from fungal fermentations.

FUTURE PERSPECTIVES

Although we have barely scratched the surface of the huge biodiversity of fungi (Suay *et al.*, 2000) and a number of important sesquiterpenes, e.g., the antibiotic terrecyclic acid (Nakagawa *et al.*, 1982) or the anti-influenza A virus merosesquiterpene stachyflin (Minagawa *et al.*, 2002), have not been discussed here, an incredible rich spectrum of sesquiterpenes has been revealed. The currently characterized sesquiterpenes from fungi comprise many different sesquiterpene skeletons, some of them only known from fungi, with a huge diversity of oxidation patterns. This structural diversity is accompanied by a large diversity of biological activities bearing great promises for applications in various drugs in agriculture and medicine.

Often fungi produce only small amounts of the bioactive compounds in fermentations. Several approaches to overcome this limitation have been described, however, there is not a single solution applicable to most fungi. Thorough optimizations are often needed to reach increases suitable for a large scale production. Progress in fungal genetics and metagenomics (Ward, 2006) will be of immense help to overcome many obstacles.

Furthermore, fungi are not only a rich source for secondary metabolites they can also be used in biotransformations for the production of new compounds from readily available starting materials. This has also been used in the field of the biotransformations of sesquiterpenes and here only the production of biologically active compounds from isolongifolol (Choudhary *et al.*, 2005) and the oxidation of aristolenepoxide to a phytotoxin (Abraham *et al.*, 1996) are mentioned. Biotransformation of sesquiterpenes can also be a valuable support for the elucidation of structure-activity relations. An example for this is the finding of the nematicidal active 2β,13-dihydroxyledol (**115**) in *Dichotomus squalens* cultures (Huang *et al.*, 2004) which could be compared with the epimeric 2α,13-dihydroxy-globulol (**116**) formed by biotransformation of globulol with *Bacillus megaterium* (Abraham *et al.*, 1992).

CONCLUSIONS

A multitude of different sesquiterpene metabolites have been characterized from fungi, subdivision Basidiomycotina. The rich variation of structures is mirrored in a rich variation of activities as well. However, as with most microorganisms still only a small number of strains has been tested for the production of secondary metabolites and screening of new isolates of rare species is urgently needed. Obviously, currently, only the minority of the compounds discussed here have been tested in biological assays. However, for some sesquiterpenes there are already interesting data on their different biological activities and some have even been identified as inhibitors of enzymes crucial in some diseases. This can give us a glimpse what still remained unknown in store. The search for novel metabolites with novel activities will be backed up by biotransformation reactions where easily available sesquiterpenes are transformed by bacteria or fungi to novel derivatives which again will be screened in various assays. A promising approach to exploit the whole biodiversity of fungi including those strains which are difficult to cultivate is the application of metagenomics. These techniques are still in its infancy but first results demonstrated already the potential of this approach. Currently, a renaissance of secondary metabolites is under way and without doubt sesquiterpenes from fungi will play here an important role.

REFERENCES

Abate D. and Abraham W-R. (1994). Antimicrobial metabolites from *Lentinus crinitus. Journal of Antibiotics* 47: 1348-1350.

Abell C. and Leech A.P. (1987). Evidence for consecutive 1,2-alkyl shifts in sterpurene biosynthesis: Isolation and characterisation of 7,12-dihydroxysterpurene. *Tetrahedron Letters* 28: 4887-4890.

Abell C. and Leech A.P. (1988). The absolute configuration of the sterpurene sesquiterpenes. *Tetrahedron Letters* 29: 1985-1986.

Abraham W-R. (2001). Bioactive sesquiterpenes produced by fungi: are they useful for humans as well? *Current Medicinal Chemistry* 8: 583-606.

Abraham W-R., Ernst L., Witte L., Hanssen H-P. and Sprecher E. (1986). New trans-fused africanols from *Leptographium lundbergii. Tetrahedron* 42: 4475-4480.

Abraham W-R. and Hanssen H-P. (1987). Fragrance compounds from fungal liquid cultures. II. New benzofuran-terpenoids from *Cystostereum murraii,* Proceedings of the 4th European Congress on Biotechnology, Amsterdam/NL, Vol. 3: 294-296.

Abraham W-R. and Stumpf B. (1987). Biotransformation of humulene by fungi and enantioselectivity of the strains used. *Zeitschrift für Naturforschung, C: Biosciences* 42: 79-86.

Abraham W-R., Hanssen H-P., and Möhringer C. (1988). Novel sesquiterpene ethers from liquid cultures of the wood-rotting fungus *Lentinus lepideus. Zeitschrift für Naturforschung, C: Biosciences* 43: 24-28.

Abraham W-R., Ernst L. and Arfmann H.-A. (1990). Rearranged caryophyllenes by biotransformation with *Chaetomium cochliodes, Phytochemistry* 29: 757-763.

Abraham W-R., Ernst L. and Stumpf B. (1990a). Biotransformation of caryophyllene by *Diplodia gossypina, Phytochemistry* 29: 115-120.

Abraham W-R., Hanssen H.-P. and Urbasch I. (1991). Lepistirones, major volatile metabolites from liquid cultures of *Lepista irina* (Basidiomycotina). *Zeitschrift für Naturforschung, C: Biosciences* 46: 169-171.

Abraham W-R., Stumpf B., Kieslich K. and Ernst L. (1992). Microbial oxidation of tricyclic sesquiterpenoids containing a dimethylcyclopropane ring. *Phytochemistry* 31: 3749-3755.

Abraham W-R, Riep A. and Hanssen H-P. (1996). Biotransformation and Phylogeny. Part 6. Microbial oxidation of aristolenepoxide to phytotoxins. *Bioorganic Chemistry* 24: 19-28.

Alvi K.A., Rabenstein J., Woodard J., Baker D.D., Bergthold J.D., Lynch J. , Lieu K.L. and Braude I.A. (2002). 14′-Hydroxymytoxin B and 16-hydroxyroridin E, two new cytotoxic trichothecenes from *Myrothecium roridum. Journal of Natural Products* 65: 742-744.

Ainsworth A.M., Chicarelli-Robinson M.I., Copp B.R., Fauth U., Hylands P.J., Holloway J.A., Latif M., O'Beirne G.B., Porter N., Renno D.V., Richards R. and Robinson N. (1995). Xenovulene A, a novel GABA-benzodiazepine receptor binding compound produced by *Acremonium strictum. Journal of Antibiotics* 48: 568-573.

Anchel M., Hervey A. and Robbins W.J. (1950). Antibiotic substances from Basidiomycetes. VII. *Clitocybe illudens. Proceedings of the National Academy of Sciences of the United States of America* 36: 300-305.

Anderson J.R., Edwards R.L., Freer A.A., Mabelis R.P., Poyser J.P., Spencer H. and Whalley A.J.S. (1984). Punctatins B and C (antibiotics M95154 and M95155): Further sesquiterpene alcohols from the fungus *Poronia punctata. Journal of the Chemical Society, Chemical Communications* 917-919.

Andina D., De Bernardi M., Del Vecchio A., Fronza G., Mellerio G., Vidari G. and Vita-Finzi P. (1980). Sesquiterpenes from *Russula sardonia. Phytochemistry* 19: 93-97.

Anke T. (1989). Basidiomycetes: a source for new bioactive secondary metabolites. *Progress in Industrial Microbiology* 27: 51-66.

Anke H., Hillen-Maske E. and Steglich W. (1989). 9β-hydroxymarasmic acid and other sesquiterpenoids from submerged cultures of a basidiomycete. *Zeitschrift für Naturforschung, C: Biosciences* 44: 1-6.

Arnone A., Nasini G., Assante G., Roeijmans H.J. and Van Eijk G.W. (1987). Sulcatine, a norsesquiterpene from the fungus *Laurilia sulcata. Phytochemistry* 26: 1739-1742.

Arnone A., Cardillo R., Di Modugno V. and Nasini G. (1988). Secondary mould metabolites. XXII. Isolation and structure elucidation of melledonals D and E and melleolides E-H, novel sesquiterpenoid aryl esters from *Clitocybe elegans* and *Armillaria mellea. Gazzeta Chimica Italaliana* 118: 517-521.

Arnone A., Cardillo R., Nasini G. and de Pava O.V. (1991). Secondary mould metabolites. Part 34. Isolation and structure elucidation of illudosin, a novel sesquiterpene from *Clitocybe illudens* using one and two dimensional NMR techniques. *Journal of the Chemical Society, Perkin Transactions I* 1787-1791.

Arnone A., Cardillo R., Nasini G. and de Pava O.V. (1991a). Secondary mould metabolites. Part 31. Isolation and structure eluci-dation of illudins A and B, and illudanenol, new sesquiterpenoids from *Clitocybe illudens. Journal of the Chemical Society, Perkin Transactions I* 733-737.

Arnone A., Nasini G., Assante G., van Eijk G. W. (1992). Three sesquiterpenes produced by the fungus *Laurilia sulcata. Phytochemistry* 31: 2047-2050.

Arnone A., Brambilla U., Nasini G. and Vajna de Pava O. (1995). Isolation and structure elucidation of tsugicolines A-D, novel protoilludane sesquiterpenes from *Laurilia tsugicola. Tetrahedron* 51: 13357-13364.

Arnone A., Nasini G. and De Pava O.V. (1997). Marasmane sesquiterpenes from the basidiomycete *Clitocybe hydrogramma. Phytochemistry* 46: 1099-1101.

Arnone A., De Gregorio C., Nasini G. and De Pava O.V. (1997a). Secondary mould metabolites: Part 53.1 Transformation of the protoilludane sesquiterpene tsugicoline A into a sterpurane derivative and its microbiological reduction. *Journal of the Chemical Society, Perkin Transactions I* 1523-1526.

Arnone A., De Gregorio C., Nasini G. and Vajna de Pava O. (1998). Isolation and structure elucidation of tsugicolines F-H, novel furosesquiterpenes, and tsugicoline I from the fungus *Laurilia tsugicola. Tetrahedron* 54: 10199-10204.

Arnone A., De Gregorio C., Meille S.V., Nasini G. and Sidoti G. (1999). Tsugicoline E, a new polyoxygenated protoilludane sesquiterpene from the fungus *Laurilia tsugicola. Journal of Natural Products* 62: 51-53.

Ayer W.A. and Browne L.M. (1981). Terpenoid metabolites of mushrooms and related basidiomycetes. *Tetrahedron* 37: 2197-2248.

Ayer W.A. and Saeedi-Ghomi M.H. (1981). The sterepolides: New isolactaranes from *Stereum purpureum. Tetrahedron Letters* 2071-2074.

Ayer W.A., Saeedi-Ghomi M.H., Van Engen D., Tagle B. and Clardy J. (1981). The sterpuric acids : A new type of sesquiterpenoid. *Tetrahedron* 37 (Suppl. No. 1): 379-385.

Ayer W.A. and McCaskill R.H. (1987). Bullerone, a novel sesquiterpenoid from *Cyathus bulleri* Brodie. *Canadian Journal of Chemistry* 65: 15-17.

Backens S., Steffan B., Steglich W., Zechlin L. and Anke T. (1984). Naematolin und Naematolon, zwei Caryophyllan-Derivate aus Kulturen von *Hypholoma*-Arten (Agaricales). *Justus Liebigs Annalen der Chemie* 1332-1342.

Battaglia R., De Bernardi M., Fronza G., Mellerio G., Vidari G. and Vita-Finzi P. (1980). Fungal metabolites. VIII. Structures of new sesquiterpenes from *Lactarius scrobiculatus. Journal of Natural Products* 43: 319-328.

Bernard A., Zink D., Dufresne C., Young K., Pelaez F., Gonzalez Del Val A., Valentino D., Graham A. and Platas G. (1997). Illudinic acid, a novel illudane sesquiterpene antibiotic. *Journal of Natural Products* 60: 188-190.

Bosetti A., Fronza G., Vidari G. and Vita-Finzi P. (1989). Norlactarane and lactarane sesquiterpenes from *Lactarius scrobiculatus. Phytochemistry* 28: 1427-1431.

Brown D.W., McCormick S.P., Alexander N.J., Proctor R.H., Desjardins A.E. (2002). Inactivation of a cytochrome P-450 is a determinant of trichothecene diversity in *Fusarium* species. *Fungal Genetics and Biology* 36: 224-233.

Burgess M.L. and Barrow K.D. (1999). Biosynthesis of illudosin, a fomannosane-type sesquiterpene, by the Basidiomycete *Omphalotus nidiformis. Journal of the Chemical Society, Perkin Transactions* 1 2461-2466.

Burgess M.L., Zhang Y.L. and Barrow K.D. (1999). Characterization of new illudanes, illudins F, G, and H from the basidiomycete *Omphalotus nidiformis. Journal of Natural Products* 62: 1542-1544.

Cane D.E., Nachbar R.B., Clardy J. and Finer J. (1977). The absolute configuration of fomannosin. *Tetrahedron Letters* 18: 4277-4279.

Cane D.E. and Nachbar R.B. (1978). Stereochemical studies of isoprenoid biosynthesis. Biosynthesis of fomannosin from [1,2-$^{13}C_2$]acetate. *Journal of the American Chemical Society* 100: 3208-3212, (1979). Corrigendum 101: 1908.

Choudhary M.I., Musharraf S.G., Nawaz S.A., Anjum S., Parvez M., Fun H.K., Atta-ur-Rahman. (2005). Microbial transformation of (-)-isolongifolol and butyrylcholinesterase inhibitory activity of transformed products, *Bioorganic and Medicinal Chemistry* 13: 1939-1944. Erratum in: (2005). *Bioorganic and Medicinal Chemistry* 13: 4589.

Clericuzio M. and Sterner O. (1997). Conversion of velutinal esters in the fruit bodies of *Russula cuprea. Phytochemistry* 45: 1569-1572.

Clericuzio M., Toma L. and Vidari G. (1999). Fungal Metabolites, 44. Isolation of a new caryophyllane ester from *Lactarius subumbonatus*: Conformational analysis and absolute configuration. *European Journal of Organic Chemistry* 2059-2065.

Comer F.W., McCapra F., Qureshi I.H., Trotter J. and Scott A.I. (1965). The structure and stereochemistry of hirsutic acid C, a novel fungal sesquiterpenoid. *Chemical Communications* 310.

Comer F.W., McCapra F., Qureshi I.H. and Scott A.I. (1967). The structure and chemistry of hirsutic acid. *Tetrahedron* 23: 4761-4768.

Daniewski W.M., Kocor M. and Thoren S. (1976). Constituents of higher fungi. Part VIII. Isolactarorufin, a novel tetracyclic sesquiterpene lactone from *Lactarius rufus. Heterocycles* 5: 77-84.

Daniewski W.M., Grieco P.A., Huffman J.C., Rymkiewicz A. and Wawrzun A. (1981). Isolation of 12-hydroxycaryophyllene-4,5-oxide, a sesquiterpene from *Lactarius camphoratus. Phytochemistry* 20: 2733-2734.

Daniewski W.M., Gumulka M., Ptaszynska K., Vidari G., Garlaschelli L., Fronza G. and Budesinsky M. (1988). 8-*epi*-pipertriol, a lactarane sesquiterpene from *Lactarius necator. Phytochemistry* 27: 3315-3317.

Daniewski W.M., Gumulka M. and Skibicki, P. (1990). Furantriol, a lactarane sesquiterpene from *Lactarius mitissimus. Phytochemistry* 29: 527-529.

Daniewski W.M., Gumulka M., Skibicki P., Krajewski J. and Gluzinski P. (1991). 2(3)-8(9)-Bisanhydrolactarorufin A and a highly oxygenated furanol from *Lactarius vellereus. Phytochemistry* 30: 1326-1328.

Daniewski W.M., Gumulka M., Ptaszynska K., Skibicki P., Jacobsson U. and Norin, T. (1992). 3-Deoxy-3-epi-lactaroscrobiculide B, a sesquiterpene from *Lactarius vellereus. Phytochemistry* 31: 3933-3936.

De Bernardi M., Fronza G., Mellerio G., Vidari G. and Vita-Finzi P. (1979). New sesquiterpene hydroxylactones from *Lactarius* species. *Phytochemistry* 18: 293-298.

De Bernardi M., Fronza G., Mellerio G., Vidari G. and Vita-Finzi P. (1980). Stereochemistry of blennin A and blennin D from *Lactarius blennius. Phytochemistry* 19: 99-101.

De Bernardi M., Fronza G., Scilingo A., Vidari G. and Vita-Finzi P. (1986). Fungal metabolites XX: Chemical correlation of lactarane and secolactarane sesquiterpenes. Absolute configuration of furosardonin A, lactaral blennin C. *Tetrahedron* 42: 4277-4283.

De Bernardi M., Garlaschelli L., Toma L., Vidari G. and Vita-Finzi P. (1993). The chemical basis of hot-tasting and yellowing of the mushrooms *Lactarius chrysorrheus* and *L. scrobiculatus. Tetrahedron* 49: 1489-1504.

Desjardins A.E., Hohn T.M. and McCormick S.P. (1993). Trichothecene biosynthesis in *Fusarium* species: chemistry, genetics, and significance. *Microbiology Reviews* 57: 595-604.

Dhar T.K., Siddiqui K.A.I. and Ali E. (1982). Structure of phaseolinone, a novel phytotoxin from *Macrophomina phaseolina,. Tetrahedron Letters* 23: 5459-5462.

Donnelly D., Sanada S., O'Reilly J., Polonsky J., Prange T. and Pascard C. (1982). Isolation and structure (X-ray analysis) of the orsellinate of armillol, a new antibacterial metabolite of *Armillaria mellea. Journal of the Chemical Society, Chemical Communications* 135.

Donnelly D.M.X., Coveney D.J., Fukuda N. and Polonsky J. (1986). New sesquiterpene aryl esters from *Armillaria mellea. Journal of Natural Products* 49: 111-116.

Dugan J.J., de Mayo P., Nisbet M., Robinson J.R. and Anchel M. (1966). Terpenoids XIV. The constitution and biogenesis of marasmic acid. *Journal of the American Chemical Society* 88: 2838-2844.

Edwards R.L., Maitland D.J. and Whalley A.J.S. (1989). Metabolites of the higher fungi. Part 24. Cytochalasin N, O, P, Q, and R. New cytochalasins from the fungus *Hypoxylon terricola* Mill. *Journal of the Chemical Society, Perkin Transactions I*, 57-65.

Edwards R.L., Maitland D.J., Poyser J.P. and Whalley A.J.S. (1989a). Metabolites of the higher fungi. Part 25. Punctaporonin G from the fungus *Poronia punctata* (Linnaeus: Fries) Fries. *Journal of the Chemical Society, Perkin Transactions I* 1939-1941.

Erkel G., Anke T., Gimenez A. and Steglich W. (1992). Antibiotics from basidiomycetes. XLI. Clavicoronic acid, a novel inhibitor of reverse transcriptases from *Clavicorona pyxidata* ,Pers. ex Fr.) Doty. *Journal of Antibiotics* 45: 29-37.

Erkel G., Becker U., Anke T. and Sterner O. (1996). Nidulal, a novel inducer of differentiation of human promyelocytic leukemia cells from *Nidula candida. Journal of Antibiotics* 49: 1189-1195.

Fabian K., Lorenzen K., Anke T., Johansson M. and Sterner O. (1998). Five new bioactive sesquiterpenes from the fungus *Radulomyces confluens* (Fr.) Christ. *Zeitschrift für Naturforschung, C: Biosciences* 53: 939-945.

Favre-Bonvin J., Gluchoff-Fiasson K. and Bernillon J. (1982). Structure du stearyl-velutinal, sequiterpenoide naturel de *Lactarius velutinus* Bert. *Tetrahedron Letters* 23: 1907-1908.

Favre-Bonvin J. and Gluchoff-Fiasson K. (1988). Structures of two glutinopallal esters, new natural sesquiterpenoids from *Lactarius glutinopallens. Phytochemistry* 27: 286-287.

Feline T.C., Mellows G., Jones R.B. and Phillips L. (1974). Biosynthesis of hirsutic acid C using ^{13}C nuclear magnetic resonance spectroscopy. *Journal of the Chemical Society, Chemical Communications* 63.

Fraga B.M. (2005). Natural sesquiterpenoids. *Natural Products Report* 22**:** 465-486.

Frank M., Kingston E., Jeffrey J.C., Moss M.O., Murray M., Simpson T.J. and Sutherland, A. (1999). Walleminol and walleminone, novel caryophyllenes from the toxigenic fungus *Wallemia sebi. Tetrahedron Letters* 40: 133-136.

Froborg J. and Magnusson G. (1978). Fungal extractives. XI. On the biogenetic interrelationship between the marsmane- and vellerane sesquiterpene skeletons. *Tetrahedron* 34: 2027-2028.

Giannetti B.M., Steffan B., Steglich W., Kupka J. and Anke T. (1986). Antibiotics from basidiomycetes. Part 24.: Antibiotics with a rearranged hirsutane skeleton from *Pleurotellus hypnophilus* (Agaricales). *Tetrahedron* 42: 3587-3593.

Giannetti B.M., Stefan B., Steglich W., Quack W. and Anke T. (1986a). Antibiotics from Basidiomycetes. Part 23. Merulidial, an isolactarane derivative from *Merulius tremellosus. Tetrahedron* 42: 3579-3586.

Grove J.F. (1988). Non-macrocyclic trichothecenes. *Natural Products Report* 5: 187-209.

Guerriero A., D'Ambrosio M., Cuomo V., Vanzanella F. and Pietra F. (1989). Novel Trinor-eremophilanes (Dendryphiellin B, C, and D), eremophilanes (Dendryphiellin E, F, and G), and branched C_9-carboxylic acids (Dendryphiellic acid A and B) from the marine deuteromycete *Dendryphiella salina* (Sutherland) Pughet Nicot. *Helvetica Chimica Acta* 72: 438-446.

Hanssen H-P., Sprecher E. and Abraham W-R. (1986). 6-Protoilludene, the major volatile metabolite from *Ceratocystis piceae* liquid cultures. *Phytochemistry* 25: 1979-1980.

Hanssen H-P. and Abraham W-R. (1988). Sesquiterpene alcohols with novel skeletons from the fungus *Ceratocystis piceae* (Ascomycotina). *Tetrahedron* 44: 2175-2180.

Hansson T. and Sterner O. (1991). Studies of the conversions of sesquiterpenes in injured fruit bodies of *Lactarius vellereus*. A biomimetic transformation of stearoylvelutinal to isovelleral. *Tetrahedron Letters* 32: 2541-2544.

Hara M., Yoshida M., Morimoto M. and Nakano H. (1987). 6-Deoxyilludin M, a new antitumor antibiotic: Fermentation, isolation and structural identification. *Journal of Antibiotics* 40: 1643-1646.

Harttig U., Anke T., Scherer A. and Steglich W. (1990). Leaianafulvene, a sesquiterpenoid fulvene derivative from cultures of *Mycena leaiana*. *Phytochemistry* 29: 3942-3944.

Hatanaka H., Kino T., Hashimoto M., Tsurumi Y., Kuroda A., Tanaka H., Goto T. and Okuhara M. (1988). FR65814, a novel immunosuppressant isolated from a *Penicillium* strain. Taxonomy, fermentation, isolation, physico-chemical and biological characteristics and structure assignment. *Journal of Antibiotics* 41: 999-1008.

Hayashi M., Kim Y.P., Takamatsu S., Preeprame S., Komiya T., Masuma R., Tanaka H., Komiyama K. and Omura S. (1996). Chlovalicin, a new cytocidal antibiotic produced by *Sporothrix* sp. FO-4649. I. Taxonomy, fermentation, isolation and biological activities. *Journal of Antibiotics* 49(7): 631-634.

Heatley, N.G., Jennings, M.A., Florey, H.W. (1947). Antibiotics from *Stereum hirsutum*. *British Journal of Experimental Pathology* 28: 35.

Heim J., Anke T., Mocek U., Steffan B. and Steglich W. (1988). Antibiotics from basidiomycetes. XXIX: Pilatin, a new antibiotically active marasmane derivative from cultures of *Flagelloscypha pilatii* Agerer. *Journal of Antibiotics* 41: 1752-1757.

Hellwig V., Dasenbrock J., Schumann S., Steglich W., Leonhardt K. and Anke T. (1998). New triquinane-type sesquiterpenoids from *Macrocystidia cucumis* (Basidiomycetes). *European Journal of Organic Chemistry* 73-79.

Huang Z., Dan Y., Huang Y., Lin L., Li T., Ye W. and Wei X. (2004). Sesquiterpenes from the mycelial cultures of *Dichomitus squalens*. *Journal of Natural Products* 67: 2121-2123.

Isaka M., Punya J., Lertwerawat Y., Tanticharoen M. and Thebtaranonth Y. (1999). Antimalarial activity of macrocyclic trichothecenes isolated from the fungus *Myrothecium verrucaria*. *Journal of Natural Products* 62: 329-331.

Ito Y., Kurita H., Yamaguchi T., Sato M. and Okuda T. (1967). Naematolin, a new biologically active substance produced by *Naematoloma fasciculare* (Fr.) Karst. *Chemical and Pharmaceutical Bulletin* 15: 2009-2010.

Jarvis B.B., Midiwo J.O., DeSilva T. and Mazzola E.P. (1981). Verrucarin L, a new macrocyclic trichothecene. *Journal of Antibiotics* 34:120-121.

Jonassohn M., Anke H., Sterner O. and Svensson C. (1994). New compounds isolated from the culture filtrate of the fungus *Merulius tremellosus*. *Tetrahedron Letters* 35: 1593-1596.

Kavanagh F., Hervey A. and Robbins W.J. (1949). Antibiotic substances from basidiomycetes. IV. *Marasmius conigenus*. *Proceedings of the National Academy of Sciences of the United States of America* 35: 343-349.

Kenfield D. Bunkers G., Wu Y.-H., Strobel G. Sugawara F., Hallock Y. and Clardy J. (1989). Gigantenone, a novel sesquiterpene phytohormone mimic. *Experientia* 45: 900-902.

Kepler J.A., Wall M.E., Mason J.E., Basset C., McPhail A.T. and Sim G.A. (1967). The structure of fomannosin, a novel sesquiterpene metabolite of the fungus *Fomes annosus*. *Journal of the American Chemical Society* 89: 1260.

Konitz, A., Bogucka-Ledochowska, M., Dauter, Z., Hempel, A., Borowski, E. (1977). The structure of isolactarorufin. *Tetrahedron Letters* 3401-3402.

Kühne B., Hanssen H.-P., Abraham W.-R. and Wray V. (1991). A phytotoxic eremophilane ether from *Hypomyces odoratus*. *Phytochemistry* 30: 1463-1465.

Lee I.K., Jeong C.Y., Cho S.M., Yun G.S., Kim Y.S., Yu S.H., Koshino H. and Yoo I.D. (1996). Illudins C2 and C3, new illudin C derivatives from *Coprinus atramentarius* ASI20013. *Journal of Antibiotics* 49: 821-822.

Ma W., Huang Y., Lin L., Zhu X., Chen Y., Xu H. and Wei X. (2004). Two new biologically active illudane sesquiterpenes from the mycelial cultures of *Panaeolus retirugis*. *Journal of Antibiotics* 57: 721-725.

Magnusson G. and Thoren S. (1973). Fungal extractives - III. Two sesquiterpene lactones from *Lactarius*. *Acta Chemica Scandinavica* 27: 1573-1578.

Magnusson G. and Thoren S. (1974). Fungal extractives - VI. Structure of lactaral, a new sesquiterpene furan-3-aldehyde from *Lactarius*, by spectroscopic methods. *Tetrahedron* 30: 1431-1435.

McDonald L.A., Barbieri L.R., Bernan V.S., Janso J., Lassota P. and Carter G.T. (2004). 07H239-A, a new cytotoxic eremophilane sesquiterpene from the marine-derived xylariaceous fungus LL-07H239. *Journal of Natural Products* 67: 1565-1567.

McMorris T.C. and Anchel M. (1965). Fungal metabolites. The structures of the novel sesquiterpenoids illudin-S and M. *Journal of the American Chemical Society*, 87: 1594-1600.

Mehta G. and Karra S.R. (1991). Polyquinanes from (R)-(+)-limonene. Enantioselective total synthesisof the novel tricyclic sesquiterpene (-)-ceratopicanol. *Journal of the Chemical Society, Chemical Communications* 1367-1368.

Mellows G., Mantle P. G., Feline T. C., Williams D. J. (1973). Sesquiterpenoid metabolites from *Stereum complicatum*. *Phytochemistry* 12: 2717-2720.

Midland S.L., Izac R.R., Wing R.M., Zaki A.I., Munnecke D.E. and Sims J.J. (1982). Melleolide, a new antibiotic from *Armillaria mellea*. *Tetrahedron Letters* 23: 2515-2518.

Mierau V., Anke T. and Sterner O. (2003). Dacrymenone and VM 3298-2 - new antibiotics with antibacterial and antifungal activity. *Zeitschrift für Naturforschung, C: Biosciences* 58: 541-546.

Minagawa K., Kouzuki S., Yoshimoto J., Kawamura Y., Tani H., Iwata T., Terui Y., Nakai H., Yagi S., Hattori N., Fujiwara T. and Kamigauchi T. (2002). Stachyflin and acetylstachyflin, novel anti-influenza A virus substances, produced by *Stachybotrys* sp. RF-7260. I. Isolation, structure elucidation and biological activities. *Journal of Antibiotics* 55: 155-164.

Moreau S., Lablache-Combier A. and Biguet J. (1980). Production of eremofortins A, B, and C relative to formation of PR toxin by *Penicillium roqueforti*. *Applied and Environmental Microbiology* 770.

Moss M.O. and Thrane U. (2004). *Fusarium* taxonomy with relation to trichothecene formation. *Toxicology Letters* 153: 23-28.

Murakami Y., Okuda T. and Shindo K. (2001). Roridin L, M and verrucarin M, new macrocyclic trichothecene group antitumor antibiotics, from *Myrothecium verrucaria*. *Journal of Antibiotics* 54: 980-983.

Nair M.S.R., Carey S.T. and Rogerson C.T. (1983). Illudoids from *Omphalotus olivascens* and *Clitocybe subilludens*. *Mycologia* 75: 920-922.

Nakagawa M., Hirota A., Sakai H. and Isogai A. (1982). Terrecyclic acid A, a new antibiotic from *Aspergillus terreus*. I. Taxonomy, production, and chemical and biological properties. *Journal of Antibiotics* 35: 778-782.

Nakamura H., Takita T., Umezawa H., Kunishima M., Nakayama Y. and Iitaka Y. (1974). Absolute configuration of coriolin, a sesquiterpene antibiotic from *Coriolus consors*. *Journal of Antibiotics* 27: 301-302.

Nasini G. and Vajna de Pava O. (1997). Protoilludane sesquiterpenes from Basidiomycetes. In: Virtual activity, real pharmacology. Different approaches to the search for bioactive natural compounds, Verotta L. Ed., Research Signpost. Trivandrum, India, 177-188.

Nozoe S., Furukawa J., Sankawa U. and Shibata S. (1976). Isolation, structure and synthesis of hirsutene, a precursor hydrocarbon of coriolin biosynthesis. *Tetrahedron Letters* 17: 195-198.

Obuchi T., Kondoh H., Watanabe N., Tamai M., Omura S., Jun-Shan Y. and Xiao-Tian L. (1990). Armillaric acid, a new antibiotic produced by *Armillaria mellea*. *Planta Medica* 56: 198-201.

Pang Z., Bocchio F. and Sterner O. (1992). The isolation of new sesquiterpene aldehydes from injured fruit bodies of *Lactarius scrobiculatus*. *Tetrahedron Letters* 6863-6866.

Poyser J.P., Edwards R.L., Anderson J.R., Hursthouse M.B., Walker N.P.C., Sheldrick G.M. and Whalley A.J.S. (1986). Punctaporonins A, D, E, and F (antibiotics M95464, M167906, M171950, and M189122), isomeric allylic alcohols from the fungus *Poronia punctata*: X-ray crystal structures of D and of E acetonide. *Journal of Antibiotics* 39: 167-169.

Pulici M., Sugawara F., Koshino H., Usawa J. and Yoshida S. (1996). Metabolites of endophytic fungi of *Taxus brevifolia*: the first highly functionalized Humulene of fungal origin. *Journal of Chemical Research (Synopses)* 378-379.

Pulici M., Sugawara F., Koshino H., Uzawa J., Yoshida S., Lobkovsky E. and Clardy J. (1996a). Pestalotiopsins A and B: New caryophyllenes from an endophytic fungus of *Taxus brevifolia*. *Journal of Organic Chemistry* 61: 2122-2124.

Quack W., Anke T., Oberwinkler F., Giannetti B.M. and Steglich W. (1978). Antibiotics from basidiomycetes. V. Merulidial, a new antibiotic from the basidiomycete *Merulius tremellosus* Fr. *Journal of Antibiotics* 31: 737-741.

Riche C., Pascard-Billy C., Devys M., Gaudemer A. and Barbier M. (1974). Structure cristalline et moleculaire de la phomenone, phytotoxine produite par le champignon *Phoma exigua* var. *non oxydabilis*. *Tetrahedron Letters* 15: 2765-2766.

Rukachaisirikul V., Tansakul C., Saithong S., Pakawatchai C., Isaka M. and Suvannakad R. (2005). Hirsutane sesquiterpenes from the fungus *Lentinus connatus* BCC 8996. *Journal of Natural Products* 68: 1674-1676.

Shirahama H., Osawa E. and Matsumoto T. (1978). Molecular mechanics studies on humulene conformations and its relations to biosynthetic route to illudoids. *Tetrahedron Letters* 19: 1987-1990.

Simon B., Anke T., Anders U., Neuhaus M. and Hansske F. (1995). Collybial, a new antibiotic sesquiterpenoid from *Collybia confluens* (Basidiomycetes). *Zeitschrift für Naturforschung, C: Biosciences* 50: 173-180.

Singh S.B., Zink D., Polishook J., Valentino D., Shafiee A., Silverman K., Felock P., Teran A., Vilella D., Hazuda D.J. and Lingham R.B. (1999). Structure and absolute stereochemistry of HIV-1 integrase inhibitor integric acid. A novel eremophilane sesquiterpenoid produced by a *Xylaria* sp. *Tetrahedron Letters* 40: 8775-8779.

Smith C.J., Morin N.R., Bills G.F., Dombrowski A.W., Salituro G.M., Smith S.K., Zhao A. and MacNeil D.J. (2002). Novel sesquiterpenoids from the fermentation of *Xylaria persicaria* are selective ligands for the NPY Y5 receptor. *Journal of Organic Chemistry* 67: 5001-5004.

Stadler M., Anke H. and Sterner O. (1994). New nematicidal and antimicrobial compounds from the basidiomycete *Cheimonophyllum candidissimum* (Berk & Curt.) sing. I. Producing organism, fermentation, isolation, and biological activities. *Journal of Antibiotics* 47: 1284-1289.

Staerk A., Anke T., Mocek U., Steglich W., Kirfel A. and Will G. (1988). Lentinellic acid, a biologically active protoilludane derivative from *Lentinellus* species (Basidiomycetes). *Zeitschrift für Naturforschung, C: Biosciences* 43: 177-183.

Sterner O., Bergman R., Franzen C. and Wickberg B. (1985). New sesquiterpenes in a proposed russulaceae chemical defense system. *Tetrahedron Letters* 26: 3163-3166.

Sterner O. (1989). The Co-formation of sesquiterpene aldehydes and lactones in injured fruit bodies of *Lactarius necator* and *L. circellatus*. The isolation of epi-piperalol. *Acta Chemica Scandinavica*, 43: 694-697.

Suay I., Arenal F., Asensio F.J., Basilio A., Cabello M.A., Diez M.T., Garcia J.B., Gonzalez del Val A., Gorrochategui J., Hernandez P., Pelaez F. and Vicente M.F. (2000). Screening of basidiomycetes for antimicrobial activities. *Antonie van Leeuwenhoek* 78: 129-139.

Sugawara F., Strobel G., Fisher L.E., van Duyne G.D. and Clardy J. (1985). Bipolaroxin, a selective phytotoxin produced by *Bipolaris cynodontis*. *Proceedings of the National Academy of Sciences of the United States of America* 82: 8291-8294.

Suri O.P., Sha R., Satti K. and Suri K.A. (1997). Russulactarorufin, a lactarane skeleton sesquiterpene from *Russula brevipes*. *Phytochemistry* 45: 1453-1455.

Sweeney M.J. and Dobson A.D. (1999). Molecular biology of mycotoxin biosynthesis. *FEMS Microbiology Letters* 175(2): 149-163.

Tabata Y., Miike N., Hatsu M., Kurata Y., Yaguchi T., Someya A., Miyadoh S., Hoshiko S., Tsuruoka T. and Omoto S. (1997). PF1092A, B and C, new nonsteroidal progesterone receptor ligands produced by *Penicillium oblatum*. I. Taxonomy of producing strain, fermentation, isolation and biological activities. *Journal of Antibiotics* 50: 304-308.

Takamatsu S., Kim Y.P., Komiya T., Sunazuka T., Hayashi M., Tanaka H., Komiyama K. and Omura S. (1996). Chlovalicin, a new cytocidal antibiotic produced by *Sporothrix* sp. FO-4649 II. Physicochemical properties and structural elucidation. *Journal of Antibiotics* 49: 635-638.

Takazawa H. and Kashino S. (1991). Incarnal. A new antibacterial sesquiterpene from Basidiomycetes. *Chemical and Pharmaceutical Bulletin* 39: 555-557.

Tamm C. (1974). The antibiotic complex of the verrucarins and roridins, Fortschritte in der Chemie Organischer Naturstoffe 31: 61-117

Tamm C. (1986). Structural studies on bioactive microbial metabolites. *Pure and Applied Chemistry* 58: 773-780.

Tanabe M., Suzuki K.T. and Jankowski W.C. (1974). Biosynthetic studies with carbon-13: The FT-^{13}C NMR spectra of the sesquiterpenoid coriolins. *Tetrahedron Letters* 15: 2271-2274.

Tanaka S., Wada K., Marumo S. and Hattori H. (1984). Structure of sporogen-AO 1, a sporogenic substance of *Aspergillus oryzae*. *Tetrahedron Letters* 25: 5907-5910.

Tirilly Y., Kloosterman J., Sipma G., and Kettens-van den Bosch J.J. (1983). A fungitoxic sesquiterpene from *Hansfordia pulvinata*. *Phytochemistry* 22: 2082-2083.

Tressl R., Kossa M. and Koeppler H. (1988). Changes of aroma components during processing of hops. *Monograph European Brewery Convention* 13: 116-129.

Urbasch I., Kühne B., Hanssen H.-P. and Abraham W.-R. (1991). Fungicidal activity of hypodoratoxide from *Hypomyces odoratus* (Ascomycotina). *Planta Medica* Suppl. 2, A 18.

Vidari G., De Bernardi M., Vita-Finzi P. and Fronza G. (1976). Sesquiterpenes from *Lactarius blennius*. *Phytochemistry* 15: 1953-1955.

Vidari G. and Vita-Finzi P. (1995). Sesquiterpenes and other secondary metabolites of genus *Lactarius* (Basidiomycetes): Chemistry and biological activities. In: Studies in Natural Product Chemistry, Atta-ur-Rahman Ed., Elsevier, Amsterdam Vol. 17: 153-206.

Vidari G., Garlaschelli L., Rossi A. and Vita-Finzi P. (1998). New protoilludane sesquiterpenes from *Lactarius violascens*. *Tetrahedron Letters* 39: 1957-1960.

Ward N. (2006). New directions and interactions in metagenomics research. *FEMS Microbiology Ecology* 55: 331-338.

Wei R., Schnoes H.K., Hart P.A. and Strong F.M. (1975). The structure of PR toxin, a mycotoxin from *Penicillium roqueforti*. *Tetrahedron* 31:109-114.

Wunder A., Anke T., Klostermeyer D. and Steglich W. (1996). Lactarane type sesquiterpenoids as inhibitors of leukotriene biosynthesis and other, new metabolites from submerged cultures of *Lentinellus cochleatus* (Pers. ex Fr.) Karst. *Zeitschrift für Naturforschung, C: Biosciences* 51: 493-499.

Xie J.-L., Li L.-P. and Dai Z.-Q. (1992). Isolation and identification of two new metabolites from silver leaf fungus *Stereum purpureum. Journal of Organic Chemistry* 57: 2313-2316.

Yang J.S., Chen Y.W., Feng X.Z., Yu D.Q. and Liang X.T. (1984). Chemical constituents of *Armillaria mellea* mycelium. I. Isolation and characterization of armillarin and armillaridin. *Planta Medica* 50: 288-290.

Yang J.S., Su Y.L., Wang Y.L., Feng X.Z., Yu D.Q., Liang X.T., He C.H., Zheng Q.T., Yang J.J. and Yang J. (1990). Chemical constituents of *Armillaria mellea* mycelium. VI. Isolation and structure of armillaripin. *Yaoxue Xuebao* 25: 353-356.

Yang J.S., Su Y.L., Wang Y.L., Feng X.Z., Yu D.Q. and Liang X.T. (1990a). Chemical constituents of *Armillaria mellea* mycelium. V. Isolation and characterization of armillarilin and armillarinin. *Yaoxue Xuebao* 25: 24-28.

Zhang J. and Feng X.Z. (1997). Sesquiterpene hydroxylactone from *Lactarius subvellereus. Phytochemistry* 46: 157-159.

16 Mycological Applications of Raman Spectroscopy

Kris De Gussem[1,*], Peter Vandenabeele[1], Annemieke Verbeken[2], Luc Moens[1]

[1]Department of Analytical Chemistry, Ghent University, Proeftuinstraat 86, B-9000 Gent, Belgium.
[2]Department of Biology, Ghent University, K.L. Ledeganckstraat 35, B-9000 Gent, Belgium.

INTRODUCTION

The classical study of mycology and taxonomical relationships among fungal species started in the 17th and 18th century with important contributors like Fries, Micheli and Persoon (Persoon, 1797; Carlile and Watkinson, 1994a; Alexopoulos *et al.*, 1996a). Their morphological concepts stressing macromorphological appearance of sporophores dominated classification systems until the mid to late 20th century. Morphological characters include amongst others the substrate, type and shape of sporophore and hymeniophore, dimensions, colour and odour. Since the invention of the microscope by van Leeuwenhoek in the 17th century, mycologists also use various microscopic characters on the arrangement of hyphae in the different tissues in the sporophore, spore formation, presence and form of structures dedicated to sexual and asexual reproduction (Camazine *et al.*, 1984; Carlile and Watkinson, 1994b; Alexopoulos *et al.*, 1996b). The inclusion of more micromorphological characters proved that the Friesian classification based mainly on sporophore and hymenophore type (e.g., Agaricales vs. Aphyllophorales) was highly artificial and contained numerous polyphyletic taxa.

Besides macro- and micromorphological characters, chemical characters have since long been used for identification and study of fungal species. Biochemical characters involve colour reactions with chemical products, such as NH_4OH, $FeSO_4$, KOH, aniline, sulfovanillin and phenol, which are used in field-tests to discriminate between fungal species (Matheis, 1972; Heilmann-Clausen *et al.*, 1998). Amyloidity, the dark blue to black staining of basidiospores or tissues in iodine reagents such as Melzer's solution, is an example of a microchemical character which has been used as a taxonomic character in fungi for nearly 150 years (Dodd and McCracken, 1972). An amyloidic spore ornamentation is reported in 35 genera of Agaricales (Singer, 1975) and at least 20 genera of Aphyllophorales (Donk, 1964).

Since a few decades, mycologists are also using molecular-genetic tools for the classification and phylogeny of fungi (Carlile and Watkinson 1994c; Alexopoulos *et al.*, 1996b; Couto *et al.* 1996; Horton and Bruns 2001; Miller *et al.* 2001; Nuytinck and Verbeken 2006). While phenotypic characters may show great variability, DNA, in which genetic information of the organisms is encoded, is highly constant and species specific. Because of the low variability and high information content in genetic material, these tools are considered as robust and reliable to investigate taxonomical relationships. In phylogeny, mitochondrial DNA is of particular interest. Because mitochondria (including their DNA) are only transferred to new generations by female individuals or through asexual reproduction, mitochondrial DNA can be seen as a molecular clock. The higher the similarities between genes, more the species are related to each other, so the similarities between genes are a measure of the relationship between taxa. Different genes are used in phylogeny, such as the genes coding for G3P-dehydrogenase or genes in the ITS-region. Widely used techniques include nucleotide sequencing, DNA-DNA hybridisation and Randomly Amplified Polymorphic DNA (RAPD). As a fingerprinting technique capable of assessing inter- and intra-species variation, Amplified Fragment Length Polymorphism (AFLP) is also of importance to verify whether closely related species could be distinguished (Bock *et al.*, 2002). As such, these molecular-genetic methods are shedding new light on taxonomy. Next to taxonomy, molecular-genetic tools are used for species identification, for ecological research as well as for studying geographical patterns in species distribution. The molecular-genetic approach requires specially trained people, and although these tools can be automated, they are mainly expensive.

The application of instrumental chemical analysis, for example, to perform extractions and enzyme electrophoresis, had sporadically been used on many different topics. Chemical analysis techniques are recently gaining importance and may prove to be a valuable alternative for time consuming, costly and highly specialized methods, such as morphological and genotypic investigations. The chemical tools which are now used in the study of fungi and their metabolites are mainly spectroscopic and chromatographic techniques. While chromatographic analysis can be optimized for the separation and selective detection of fungal metabolites, different spectroscopic techniques offer additional and complementary information. Gas chromatography (GC) is generally used for the detection of volatile compounds and is often coupled to mass spectrometry (MS) for selective detection of low concentrations. Brondz *et al.,* (2004) studied the fatty acid spore composition by GC-MS as a possible and novel chemotaxonomic marker. NMR spectroscopy is a spectroscopic method of analysis, especially capable of structural analysis of pure compounds. The technique was applied by Himmelreich *et al.,* (2003) for identification of *Candida* species. Several authors also used NMR-spectroscopy to investigate *Lactarius* latex composition and its sesquiterpenoid origin (Camazine and Lupo, 1984; Sterner *et al.*, 1985). Vibrational spectroscopic methods are of further importance, because of their advantageous properties. Vibrational spectroscopy includes infrared and Raman spectroscopy which are both non-destructive and fast analysis techniques, capable of analysing complex mixtures. Applications include, for example, characterisation and identification purposes as well as single cell analysis (Kummerle *et al.*,1998; Mohaček-Grošev *et al.*, 2001; Maquelin *et al.*, 2002; Maquelin *et al.*, 2003; Huang *et al.*, 2004; Rösch *et al.*, 2005).

Raman spectroscopy is in general a promising technique for chemical analysis in biology, microbiology and mycology. Its properties include amongst others limited sample preparation, high spectral resolution and the ability to analyse aqueous solutions (McCreery, 2000). Rapid analysis or identification of samples is possible with such an analytical spectroscopic technique and Raman spectroscopy may prove to be a valuable alternative for, or complement to, phenotypic and genotypic methods. Fingerprinting capability, non-destructiveness, narrow Raman bands and low water background make the technique well-suited for the analysis of biological samples. Opposite to other taxonomic tools (Fatty Acid Methyl-Ester analysis (FAME), AFLP, enzyme electrophoresis ...), which focus on one particular group of taxonomic discriminative molecules, Raman spectroscopy offers a complete phenotypic analysis because information on the complete chemical composition of the cell is gathered. As such, it may be used to analyse the chemical composition of specific parts of a sporophore, to distinguish and identify different taxa and to monitor biochemical reactions. Although Raman spectroscopy is not yet used routinely for biological research so far, mycological studies implying Raman spectroscopy reported on topics with either biochemical, taxonomic, medical or pharmaceutical aspects.

Raman spectroscopy has great potential due to its inherent properties (McCreery, 2000) in mycological research. Due to several instrumental improvements of Raman instrumentation during the 1970's and 1980's, Raman spectroscopy became available for the demanding analysis of complex biological materials. Therefore, because of this ongoing progression, new mycological applications of Raman spectroscopy are still in development. This evolution may eventually lead to mycotechnological and industrial applications as well as to applications for the identification of fungi.

RAMAN SPECTROSCOPY FONT SIZE

Although the Raman effect was already predicted by Smekal in 1923, it lasted until 1928 when Sir Chandrasekhara Venkata Raman discovered this effect, by using solar irradiation (Smekal, 1923; Raman and Krishnan, 1928; Gardiner, 1989). Because monochromatic light sources were not available in the 1920's and 1930's, Raman spectra were of low quality and it took over 24 hours to obtain a spectrum. Even when mercury arc light sources became available, Raman spectroscopy remained a niche technique in the 1940's. Electron multipliers improved accuracy, but it was only with the invention of the laser in the 1960's and the massive progression in instrumentation in the mid-1980's that Raman spectroscopy became widely available. This includes automation, miniaturisation of instrumentation, usage of optical fibres and personal computers. Progress in the accuracy and response times of Raman spectrometers through better lasers and the invention of charge-coupled devices (CCD's) maximised the range of applications and enabled the investigation of biomaterials with Raman spectroscopy. As biomaterials mostly contain water and have a complex composition, one needs a reliable and accurate tool of analysis. Due to its inherent properties and the increase of sensitivity since the mid-1980's, Raman spectroscopy, although relying upon a weak effect, offers opportunities towards biological research.

THE RAMAN EFFECT

When a monochromatic light beam encounters a sample, most photons will be scattered at the same frequency, although there is a small probability that the photons will have a different frequency after the scattering process. This Raman effect is inherently weak: only 1 out of 10^6 to 10^8 scattered photon undergoes Raman scattering. Molecules in a sample usually only populate the electronic ground state, but can occupy different vibrational energy levels. A photon coincides with a sample molecule and excites it into a virtual energy level. This virtual energy state is an unstable situation and can be considered as a short lived distortion of the molecule's electron cloud perturbed by the light beam's electric field. Molecules in this unstable state will relax through the emission of a secondary photon, which is shown in Fig. 16.1.

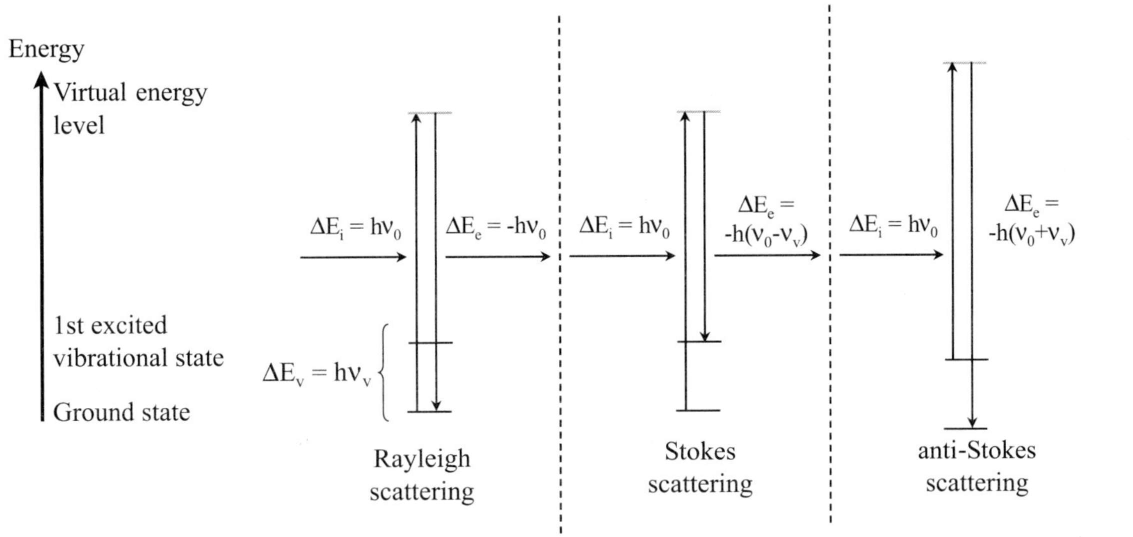

Fig. 16.1. Scattering of incident monochromatic laser light and different types of scattering. From the temporary virtual energy level, a molecule may relaxes to the same vibrational energy level (Rayleigh scattering), a higher vibrational energy level (Stokes scattering) or a lower vibrational energy level (anti-Stokes scattering).

The most probable situation is the emission of a photon of equal energy compared to the initial light beam and hence molecules relax to the ground state (Rayleigh scattering). The electron cloud of the molecule is not only perturbed by the electric field of the light beam, but is also perturbed by the molecular vibrations and therefore both perturbations may eventually interact. The molecule can thus after the process alternatively be found in a vibrational level below (anti-Stokes scattering) or above (Stokes scattering) the initial vibrational energy level. A photon of different energy relative to the incident radiation is thus scattered and the sample molecule effectively exchanged a quantum of energy equal to the energy difference between the vibrational energy levels. These energy differences are recorded in a Raman spectrum as a wavelength shift from the incident radiation. As vibrational energy levels are product specific, a Raman spectrum can be considered as a vibrational signature of the molecule, which is a highly characteristic and structurally sensitive fingerprint.

The intensity of a Raman band is dependent on the population of the different vibrational energy levels and is, therefore, directly related to the Boltzmann-distribution. Molecules are mostly found in the lowest vibration states. As a result, Stokes scattering is more intense than anti-Stokes scattering. Because Stokes and anti-Stokes scattering offer the same information with different signal-to-noise-ratios, Stokes scattering is often generalised in discussions as Raman scattering.

PROPERTIES OF RAMAN SPECTROSCOPY

Raman bands are characteristic for functional groups, so different regions in a Raman spectrum correspond with different types of functional groups. A comprehensive list of Raman frequencies is compiled by Socrates (2001). Despite the fact that all functional groups can lead to signals in a Raman spectrum, the symmetry of the irradiated molecule imposes certain limits on the presence of Raman bands (McCreery, 2000). Signals in a Raman spectrum are only observed if the polarisability of the molecule is changed. As the intensity of Raman bands is proportional to the polarisability of the different functional groups, Raman spectroscopy is especially useful in the characterisation of functional groups with loosely bound electron clouds (Ferraro *et al.,* 2003). These functional groups include carbonyl, COOH, alkenes, aromatic rings and sulphur containing groups.

In Raman spectroscopy, signal is normally collected at an angle of 90 degrees or 180 degrees (back-scattering) relative to the incident beam (McCreery, 2000). By using a transparent window relative to the laser light and by adapting focus distance, one can monitor a sample within a closed container (Pelletier, 1999). This opens possibilities, amongst others, for the investigation of pathogens without risk of infection, as well as it allows controlling industrial fermentation processes. Optical fibres can additionally be used. Through transport of laser light and Raman scattered photons, analysis is possible at several metres up to one hundred metres from the analytical equipment by using an automatic probe head.

ADVANTAGES OF RAMAN SPECTROSCOPY

Raman spectroscopy has some obvious advantages for the analysis of biomaterials. The technique is non-destructive, while only a limited preparation of the sample is necessary, if necessary at all (McCreery, 2000). Moreover, only small amounts of sample are necessary for Raman spectroscopic analysis. Compared to other spectroscopic techniques, Raman spectroscopy offers a high spectral resolution and narrow spectral bands. This is important in the analysis of complex mixtures and simplifies signal detection and interpretation without using advanced mathematical techniques. When water is irradiated using an intense light beam, the polarisability of water molecules changes only slightly. This means Raman spectra of aqueous solutions have a low water background, which makes Raman spectroscopy water compatible. This is an important factor for the analysis of biological materials, as biomaterials in general contain considerable amounts of water.

Raman spectroscopy is a fast and non-destructive tool of analysis, which offers the opportunity to measure cell composition *in vivo* or to monitor a changing sample composition in function of the time. This is made possible as small sample volumes could be measured by using confocal Raman spectrometers, while the spectroscopist can laterally focus on surfaces at micrometer level. At the most extreme, even single cell analysis was already performed (Huang *et al.*, 2004; Rösch *et al.*, 2005).

DISADVANTAGES OF RAMAN SPECTROSCOPY

Fluorescence background can be very intense in Raman spectra and is often the main interference (McCreery, 2000). It is a physical phenomenon in which the virtual energy level has energy comparable to an electronic level of the sample molecule and as such the molecule is excited to a higher electronic state. Intermediary relaxation, accompanied with heat production, leads to many possibilities for the molecule to return to the electronic ground state. The sample molecule in the higher electronic state is able to relax to all vibrational energy levels of the electronic ground state, which results in a lot of overlapping signals and because of the efficiency of the fluorescence process, the fluorescence background can be an intense and continuous signal. The amount of fluorescence depends on the laser wavelength; in general fluorescence is less intense with increasing laser wavelength. 1064 nm lasers are often used, at the expense of sensitivity.

A second phenomenon which might delimit the application of Raman spectroscopy is the absorption of laser light. Obtaining a Raman spectrum of dark coloured samples might be problematic, especially when the colour of the laser light is complementary to the sample colour. The intense laser bundle will then be absorbed by the sample, which causes heating and eventually sample damage. Absorption can often be dealt with by selecting an appropriate laser for the selected application.

Due to the weakness of the Raman effect, the detection limit of ordinary Raman spectroscopy, which is typically a few percent, might be another important issue. Resonance Raman spectrocopy as well as the SERS-technique (Surface Enhanced Raman Scattering) enhance sensitivity, but are not always applicable.

COMPARISON OF RAMAN SPECTROSCOPY WITH OTHER TECHNIQUES

Raman spectroscopy and infrared spectroscopy both probe molecular vibrations, although Raman and IR-spectroscopy are subject to different selection rules (McCreery, 2000). Furthermore, IR-signals are only observed when polarity changes during the vibrational transition, compared to a change in polarisability during a Raman transition. Polarity and polarisability are two different physical phenomena. This results in complementary information which makes it interesting to combine Raman and infrared spectroscopy, so infrared and Raman spectroscopy are used for similar applications, such as in mycological research (Maquelin *et al.*, 2000; Wenning *et al.*, 2002). However, as Raman spectra are mainly composed of narrow bands, spectral analysis is simplified. On the contrary, low vibrational modes are generally not observable through infrared spectroscopy. Disadvantages of infrared spectroscopy compared to Raman spectroscopy, also include a larger spot size of the light beam as well as a complicated and elaborate sample preparation.

Raman and infrared spectroscopy are often interchangeable and could replace each other, in case one technique fails. In contrast with Raman spectroscopy, infrared spectroscopy is water-sensitive and can thus, besides for classification, also be used for studying water capacity and water absorption (Decker *et al.*, 2005). Unlike Raman spectroscopy, IR spectroscopy is not subject to fluorescence, and is thus preferred in cases of very strong fluorescence. Where strong interference due to water absorption occurs, Raman spectroscopy is preferred to IR spectroscopy.

As stated before, both morphological and molecular tools are used for fungal taxonomy. Raman spectroscopy can be used for chemotaxonomic purposes as well, although Raman spectroscopy should not be considered as a substitute for either these genetic tools or the classical morphological investigations. It is an additional and alternative tool, offering a lot of potential due to its non-destructive nature and fast analysis.

CALIBRATION AND THE USE OF CHEMOMETRICS

The main interferences in Raman spectroscopy were discussed in the previous section. As fungal materials, like other biological materials, in general have a complex composition, their Raman spectra often have several overlapping bands. These bands often have valuable shoulders, characteristic for specific functional groups. In addition, biomaterials frequently consist of many different molecules of a similar structure. The frequencies of the Raman bands of such molecules only differ slightly, as frequencies are mainly determined by the functional group and to a smaller amount by the rest of the molecular structure. All these facts, as well as the presence of spectral fluctuations, may hinder the spectral interpretation and the assignment of bands to specific structures. The requirements to the calibration depend on the application in mind. When a qualitative analysis of simple fungal extracts is needed, calibration is less stringent. On the contrary, when direct data comparison is aimed or when a spectral database needs to be built, then a very accurate calibration protocol has to be followed.

Extraction of useful information from Raman spectra of complex fungal materials is mainly a three stage process. As a first requirement, the Raman spectrometer needs to be calibrated adequately for the intended application. Furthermore, all interferences and background effects still present in the Raman spectra need to be eliminated by mathematical preprocessing of the data. Finally, the information is drawn from the corrected spectra and conclusions are deduced.

ACCURATE CALIBRATION OF A RAMAN SPECTROMETER

In order to relate bands and shoulders to specific molecular groups, the Raman shift axis needs to be very accurate. Hutsebaut *et al.* (2005) presented a well tested calibration protocol to solve this issue. By measuring 7 calibration standards (4-acetamidophenol, acetonitrile / toluene, 1,4-bis (2-methylstyryl) benzene, cyclohexane, naphthalene, polystyrene and sulphur) during the day, the Raman shift axis can be calculated. To correct for differences in detector-response, the intensity values on the y-axis are divided by a function which correlates with the quantum efficiency of the detector. This function is measured by using a calibrated tungsten bulb, operating at a well-known voltage and electric current.

When one wants to use Raman spectroscopy for identification of samples, a spectral database needs to be built through repeatedly and randomly remeasuring many samples. The major problems to obtain a stable database are laser shifts occurring over a longer period, which are as well addressed by the procedure of Hutsebaut *et al.*, (2005). By collecting each hour a neon spectrum and an e-caprolacton spectrum to correct for intra-day laser shifts, long-time spectral differences can be as small as 0.2 cm^{-1}.

PRE-TREATMENT OF DATA

Background signals differ from spectrum to spectrum. In many cases, a fluorescence background of biological samples can be modelled mathematically by fitting a fifth order polynomial through the spectra (Lieber *et al.*, 2003). Other variation in the Raman spectra may be due to differences in scattering, focus or water content, etc. (Maquelin *et al.*, 2000). This is not problematic for monitoring of a well-known sample or for a simple check on the presence of specific molecular groups. However, for more demanding applications, such as chemotaxonomy or product identification using spectral databases, one needs to correct for these issues in order to be able to use these spectra adequately. Otherwise, long-term drifts in the spectrometer's output will be the main variation in the data, instead of sample specific information. Thus, not only a careful calibration routine is needed, but a proper chemometric preprocessing protocol, to eliminate all unnecessary spectral variation, is necessary as well.

Interferences in a Raman spectrum can be of both physical and chemical origin. Baseline fluctuations are amongst the most important obstacles a Raman spectroscopist needs to be aware of. Major causes of these physical interferences are fluctuations in scattering, focussing and density changes. Mathematical techniques such as vector normalisation are used for intensity corrections, while the Standard Normal Variate (SNV) procedure is often applied to take care of scattering fluctuations (Barnes *et al.*, 1989). In addition, the Savitsky-Golay algorithm is used to correct for baseline shifts between sample spectra (Savitsky and Golay, 1964). The effect of this mathematical preprocessing technique on spectra is represented in Fig. 16.2. In Fig. 16.2A two *Lactarius lacunarum* spore spectra are shown, which show quite a few spectral differences. The main spectral difference is a change in slope, which is handled effectively by taking the first derivative. Fig. 16.2B shows the same spectra on which the Savitsky-Golay algorithm is applied (window width of 9 data points, second order polynomial for fitting and first derivative; derivative spectra are 4 times magnified and shown with an offset of -4). The resulting derivative spectra can be used to build a more stable spectral database.

Another category of preprocessing routines includes techniques which enable a Raman spectroscopist to process spectral fluctuations mathematically using a model trained with a reference

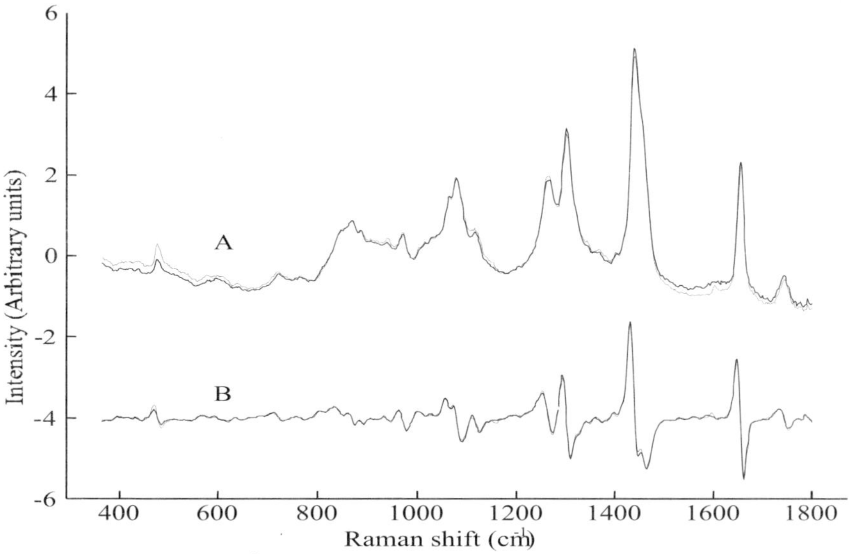

Fig. 16.2. Effect on spectral backgrounds of applying the first derivative using the Savitsky-Golay algorithm. A) The 2 top spectra of *Lactarius lacunarum* spores have a different slope. B) The same spectra with the Savitsky-Golay algorithm applied.

set of spectra. A mathematical model of the data is built using a suggested fluctuation pattern: in a first stage, the model is trained by using a training set to obtain the parameters of the mathematical routines. In a second stage, this model is applied to other spectra in order to correct for the interference under investigation. Multiplicative Signal Correction (MSC) (Martens and Naes, 1989) and Extended Multiplicative Signal Correction (Martens and Stark, 1991; Martens *et al.*, 2003) are representatives of these trained model techniques. Next to these techniques to take care of physical interference, chemical interference can be handled by using vector subtraction (Maquelin *et al.*, 2000) as well as by EMSC.

DATA ANALYSIS

Common chemometric tasks include estimation of concentrations, sample classification and identification. The chemical composition of biomaterials is considered to be sample or species dependant for classifying purposes. However, the major composition of related fungal samples is often comparable, with only minor compositional differences, thus resulting in highly similar Raman spectra.

The number of variables in these high-dimensional spectra is reduced for data analysis in almost all cases, while the information is retained in only a few latent variables. Principal component analysis (PCA) is one of the most frequently used techniques to reduce the amount of data, while retaining information. In order to perform concentration estimation, a subsequent multiple linear regression (this is also called principal components regression or PCR) is performed. Partial least squares (PLS) analysis is alternatively executed, often yielding better results. In other cases, sample classification such as species identification is needed and PCA is followed by cluster analysis, to reveal sample relationships, or linear discriminant analysis, to stress spectral differences between samples.

APPLICATIONS OF RAMAN SPECTROSCOPY IN MYCOLOGY

As Raman spectroscopy is a vibrational spectroscopic method of analysis, it measures signals characteristic for functional groups of the sample molecules. The inherent properties of Raman spectroscopy show the technique's possibilities and opportunities for the analysis of fungi or fungal materials. The lack of sample pretreatment makes Raman spectroscopy directly applicable for chemical analysis of fungi, especially for the analysis of sample composition and chemical reaction monitoring. Since the progress in Raman instrumentation, papers on studies with very different goals have been published, including clinical, taxonomic or other scientific aspects. In the majority of these studies, researchers investigated lower fungi such as moulds and yeasts.

CLINICAL APPLICATIONS

In case of problematic infections, there is no time to wait for mycelial cultures in order to identify the pathogen and thus a fast method of analysis is necessary in order to accurately counter infectious diseases and is especially of importance in intensive care units to reduce mortality. Raman spectroscopy has perspectives here, as was shown by Maquelin *et al.,* who used Raman spectroscopy to identify and discriminate between *Candida* species (Maquelin *et al.*, 2000; Ibelings *et al.*, 2005). They also obtained a prediction accuracy of 90% of *Candida* infections in peritonitis patients. This approach had a turnover time of 1 day compared to 3 to 4 days for classical microbiological tests. Such a fast and non-destructive identification is especially handy leading to better treatment of the *Candida* infection and in general lower drug resistance. Maquelin *et al.*, (2000) also investigated other clinical aspects of Raman spectroscopy, such as the detection of pathogens in blood cultures (Maquelin *et al.*, 2003).

INVESTIGATION OF LICHENS

Lichens have been intensively studied with Raman spectroscopy (Russell *et al.*, 1998; Edwards *et al.*, 2003; Edwards *et al.*, 2003b; Villar *et al.*, 2004). The lichens do not need to be detached from their substrate prior to analysis, on the contrary to the destructiveness of commonly used chromatographic techniques. The lichens and their substrata were collected in the field and analysed afterwards in the lab. With an optimized Raman spectrometer, spectra were collected from different sample layers at micrometer level and the organic content of the lichen's centre and the outer photosynthesizing layers could be measured separately. The fungus could be clearly differentiated from its photosymbiont and different lichens as well could be recognized from each other. The ability to analyse organic and inorganic molecules in the same experiment, was used to study biodeterioration and product formation. Lichens, which fungi feed on the sugars produced by their photosymbionts, produce acidic compounds as they grow, while calcareous substrates are degraded by these acidic products. This leads to the leakage of chalk ($CaCO_3$) and to the formation of oxalates. This oxalate formation may cause the degradation of monuments and the deterioration of their structure. Raman spectroscopic investigation of such monuments not only help in determining the degree of degradation, but also of interest for the identification of pigments in infected wall paintings. This ability makes Raman spectroscopy an interesting tool in the restoration of wall paintings and monuments.

RAMAN SPECTROSCOPIC ANALYSIS OF HIGHER FUNGI

Vibrational spectroscopic classification of fungal samples is a major research topic in literature. Mohaček-Grošev *et al.,* (2001) intensively investigated different genera like *Amanita*, *Lactarius* (Russulales), *Macrolepiota* (Agaricales) species with infrared spectroscopy and compared some of them with their corresponding Raman spectrum. Their research was a broad prospection rather than an in-depth study. They found, however, that specimens could be recognized at generic level based on infrared spectra of the spores. De Gussem *et al.,* (2005) analysed *Lactarius* spore composition by using Raman spectroscopy, prior to further research to investigate the possibility of using basidiomycetous spores for chemotaxonomic purposes. Vibrational spectroscopy as a taxonomical tool is rather new and up to now mainly done for lower fungi or on a very broad taxonomic scale. Edwards *et al.,* (1995) for example compared the spectra of Zygomycota such as *Mucor* (Mucorales), *Mortierella* (Mortiellales) and Basidiomycota such as *Agaricus* (Agaricales).

OTHER APPLICATIONS

Raman spectroscopy can be used to monitor life cycles and biological processes (Huang *et al.*, 2003; Huang *et al.*, 2004), e.g., to control metabolite production or screen for metabolic activity. Monitoring of antibiotic production has been demonstrated by Clarke *et al.*, (2005), by using surface enhanced Raman spectroscopy. An extreme case of this monitoring capacity is single cell analysis. Rösch *et al.*, (2005) used Raman spectroscopy for single cell yeast identification. For each sample, they collected 10 spectra of different compartments in a yeast cell and concluded that the average spectrum of the different cell compartments could be a chemotaxonomic marker for yeast identification. Huang *et al.*, (2003) studied the mitosis process of yeasts by using Raman spectroscopy. With a confocal Raman spectrometer a time resolution of 100s and a lateral resolution of 250 nm was achieved. These results demonstrate the favourable features of Raman spectroscopy, because other biochemical methods usually lack detailed time and space information, while molecular information is time and space averaged. During the mitotic division of yeasts, different signals in the Raman can be observed and attributed to proteins, lipids, phospholipids and carbohydrates.

As a last example, tissue characterisation and pathogen detection on foods can be mentioned. Yang *et al.*, (2003) applied Raman spectroscopy to detect microorganisms on apples, in order to screen for pathogens. Real-time screening and detection of pathogens by an imaging spectrometer can be used in the food industry to further minimize infections and rise the degree of food safety.

FUTURE PROSPECTS OF RAMAN SPECTROSCOPY IN MYCOLOGY

The evolution in Raman instrumentation in the past has given an impression of the way in which Raman spectroscopy may evolve. During the last two decades there were huge improvements in instrument stability and sensitivity. Raman spectrometers also became smaller and easier to operate and this trend is expected to continue. These improvements lead to the development of mobile spectrometers, which can be used for fast on-site analysis. This on-site analysis results in many more applications of Raman spectroscopy, because some samples for instance may not be transported to a laboratory because they are too heavy or because they can not be collected due to sample degradation. The importance of personal computers to operate Raman instrumentation can not be neglected. Computers improve the ease of use of Raman instrumentation, allow reliable mapping experiments and are more and more used for the mathematical treatment of spectra. Increased storage capacity and processor power will be used to deal with the huge amount of data in mapping experiments, while the application of more refined mathematical algorithms results in a more robust data analysis. Such algorithms of interest are EMSC, including its extensions such as Spectral Interference Subtraction (Martens and Stark, 1991; Martens *et al.*, 2003), and 2D-Raman spectroscopy (Noda and Ozaki, 2004).

Fluorescence is one of the major interferences in Raman spectroscopy. Fluorescence backgrounds can be dealt with by choosing an appropriate laser wavelength. Unfortunately, the laser wavelength can not be chosen arbitrarily. Tuneable lasers are in development but they are unstable and have a low laser light intensity to date. This eventually may be solved to reduce fluorescence backgrounds in Raman spectra of biomaterials.

As outlined in the previous section, clinical applications of Raman spectroscopy include a fast spectroscopic identification of pathogens in order to reduce mortality in hospitals. A fast identification could also be used to counter more effectively fungal infections, while the usage of Raman spectroscopy in agriculture may save crops. Mobile Raman spectrometers could be used in the field to screen for pathogens on crops.

Current research in Raman spectroscopy will certainly lead to industrial applications in this field. Due to its ability to obtain spectra through transparent windows, reaction monitoring by Raman spectroscopy is a realistic example. In addition, the usage of optical fibres permits the analysis of chemical content at large distances from the bioreactor. Monitoring of yeast fermentation has been examined in literature by Shaw *et al.,* (1999). On-line spectroscopic analysis assists the optimization of biotransformations for efficient production of green chemicals starting from raw biomaterials. In this way Raman spectroscopy as well as other spectroscopic methods contribute to the applicability of sustainable chemistry.

CONCLUSIONS

Mycologists are familiar with morphological characters and phenotypic tools in scientific research, while chemical tools are used to a lesser extent. Raman spectroscopy, being a vibrational molecular spectroscopic method, became recently available as a chemical tool of analysis in biological research. The properties of Raman spectroscopy highlight its opportunities in mycology. Compared to other analytical techniques, Raman spectroscopy has some unique medical and industrial opportunities, due to its monitoring capacity and also partly due to its non-destructive nature and speed of analysis. However, as Raman spectroscopy for high-demanding applications, such as the analysis of fungal biomaterials, only became available in the past 10-15 years, the technique is still under development. Potential of Raman spectroscopy for mycological applications still is a growing field and scientists are working on new opportunities before mycological applications of Raman spectroscopy may be developed into commercial products.

REFERENCES

Alexopoulos C. J., Mims C.W. and Blackwell M. (1996). *Introductory Mycology, 4th Edition*, John Wiley and sons, New York, 1-25, 61-75.

Barnes R.J., Dhanoa M.S. and Lister S.J. (1989). Standard normal variate transformation and de-trending of near-infrared diffuse reflectance spectra. *Applied Spectroscopy*, 43(5): 772-777

Bock C.H., Thrall P.H., Brubaker C.L. and Burdon J.J. (2002). Detection of genetic variation in *Alternaria brassicicola* using AFLP fingerprinting. *Mycological Research* 106(4): 428-434.

Brondz I., Hoiland K. and Ekeberg D. (2004). Multivariate analysis of fatty acids in spores of higher basidiomycetes: a new method for chemotaxonomical classification of fungi. *Journal of Chromatography B*, 800(1-2): 303-307

Camazine S. and Lupo A.T. (1984). Labile toxic compounds of the Lactarii - the role of the laticiferous hyphae as a storage depot for precursors of pungent dialdehydes. *Mycologia*, 76(2): 355-358.

Carlile M.J. and Watkinson S.C. (1994). *The fungi*, Academic press, London, 1-8.

Carlile M.J. and Watkinson S.C. (1994). *The fungi*, Academic press, London, 9-76.

Carlile M.J. and Watkinson S.C. (1994). *The fungi*, Academic press, London, 203-248.

Clarke S.J., Littleford R.E., Smith W.E. and Goodacre R. (2005). Rapid monitoring of antibiotics using Raman and surface enhanced Raman spectroscopy, *Analyst*, 130(7): 1019-1026.

Couto M.M.B., Eijsma B., Hofstra H., Huis in't Veld J.H.J. and van der Vossen J.M.B.M. (1996). Evaluation of molecular typing techniques to assign genetic diversity among *Saccharomyces cerevisiae* strains. *Applied and Environmental Microbiology*, 62(1): 41-46

Decker M., Nielsen P.V. and Martens H. (2005). Near-infrared spectra of *Penicillium camemberti* strains separated by extended multiplicative signal correction improved prediction of physical and chemical variations. *Applied spectroscopy*, 59(1): 56-68.

De Gussem K., Vandenabeele P., Verbeken A. and Moens L. (2005). Raman spectroscopic study of *Lactarius* spores (Russulales, Fungi). *Spectrochimica Acta A*, 61(13-14): 2896-2908

Dodd J.L. and McCraken D.A. (1972). Starch in fungi. Its molecular structure in three genera and an hypothesis concerning its physiological role. *Mycologia,* 64: 1341-1343.

Donk M.A. (1964). A conspectus of the families of Aphyllophorales, *Persoonia*, 3: 199-324.

Edwards H.G.M., Newton E.M., Wynn-Williams D.D. and Lewis-Smith R.I. (2003). *Spectrochimica acta part A*, 59(10): 2301-2309

Edwards H.G.M., Russell N.C., Weinstein R. and Wynn-Williams D.D. (1995). Fourier-transform Raman-spectroscopic study of fungi. *Journal of Raman Spectroscopy*, 26(8-9): 911-916.

Edwards H.G.M., Seaward M.R.D., Attwood S.J., Little S.J., de Oliveira L.F.C. and Tretiach M. (2003). FT-Raman spectroscopy of lichens on dolomitic rocks: an assessment of metal oxalate formation. *Analyst*, 128(10): 1218-1221.

Ferraro J.R., Nakamoto K., Brown C.W. (2003), *Introductory Raman spectroscopy*, Academic Press, 1-94.

Gardiner D.J. (1989). *Practical Raman spectrocopy*, Springer-Verlag Berlin Heidelberg, ISBN 3-540-50254-8, 1-12.

Heilmann-Clausen J., Verbeken A. and Vesterholt J. (1998). *The genus Lactarius - fungi of Northern Europe, vol. 2*, Svampetryk, Mundelstrup Denmark, 8-30.

Himmelreich U., Somorjai R.L., Dolenko B., Lee O.C., Daniel H.M., Murray R., Mountford C.E. and Sorrell T.C. (2003). Rapid identification of *Candida* species by using nuclear magnetic resonance spectroscopy and a statistical classification strategy. *Applied and Environmental Microbiology*, 69(8): 4566-4574.

Horton T.R. and Bruns T.D. (2001). The molecular revolution in ectomycorrhizal ecology: peeking into the black-box. *Molecular Ecology*, 10(8): 1855-1871

Huang Y.S., Karashima T., Yamamoto M. and Hamaguchii H. (2003). Molecular-level pursuit of yeast mitosis by time- and space-resolved Raman spectroscopy. *Journal of Raman Spectroscopy*, 34(1): 1-3.

Huang Y.S., Karashima T., Yamamoto M., Ogura T. and Hamaguchi H. (2004). Raman spectroscopic signature of life in a living yeast cell. *Journal of Raman Spectroscopy*, 35(7): 525-526.

Hutsebaut D., Vandenabeele P. and Moens L. (2005). Evaluation of an accurate calibration and spectral standardization procedure for Raman spectroscopy. *Analyst*, 130(8): 1204-1214.

Ibelings M.S., Maquelin K., Endtz H.P., Bruining H.A. and Puppels G.J. (2005). Rapid identification of *Candida* spp. in peritonitis patients by Raman spectroscopy. *Clinical microbiology and infection*, 11(5): 353-358.

Kummerle M., Scherer S. and Seiler H. (1998). Rapid and reliable identification of food-borne yeasts by Fourier-transform infrared spectroscopy. *Applied and Environmental Microbiology*, 64(6): 2207-2214.

Lieber C.A. and Mahadevan-Jansen A. (2003). Automated method for subtraction of fluorescence from biological Raman spectra. *Applied Spectroscopy*, 57(11): 1363-1367.

Maquelin K., Choo-Smith L.P., Endtz H.P., Bruining H.A. and Puppels G.J. (2002). Rapid identification of *Candida* species by confocal Raman micro spectroscopy. *Journal of Clinical Microbiology*, 40(2): 594-600.

Maquelin K., Choo-Smith L.P., van Vreeswijk T., Endtz H.P., Smith B., Bennett R., Bruining H.A. and Puppels G.J. (2000). Raman spectroscopic method for identification of clinically relevant microorganisms growing on solid culture medium. *Analytical Chemistry*, 72(1): 12-19.

Maquelin K., Kirschner C., Choo-Smith L.P., Ngo-Thi N.A., van Vreeswijk T., Stammler M., Endtz H.P., Bruining H.A., Naumann D., Puppels G.J. (2003). Prospective study of the performance of vibrational spectroscopies for rapid identification of bacterial and fungal pathogens recovered from blood cultures. *Journal of Clinical Microbiology*, 41(1): 324-329.

Martens H., Hoy M., Wise B.M., Bro R., Brockhoff P.B. (2003). Pre-whitening of data by covariance-weighted pre-processing. *Journal of Chemometrics*, 17(3): 153-165.

Martens H. and Naes T. (1989). *Multivariate calibration*, John Wiley and Sons, 314-354.

Martens H. and Stark E. (1991). Extended multiplicative signal correction and spectral interference subtraction: New preprocessing methods for near infrared spectroscopy. *Journal of Pharmaceutical & Biomedical Analysis,* 9(8): 625-635.

Matheis W. (1972). Chemische Reagenzien in der Hand des Mykologen. *Zeitschrift für Pilzkunde*, 38: 33-47.

McCreery R.L. (2000). *Raman spectroscopy for chemical analysis (Chemical analysis series vol. 157)*, John Wiley and Sons, New York, 1-14.

Miller S.L., McClean T.M., Walker J.F. and Buyck B. (2001). A molecular phylogeny of the Russulales including agaricoid, gasteroid and pleurotoid taxa. *Mycologia*, 93(2): 344-354.

Mohaček-Grošev V., Boňac R. and Puppels G.J. (2001). Vibrational spectroscopic characterization of wild growing mushrooms and toadstools. *Spectrochimica Acta A*, 57(14): 2815-2829.

Noda I. and Ozaki Y. (2004). *Two-Dimensional correlation spectroscopy—applications in vibrational and optical spectroscopy*, John Wiley & Sons Ltd. Chichester, ISBN: 0-471-62391-1.

Nuytinck J. and Verbeken A. (2006). Species delimination and phylogenetic relationships in *Lactarius* sect. *Deliciosi* in Europe. *Mycological research* (accepted for publication).

Pelletier, M.J. (1999), *Analytical applications of Raman spectroscopy*, Blackwell Science Ltd., Oxford, 106-126.

Persoon C.H. (1797). *Tentamen dispositiones methodicae fungorum in classes, ordines, genera et familias,* Lipsiae, [Leipzig].

Raman C.V. and Krishnan K.S. (1928). A new type of secondary radiation. *Nature*, 121: 501-502.

Rösch P., Harz M., Schmitt M. and Popp J. (2005). Raman spectroscopic identification of single yeast cells. *Journal of Raman Spectroscopy*, 36(5): 377-379.

Russell N.C., Edwards H.G.M. and Wynn-Williams D.D. (1998). FT-Raman spectroscopic analysis of endolithic microbial communities from Beacon sandstone in Victoria Land, Antarctica. *Antarctic Science* 10(1): 63-74.

Savitsky A. and Golay M.J.E. (1964). Smoothing and differentiation of data by simplified partial least squares. *Analytical Chemistry*, 36(8): 1627-1639.

Shaw A.D., Kaderbhai N., Jones A., Woodward A.M., Goodacre R., Rowland J.J. and Kell D.B. (1999). Noninvasive, on-line monitoring of the biotransformation by yeast of glucose to ethanol using dispersive Raman spectroscopy and chemometrics. *Applied Spectroscopy*, 53(11): 1419-1428.

Singer R. (1975). The Agaricales in modern taxonomy. 4th ed., Koeltz scientific books, Koenigstein, Germany.

Smekal A. (1923). Zur Quantumtheorie des Dispersion. *Naturwissenschaften*, 11: 873-875.

Socrates G. (2001). Infrared and Raman characteristic group frequencies - Tables and charts, John Wiley and Sons, Chichester, ISBN 0-471-85298-8.

Sterner O., Bergman R., Franzen C. and Wickberg B. (1985). New sesquiterpenes in a proposed Russulaceae chemical defense system, *Tetrahedron Letters*, 26(26): 3163-3166.

Villar S.E.J., Edwards H.G.M. and Seaward M.R.D. (2004). Lichen biodeterioration of ecclesiastical monuments in northern Spain. *Spectrochimica Acta A*, 60(5): 1229-1237.

Wenning M., Seiler H. and Scherer S. (2002). Fourier-transform infrared microspectroscopy, a novel and rapid tool for identification of yeasts. *Applied and Environmental Microbiology*, 68(10): 4717-4721.

Yang H. and Irudayaraj J. (2003). Rapid detection of foodborne microorganisms on food surface using Fourier transform Raman spectroscopy. *Journal of Molecular Structure*, 646(1-3): 35-43.

Index